ADVANCED MACHINE TOOL TECHNOLOGY AND MANUFACTURING PROCESSES

C. Thomas Olivo

C. Thomas Olivo Associates

DEDICATION

This book is dedicated to my wife and constant teacher, Hilda G. Olivo. She shared generously of her unique editorial talents, provided inspiration, and contributed significantly to this work from the initial research, analyses, and organization stages to this final publication.

Copyright © 1990 by C. Thomas Olivo Associates
Albany, New York, 12206

Printed in the United States of America

10   9   8   7   6   5   4   3   2   1

Library of Congress Catalog Card Number: 89–64000

Olivo, C. Thomas

   Advanced machine tool technology and manufacturing
   processes

   Includes handbook tables and index.

   1. Machine tool technology. 2. Machine shop practice.
   3. Tool and die making.   I. Title.

       1990   621.9'02   89–64000

ISBN 0–938561–04–9

ISBN 0–938561–05–7 (laboratory manual)

ISBN 0–938561–06–5 (instructor's guide)

Cover credits. . . *Design and composite art*, The Drawing Board
Cutting tools, Schuyler Photography
Automated turning/milling (screw machine), Brown & Sharpe Manufacturing Company
Cylindrical grinding, Pratt & Whitney Company, Inc., Precision Grinding Division

(Back cover) *CNC machining center control panel*, Warner & Swasey Turning Division, a Cross & Trecker Company

PREFACE

Advanced Machine Tool Technology and Manufacturing Processes is a state-of-the-technology, fascinating study of the performance skills and the related technology upon which industry depends for craftspersons and technicians; design, programming, and engineering support personnel, and other high-skill workers.

This newly titled book (formerly, **Advanced Machine Technology**) reflects changes in organization and the extended scope and contents of this third edition. The new dimensions are the end result of three parallel studies, as follows. (1) Analyses of performance skills, technology skills, and related technical subjects skills required by occupationally competent workers, projected to identify evolving industry developments. (2) Studies of courses and programs conducted by educational institutions; industry, labor, and military organizations; and by special training agencies. (3) Evaluations of systems, processes, and products used to measure occupational competency skills.

PRACTICAL APPLICATIONS

This book is crafted for students/trainees and employed workers who relate to programs or occupations such as: machine shop practice; tool and die making; skilled machine operatives; machine and systems design, programming, and industrial engineering personnel; machine tool technology and manufacturing specialists, and others in the machining industry.

This highly practical book is designed for use at the technical institute and community college level; in area vocational-technical schools; in college level industrial engineering courses; in apprenticeship and other on-the-job or cooperative school/industry/military occupational programs; and for adult training and upgrading courses. Such courses/programs may be under public or private administration. The book is also adaptable to directed self study.

SPECIAL FEATURES

The following special features are incorporated within the structure and contents.

√ AUTHENTICITY. The contents are written by a highly qualified, occupationally competent writer. As stated, the contents are based on *valid analyses* of industry requirements, training and *curriculum* development *patterns*, and occupational *competency assessment* systems.

√ COMPREHENSIVENESS. The contents reflect the major areas of training in which an individual must develop performance skills, technology skills, and related technical subjects skills for successful employment and advancement.

√ INTERLOCKED PERFORMANCE SKILLS AND TECHNOLOGY. Equal emphasis is placed on essential manipulative "how to" skills and companion "why-to" technology skills. These skills are further interlocked with problem-solving mathematics skills, applications of physical science skills, and blueprint reading communication skills. . .as required at the workplace.

√ FLEXIBLE UNIT ORGANIZATION. The structure of twelve *Parts* and *Sections* and 33 instructional *Units* provides a practical sequence to follow or to rearrange to meet other particular training requirements.

√ STRUCTURE WITHIN UNITS. The instructional contents flow from one Unit to the next higher level of learning. Each Unit starts with an *introductory overview* followed by carefully identified *Unit Objectives*. New *Technical Terms* are defined and applied within the Unit contents and are summarized at the end. The content *Summary* provides another opportunity to bring together all of the learning experiences within each unit.

√ MODERN, ACCURATE ILLUSTRATIONS.   Specially selected, relevant, accurate art work and callouts complement the contents.

√ HANDBOOK TABLES. Appropriate *Handbook Tables* are content referenced within the units. Additional complete tables appear in the Appendix. Practical test items require the use of handbook data.

√ CONTINUOUS TESTING. End-of-unit *Review and Self-Test Items* (with additional workplace-simulated problems provided in the Laboratory Manual) provide the most comprehensive continuous testing resource materials available.

√ PERSONAL, MACHINE, AND PRODUCT SAFETY. A positive, planned *Safety Program* threads throughout the unit contents. Test items are contained in each unit.

√ PRACTICAL AESTHETIC QUALITIES. *Two colors* are used on every page to emphasize important components, features, and concepts within each unit. Page layout, color, and physical properties are planned to enhance learning.

√ COMPLEMENTARY LABORATORY MANUAL. A student/trainee manual provides *Guidelines for Study* that parallel each textbook unit. Additional shop/laboratory problems are included for individual competency enrichment.

√ INSTRUCTOR/TRAINER GUIDE.   s complementary teacher/teaching resource suggests unit-by-unit *Teaching Plans* and *solutions* to Review and Self-Test Items in both the textbook and the Laboratory Manual. The Guide is designed to maximize teaching/learning effectiveness.

*   *   *   *   *

The twelve major content areas (Parts) of the book follow. Detailed Section and Unit content information is provided in the Table of Contents.

- 1   Advanced Precision Measurement and Inspection Processes
- 2   Surface Technology and Statistical Process Control (SPC)
- 3   Turning Machines: Advanced Technology and Processes
- 4   Milling Machines: Advanced Technology and Processes
- 5   Vertical Band Machines: Technology, Setups, and Processes
- 6   Shaping, Slotting, Planing, and Broaching Machines and Processes
- 7   Abrasive Machining: Technology and Processes (including Superabrasives)
- 8   Precision Machine Tools, Production Tooling, and Tool Design
- 9   Numerical Control (NC), CNC, CAM, and Flexible Manufacturing Systems (FMS)
- 10   Particle and Powder Metallurgy, Coating Materials, and Hardness Testing
- 11   New Manufacturing Processes: Laser Beam and Electrical Discharge (Nontraditional) Machining
- 12   Flexible Manufacturing Systems (FMS) and Robotics

About the Author

Dr. C. Thomas Olivo ("Dr. Tom") is a distinguished technical writer and foremost authority on institutional and training-within-industry programs related to machine tool technology, machine shop practice, tool and die making, and metal products manufacturing.

After serving an apprenticeship, Dr. Olivo achieved journeyman and master status as a tool and die maker. On the training scene, Dr. Tom served successively responsible positions as a teacher, training supervisor, technical institute director, branch chief, Wharton (England)-American military occupational technical school, State Director of industrial-technical education, and industry training consultant.

CONTENTS

Part 1 ADVANCED PRECISION MEASUREMENT AND INSPECTION

Section One MECHANICAL MEASURING INSTRUMENTS AND GAGES

Section Two PRECISION MEASUREMENT: NONMECHANICAL INSTRUMENTS

Part 4 MILLING MACHINES: ADVANCED TECHNOLOGY AND PROCESSES

Section One VERTICAL MILLING MACHINES

Section Two ADVANCED HORIZONTAL MILLING MACHINE PROCESSES

Part 5   VERTICAL BAND MACHINES

Section One   BAND MACHINE TECHNOLOGY AND BASIC SETUPS

Section Two BAND MACHINING: TECHNOLOGY AND PROCESSES

Part 6 SHAPING, SLOTTING, PLANING MACHINES AND BROACHING

Section One MACHINE TOOL TECHNOLOGY AND PROCESSES

Part 7 ABRASIVE MACHINING: TECHNOLOGY AND PROCESSES

Section One GRINDING MACHINES, ABRASIVES WHEELS, AND CUTTING FLUIDS

x • CONTENTS

Part 10 NEW METALLURGY AND HARDNESS TESTING

Section One NEW MATERIALS AND HARDNESS TESTING: TECHNOLOGY AND PROCESSES

Part 11 NEW MANUFACTURING PROCESSES

Section One NONTRADITIONAL MACHINING

Part 12 FLEXIBLE MANUFACTURING SYSTEMS (FMS) AND ROBOTICS

Section One INTERFACED SUBSYSTEMS IN FLEXIBLE MANUFACTURING SYSTEMS (FMS)

APPENDIX

HANDBOOK TABLES

ACKNOWLEDGMENTS

All advancements since the machining of single parts on the Wilkinson's lathe of the 1790's to current computer numerically controlled machining centers and totally flexible integrated manufacturing processes and systems, are founded on "sustained incremental contributions" of leaders. So, too, the quality, relevance, and authoritativeness of this book depends on resources provided by a broad spectrum of leaders, undergirded by other contributions from the institutions and organizations wherein they serve.

The spectrum of individuals who provided resources include: lead teachers serving in institutional programs, trainers from within industry and union sponsored programs, military occupational specialty training personnel, human resource training supervisors, engineers, and others. Marketing directors supplied important up-to-the-minute product information, technical data, and excellent illustrations. The scope represented by this cadre of leaders reached across many different types of training programs and training levels; semiskilled to high skill craftspersona and technicians, to product design, tooling, and programming specialists, to engineering support persons.

The *Courtesy of* line that appears in the text above an illustration sincerely acknowledges each company that provided technical assistance. A personal "thank you" is expressed to each contributing individual from the following companies.

Amada Laser Division; ACT, Inc.
AMATROL, Inc.
American Drill Bushing Company
American Precision Museum Association, Inc.
American Tool & Grinding Company, Inc.
Ames (B.C.) Company
Apex Tool & Cutter Company
Arcadia Supply, Inc.
Automation and Measurement Division, Bendix Corporation
Balzers Tool Coating, Inc.
Barber-Colman Company; Machine Tools Division
Bausch & Lomb Inc.
Bay State Abrasives; Dresser Industries, Inc.
Blanchette Tool & Gage Mfg. Company
Bliss (E.W.) Company
Bridgeport Machines, Inc.
Brown & Sharpe Manufacturing Company
Carpenter Technical Corporation
Cincinnati Milacron, Inc.
Clausing Industries, Inc.
Cleveland Twist Drill; an Acme-Cleveland Company
Coherent General, Inc.
Crucible Service Centers; Crucible Materials Corporation
Cushman Industries, Inc.
Danly Die Set; Division of Connell Ltd. Partnership

Darex Corporation
DeVlieg Machine Company; Microbore Division/Sundstrand Division
DoALL Company
Dorsey Gage Company, Inc.
duMont Corporation (The)
Dunham Tool Company, Inc.
Eimeldingen Corporation
Elcometer, Inc.
Elox Corporation
Ex-Cell-O Corporation
Federal Products Corporation
General Electric Company; Carboloy Systems; Superabrasives Department
Gleason Works; Division of Gleason Corporation
Grinding Wheel Institute
Hammond Machinery, Inc.
Hardinge Brothers, Inc.
Heald Machine Division; Cincinnati Milacron, Inc.
Hitachi Magna-Lock Corporation
Illinois/Eclipse; Division of Illinois Tool Works, Inc.
J & S Tool Company, Inc.
Jones & Lamson Division; Textron, Inc.
Kearney & Trecker(K-T)-Swasey Company
Kennametal, Inc.
LeBlond Makino Machine Tool Company
Lewis & Judge Machine Tools Company, Inc.
Light Machines Corporation

Litton Industrial Automation Systems; Landis Grinding Division

Lodge & Shipley; Division of Manuflex

Metcut Research Associates, Inc.; Manufacturing Technology Division

Micromatic Hone Corporation

Modern Machine Company

Modern Machine Shop; Gardner Publications, Inc.

Monarch Machine Tool Company; Monarch Cortland and Monarch Sidney Divisions

Moore Special Tool Company, Inc.

National Broach & Machine Company

National Twist Drill Division; Lear Siegler, Inc.

Norton Company

Pratt & Whitney Company, Inc.; Precision Grinding and Machine Tool Divisions

Rams-Rockford Products, Inc.

Rockford Engineered Products Company

Rockford Machine Tool Operations; Ex-Cell-O Corporation

Schlumberger CAD/CAM Division, Schlumberger Ltd.

Sheffield Measurement Division; Warner & Swasey, A Cross & Trecker Company

Sheldon Machine Company, Inc.

Shore Instrument and Manufacturing Company

SMW Systems, Inc.

Sodick Inc.

South Bend Lathe Company, Inc.

Starrett (L.S.) Company; Webber Gage Division

Systems 3R USA, Inc.

Taft-Peirce Manufacturing Company; Division of American Machines

TE-CO Tooling Components Company

Thompson Grinder Products; Waterbury Farrel Division of Textron, Inc.

Tinius Olsen Testing Machine Company

Trumpf Industrial Lasers, Inc.

TRW Geometric Tool Division

Universal Vise & Tool Company

Vermont Gage Division; Vermont Precision Tools, Inc.

Vermont Tap & Die Division; Vermont American Corporation

Warner & Swasey, A Cross & Trecker Company

White-Sunstrand Machine Tool Company

Wilson Instruments Company

Special recognition is made of Doyle K. Stewart, Coordinator of Manufacturing Processes, Delaware Technical and Community College, and to Dr. Ralph D. O'Brien, Professor and Chairman, Department of Industrial Technology, Northern Kentucky University, for their capable assessment of organization and scope and meticulous technical review of contents; to Noble Stuart, Coordinator of Machine Trades, Regional State Vocational-Technical School at Bowling Green; to Robert Carlton, Product Engineer, MDC; Metcut Research Associates, Inc., Manufacturing Technology Division, for technical assistance and resource materials; and to Joseph Reynolds, Vice President of Marketing and Sales, Delmar Publishers, Inc. for friendly assistance relating to market assessment and product marketing and communications.

Recognition is made to Hilda G. Olivo for developmental assistance and editorial expertise and to Judith Ellen Forward for skillful preparation of the functional Index. Thanks is expressed to Jean A. LeMorta of Thoroughbred Graphics for production expertise and quality composing services and to John R. Orozco of The Drawing Board for the unusually fine layout and preparation of the cover and other art work.

Special commendation is made to Judi S. Orozco recognizing her talents and contributions in converting specifications into excellent product layout designs, mechanicals of the contents, art work, and overall supervision. Finally, appreciation is expressed to Frederick J. Sharer, Director of Manufacturing, Delmar Publishers, for helpful advice and counsel in the areas of production and manufacturing.

The aesthetic, physical properties of this book and the quality, relevance, and value of the contents, are a tribute to each individual and to each organization that gave so generously and shared so graciously.

C. Thomas Olivo

PART 1 Advanced Precision Measurement and Inspection

SECTION ONE

Mechanical Measuring Instruments and Gages

Advanced machine technology and manufacturing processes require the use of more precise and accurate measuring instruments, indicators, gages, and test equipment than are needed for basic layout, measurement, and inspection.

This Section deals with technology and processes as related to mechanical measuring instruments, gages, and accessories. These are covered in detail in three separate Units: (1) Vernier Measuring and Dial Indicating Instruments, (2) Advanced Linear Measurement Using Gage Blocks and Accessories, and (3) Precision Layout and Measurement Practices for Surface Plate Work.

Vernier Measuring and Dial Indicating Instruments

OBJECTIVES

After satisfactorily completing this unit, you will be able to:

- Identify the main features and apply principles relating to vernier measuring and dial indicating instruments.
- Perform the following processes:
    - Read a Vernier Caliper Measurement.
    - Measure with a Vernier Protractor.
- Follow *Safe Practices* when using vernier measuring and dial indicating instruments.
- Apply *Technical Terms* that relate to precision linear and angular measuring instruments.
- Solve the problems assigned from the *Review and Self-Test Items*.

Measurements in thousandths of an inch or 0.02mm are the most widely used measurements in industry. Finer measurements are obtained by applying the vernier principle to metric and inch-standard micrometers and other instruments. Vernier scale graduations extend the range of fineness of measurement to 0.0001″ (one ten-thousandth of an inch) and 0.002mm (the metric equivalent of 0.00008″).

1

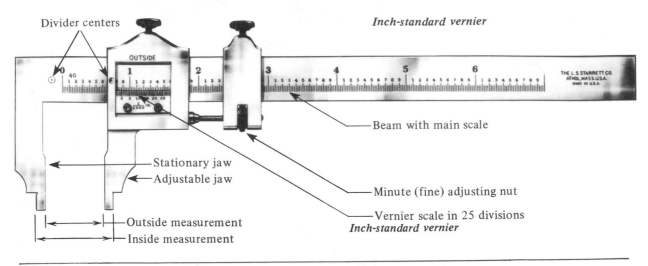

Figure 1–1 Principal Parts of a Vernier Caliper

Vernier measuring instruments incorporate the vernier principle with special features of measuring tools. The vernier measuring instruments described in this unit are the:

- Vernier caliper (inside and outside),
- Vernier depth gage,
- Vernier height gage,
- Vernier bevel protractor.

The vernier principle is also applied to graduated handwheels and tables on machine tools. In each of these applications, the vernier scale increases the precision to which a machine or a workpiece may be set for machining operations.

Two groups of precision measuring and layout instruments also covered in this unit are:

- *Test indicators* and *dial indicators*. These instruments provide for accurate, direct measurements by multiplying minute movements through a system of levers or gears;
- *Gage blocks*. These standards are combined with vernier measuring and dial indicating instruments for both measurement and layout work. Gage blocks are made to extremely close tolerances. These blocks permit measurements to within two one-millionths of an inch (0.000002″) or the equivalent metric value.

PARTS AND SCALES OF VERNIER INSTRUMENTS

In the mid 1800s Joseph R. Brown (a cofounder with his father of the Brown & Sharpe Manufacturing Company) applied the vernier principle to the slide caliper. The series of measuring instruments that resulted from Brown's application are identified by names that incorporate the term *vernier* with the name of the type of instrument. For example, the combination of *vernier* and *height gage* identifies the vernier height gage. Similarly, the names of the vernier depth gage and vernier bevel protractor incorporate the main features of the instruments with the vernier term.

The main parts and scales of a vernier measuring instrument are shown on the vernier caliper in Figure 1–1. A *vernier scale* is generally mounted on the adjustable jaw of the instrument. A *beam*, or frame, with a stationary jaw contains the *main scale*. The vernier and main scales are designed in a straight line, as exemplified by the vernier caliper (Figure 1–1), vernier height gage, and vernier depth gage. These scales also may be imprinted along a curved surface. The vernier bevel protractor is an example.

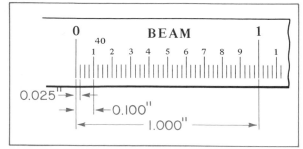

Figure 1–2 Inch, Subdivided into 40 Equal Parts, on the Main Scale of an Inch-Standard Vernier Caliper

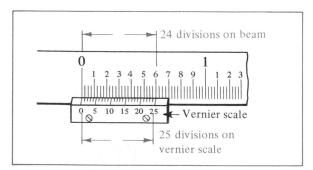

Figure 1–3 Difference between Main Scale Graduations and Vernier Scale Graduations (0.001″)

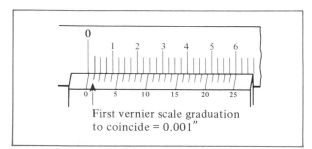

Figure 1–4 Vernier Scale Reading of 0.001″

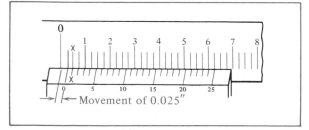

Figure 1–5 Zero (0) on the Vernier Scale Coinciding with a 0.025″ Graduation on the Main Scale

VERNIER PRINCIPLE AND VERNIER CALIPER MEASUREMENT

The vernier caliper differs from the vernier micrometer in construction and in method of taking a measurement. The vernier caliper is used for *internal* (inside) and *external* (outside) linear measurements. The following explanation of the main and vernier scales is made in relation to the inch-standard vernier caliper. The common 25-*division vernier caliper* is used to describe the vernier principle. The vernier principle also applies to the metric vernier caliper. However, measurements are taken in decimal parts of a millimeter.

THE INCH-STANDARD VERNIER CALIPER

The main scale of the beam of the inch-standard vernier caliper is divided into inches (Figure 1–2). Each inch is subdivided into 40 equal parts. The distance between graduations is 1/40″, or 0.025″. A second scale on the vernier caliper, the vernier scale, is imprinted on the movable jaw and therefore moves along the main scale. The 25 divisions on the vernier scale correspond in length to 24 divisions on the main scale (Figure 1–3). The difference between a main scale graduation and a vernier scale graduation is 1/25 of 0.025″, or 0.001″.

When the adjustable jaw of the vernier caliper is moved until the first vernier scale graduation coincides with the first graduation on the beam, the jaw is opened 0.001″. The vernier scale reading is illustrated in Figure 1–4. As the movement is continued until the second graduations coincide, the opening represents a distance of 0.002″. When the jaw has moved 0.025″, the *zero index line* on the vernier scale and the first graduation (0.025″) on the beam coincide (Figure 1–5). This represents a measurement of 0.025″.

A vernier caliper reading consists of (1) the number of inches (for dimensions larger than one inch) and (2) decimal parts of an inch. The measurement is read from the beam and vernier graduations. The decimal values on the beam (0.025″ graduations) and on the vernier scale (0.001″ graduations) are added to any whole-inch values.

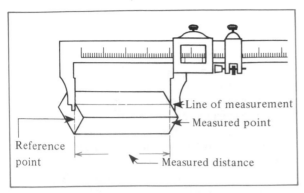

Figure 1–6 Position of the Vernier Caliper
for an Accurate Measurement

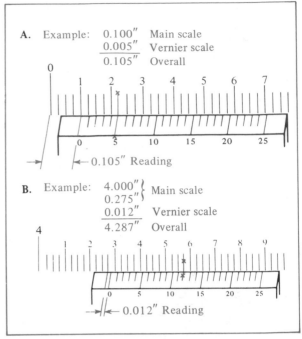

Figure 1–7 Steps in Reading a
Vernier Caliper Measurement

Two features of the vernier caliper are that (1) inside and outside measurements may be taken directly and (2) an adjustment screw permits the movable jaw to be minutely adjusted to a measurement (the jaw is then locked in position by a clamping screw). Because a vernier instrument is often more difficult to use than a micrometer, the worker must establish which instrument is more practical for the job. Figure 1–6 shows the positioning of a vernier caliper for an accurate measurement.

How to Read a Vernier Caliper Measurement

Taking a Measurement

STEP 1 Check the workpiece and the vernier caliper. They must be clean and free of burrs.

STEP 2 Check the vernier caliper in the fully closed position. The reading should be zero.

STEP 3 Loosen the clamping screws. Slide the movable jaw. The distance from the stationary jaw should be slightly smaller than the inside measurement to be taken. The jaw opening should be a little larger for an outside measurement.

STEP 4 Bring the vernier caliper to the feature of the workpiece that is to be measured. Make "rough" adjustments, if needed; then lock the clamping screw.

STEP 5 Hold the stationary jaw to the reference point. The beam is brought to a position parallel to the line of measurement.

STEP 6 Bring the movable jaw into position at the measured point. Turn the fine adjusting nut until the caliper jaw just slides over the measured point. Lock the movable jaw at this point. Recheck the feel of the measurement.

Reading a Measurement

STEP 1 Read the 0.025″ graduation(s) on the beam (main scale). The graduations appear to the left of the "0" on the vernier scale. Each graduation equals 0.025″. Multiply 0.025 by the number of graduations. The example in Figure 1–7A shows a beam (main scale) reading of 0.100″.

STEP 2 Sight along the vernier scale to establish which vernier line (graduation) coincides with a line (graduation) on the beam. In the example in Figure 1–7A, the vernier scale reading is 0.100″ + 0.005″, or 0.105″.

STEP Check the whole-inch number on the
3 beam scale (for measurements of one
inch and larger).

STEP Establish the overall vernier measurement.
4 Add the whole-inch number, the 0.025"
reading on the beam scale, and the thousandths reading indicated on the vernier
scale. (In the example given in Figure
1–7B, the overall reading is 4.287".)

STEP Retake the measurement and recheck the
5 reading for accuracy. Then clean the instrument and safely store it.

METRIC VERNIER CALIPER MEASUREMENTS

The applications of metric and inch-standard
vernier calipers are the same. The major difference is that measurements are taken in metric
units. The beam of the metric vernier caliper is
graduated in millimeters. A graduation on the
beam, illustrated in Figure 1–8 equals 1mm.
Every tenth graduation is marked with a numeral.
The "80" graduation on the beam scale thus
represents 80.00mm.

The illustration in Figure 1–8 shows a
beam with two scales. The top scale is offset
horizontally from the bottom scale to compensate for the width of the vernier caliper jaws.

Internal (inside) measurements are read on the
top scale. The bottom scale has the same
graduations as the top scale has. These graduations are also 1mm apart. The bottom scale is
used for direct external (outside) measurements.
On the model in Figure 1–8, inside and outside
measurements are read on the same side. Some
vernier calipers are available with only inch-
standard graduations. Others have one scale for
metric and a second scale for inch-standard
measurements.

On the all-metric vernier caliper in Figure
1–8, the top and bottom vernier scales are
graduated in 0.02mm. Every fifth graduation is
numbered. The "10" graduation on the vernier
scale thus represents 0.10 mm, "20" is 0.20mm,
and so on.

The metric vernier caliper is read in the same
manner as the inch-standard vernier caliper. The
decimal fraction of a millimeter on the vernier
scale is added to the whole number of millimeters on the beam. The decimal value is established at the point where a vernier graduation
coincides with a beam graduation.

The internal measurement indicated on
both the top and vernier scales in Figure 1–8
is 78.08mm. This measurement consists of a
beam reading of 70, or 78.00mm. The vernier
reading is the "4" graduation × 0.02mm, or
0.08mm. The correct measurement is thus
78.08mm. The external measurement, taken
from the bottom scale, is 80.08mm.

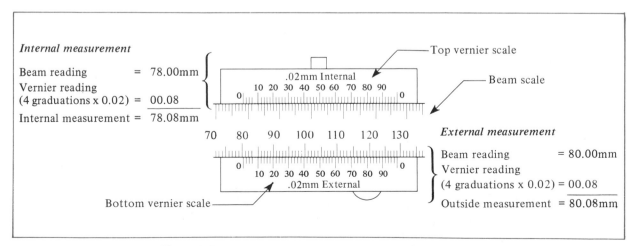

Figure 1–8 Beam and Vernier Scales with Metric Graduations

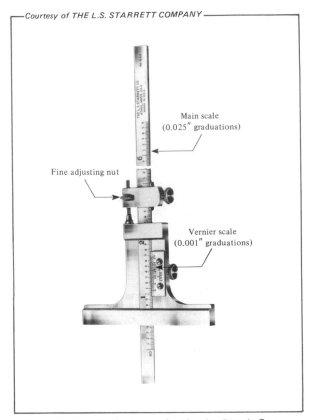

Figure 1–9 Main Features of a Vernier Depth Gage

The three main parts are the base, the column (beam), and the slide arm. The main scale is on the column. The vernier scale is attached to the slide arm.

Measurements are taken in the same manner as with other vernier measuring instruments. Whole-inch and 0.025″ (or centimeter and fractional millimeter) readings are obtained from the graduations on the beam column. The additional decimal values in multiples of 0.001″ (or 0.02mm) are read on the vernier scale. These readings are taken at the two graduations that coincide. The vernier height gage measurement consists of the (1) whole-inch number, (2) 0.025″ graduations to the left of the zero index line on the vernier scale, and (3) vernier scale reading in thousandths (0.001″).

There are a number of attachments used with the vernier height gage. A flat scriber or an offset scriber may be secured to the slide arm. These attachments are used in layout work or for making linear measurements. Depth measurements may be taken with a depth gage attachment or offset attachment. Many other special attachments are used for different measurements.

MEASURING WITH THE VERNIER DEPTH GAGE

The vernier depth gage (Figure 1–9) is essentially a regular depth gage with locking screws, an adjusting nut, a rule, and a vernier scale added. The rule in the inch-standard system is graduated in 0.025″.

Measurements are taken with the same care as for a regular depth gage with a line-graduated rule or a depth micrometer. The adjusting screw permits the blade to be brought against the work surface and locked in position. However, taking a precise accurate measurement is difficult.

MEASURING WITH THE VERNIER HEIGHT GAGE

The vernier height gage (Figure 1–10) is widely used in layout work and for taking measurements to an accuracy of 0.001″ or 0.02mm.

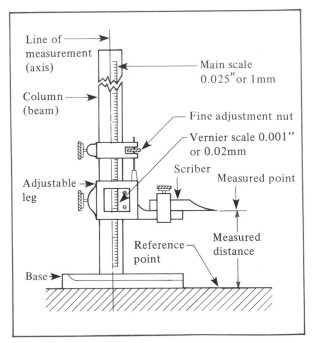

Figure 1–10 Features of the Vernier Height Gage

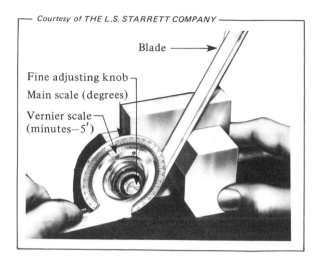

Figure 1–11 Features of a Vernier Bevel Protractor

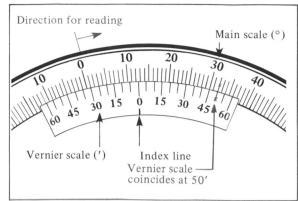

Figure 1–12 Example of 12° 50′ Readings

THE UNIVERSAL VERNIER PROTRACTOR

The vernier principle is also applied in making angular measurements. The vernier bevel protractor is an instrument used to measure angles to an accuracy of 5 minutes (5′), or 1/12 of one degree. The universal vernier protractor includes a number of attachments that make it possible to make a wide range of measurements. All measurements are within the same degree of accuracy as the standard vernier protractor.

Figure 1–11 shows the main features of a vernier bevel protractor. The protractor main scale (dial) is graduated in whole degrees. On some instruments this scale is on the body of the instrument itself. On other instruments the graduations are on the rotating turret. Depending on the manufacturer, the vernier scale is part of either the rotating turret or the body. Some main scales are arranged in four 90° quadrants (0° to 90° to 0° to 90° to 0°).

The vernier scale has 24 divisions; 12 divisions are on each side of the zero index line. The 24 vernier scale divisions are numbered from 60 to 0 to 60, as illustrated in Figure 1–12. Each vernier scale graduation represents 5′. The graduations on the vernier scale are numbered on *both* sides of the zero index line to make it easy to read the minute values. Some vernier scales are marked at every third graduation (60,

45, 30, 15, 0, 15, 30, 45, 60). Again, each line (graduation) on the vernier scale represents 5′.

In taking angular measurements, it is important to read the vernier scale in the same direction from zero as the main scale reading. When the angle is an exact whole number of degrees, the index line (0) on the vernier scale coincides with a whole-degree graduation on the main scale. However, if the angle is more than an exact whole number of degrees, the fractional value (in 5′ increments) is read on the vernier scale. In Figure 1–12 the index line (0) has moved past 12° on the main scale. The 50′ line on the vernier scale coincides with a graduation (32°) on the main scale. When the vernier scale is read in a clockwise direction, the angle represented on the vernier protractor in Figure 1–12 is 12°50′.

How to Measure with the Vernier Protractor

STEP 1 Position the vernier protractor and the workpiece on a surface plate.

STEP 2 Adjust the blade of the vernier protractor with the sensitive adjusting screw to the angle of the workpiece.

STEP 3 Remove the vernier protractor. Read the number of the whole degrees (12°) indicated on the protractor main scale.

STEP 4 Continue reading in the same direction. Determine which graduation on the vernier scale coincides with a graduation on the protractor scale. The vernier scale 50′ graduation coincides with the main scale 32° graduation (Figure 1–12).

STEP 5 Add the vernier scale reading (in minutes) to the whole-degree reading on the main scale. The reading in whole and fractional degrees is 12°50′ (Figure 1–12).

SOLID AND CYLINDRICAL SQUARES

SOLID SQUARE

The solid, hardened steel square is a precision layout and angle-measuring tool. The solid square may have either a flat or a knife edge. This square is widely used for checking the squareness of a part feature that is 90° from a reference surface. The square is brought into contact with the feature. Any variation from a true 90° may be detected by placing a white paper behind the square and observing whether any white is visible.

CYLINDRICAL SQUARE

The cylindrical square, shown in Figure 1–13, is another tool that is used to make extremely accurate right-angle measurements. Markings on the cylindrical square show deviations from a true square condition in steps of 0.0002″.

The cylindrical square and the workpiece usually are placed on a surface plate. The part surface to be measured is moved into contact with the cylindrical square. The cylindrical square is rotated until all light between the part feature and the square is shut out. The deviation from squareness is read directly from the 0.0002″ graduations on the cylindrical square. Figure 1–13 shows a common application of the cylindrical square.

TEST INDICATORS AND DIAL INDICATING INSTRUMENTS

Two groups of indicators are commonly used in shops and laboratories. Test indicators and dial indicators are *comparison instruments*. Unlike direct measuring instruments, which incorporate

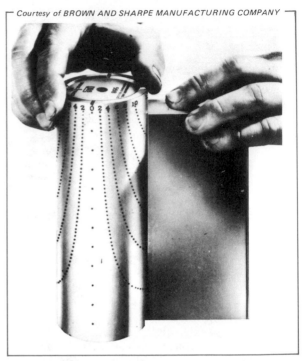

Figure 1–13 Measuring Squareness with a Cylindrical Square

a standard unit of length, test indicators and dial indicators first must be set to a reference surface. The unknown length of the part to be measured is then *compared* to a known length. Test indicators and dial indicators are positioned from a stand or adjustable holder to which they may be secured.

TEST INDICATORS

Test indicators (Figure 1–14) are used for trueing and aligning a workpiece or a fixture in which parts are positioned for machining. Once the workpiece or fixture is aligned, machining, assembling, and other operations are performed.

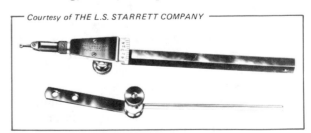

Figure 1–14 A Test Indicator and Attachment

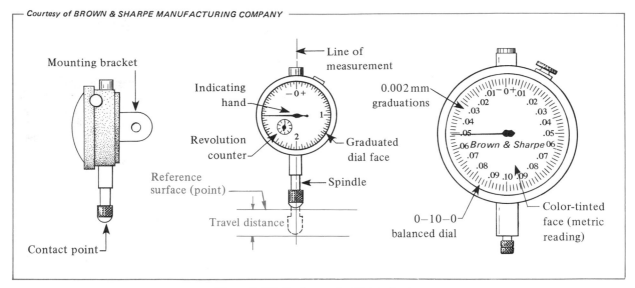

Figure 1–15 Features of a Dial Indicator

Test indicators are also used to test workpieces for roundness or parallelism or for comparing a workpiece dimension against a measurement standard.

The test indicator is held securely in a stationary base. The ball end of the test indicator is usually brought into contact with the workpiece surface. As the workpiece is moved, any deviation from the dimension to which the test indicator is set is indicated by movement of the indicator. The range of measurements is limited. Most test indicators employ a lever mechanism that multiplies any minute movement of the measuring (ball) end.

DIAL INDICATING INSTRUMENTS

The dial test indicator (Figure 1–15) is commonly called the dial indicator. It contains a mechanism that multiplies the movement of a contact point. This movement is transmitted to an indicating hand. The amount of movement is read on a graduated dial face. Because the dial indicator is more functional and more accurate than the test indicator, it has a wider range of applications.

The dial indicator serves two major functions:

• To measure a length (the distance between a standard dimension to which the dial indicator is set and the length of the part being measured),

• To measure directly how much a feature of the workpiece is *out-of-true*.

Inch-standard dial indicators are commercially produced to measure to accuracies of 0.001″, 0.0005″, and 0.0001″. Metric dial indicators are accurate to 0.02 mm and 0.002mm. Dial face graduations are marked to indicate the degree of accuracy.

Some dials are graduated to show + and − variations from 0 to 0.025″ or from 0 to 0.0005″. These dials are known as *balanced dials*. Examples of the many types of balanced dials are: 0–5–0 (0–0.0005″–0), 0–50–0 (0–0.050″–0), and 0–100–0 (0–0.100″–0).

Continuous dials are graduated from 0 to 0.100″. Readings beyond 0.100″ are made by multiplying the number of complete dial hand revolutions by 0.100″. The revolutions then are added to the reading of the indicating hand.

Long-range dial indicators are available for measuring dimensions up to 10″. These dial indicators are equipped with revolution counters. The long-range indicator in Figure 1–16 has a 0–2.000″ range, a 0–0.100″ dial, a revolution counter (0–1.000″), and an inch counter (0–1.000″–2.000″).

Common Applications of Dial Indicators. The dial indicator is constructed so that the

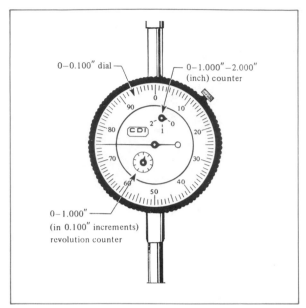

0-0.100″ dial

0-1.000″-2.000″ (inch) counter

0-1.000″ (in 0.100″ increments) revolution counter

Figure 1-16 Long-Range Dial Indicator with Revolution Counters (Balanced Type Dial)

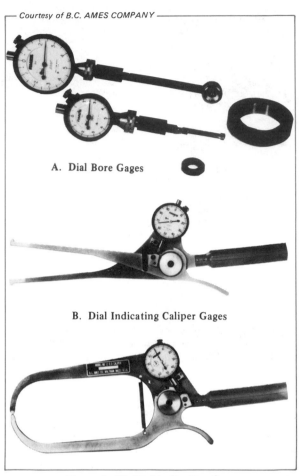

A. Dial Bore Gages

B. Dial Indicating Caliper Gages

Figure 1-17 Dial Indicating Gages

indicator mechanism may be used alone. It may be secured to a holding device and positioned to take a measurement. For example, in applying the dial indicator to check the out-of-roundness of a part on a lathe, the indicator is clamped in a holder. The indicator holder, in turn, is mounted in a lathe tool holder. As the workpiece is slowly rotated, any roundness variation causes the contact end of the indicator to move up or down.

Another common application is checking the accuracy of a lathe spindle. A precision-machined test bar, which is inserted in the taper of the spindle, is tested for runout (levelness) with the dial indicator.

Dial Indicating Gages. The dial indicator often is combined with other gaging devices and measuring tools. A few examples are given in Figure 1-17 to show the versatility of the instrument.

When the dial indicator is combined with a depth gage, it is called a *dial indicating depth gage*. When the dial indicator is adapted to testing holes for size and out-of-roundness or other surface irregularities, the instrument is called a *dial indicating hole gage*, or *dial bore gage*. The dial indicator may also be adapted for taking inside or outside measurements. Instruments called *dial indicating caliper gages* have revolution counters. The counter permits measurement of dimensions to 3″ (75mm).

Dial indicating snap gages are another adaptation of the dial indicator. These gages are used for linear measurements as well as for measuring diameters.

Dial Indicating Micrometers. Inch-standard *dial indicating micrometers* (Figure 1-18) can measure dimensional variations in one ten-thousandth of an inch (0.0001″) and one one-thousandth of an inch (0.001″). Metric dial indicating micrometers are used to measure to accuracies of 0.02mm and 0.002mm.

The lower anvil of the dial indicating micrometer is a sensitive contact point. It is

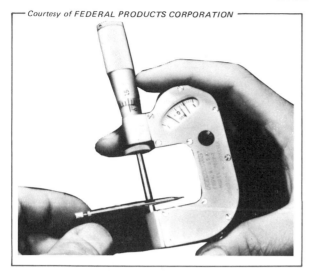

Courtesy of FEDERAL PRODUCTS CORPORATION

Figure 1–18 Dial Indicating Micrometer (±0.0001")

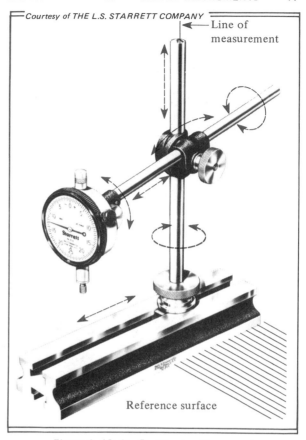

Courtesy of THE L.S. STARRETT COMPANY

Line of measurement

Reference surface

Figure 1–19 Six Positioning Movements of a Dial Indicator

connected directly to a dial indicator that is built into the instrument. When used as a micrometer, the thimble is set to the one one-thousandth (0.001") setting nearest the required measurement and is locked in position. As the dial indicating micrometer is brought into contact with the workpiece, it serves as a gage. Any variation from the required size appears on the dial indicator. The dial is graduated in increments of 0.0001", with a range up to 0.001".

On other model dial indicating micrometers with vernier thimble readings of 0.0001", the dial graduations are 0.00005"; the dial range, ±0.0025" (or metric equivalent).

Universal Dial Indicator Sets. The term *universal dial indicator set* denotes that there are a number of attachments for holding and positioning the dial indicator. For instance, to check the height of a machined surface, the dial indicator head may be mounted on a T-slot base (Figure 1–19). The indicator is set at a desired height. It is then moved along a surface plate and positioned over the part of the workpiece that is to be measured.

Dial indicator applications include checking the out-of-roundness, runout, parallelism, or face alignment of a workpiece. In these applications the dial indicator may be held in a tool post holder attachment on a lathe or other machine. The dial indicator may be positioned to measure along any one, or combination of three axes (Figure 1–19).

Safe Practices in Using Precision Measuring and Layout Instruments

- Check the workpiece to see that it is free of burrs and foreign particles. A clean workpiece helps to ensure the accuracy of all measurements.
- Clean all contact surfaces of vernier measuring and dial indicating instruments and related accessories.
- Check each instrument for accuracy. For example, in the fully closed position, the inside and outside readings of the vernier caliper must be zero (0.000" or 0.00mm).
- Bring the jaws or other instrument surfaces lightly into contact with the workpiece. Any force applied to the instrument will cause springing and inaccurate measurements.

- Take measurements as close to the instruments as possible. The longer the distance a measurement is taken from the base or frame of the instrument, the greater the probability of error.
- Check the alignment of the instrument, the condition of the reference surface, and the position of the workpiece before taking a measurement.

- Recheck each measurement to ensure that it is accurate.
- Clean each precision tool thoroughly. Instrument moving parts must be lubricated with a thin, protective film. All instruments should be carefully placed and stored in appropriate containers.

MEASUREMENT TERMS FOR VERNIER AND DIAL INDICATING INSTRUMENTS

Vernier instruments	Precision measuring instruments, layout measuring tools, and machine parts having a basic (main) measuring scale and a vernier scale.
Vernier scale	A secondary scale attached to a movable jaw or turret. (The vernier scale usually contains one less graduation than the number on the main scale of a fixed beam or head.)
Main (beam) scale	The graduations on the stationary frame or body of a measuring instrument. (The graduations on the scale are in inch or metric units and a decimal part of the unit.)
Coinciding lines	The alignment of two scale graduations. A graduation on the vernier scale that aligns with one on the stationary part of the instrument.
Vernier reading	An additional decimal value of a measurement. (The value is read on the vernier scale at the point where the vernier graduation coincides with a graduation on the main scale.)
Cylindrical square	A precision-finished cylinder that is graduated to show deviations in the squareness of a workpiece. (The variation is read in steps of 0.0002″ from a perfectly square condition.)
Alignment	The exact relationship between an instrument and a workpiece. A correct positioning of an instrument. Taking a measurement along the line of measurement between the measured points.
Comparison measurement	Setting a measuring instrument to a known length or measurement standard. (The unknown length is compared with the instrument setting.)
Dial indicator	An instrument that multiplies the contact point movement on a part to be measured. (The linear movement may be read directly. The indicator dial may be graduated in decimal values of the standard inch or millimeter.)
Test indicator	A measuring instrument in which movement is multiplied through a system of levers or gears. (A minute variation in movement of the contact end is magnified and observable on a scale at the measurement end.) An instrument for testing parallelism, alignment, trueing workpieces for machining, and making comparison measurements.

―――――――――――――――――――SUMMARY――――――――――――――――――

■ Vernier measuring tools are named according to the basic function of
the tool. The vernier height gage, the vernier bevel protractor, and
the vernier caliper are examples.

■ The vernier scale on inch-standard instruments extends the accuracy
range from 0.001″ to 0.0001″. Similarly, on metric standard instruments the basic 0.02mm accuracy is increased to 0.002mm.

■ A vernier measurement includes the reading on the main scale of the
instrument plus the reading on the vernier scale.

■ The vernier protractor applies the vernier principle to angular measurements. The vernier scale extends the accuracy of the angular reading
to within 5 minutes (5′).

■ Test indicators usually have a limited range of measurement. They
are particularly useful in checking alignment, parallelism, or out-of-
roundness.

■ Dial indicators are used to measure linear movements. When set to a
particular standard, it is possible to measure any variation on either
the + or − side of the standard measurement.

■ Solid and cylindrical squares are used to check squareness. The
cylindrical square measures any deviation from a true square condition to within 0.0002″.

■ Dial indicators are calibrated to show measurements over a wide range
of sizes. The range of accuracy is within 0.0001″ and 0.001″. In the
metric system the accuracies are 0.002mm and 0.02mm.

―――――――――――――UNIT 1 REVIEW AND SELF-TEST――――――――――――

1. Indicate the functions of (a) the beam containing the main scale of a vernier
measuring instrument and (b) the vernier scale.

2. Use the inch-standard 25-division vernier caliper to explain how to read a
measurement of 4.377″.

3. State how the accuracy of a metric vernier height gage may be checked.

4. Give three major differences between a universal bevel protractor and a standard bevel protractor of a combination set.

5. State two distinguishing design features between a solid square and a cylindrical square.

6. Indicate the different purposes that are served by test indicators and dial
indicating instruments in relation to vernier instruments.

7. Give four applications of dial indicators.

8. List three examples of instruments that combine the measurement features
of dial indicators with other measurement instrument functions.

Advanced Linear and Angular Measurement

Most linear measurements in the machine trades and metal products manufacturing industries are held to tolerances of ±0.001″ to ±0.0001″ or ±0.02mm and ±0.002mm. Similarly, angular measurements are made to within ± five minutes of arc (±5′). These dimensional accuracies are practical and efficient to produce and are reproducible anywhere in the world.

More precise limits are also common for machining, assembling, and measuring other workpieces. Linear measurement accuracies ranging to the millionth part of an inch and 0.000025mm are used regularly. Also, angular dimensions need to be measured to such degrees of fineness as one minute of arc.

Instruments and gages for highly precise measurements are normally made of metal. However, design improvements in measurement and layout tools, instruments, and accessories include the use of granite.

Surface plates, angle plates, universal right angles, parallels, and sine plates (described in this unit) are all available in granite. The advantages and applications of granite as a base material for linear and angular layout and measuring instruments also are covered in detail.

OBJECTIVES

After satisfactorily completing this unit, you will be able to:

- Apply technical information about black granite gage, measurement, and layout accessories.
- Identify features and applications of granite precision plates, universal right angles, parallels, etc. and a general-purpose bench comparator.
- Deal with factors influencing angular layouts, measurement, inspection, and machining setups.
- Translate design features of sine bars, sine blocks, and sine plates for single and compound angles.
- Use the sine formula, natural trigonometric tables, and a table of constants for sine bar measurements.
- Perform the following angle setups and measurement processes:
  - Set Up a Gage Block Measurement.
  - Make Direct Calculations for Sine Bar Setups.
  - Set Up a Sine Plate for Layout and Inspection.
  - Set Up a Permanent Magnetic Sine Plate.
- Analyze *Safe Practices* for protecting precision instruments.
- Apply each new precision measurement *Term*.

PRECISION GAGE BLOCKS

Gage blocks are blocks of steel that are heat treated to a high hardness. Gage blocks are machined, ground, and finished to precise limits of dimensional accuracy. The heat-treating process makes it possible to retain accuracy and dimensional stability. The hardness produced by heat treating helps prevent wear and damage to the block surfaces.

Parallel gage blocks are available in three general grades: (1) *master blocks*, which are accurate to within 0.000002″ or 0.00005mm; (2) *inspection blocks*, which are accurate to within 0.000005″ or 0.0001mm; and (3) *working blocks*, which are accurate to within 0.000008″ or 0.0002mm. Master and inspection blocks are usually used in temperature-controlled laboratories.

The dimensional accuracy of parallel gage blocks relates to flatness, parallelism, and length. Some sets include two additional *wear blocks* (0.050″), which are used as *end blocks* to prevent wear on the other blocks.

CARE OF GAGE BLOCKS

Gage blocks are furnished in specially designed cases. The size of each block is easily identified. The cover of each case lies flat. It serves as a tray in selecting and holding gage block combinations.

Lint or dust should be brushed away from a gage block with a soft camel's hair brush. The blocks should be wiped clean before and after use with lint-free wiping tissues. An aerosol-propelled cleaning fluid serves as a convenient way to clean blocks as they are selected. After they are used and wiped, blocks should be coated with a fine wax-base film and then returned to their appropriate cases.

Microscopic nicks and burrs prevent *perfect wringing* of gage blocks. A deburring stone is used to remove the fine imperfections. Blocks are selected and handled during rustproofing with plastic-tipped forceps.

PARALLEL GAGE BLOCK SETS AND APPLICATIONS

Practical gage block applications include:

- Checking precision measuring instruments and gages;
- Setting other instruments such as dial indicators and height gages to make comparison measurements;
- Laying out machined surfaces where highly accurate linear measurements are required.

Gage blocks are furnished in many combinations. Two common sets have 83 or 35 gage blocks each. The sizes of the gage blocks in the 83-piece inch-standard set are given in Table 2-1. Note that in the 9-block 0.0001″ series each block increases in size by an increment of 0.0001″. The increment in the 49-block 0.001″ series is 0.001″. In the 0.050″ series there are 19 blocks varying in size from 0.050″ to 0.950″ in increments of 0.050″. There are 4 blocks in the 1.000″ series, with sizes from 1.000″ to 4.000″.

Over one-hundred thousand different measurements may be made with the 83-piece set by combining different gage blocks. The gage block surfaces are so precisely finished that

Table 2-1 Series and Sizes of Gage Blocks—83 Piece Set

0.0001 Series (9 Blocks)								
.1001	.1002	.1003	.1004	.1005	.1006	.1007	.1008	.1009

0.001 Series (49 Blocks)												
.101	.102	.103	.104	.105	.106	.107	.108	.109	.110	.111	.112	.113
.114	.115	.116	.117	.118	.119	.120	.121	.122	.123	.124	.125	.126
.127	.128	.129	.130	.131	.132	.133	.134	.135	.136	.137	.138	.139
.140	.141	.142	.143	.144	.145	.146	.147	.148	.149			

0.050 Series (19 Blocks)											
.050	.100	.150	.200	.250	.300	.350	.400	.450	.500	.550	.600
.650	.700	.750	.800	.850	.900	.950					

1.000 Series (4 Blocks)			
1.000	2.000	3.000	4.000

0.050 Wear Blocks (2 Blocks)

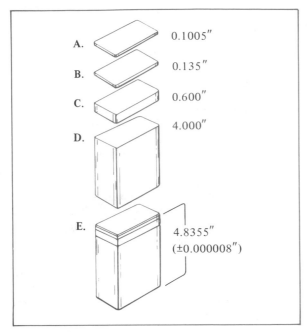

A. 0.1005″

B. 0.135″

C. 0.600″

D. 4.000″

E. 4.8355″
(±0.000008″)

Figure 2–1 Selection of Gage Blocks
for a Measurement of 4.8355″

when perfectly clean, they may be carefully *wrung together* to become almost one gage block.

Gage blocks are assembled by overlapping one clean block on another. A gentle force is applied while sliding the blocks together. The gage blocks may be taken apart by reversing the assembly process. In all gage block applications, it is important to use the least number of blocks possible.

Gage blocks are selected starting with a block in the 0.0001″ series which corresponds with the fourth place decimal value in the required measurement. This is followed by one or more blocks in the 0.001″ series, then the 0.050″ series, and the 1.000″ series, as required. Figure 2–1 provides an example of a gage block combination for a measurement of 4.8355″ (±0.000 008″).

METRIC GAGE BLOCKS

Sets of metric gage blocks are available in series. Increments within a common series range from 0.001mm to 25.0mm. The lengths of the blocks in the series range from 0.50mm through 100.0mm.

The number of blocks in a set may vary up to 112 pieces.

How to Set Up a Gage Block Measurement

Note: Refer to Figures 2–1 and 2–2 in attempting to make a gage block setup for measuring 4.8355″ ±0.000008″. A set of *working gage blocks* will give the degree of accuracy specified.

Selecting the Gage Blocks

STEP 1 Determine the last-place decimal value (number) in the required setup (.— — — 5″).

STEP 2 Select a gage block in the 0.0001″ series that ends in 5. In this example it is 0.1005″ (Figure 2–1A).

STEP 3 Select a gage block in the 0.001″ series. The size must have the same second- and third-place decimal value (.—35) as the required dimension. A practical size for the example is 0.135″ (Figure 2–1B).

STEP 4 Select a gage block in the 0.050″ series. This block must equal the remaining first-place decimal value of (.6— —). The 0.600″ block is selected for the example (Figure 2–1C).

STEP 5 Select a gage block in the 1.000″ series that is equal to the required whole number (4.— — —). The 4.000″ block is used for the example (Figure 2–1D).

STEP 6 Add the values of the gage block combination. Recheck it against the required measurement. The combination should contain the least number of blocks possible. The four gage blocks selected for the example measurement of 4.8355″ ±0.000008″ are shown combined in Figure 2–1E.

STEP 7 Check each block to see that it is clean.

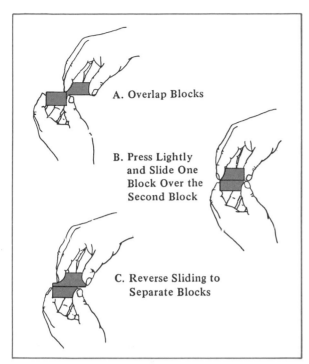

A. Overlap Blocks

B. Press Lightly and Slide One Block Over the Second Block

C. Reverse Sliding to Separate Blocks

Figure 2–2 Wringing and Separating Gage Blocks

Assembling (Wringing) Gage Blocks

STEP 1   Start with the largest blocks. Overlap each block as illustrated in Figure 2–2A.

STEP 2   Press the blocks together by hand. Carefully slide each block into position (Figure 2–2B).

STEP 3   Repeat the overlapping, pressing, and sliding steps with each successive set of blocks.

Disassembly

STEP 1   Reverse the assembly process (Figure 2–2C). Start with the smallest block and slide it off the next block combination.

STEP 2   Continue to slide each successive block off the combination.

STEP 3   Rewipe the gage blocks. Insert each block in its proper storage position.

TUNGSTEN CARBIDE GAGE BLOCKS

Steel gage block sets are commercially available in round, rectangular, and square shapes. Other gage blocks are made of tungsten carbide with a hardness of 70 R_C. Tungsten carbide gage blocks have increased stability and durability. Also, the fine grain structure permits surface finishing to extreme accuracy (0.2 millionths or 0.0005 metric microns) to further reduce wear. Since tungsten carbide gage blocks have a high resistance to corrosion and a low coefficient of expansion, these properties permit greater measurement accuracy.

GAGE BLOCK ACCESSORIES SYSTEM

Applications of gage blocks are extended by using accessories. For example, assembled gage blocks and accessories may be used for precision layouts, gaging, and the taking of internal and external measurements. When a number of accessories are included in a set, they are referred to as a *gage block accessory system*.

Figure 2–3 shows a number of components in a rectangular gage block accessory system. Similar parts are available in both inch and metric measurement standards. The inside and outside *caliper bars* shown in the figure each have a flat surface lapped on one side. This surface is used when outside linear measurements are taken. The second side of each caliper bar has a radius. This radius permits the taking of inside linear measurements.

The gage block accessory system may also serve as a snap gage, as shown in Figure 2–4. The caliper or gage block is assembled in a component called a *channel* (Figure 2–3). The channel is closed at one end. Accessories like the combination caliper and gage blocks drop easily against the back of the channel. The gage blocks are automatically aligned.

A *micro-clamp* (Figure 2–3) is used at the end of the buildup. The clamp exerts the proper force for the stack of gage blocks and accessories.

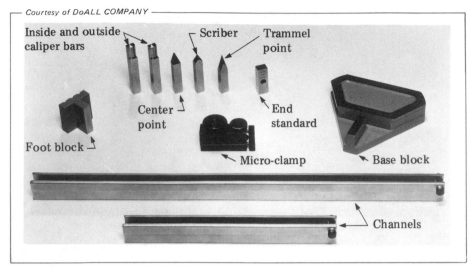

Figure 2–3 Rectangular Gage Block Accessory System

The micro-clamp consists of a clamp screw, a pusher screw, ball, and block. The clamping force applies in-line pressure. This pressure assures rigidity, parallelism, and accuracy. Surface wear on the gage blocks is thus minimized. Wringing the blocks together is eliminated in a micro-clamp setup. An *end standard* (Figure 2–3) may be assembled with gage blocks in a channel.

A *scriber* (Figure 2–3) is an accessory that may be set to permit scribing accurate lines at required precision height for a layout application. A height gage assembly for a measurement application is shown in Figure 2–5. Gage blocks, the outside (flat) face of a combination caliper, a channel, and a micro-clamp are used in the setup. While accessories and combinations have

Figure 2–4 Snap Gage Assembly of Gage Block Accessory System

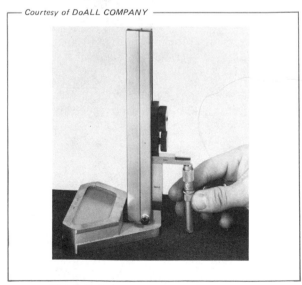

Figure 2–5 Height Gage Assembly Using Outside Face of Caliper Bar

been described for rectangular gage block applications, a similar system is available for square gage block sets.

BLACK GRANITE SURFACE PLATES

Granite surface plates are used in many laboratory, inspection, and shop applications. They provide high degrees of flatness and accuracy required in precision layout and machining processes.

Black granite surface plates are manufactured in three standard grades: AA, A, and B. *Grade AA* is used in the laboratory. Grade AA surface plates are produced commercially in basic sizes of 8" × 12" × 3" thick or 200mm × 300mm × 75mm. The overall accuracy is 50 millionths of an inch (0.000050") or 12 ten-thousandths of a millimeter (0.0012mm). One of the largest commercial sizes of black granite surface plates is 72" × 144" × 18" thick (1800mm × 3600mm × 450mm). The laboratory quality of this size of surface plate has an overall accuracy of 0.0011" or 0.028mm.

Grade A is generally used for inspection work. Grade A surface plate sizes begin at the same basic size as grade AA plates—that is, 8" × 12" (200mm × 300mm). However, the surface plate thicknesses are slightly smaller for each size. The overall accuracy of grade A plates is not as precise as the AA laboratory grade. For example, the accuracy of the work surface of the 8" × 12" × 2" thick (200mm × 300mm × 50mm) basic plate is 0.0001" (0.0025mm). The largest size of plate, 72" × 144" × 16" (1800mm × 3600mm × 400mm) is accurate to within 0.0022" (0.056mm).

Grade B surface plates (for the same range of sizes) are the least accurate of the three grades. The 8" × 12" surface plate is lapped to an accuracy of 0.0002" (0.005mm); the 72" × 144", to within 0.0044" (0.011mm).

ADAPTATIONS OF GRANITE SURFACE PLATES

Toolmaker's Flats. Handy, lightweight, and portable granite plates, called *flats*, are used by toolmakers, diemakers, inspectors, and machinists. These plates provide uniform flatness standards. Square or rectangular surface plates are used for layout, assembly, and inspection processes on the bench.

Master Flats. Super-accurate master flats are available for laboratory applications. *Master flats* are usually cylindrical granite plates. Standard sizes range from 6" to 48" (150mm to 1200mm) in diameter. The thickness varies from 2" to 16" (50mm to 400mm), respectively. The flat surface is lapped to provide for repeat measurements to 10 millionths of an inch. This degree of accuracy is compatible with laboratory-grade gage blocks

Modified Surface Plates. Granite surface plates are available with holes, grooves, slots, and metal inserts. Holes may be drilled into a surface plate. The holes permit the layout and measurement of a part that has one or more projecting lugs. A stainless steel threaded insert is useful in clamping a part to the face of the plate. T-slots that conform to dimensions and tolerances approved by the American National Standards Institute (ANSI) may be machined. Metal T-slot inserts may be fitted to a granite plate. Granite surface plates may also be machined with U-slots or to any contour shape. Three of these common design features are shown in Figure 2-6.

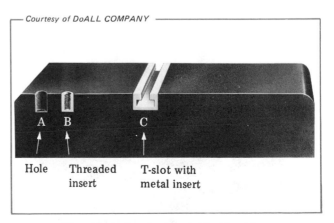

Courtesy of DoALL COMPANY

Figure 2-6 Granite Block Machined for Layout and Machining Processes

GRANITE PRECISION ANGLE PLATES

Granite *precision angle plates*, such as the plates shown in Figure 2–7, are available for production or for inspection. The difference between production and inspection plates is in the degree of accuracy. Production plates are accurate to within ±0.00005″ in 6″ (0.001mm in 150mm). The accuracy of inspection, or master, plates is ±0.000025″ in 6″ (0.0006mm in 150mm). The accuracies relate to flatness and squareness. Angle plates are lapped on either two surfaces (base and face) or four surfaces (base, face, and two ends). These surfaces are finished square and flat.

Angle plates are available with *metal inserts.* Examples of these plates are shown in Figure 2–7. Some inserts are recessed into one side and the main gaging face. These inserts are used for magnetizing purposes. Threaded inserts are available for clamping applications.

The term *universal* indicates that all faces on a *universal right angle plate* may be used. All six faces of the universal right angle shown in Figure 2–8 are square and parallel.

PARALLELS AND OTHER FORMS OF GRANITE ACCESSORIES

Parallels are matched in pairs. Granite parallels serve the same purposes as ground steel parallels. They are machined and finished for production and inspection (master) standards of accuracy.

Either the top and bottom or the top and bottom and two sides may be finished with parallel surfaces.

Black granite precision box parallels are matched pairs of parallels. The base area is smaller in comparison to the height than it is on regular parallels. Workpieces that are difficult to layout and inspect are usually elevated from a surface plate with box parallels. These parallels provide an accurate working surface that is parallel to the surface plate.

The granite *straight edge* has one precision finished plane face. If it is of inspection grade, the plane face is accurate to ±0.00005″ (1.0 μm) over 12″. The more precise laboratory-grade straight edge is finished to ±0.000025″ (0.6 μm) accuracy over 12″. Straight edges are used for such purposes as checking the straightness and parallelism of flat surfaces and the beds of machine tools.

Granite *precision step blocks* have flat, parallel, and square surfaces. These surfaces are held to close tolerances for flatness and parallelism.

A *cube* is really a combination of three sets of parallels. The opposite faces are parallel and square. The faces provide perpendicular, flat, and parallel reference surfaces in relation to the plane of a surface plate.

V-blocks support cylindrical workpieces during layout, manufacture, or inspection. The 90° V is centered and parallel to the bottom face. Some V-blocks are produced with the 90° V parallel to the bottom and two sides. The V is machined square (perpendicular) to the two ends.

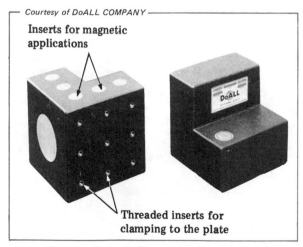

Inserts for magnetic applications

Threaded inserts for clamping to the plate

Figure 2–7 Angle Plates with Metal Inserts

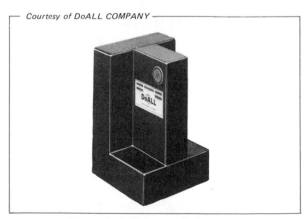

Figure 2–8 Universal Right Angle

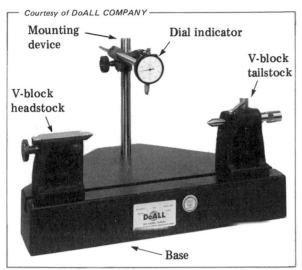

Mounting device

Dial indicator

V-block tailstock

V-block headstock

Base

Figure 2–9 Bench Comparator Equipped with V-Block Headstock, Tailstock, and Indicator

A *bench comparator* consists of a perfect plane surface (surface plate), a vertical post, a swivel arm, and a fine-feed adjustment with a dial indicator. The surface plate (plane reference surface base) may be produced from metal or black granite. The base may be slotted. Its shape may be modified to accommodate layout, inspection, and measurement accessories such as a headstock and an adjustable tailstock, V-blocks, or bench centers. An example of a V-block headstock and tailstock combination with a dial indicator is shown in Figure 2–9. The base, headstock, and tailstock are made of black granite.

Bench comparators are highly accurate for gaging roundness, flatness, parallelism, and concentricity. Measurements of length, width, height, steps, depths, and angles may be compared against linear standards.

PRECISION ANGLE MEASUREMENTS

INSTRUMENTS AND CONSIDERATIONS

Angle measurements to an accuracy within five minutes of arc are usually measured with a universal vernier bevel protractor. While the dial of this protractor may be rotated through 360°, part of the base and blade are covered. The range

of use of the protractor is thus limited. An angle attachment is used to increase the range. Since an attachment is needed for certain acute angles, it is known as an *acute angle attachment*. Two models of universal bevel protractors using the acute angle attachment are pictured in Figure 2–10. A magnifier lens on the protractor head of some models permits fast, easy angle readings.

Accurate, reliable angle measurements depend on three groups of considerations: mechanical, positional, and observational. These considerations are summarized in Table 2–2.

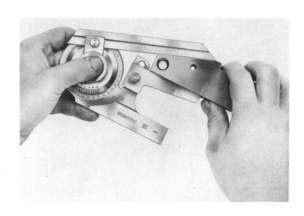

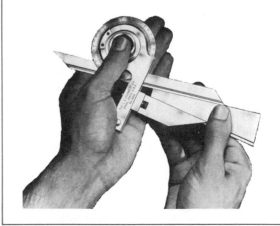

Figure 2–10 Measurements of Small Angles with Two Models of Universal Bevel Protractors Using the Acute Angle Attachment

Table 2-2 Mechanical, Positional, and Observational Considerations in Precision Angle Measurement

Mechanical considerations	• The surfaces to be measured are free of burrs, dust, and other foreign particles.
	• Surface roughness is reduced. The base and blade are brought into close contact with the angle surfaces to be measured.
	• The movable parts of the instrument have freedom of movement.
	• The accuracy of the instrument has been checked against a known measurement standard.
Positional considerations (relation between the instrument and the angle to be measured)	• The vertical (X) axis of the vernier bevel protractor falls in the vertical reference plane of the angle.
	• The horizontal (Y) axis of the instrument base is parallel to the base (surface) of the part to be measured.
	• The (Z) axis of the instrument is parallel to the (Z) plane of the angle.
Observational considerations	• The vernier reading is taken without parallax error.
	• Personal measurement bias is not read into the bevel protractor reading.
	• The angular reading is correct in terms of the required complement or supplement of the angle being measured.

SINE BAR, SINE BLOCK, AND SINE PLATES

Sine bars, sine blocks, and simple and compound sine plates and perma-sines are commonly used in layout, inspection, assembly processes, and measurement.

Angular surfaces of workpieces or assemblies are often layed out, machined, or inspected to a degree of accuracy within *one minute of arc*. This is a finer measurement than can be taken with a vernier bevel protractor. A sine bar, sine plate, sine block, and compound (universal) sine block are four precision angular layout and measurement instruments. Each of these instruments may be set to a required angle by applying trigonometry.

SINE BAR

A *sine bar* (as displayed in Figure 2-11) is a precise layout and measuring accessory. It consists of two rolls mounted on opposite ends of a notched steel or granite parallel.

All metal parts are hardened, machined to precise limits, and accurately positioned. The center distance between the rolls is fixed at lengths that are a multiple of 5″ (127mm).

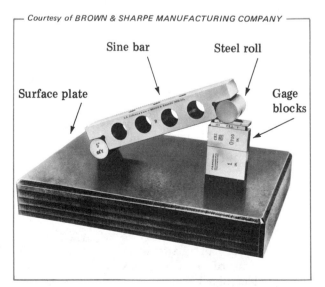

Courtesy of BROWN & SHARPE MANUFACTURING COMPANY

Sine bar   Steel roll

Surface plate   Gage blocks

Figure 2-11 Typical Angle Setting Using a 5″ Sine Bar, Gage Blocks, and Surface Plate

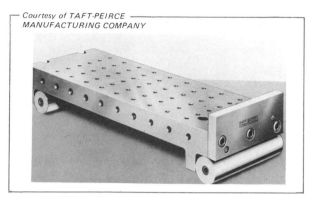

Figure 2–12 Sine Block with Tapped
Holes and End Plate (Stop)

Figure 2–13 Compound Sine Plate

SINE BLOCK

A *sine block* is essentially a wide sine bar (Figure 2–12). A sine block (or modification) provides maximum efficiency for layout, inspection, and angle measurements. The exact angular degree that a 5″ or 10″ sine block makes with a plane reference surface is determined by a precise vertical height (combination of gage blocks). These blocks are placed under the elevated end of the sine block.

SIMPLE SINE PLATE

Single angles, which lie in one plane, are usually set by using a *simple sine plate*. This sine plate has a precision base, a movable sine plate, and two hardened precision rolls. The base and elevating sine plate are hinged.

The simple sine plate is set at the required angle by establishing the overall measurement of a gage block combination.

COMPOUND SINE PLATE

Compound angles are angles formed by edges of triangles that lie in different planes. Compound angles may be layed out, inspected, or measured by holding a workpiece on a *compound sine plate* (Figure 2–13). A series of tapped holes in the top, sides, and end permit the direct strapping and clamping of workpieces.

A typical compound sine plate has an intermediate hinged plate. Attached to it is the top

sine plate. The design of this plate includes the same features as the features on a simple sine plate.

The intermediate plate and the top sine plate are each set at a required angle in one of two planes. The two angles make up the compound angle. The gage block combination for each angle is established. These combinations represent the heights to which the intermediate plate and sine plate are set.

PERMANENT MAGNETIC SINE PLATE (CHUCK): PERMA-SINE

A *permanent magnetic sine plate* is a simple or compound angle sine plate (Figure 2–14) in

Figure 2–14 Simple Angle, Permanent Magnet
Sine Plate (Perma-Sine)

which a permanent magnetic chuck forms the working face. The name is shortened so that the combination is often referred to as a *perma-sine* or a *sine chuck.*

Perma-sines are especially adapted to surface grinder work. The workpiece may be layed out, inspected, or machined. Angles produced are within one minute of arc accuracy. The fine magnetic pole spacing on the compound angle plate permits holding small and large parts securely. The magnetic force is turned on for holding and off for releasing by moving a simple lever.

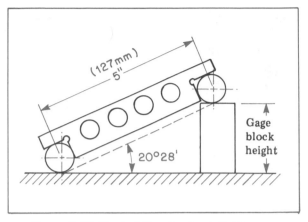

Figure 2–15 Typical Sine Bar Angle Setting

COMPUTING THE SINE VALUE (GAGE BLOCK HEIGHT)

The name *sine bar* is derived from the use of triangles and trigonometric functions. The *sines* of required angles are used to calculate the height of a triangle. The triangle formed consists of the hypotenuse of a known length and the required angle.

Figure 2–15 shows a typical angle setting of a sine bar. Gage blocks and a surface plate are used. The center-to-center distance between the hardened steel rolls of equal diameter is a multiple of 5″. The basic center distances are 5″, 10″, or 20″.

Example: Determine the gage block measurement (height) required to set a workpiece at an angle of 20°28′. A 5″ sine bar and surface plate setup are required. The sine value of 20°28′ must first be established. This value may be read directly in a *table of natural trigonometric functions.* The sine value for 20°28′ is 0.34966″.

COMPUTING INCH- OR METRIC-STANDARD GAGE BLOCK COMBINATIONS

After the sine value of a required angle has been established, the required gage block combination may be found. The sine formula is used. The values from the previous example are used in the formula as follows:

$$\frac{\text{sine of angle}}{(20°28')} = \frac{\text{height (gage block combination)}}{\text{length of sine bar (5'')}}$$

height = sine of angle (0.34966) × length of sine bar (5″)

= 1.7483″

If metric-standard gage blocks are to be used, the problem may be solved according to the same trigonometric function. The 1.7483″ gage block height may be converted to the metric equivalent of 44.407mm. Metric gage blocks of this height will produce an angle of 20°28′ when a 127mm (5″) sine bar is used.

How to Make Direct Calculations for Sine Bar Setups

Computing the Gage Block Height (5″ Sine Bar)

Note: The mathematical process of determining a gage block combination (height) for setting a sine bar or sine plate may be simplified. The use of a 5″ sine bar is assumed in the following procedure.

STEP 1 Look up the natural trigonometric function value of the required angle. In the earlier example, the sine value of 20°28′ is 0.34966.

STEP 2 Move the decimal point one place to the right (3.4966).

STEP 3 Divide by 2. The result represents the height of the gage block combination. In the example, 3.4966 ÷ 2 = 1.7483″.

> Note: The result represents the height to which the rear roll must be set on a sine bar, sine block, or sine table.

Determining the Angle from a Height Setting (5″ Sine Bar)

> Note: The angle of sine bar setup may be determined by reversing the mathematical steps used in computing gage block height. The 1.7483″ gage block combination is again used as an example.

STEP 1 Move the decimal point in the 1.7483″ measurement one place to the left (0.17483).

STEP 2 Multiply by 2. The product represents the natural trigonometric function of the required angle. In the example, 0.17483 × 2 = 0.34966.

STEP 3 Refer to a table of natural trigonometric functions. Locate the 0.34966 value in the sine column.

STEP 4 Read the angle corresponding to the 0.34966 value.

ESTABLISHING A HEIGHT SETTING FOR DIFFERENT LENGTH SINE BARS

Sine bars with center distances of 5″ and 10″ are common. For larger layout, inspection, and measurement operations, a 20″ sine bar is used. The basic 5″, 10″, and 20″ lengths were espe-cially selected in this unit to simplify the mathematical computations and setups. The height to which a 10″ sine instrument is set is established by simply moving the decimal point (in the natural trigonometric function value of the required angle) one place to the left. For example, with the 20°28′ angle, by moving the decimal point in the trigonometric sine value of 0.34966, the height becomes 3.4966. The value represents the gage block combination height.

TABLES OF CONSTANTS FOR SINE BARS AND PLATES

The machine tool industry has further simplified layout, inspection, and machining practices with sine bars and sine plates. *Tables of constants* are available in handbooks or are furnished by product manufacturers. Part of a table of constants is reproduced in Table 2-3. Note that values of degrees and minutes are given with the accompanying constant (inch value).

The values in the Table apply to 5″ sine bars or plates. Settings for 10″ instruments are found by multiplying by 2. For example, in the 20°28′ setting, the constant value shown in a complete table is 1.7483″ for a 5″ sine bar. A 10″ sine bar is set at 1.7483″ × 2, or 3.4966″. For a 20″ sine bar, the constant value is multiplied by 4. By contrast, the constant value in the table is divided by 2 when a 2 1/2″ sine bar is used.

The same table of constants is used to determine the degrees and minutes in an angle. For a 5″ setup, the height (gage block measurement) is located in the inch values section of the table. The degrees and minutes corresponding to the inch value represent the angle of the setup.

Table 2-3 Partial Table of Constants for Setting 5″ Sine Bars or Sine Plates (16°0′ to 23°5′)

Minutes	16°	17°	18°	19°	20°	21°	22°	23°
0	1.3782	1.4618	1.5451	1.6278	1.7101	1.7918	1.8730	1.9536
1	.3796	.4632	.5464	.6292	.7114	.7932	.8744	.9550
2	.3810	.4646	.5478	.6306	.7128	.7945	.8757	.9563
3	.3824	.4660	.5492	.6319	.7142	.7959	.8771	.9576
4	.3838	.4674	.5506	.6333	.7155	.7972	.8784	.9590
5	1.3852	1.4688	1.5520	1.6347	1.7169	1.7986	1.8797	1.9603

How to Set Up a Sine Plate for Layout and Inspection

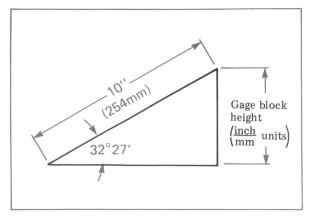

Figure 2–16 Known and Required Measurement (Inch-Standard Units)

Note: The following procedure requires the setting up of a 10″ sine plate to inspect a workpiece that is finished to an angle of 32°27′. Wear blocks are to be used.

STEP 1   Move the hand carefully over the working plane of the surface plate. Check for and remove any burrs and nicks.

STEP 2   Wipe the surface clean so that it is free of all dust.

STEP 3   Check and clean the two faces of the base, the top face of the sine plate, and the rear roll.

STEP 4   Position the workpiece on the sine plate. It should be square with the sides and end of the top plate.

STEP 5   Clamp the workpiece securely to the top plate.

STEP 6   Compute the gage block height for the 32°27′ angle (Figure 2–16):

$$\text{sine } 32°27' \ (0.53656) = \frac{\text{height}}{\text{hypotenuse (10″)}}$$

$$\text{height} = 0.53656 \times 10″$$

$$= 5.3656″$$

Note: The height may also be established by using a table of sine bar constants. The constant for a 5″ sine bar is 2.6828″, which is multiplied by 2 for a 10″ sine bar. The height, again, is 5.3656″.

STEP 7   Select a gage block combination for the 5.3656″ height. Use protective wear blocks in the combination.

Note: If metric gage blocks are to be used, multiply the natural trigonometric value by the hypotenuse—that is, by the center distance of the rolls in millimeters.

$$\text{sine } 32°27' \ (0.53656) = \frac{\text{height}}{\text{hypotenuse (254mm)}}$$

$$\text{height} = 0.53656 \times 254$$

$$= 136.28624\text{mm}$$

The five decimal places are rounded off to the nearest 5 ten-thousandths of a millimeter. In this example, the metric gage block measurement is 136.286mm.

With wear blocks, one possible combination of metric gage blocks for the 136.286mm height is as follows: 2.00 (wear block), 2.006, 2.28, 18.00, 50.00, 60.00, and 2.00 (wear block).

STEP 8   Clean each gage block. Use a camel's hair brush to remove any dust or lint.

STEP 9   Wring each gage block until the 5.3656″ (or 136.286mm) combination is formed.

STEP 10   Slide the gage block combination carefully onto the sine plate base.

STEP 11   Lower the top plate gently until the roll rests on the wear block of the gage block combination.

STEP 12   Set a dial indicator, vernier height gage, or other layout instrument to the required dimension. Perform any subsequent layout, gaging, or measurement operations. The sine bar and workpiece in this example are set at 32°27′.

How to Set Up a Permanent Magnetic Sine Plate (Perma-Sine)

Setting a Compound Angle

STEP 1 Examine the base of the perma-sine and face of the surface plate or machine table. Remove any burrs or nicks.

STEP 2 Strap the base of the perma-sine to the machine table. Flanges on the base permit clamping.

STEP 3 Wipe and use a fine bristle brush on the base, sine plate, and rolls. They must be clean and free of dust and foreign particles.

STEP 4 Determine the required gage block combination for the first angle and then for the second angle.

STEP 5 Select and wring the gage blocks together for the first setup.

STEP 6 Place the gage block setup between the base and the first hinged plate.

STEP 7 Lower the hinged plate and other parts of the perma-sine. The hardened roll should just set on the wear block of the gage block setup.

STEP 8 Repeat the preceding two steps for the second gage block combination.

STEP 9 Remove any burrs from the workpiece. Clean the workpiece and face of the magnetic chuck (sine plate).

STEP 10 Position the workpiece in relation to the sides of the sine plate. Use parallels between the end plates and the workpiece.

STEP 11 Turn the magnet lever on. The workpiece is ready for layout, inspection, or machining at the compound angle.

SAFE PRACTICES WITH PRECISION LINEAR AND ANGULAR MEASURING ACCESSORIES AND INSTRUMENTS

- Remove dust and foreign particles from gage blocks and microfinished surfaces with a fine, soft camel's hair brush.
- Check along microfinished surfaces that are to be in contact in order to detect any nicks or burrs. Remove all imperfections properly.
- Allow time for measuring instruments and workpieces to normalize to the same temperature.
- Avoid working near heat sources that produce varying temperatures of a workpiece and measurement tool.
- Check gage blocks, measurement surfaces, and instruments against master standards to ensure that correct measurements can be established.
- Place granite angle blocks, parallels, gage blocks, and other precision measurement accessories and instruments in their appropriate cases.
- Apply the amount of force on straps and clamps that is needed to hold parts securely. Any excess force tends to spring or bend an assembly and thus produces inaccurate measurements.
- Clean measuring instruments and dry them thoroughly.
- Lubricate moving parts. After an instrument is used and cleaned, apply a fine corrosive-resistant lubricant to it and then store the instrument in an appropriate container.
- Make sure that the dimensions, shape, and materials of a workpiece permit it to be drawn securely to a perma-sine.

TERMS USED IN PRECISION LINEAR AND ANGULAR MEASUREMENT

Dimensional stability (precision instruments) — The relief of stresses within an instrument to ensure continuing accuracy of measurement.

Black granite — A natural black-colored granite with great strength, hardness, rigidity, thermal stability, and a low (water) absorption rate.

Modified (granite) surface plate	A granite precision surface plate with holes or grooves cut into the face and/or with steel slides or threaded inserts.
Universal right angle	A precision ribbed steel or granite layout, measurement, and holding block. A right angle block with six faces that are accurately finished at right angles and are parallel.
Box parallel	A metal or granite parallel used to elevate a workpiece so that it may be accurately measured and layed out.
Precision step block	A block with stepped areas that are precisely machined for flatness and parallelism; used singly or in pairs.
Bench comparator	A machine having a precisely finished base, vertical post, and universal dial indicator that may be adjusted to a linear dimension.
One minute of arc	A precision angular measurement that is accurate within one minute of an arc of one degree.
Sine bar	A combined precision parallel and two hardened and finely finished rolls that are set apart at a fixed center distance.
Compound sine plate	Two sine blocks in different planes that are adjustable on a single base and may be set at a required compound angle.
Simple (single angle) sine plate	A combination adjustable sine block and attached base.
Table of constants for sine bars and plates	Tables for directly reading an angle or an inch or metric value that represents the measurement needed to produce a required angle setting when using a particular size sine bar.
Perma-sine	A trade designation for a permanent magnetic sine plate.

SUMMARY

- Steel gage blocks are produced to accuracies within one-millionth of an inch and 0.000025mm.
  - Metric tolerances in the equivalent of millionths of an inch are often given as microns (μm). For example, a tolerance of ±0.00005mm = ±0.05μm.
- Protective wear blocks are usually made of tungsten carbide. Wear blocks prolong the accuracy and life of gage blocks.
  - Precision scribers, inside and outside caliper bars, and other accessories extend the range of use of gage blocks.
- Some of the desirable properties of granite are: rigidity, low water absorption rate, thermal stability, fine texture (which permits the production of a precise lapped surface to which gage blocks and other precision tool surfaces do not wring), and hardness.
  - Accurate angular measurements depend on mechanical, positional, and observational factors.

- The use of sine bars, blocks, and plates requires trigonometry, computing gage block heights, and the use of formulas and tables of constants.

    - Toolmaker's flats and super-accurate cylindrical master flats provide precision plane surfaces.

- Angle plates; universal right angles; adaptations of parallels such as cubes, box parallels, and step blocks; and V-blocks are available in either metal or granite.

    - The bench comparator is a highly accurate precision instrument. It is adapted to gaging operations or for comparing linear measurements with fixed standards.

- Sine bars and accessories used in precision layout, inspection, machine setups, and measurement, permit accuracies to one minute of arc.

    - Simple or compound sine plates are used for work setups with angles in one or two planes, respectively.

- Perma-sines provide a permanent magnetic chuck mounted on a sine plate.

UNIT 2 REVIEW AND SELF-TEST

1. List three advantages of using granite over similar cast iron and steel layout and measuring instruments and accessories.

2. Tell briefly why the material in a precision gage block must be dimensionally stable.

3. State one reason for "burr proofing" the edges of gage blocks.

4. List the parts from a gage block accessory system to use for the following two setups: (a) an inside diameter snap gage and (b) a height gage assembly.

5. a. Name three properties of black granite.
   b. State briefly the importance of each property to precision measurement and inspection.

6. a. Refer to a table of constants for 5″ sine bars (blocks).
   b. Determine the required gage block height for each of the following 5″ sine block angle settings: (1) 16°, (2) 18°5′, and (3) 22°57′.

7. a. State four main uses of compound sine plates.
   b. Indicate the design features of a compound sine plate that make compound angle settings possible.

8. State two precautions to take to safely store precision linear and angular measuring instruments and gages.

Precision Layout, Measurement, and Inspection Practices for Surface Plate Work

The *surface plate* provides an accurate reference plane that relates to layout, inspection, and measurement processes (called *surface plate work*). In surface plate work, workpieces are inspected for parallelism, squareness, roundness, and concentricity. General measurements are taken in relation to length, height, hole locations, angles, and tapers.

Ordinate dimensioning is used on drawings to represent the design or measurement features of a piece part. These features are referenced from a fixed point or surface called a datum. Principles of drafting and practices in interpreting typical shop and laboratory drawings that require ordinate measurements are also treated in this unit.

OBJECTIVES

After satisfactorily completing this unit, you will be able to:

* Relate the importance of perfectly flat reference planes on surface plates to precision measurements and inspection practices.

* Comprehend principles for checking flatness, parallelism, squareness, roundness, concentricity, and axial runout.

* Deal with new linear measuring instruments requiring gage block stacking, a micrometer head, and a digital readout.

* Deal with angle layout, measurement, and inspection techniques using precision angle gage blocks or a granite sine bar bench center and electronic height gage.

* Interpret and apply technical information related to design features and workpiece specifications from drawings on which datums and ordinate dimensioning are used.

* Perform the following layout and measurement processes.
  * Check Parallel Surfaces Using a 0.0001″ (0.002mm) Dial Test Indicator.
  * Check for Squareness and Flatness Using a Transfer Gage and Dial Indicator.
  * Check for Concentricity, Axial Runout, and Roundness.
  * Set Up and Check Angles with Precision Angle Gage Blocks.

* Analyze *Safe Practices* for protecting precision instruments.

* Apply each new precision measurement *Term*.

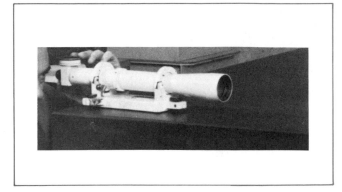

Figure 3-1 Calibrating a Flat (Plane) Surface
with an Autocollimator

ACCURACY OF SURFACE PLATES

Fine, precise reference planes have been produced commercially for many years. Some early surface plates were made by hand-filing and hand-lapping and then later, by hand-scraping. One technique for producing flat surfaces dates back to Henry Maudslay. In 1797, he used the three-plate method to hand-file and hand-lap the finished face of surface plates. Later, in 1840, Sir Joseph Whitworth replaced lapping with abrasive particles by hand-scraping.

Today, many design, development, and production contracts require part specifications that conform to both national and international measurement and design standards. Some manufacturers provide a *certificate of accuracy*. This certificate indicates that the product conforms with government specifications. The calibration is traceable to the National Bureau of Standards. The Federal Specification accuracies to which a surface plate conforms are certified.

Expressed in technical terms, "all points of the work surface (plane) shall be contained between two parallel planes that are separated by a distance no greater than that specified for its respective grade." As stated in Unit 2, there are three general grades of surface plates. *Grade AA* provides for extremely precise measurements, such as measurements required in the laboratory. The tolerance per two square foot area is ±0.000025"

(0.0006mm). The flatness tolerance of the *grade A* inspection-practice surface plate is ±0.000050" (0.0012mm). The *grade B* shop accuracy is ±0.0001" (0.0025mm) per two square foot area.

CALIBRATING FLAT (PLANE) SURFACES

The term *calibration*, as used in this unit, relates to the inspection, measurement, and reporting of size, flatness, and parallelism errors of plane surfaces. The degree of accuracy and method of calibration depend on the nature and use of the measuring equipment. For example, federal specifications suggest an annual check of ground master- and laboratory-grade gage blocks. Semi-annual checks of inspection gage blocks and surface plates are common practice.

Under normal conditions, the accuracy of a shop or laboratory surface plate may be checked easily by a simple test for the *repeatability* of a measurement. That is, the same part is measured under the same conditions on a number of different areas of the surface. The plane surface is accurate if the measurement on all areas is the same.

Large surface plates and plane surfaces are commercially calibrated and periodically checked. Their accuracy is checked with a precision instrument called an *autocollimator* (Figure 3-1). This

instrument uses lenses, light rays, and micrometer readings. The light source is built into the
autocollimator. The lens system causes light
rays to leave the instrument in parallel paths.
The light rays are directed at a target mirror.
The rays are reflected back into the instrument.
The rays are viewed into the eyepiece. A corner
target mirror and a reflecting target mirror are
required.

A micrometer on the autocollimator is used
to read any deviation in the right-angle setting of
the mirror target. Errors to within one-fifth second of arc may be read on the micrometer. Readings are taken (1) around the four edges of the
surface plate, (2) along the two diagonals, and
(3) across the two centerlines. A number of other
readings are also taken along each line. Each
reading is recorded and plotted on a graph. The
readings form a profile along the eight lines of
sighting.

The time required for calibrating a surface
plate with an autocollimator may be reduced
substantially by using a laser-powered surface
contour projector.

ADVANTAGES OF GRANITE PLANE SURFACES

In addition to the qualities of granite stated
earlier, consideration should be given to the following advantages over similar metal items.

- *Hardness:* Black granite has an 80 Rockwell
  C scale hardness by test. It is harder than
  tool steel. Black granite is also tough
  enough to resist chipping and scratching.
- *Rigidity:* Black granite surface plates have
  extremely low deflection under load. They
  are, therefore, almost distortion free and
  retain their original accuracy.
- *Density:* The uniform density of black
  granite makes it possible to produce a fine-
  textured, lapped, and frosted surface. The
  surface finish prevents gage blocks and precision surfaces from wringing.
- *Low reflectivity:* A black granite surface
  provides a dark nonglare finish. Light reflected from illumination, the surface
  plate, tools, instruments, and workpieces
  is limited. The dark surface is also an ex-

cellent background for layout lines that are
scribed with a brass rod. Such lines may
be removed with an ordinary eraser.
- *Porosity:* The uniform grain structure of
  black granite gives it a low water (moisture)
  absorption rate. A black granite surface is
  not affected by acids, chemical fumes, dust,
  or dirt. Surface plates and gages are, therefore, rustproof, and rusting and pitting of
  precision gages that are left on the plates
  are prevented.
- *Thermal (heat) stability:* Temperature
  changes have extremely limited effect on the
  accuracy of black granite gaging products.
- *Easy care and maintenance:* Black granite
  surface gages and accessories may be cleaned
  easily. Soap and water, detergent, or a
  manufacturer's solvent may be used. The
  surface will not stain or discolor.
- *Low cost:* The initial, maintenance, and
  relapping costs of a granite plate are lower
  than for a metal surface plate.

TECHNIQUES FOR CHECKING PARALLELISM, SQUARENESS, ROUNDNESS, AND CONCENTRICITY

Extremely precise checking for parallelism,
squareness, roundness, and concentricity may be
done with standard measuring instruments such
as a vernier height gage and/or a dial test indicator. Greater sensitivity is achieved by using electronic and pneumatic gages and instruments.

CHECKING FOR PARALLELISM

To check for *parallelism*, the workpiece is
placed with one of the parallel surfaces resting
flat on the face of the surface plate. If the part
has a projection on the bottom surface, precision
parallels may be used. These parallels accurately
elevate the surface to permit accurate measurement.

A test indicator is then selected with either a
front- or top-mounted dial. Whether a swivel-
point test indicator or a regular dial indicator
(with plunger-type stem) is used depends on the
location of the surface to be checked. A dial indicator with a magnet base is especially adapted
for rigid mounting on any ferrous surface (Figure
3–2). An indicator with a regular base is adapted
to either black granite or metal surface plates.

Courtesy of BROWN & SHARPE MANUFACTURING COMPANY

Figure 3–2 Dial Indicator and Magnet Base Set

Courtesy of BROWN & SHARPE MANUFACTURING COMPANY

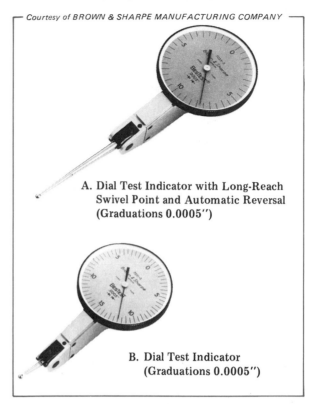

A. Dial Test Indicator with Long-Reach
Swivel Point and Automatic Reversal
(Graduations 0.0005″)

B. Dial Test Indicator
(Graduations 0.0005″)

Figure 3–3 Two General-Purpose Dial Test Indicators

Two general-purpose dial test indicators are pictured in Figure 3–3. The swivel-point model is available with a short- (1/2″) or long-range (1 7/16″) contact point. The point may be swiveled on some models through a 210° arc. An automatic reversal feature permits the dial test indicator to measure over, under, outside, or inside a surface (Figure 3–3A).

The smallest division on test indicators such as the fine-measurement model shown in Figure 2–3B is 0.0005″. A comparable metric-reading indicator has the smallest division of 0.02mm. The range of movement (measurement) of dial test indicators is limited. The dials, mechanisms, and contact points on regular sizes of dial test indicators may range from 0.008″ (0–4–0) to 0.030″ (0–15–0).

Like the dial indicator, the dial test indicator (Figure 3–3) is usually held in an attachment from a universal set. The dial test indicator is mounted on a column, magnetic base, or height gage. The contact point of the indicator is slid into position over one corner of the workpiece. The readings on the indicator dial face show any deviation in parallelism. As in all precision measurement work, the readings are rechecked.

How to Check Parallel Surfaces by Using a 0.0001″ (0.002mm) Dial Test Indicator

STEP 1 Check the cleanliness of all tools and instruments and of the workpiece. Remove all burrs.

STEP 2 Position the workpiece on a surface plate by using parallels.

STEP 3 Select a 0.0001″ dial test indicator with a swiveling point that is long enough to reach the areas of the surface that are to be checked.

STEP 4 Mount the indicator on a permanent magnet or adjustable base. The accessory used should permit the point to be adjusted.

STEP 5 Bring the dial test indicator point down over the work surface. Adjust the point to a zero reading.

Note: If the dial test indicator is brought to the work, the point should be moved carefully by hand and set on the work surface. The range of the dial test indicator must provide the movement needed to take complete measurements.

STEP 6 Move the dial test indicator setup over each corner of the work surface. Check for variations in the readings.

Note: Readings are usually taken near the corners of square and rectangular workpieces. Similar readings are noted near the outer edge (periphery) of round-shaped workpieces.

CHECKING FOR SQUARENESS AND FLATNESS

Usually the squareness of two surfaces is given in terms of *deviation per six inches of length*, particularly when stating the accuracy of measuring tools. One common practice in determining the amount that a workpiece or other part is out-of-square is to use a master cylindrical square. The cylindrical square is carefully brought against the surface to be checked. The cylindrical square is rotated until no light is visible. Any variation within the range of the cylindrical square may be read directly from the graduations.

A transfer gage is sometimes used with a master cylindrical square. The transfer gage has a button on the front of the base. This button is positioned against the square. The dial indicator is then adjusted to a zero reading while the button is in contact with the square.

The transfer gage is then moved carefully to the face of the squared surface to be measured. The button on the transfer gage is brought into contact with the workpiece. The dial indicator shows any difference in reading from the original setting.

How to Check for Flatness and Squareness by Using a Transfer Gage and Dial Indicator

STEP 1 Mount a dial indicator on a transfer gage.

STEP 2 Move the transfer gage until the button just touches the base of a master cylindrical square (or a universal square).

STEP 3 Set the indicator dial at zero at a height that will be near the top of the workpiece.

STEP 4 Remove from this setting. Place the button against the workpiece that is to be checked for squareness.

STEP 5 Record the reading on the dial indicator. Any variation from the original zero setting indicates the amount the workpiece is out-of-square.

STEP 6 Repeat the process at a few places.

STEP 7 Continue to test for squareness by resetting the dial indicator at midpoint of the workpiece. The indicator point is again set at zero against a cylindrical square.

STEP 8 Check the workpiece with this new setting. The dial indicator readings are compared with the first set of readings obtained in steps 5 and 6. Any difference in readings indicates whether the workpiece is dished or bowed. The amount the workpiece varies from a flat plane surface is found in 0.001″ or 0.0001″ (0.02mm or 0.002mm), depending on the precision of the dial indicator.

STEP 9 Disassemble the setup. Clean and prepare the instruments for storage. Return each instrument to its proper container.

CHECKING FOR ROUNDNESS

Roundness refers to the uniform distance of the periphery of a plug, ring, or hole in relation to the axis. Roundness is a measurement of a true diameter or radius. V-blocks, a surface plate, a dial indicator, and a holding device are generally used for checking roundness.

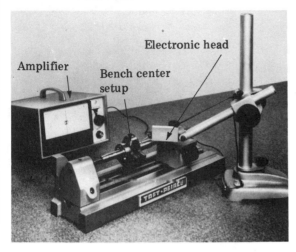

Figure 3–4 Measuring Concentricity by Using a Precision Bench Center and Electronic Height Gage and Amplifier

The workpiece is placed in the V-block. The contact point of a dial indicator (that is mounted securely) is moved onto the outside diameter. The indicator is positioned at the point where the highest reading is obtained—that is, at the centerline of the workpiece.

The dial face is turned until the indicator reading is zero. The part is then rotated carefully so that the position of the V-block, part, and original indicator setting are not changed. Any out-of-roundness from the true diameter shows up on the direct dial readings of the indicator.

CHECKING FOR CONCENTRICITY

Concentricity relates to the periphery (outside surface) of a cylindrical part and an axis of rotation. A cylindrical surface is said to be concentric when every point is equidistant from the imaginary axis. The measurement of concentricity is often called *total indicated runout (TIR)*. Testing and measuring for runout are usually done on a metal or black granite surface plate. A bench center with two dead centers and a dial test indicator are generally used. Higher accuracy of measurement is obtained with an electronic height gage and amplifier unit (Figure 3–4).

How to Check for Concentricity, Axial Runout, and Roundness

Checking the Concentricity of a Centered Workpiece

STEP 1 Secure a bench center. Check the centers to be sure that they are clean and burr-free.

STEP 2 Examine the centers of the workpiece. Remove any burrs or foreign particles. Scraping or lapping the centers may be necessary.

STEP 3 Mount the workpiece between the centers. They should be free to turn, but without end play.

STEP 4 Position the dial test indicator. The contact point must be on dead center. Turn the dial to read zero.

STEP 5 Rotate the workpiece one turn and note any movement of the contact point.

Note: The dial reading at the highest point represents the total indicated runout (TIR).

Checking the Concentricity of a Noncentered Part

STEP 1 Place the workpiece on V-blocks. Whether a single block or pair of blocks is used depends on the length and diameter of the part.

STEP 2 Position a dial indicator on a stand. The contact point must be able to reach the top of the workpiece.

STEP 3 Move the dial indicator over the outside diameter. Set the contact point reading at zero.

STEP 4 Turn the workpiece slowly and carefully one complete revolution. Note any variation (+ or –) from the zero setting. The difference indicates the amount the workpiece is out-of-round.

Note: The steps are repeated to recheck the accuracy.

STEP 5 Clean and replace the instruments in their proper storage containers.

CHECKING FOR AXIAL RUNOUT

Axial runout relates to the amount a shoulder or flange on a cylindrical part is out of alignment with the axis of rotation. A bench center, plane surface, and mounted dial test indicator are needed to test for axial runout.

The centered workpiece with a shoulder or flange is placed between centers on the bench center. The bench center is placed vertically on the surface plate. The part is turned one revolution. Any runout is indicated by the movement of the contact point. The amount of runout is read directly from the readings on the dial test indicator face.

GENERAL PRECISION LAYOUT AND MEASUREMENT PRACTICES

TAKING LINEAR MEASUREMENTS (LENGTH AND HEIGHT)

Linear measurements may also be checked, or measuring instruments may be set, with a combination gage block/height gage instrument. This instrument (Figure 3–5) consists of a base and post. The post supports a carrier with 1.0000″ (in the case of a metric instrument, 25mm) thick precision blocks. These blocks are arranged so that the top of one block is on the same plane as the bottom of an adjacent block. On some models, the lower gage block in the column is 0.090″ thick. This size permits measurements of 0.100″ height to be checked.

The 12″ instrument in Figure 3–5 has a super-accurate micrometer head. The thimble graduations may be read directly in 0.0001″. A metric micrometer head has readings from the spindle in increments of 0.002mm. The two scales on the sides of the precision blocks show each 0.100″ of movement. The direct readings are further simplified by a digital readout.

LAYING OUT HOLE LOCATIONS

Measuring the exact locations of holes is another common surface plate process. Hole locations within ±0.0001″ (±0.002mm) may be layed out with a gage block accessory set. Gage blocks of the required height and the precision scriber point combination are applied.

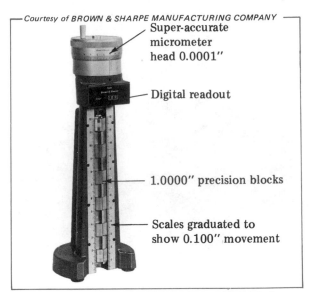

Super-accurate micrometer head 0.0001″

Digital readout

1.0000″ precision blocks

Scales graduated to show 0.100″ movement

Figure 3–5 Precision Height Gage with Digital Readout and 1.0000″ Precision Blocks

INDIRECT MEASUREMENT OF A HOLE POSITION

Instruments for laying out and measuring hole locations are also used to measure the location of holes that are already formed. One major difference is that the hole location is not established directly. The measurement is made to the bottom of the hole or to the top of a plug inserted in the hole. Many parts in punch and die, jig and fixture, and job shop work require that the location of one or more holes be measured. The workpiece is positioned on a surface plate and strapped to a precision angle block with a clamp. A dial test indicator is secured to a vernier height gage.

The contact point of the indicator is set at the required height (the centerline distance less one-half the diameter of the hole). The measurement is made with a gage block combination. The hole position is then checked against the setting of the dial test indicator.

MEASURING SMALL HOLE LOCATIONS

It is not always possible to position the indicator contact point in a hole. If the diameter of the hole is too small, a press or push fit pin or plug is inserted temporarily. The indicator contact point is set at zero at the required gage block height. This height equals the center distance plus the radius of the pin or plug. Any variation from the zero setting indicates the amount the hole position varies from the required dimension.

PRECISION ANGLE MEASUREMENT WITH ANGLE GAGE BLOCKS

Angle gage blocks permit the measurement of angles within limits of a fraction of a second of arc. Angle gage blocks are long, tapered rectangular blocks.

Angle gage blocks are available in three accuracy grades. *Toolroom* angle gage blocks are used for angle measurements within 1-second accuracy. *Inspection grade* angle gage blocks are applied to measurements within 1/2-second accuracy. The most precise set is called the *laboratory master*. Laboratory masters are used for measurements within 1/4-second accuracy. The surface finish of the gaging surfaces also varies from 0.6 microinch for the toolroom grade to 0.1–0.3 microinch for the laboratory grade.

A standard 16 angle gage block set has a range of measurement from 0° to 99° in steps of one second. The number of blocks in this set and the angles of each block appear in Table 3–1. It is possible, with a laboratory grade set of 16 blocks, to measure 356,400 angles in steps of one second to an accuracy of ±1/4 second. In addition to the angle gage blocks, a 6″ parallel and a 6″ knife edge (form of parallel having one edge that is narrowed to a smaller area) are available.

Table 3–1 Angles of the Gage Blocks in a 16-Block Set

Number of Blocks	Angles
6	1, 3, 5, 15, 30, 45 degrees
5	1, 3, 5, 20, 30 minutes
5	1, 3, 5, 20, 30 seconds

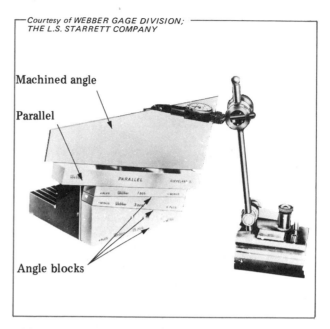

Courtesy of WEBBER GAGE DIVISION; THE L.S. STARRETT COMPANY

Machined angle

Parallel

Angle blocks

Figure 3–6 Angle Gage Blocks and Dial Test Indicator Setup to Measure a Required Machined Angle on a Workpiece

The angle gage blocks are designed to be combined in plus and minus positions. One end of a gage block is marked *plus*. The other end is marked *minus*. For example, Figure 3–6 shows a workpiece that is being checked with a dial test indicator and angle gage blocks. The three angle gage blocks produce an angle of 13°. The workpiece is resting on a parallel block that is wrung to the angle blocks. The setup is aligned at a right angle by positioning the parts against an angle plate. The 13° angle requires a 15° angle gage block, minus a 3° block, plus a 1° block.

Angle combinations are easy to establish. Figure 3–7 shows three angle gage blocks and a parallel block. Toolroom grade blocks are used in the setup with a dial test indicator. The face of a magna-sine is being set to an angle of 38° ±1 second. The angles of the three blocks are (+30°), (+5°), and (+3°). The addition of the three angles of the blocks produces the 38° angle. If an angle of 38°5′ were required, the five-minute block would be used in addition. The plus side would be added to the previous 38° setup.

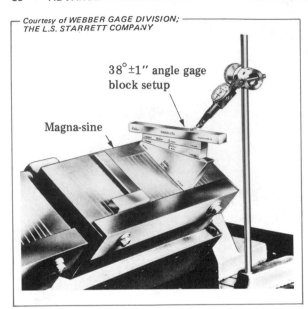

38°±1″ angle gage block setup

Magna-sine

Figure 3–7 Setting a Magna-Sine at 38° ±1″ with Angle Gage Blocks and a Dial Test Indicator

How to Set Up and Check Angles With Angle Gage Blocks

Note: The following procedure requires the setting up of a magnetic sine plate to an angle of 12°20′30″ ±1″ of accuracy. Angle gage blocks are to be used.

STEP 1   Determine the angle gage blocks to use.

Note: One combination includes the 15°, 3°, 20′, and 30″ angle gage blocks. A parallel block may also be used.

STEP 2   Wipe, clean, and check each block.

STEP 3   Slide the 15° block and the 3° block so that they are wrung together. The minus side of the 3° block matches the plus side of the 15° block.

Note: The two blocks, when wrung together, produce the 12° angle.

STEP 4   Wring the 20′ block to the combination. The plus side is matched with the 12° of the two wrung blocks.

Note: The combination produces an angle of 12°20′.

STEP 5   Add 30″ to the setup by using the plus side of the 30″ gage block.

Note: The overall angle is 12°20′30″.

STEP 6   Wring the parallel gage block on the total combination.

STEP 7   Place the 15°20′30″ angle gage block setup on the cleaned and burr-free surface of the magna-sine.

STEP 8   Place a dial or test indicator on a stand or holder.

STEP 9   Bring the contact point onto the top of the gage blocks at one end. Set the dial at zero.

STEP 10   Move the indicator across the face of the gage block combination. Note any variation in the dial reading.

STEP 11   Adjust the angle of the sine plate until a zero reading is obtained across the entire face of the gage blocks.

STEP 12   Lock the magna-sine at the required angle. Recheck to ensure that the angle position has not changed.

STEP 13   Disassemble the angle gage block combination. Start with the parallel block. Place each gage block on the protective cover.

STEP 14   Clean, wipe, and prepare each block for storage in the case.

MEASURING TAPER ANGLES

Tapers that have been machined between centers may be measured with bench centers. These centers are sometimes mounted on a sine bar, as shown in the setup in Figure 3–8. The required gage block height combination is determined for the required angle. The part is carefully inserted between the bench centers. The sine bar is then raised to the height of the gage blocks.

While the figure shows an application of an electronic height gage, it is common practice to use a dial test indicator. The dial test indicator

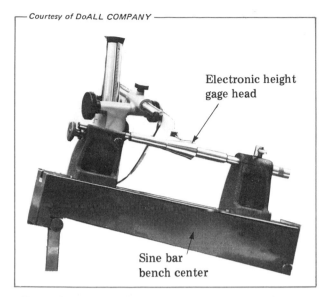

Electronic height gage head

Sine bar bench center

Figure 3-8 Measuring a Taper Angle by Using a Granite Sine Bar Bench Center and Electronic Height Gage

is moved across the workpiece at one end of the taper. The highest reading indicates the center-line. The dial face is set at zero at this point. The indicator is then moved across the part at the other end of the taper. Any variation in readings across the small end and the large end of the taper indicates the amount the workpiece varies from a required measurement.

DRAWINGS WITH DATUMS AND ORDINATE DIMENSIONS

Important dimensioning for layouts and measurement is provided on drawings and prints by using datums and zero coordinates. Linear, angular, and circular surfaces may be located and measurements may be taken in relation to a datum and zero coordinates.

DATUMS

The term *datum* means an exact point, line, plane, or other reference surface. Two or three datums are usually required to derive all essential information for dimensioning, machining, and inspection. One advantage of using datums is that a series of measurements may be taken from the same reference point. Thus, the accuracy of one dimension does not depend on another. That is, dimensional errors are not cumulative; one error does not add on to another. Dimensions governing each feature are layed out and measured from the datum and not from other features.

DATUM PLANES AND ORDINATE DIMENSIONS

Two dimensions are normally required to establish the height, depth, or length of a feature. When datums are used, two or more mutually related *datum planes* are needed. These planes are called *zero coordinates*. The zero coordinates of the two datum planes in Figure 3-9 are easily identified. One zero coordinate is the vertical datum plane; the other is the horizontal datum plane. Note on the drawing that the centerline distances from the zero coordinates are represented by extension lines. There are no dimension lines or arrowheads.

Each center for holes **A** through **G** is located at the intersection of two datum dimensions. *Ordinate dimensioning*, rectangular datum dimensioning from zero coordinates, is used in Figure 3-9. An *ordinate table*, the simple table accompanying the drawing, is included.

Many features or dimensions on a conventional drawing are often difficult to read and cause layout, measurement, and machining errors. Ordinate dimensioning and tables simplify a drawing.

The machine operator is able to accurately position the workpiece for each center. Horizontal, cross feed, and vertical feed handwheels may be moved a specified distance to position the workpiece for each dimensioned location. The part is moved a specified distance in a required direction. The amount of movement (dimension) is read on the graduated micrometer collar of the feed handwheel. Drawings that use datums and ordinate dimensioning are particularly useful in programming numerically controlled machine tools.

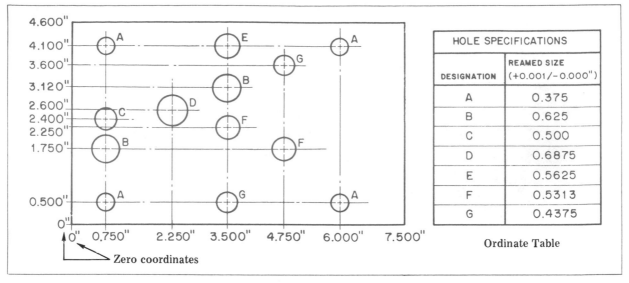

Figure 3-9 Locating Hole Positions from Zero Coordinates by Using Ordinate Dimensions

In Figure 3-9, the center of one hole **A** is located 0.500″ up from the zero horizontal coordinate and 0.750″ from the zero vertical coordinate. Four holes on the drawing are designated **A**. As shown in the table, each of these holes is reamed to 0.375″ +0.001/-0.000. One **B** hole falls along the same 0.750″ dimension and is 1.750″ from the zero horizontal coordinate. The two **B** holes on the drawing are each reamed to 0.625″ +0.001/-0.000. However, the second **B** hole is located at distances of 3.500″ and 3.120″ from the zero coordinates. The locations of the remaining holes are measured in a similar manner.

SAFE PRACTICES IN PRECISION LAYOUT, MEASUREMENT AND INSPECTION

- Protect the accuracy of a surface plate by not laying unnecessary tools, instruments, and workpieces on it. All items used on a surface plate should be checked for burrs, scratches, and nicks. Remove all imperfections before the items are used.
- Place only the workpiece and the measuring and layout tools and accessories required for a job on a surface plate.

- Wipe a surface plate frequently. It should be kept clean and free of dust, dirt, and oil. Use the commercial cleaner recommended by the plate manufacturer.
- Spray a dust inhibitor over a cast-iron surface plate when the plate is not in use.
- Check the temperature of instruments and workpieces. The heat absorbed from handling must have a chance to dissipate.
- Bring the contact point of an indicator down on a surface, where possible. If the contact point is to be carefully swung onto a work surface, raise the point gently before making contact. Select a movement range that permits the indicator to cover any variation of the workpiece.
- Elevate parts with projecting lugs, arms, and other surfaces. These surfaces should clear the face of a surface plate. Use protective materials between a rough-surfaced workpiece and a precision-finished surface.
- Recheck all measurements for repeatability of measurement.
- Lap work centers if any burrs or nicks exist on the angular surfaces.
- Follow the practice of noting measurements, particularly where a number of measurements are to be taken.

TERMS USED IN PRECISION SURFACE PLATE WORK

Three-plate method of producing flat surfaces	A technique of scraping flat surfaces. Each surface forms an accurate plane in relation to the two other plane surfaces.
Certificate of accuracy (surface plate)	A written statement indicating that the accuracy of the plane surface conforms to federal specifications for flatness tolerances.
Repeatability of measurement	The accuracy of a plane surface. Providing the same measurement under the same conditions at different locations on a plane surface.
Calibration (surface plates)	The inspection and measurement of size and flatness errors. The reported accuracy of a plane surface.
Autocollimator	An instrument designed to use light rays, lenses, and micrometer to check flatness errors to within one-fifth second of arc.
Swivel-point test indicator	A standard short (1/2″) or long (1 7/16″) contact-point test indicator that may be swiveled through a 210° arc.
Roundness	A true, closed cylinder where every point on the periphery is equidistant from its center.
Concentricity	The trueness of the periphery of a cylinder in relation to its axis.
Axial runout	The amount a shoulder varies from a true square in relation to its axis.
Angle gage blocks	Heat-treated, dimensionally stabilized, long-tapered, rectangular, precision gage blocks.
Bench centers	Precision-machined centers that may be mounted and positioned on a bed or angle sine bar.
Datum (datum plane)	An exact line, surface, or reference point.
Ordinate dimensions	Dimensioning from a datum (zero coordinates). A basic dimensioning system used for numerical control.

SUMMARY

- The generating of standard planes by the three-plate method dates back to Henry Maudslay in 1797.

    - Tolerances of surface plates are given as ±0.000025″ (0.0006mm) per two square foot area for grade AA laboratory surface plates. The flatness tolerance for grade A inspection-practice plates is ±0.000050″ (0.0012mm). Grade B shop plates are accurate to ±0.0001″ (0.0025mm) over a two square foot area.

- Surface plates and other plane surfaces may be calibrated commercially by using an autocollimator. Applications of light rays, lenses, and a micrometer are involved in calibrating to within one-fifth second of arc.

    - Precision inspection and measurement of flatness, parallelism, squareness, roundness, and concentricity are typical surface plate processes.

■ The plunger-type dial indicator and the swivel-point test indicator are common inspection and measurement instruments. These indicators are used with other indicator attachments, a cylindrical square, transfer gage, precision V-blocks, and accessories.

■ Bench centers are used with indicators to measure total indicated runout (concentricity) and axial runout.

■ Gage blocks, stacked in a post that is supported on a base, permit taking precise linear measurements. The instruments have a super-accurate micrometer head from which direct readings are taken. Digital readouts further simplify readings.

■ Angle gage blocks make setting and taking angle measurements to within 1/4 second possible. The three grades of accuracy are toolroom (within one second), inspection (1/2 second), and laboratory master (1/4 second).

■ Taper angles may be measured with angle gage block setups. A bench center and indicator may also be used on the surface plate.

■ Precision layouts and numerically controlled programming and machine processes involve the interpretation of drawings with datums and ordinate dimensions.

■ Temperature control of workpieces and instruments is necessary. Temperature variations affect dimensional accuracy.

UNIT 3 REVIEW AND SELF-TEST

1. Explain the importance of a perfectly flat plane surface to precision layout and inspection processes.

2. List the steps required to check the flat faces of a hardened and ground rectangular fixture plate for parallelism (within ±0.0001″).

3. Explain the difference between the *roundness* and the *concentricity* of a turned part.

4. List the steps to lay out three horizontal lines when a vernier height gage is used. The lines are located 25.4mm, 38.1mm, and 50.8mm from the same ground edge.

5. a. Indicate the degree of accuracy to which angle measurements may be held when (1) laboratory master, (2) inspection, and (3) toolroom grades of angle gage blocks are used.
   b. Cite two advantages of using angle gage blocks for laying out and inspecting angle surfaces in preference to regular gage blocks and a vernier height gage setup.

6. a. Define the term *datum* as used on a shop drawing.
   b. Make two statements about the advantages of using datums for layouts and measurements.

7. State two precautions to take in precision layout and inspection work.

SECTION TWO

Precision Measurement: Nonmechanical Instruments

All of the measuring instruments covered thus far are mechanical. The range of accuracy has been from ±1/64″ (±0.5mm) using a steel rule to ±0.0001″ (±0.002mm) using a vernier-scale instrument. More precise measurements to higher degrees of accuracy are obtainable with *optical flats* and *microscopes*. These instruments depend on physical science principles related to light waves and optics. Still other *high amplification comparators* are treated in the next unit. The comparators depend on electrical, electronic, and pneumatic principles, often in combination with mechanics and optics.

Measurements related to flatness, parallelism, precise dimensional, and form differences are treated in terms of quality control and surface texture.

Measurement with Optical Flats and Microscopes

OBJECTIVES

After satisfactorily completing this unit, you will be able to:

- Apply principles of optical measurement to lenses, industrial magnifiers, and shop and laboratory microscopes in relation to linear, angle, and radius measurement.
- Relate optical flats to measurement of surface flatness, parallelism, size, and finish.
- Establish dimensions using optical flats and a lightwave conversion table.
- Interpret interference optical band patterns for flat and round surfaces.
- Relate tabular dimensioning and data from coordinate charts to layout and machining processes.
- Perform the following processes:
  - Use a Toolmaker's Microscope.
  - Measure a Workpiece with an Optical Flat.
- Analyze the reasons for following each recommended *Safe Practice.*
- Describe each *Term* as related to each new principle, instrument, and process.

MEASUREMENT WITH OPTICAL FLATS AND MICROSCOPES

SHOP AND TOOLMAKER'S MEASURING MICROSCOPES

Two common optical instruments found in the shop and laboratory are the *shop microscope* and the *toolmaker's measuring microscope*. Practical measurement and inspection procedures may be carried on with these microscopes.

BINOCULAR-HEAD MICROSCOPE

Another microscope, which is not identified as a shop or toolmaker's microscope, combines the advantages of a *binocular* (two-eyepiece) head design and a *zoom* feature. The head design and zoom feature deliver a bright, three-dimensional stereo image. The image keeps the area being viewed continuously in focus at a constant working distance up to 10.5″ (267mm).

The magnification knob is turned to zoom the image to the exact size for optimum viewing. Zoom microscopes generally use a *flat field* optical system which permits the image to remain clear from one edge to the opposite edge. This type of microscope minimizes operator fatigue.

OPERATION OF THE SHOP MICROSCOPE

A shop microscope of simple design is pictured in Figure 4–1. The main features include an adjustable ocular lens and tube, a column and stand, an adjusting ring with a graduated scale, and a light source. The microscope is mounted in a stand. The tube may be raised or lowered by a clamp screw.

The ocular lens is adjusted by turning it to the right or left. Adjustment brings the object into proper focus. The light source in the column is directed to the object through the relieved area in the base. The degree of magnification of the microscope is the product of the magnifications of the ocular lens and the objective lens. For example, the 40X shop microscope has a power of magnification of 40 (10X ocular lens with a 4X objective lens).

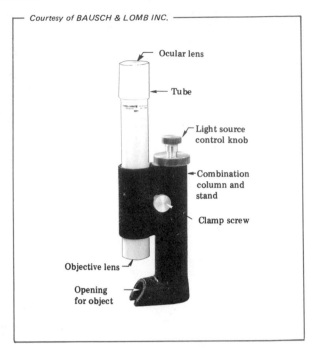

Courtesy of BAUSCH & LOMB INC.

Figure 4–1 40X Shop Microscope with 0.150″ Scale Graduated in 0.001″

A micrometer scale is mounted within the ocular. The 150-division scale represents 0.150″. Measurements may be made to 0.001″. Subdivisions of 0.001″ (0.02mm) may be estimated to within 0.00025″ (0.005mm), Figure 4–2. The instrument is thus capable of a high degree of accuracy.

One advantage of the shop microscope is that it is small and convenient to carry to a required location. A disadvantage is that it has a limited

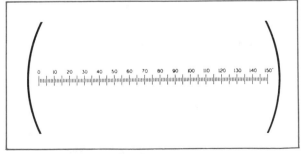

Figure 4–2 150-Division (0.150″) Scale Graduated in 0.001″ (Mounted within a Shop Microscope)

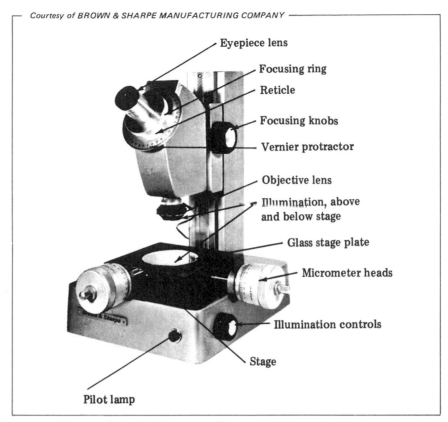

Eyepiece lens

Focusing ring

Reticle

Focusing knobs

Vernier protractor

Objective lens

Illumination, above
and below stage

Glass stage plate

Micrometer heads

Illumination controls

Stage

Pilot lamp

Figure 4–3 Main Features of a Toolmaker's Measuring Microscope

field. *Field* refers to the diameter of the area that may be observed in a single position. The model shop microscope, as illustrated in Figure 4–1, has a field of 7/32" (5.5mm) and magnifies 40X.

GENERAL APPLICATIONS OF THE SHOP MICROSCOPE

General applications of the shop microscope are as follows:

- Measuring the diameter of small holes having tolerances of ±0.001" (0.02mm) (such holes often require measurement and inspection during machining);
- Measuring small dimensions;
- Checking the thickness of the case on hardened parts;
- Measuring the diameter of impression made during hardness testing;
- Checking mechanical parts quickly for wear (welds, for example, may be inspected easily);

- Examining surfaces on-the-spot to detect cracks and flaws (the condition of plated, polished, and painted surfaces may be checked quickly).

TOOLMAKER'S MEASURING MICROSCOPE

The toolmaker's measuring microscope is a ruggedly designed microscope with high-power magnification. It is adaptable to taking linear and angular measurements. The microscope is especially valuable for inspecting small parts and tools. The main features of the microscope are labeled in Figure 4–3.

A number of attachments permit the microscope to be used as a comparator. A thread form or other tool contour may be compared with a charted outline that is superimposed directly upon the virtual image of the object. The contour of the object may thus be checked to within 0.0001" (0.002mm).

Angles may be measured to an accuracy of five minutes of arc on the toolmaker's measuring microscope. The microscope in Figure 4–3 has a *protractor eyepiece* (eyepiece and lens assembly, combined with a vernier protractor) that is used to make angular measurements.

The toolmaker's microscope may also be used to measure pitch diameter, major and minor diameters, pitch, and lead. Lead-measuring and thread-measuring attachments extend the practical applications of this microscope. Some advanced models include a projection attachment that permits making contour and other precision measurements normally performed on comparators.

OPTICAL HEIGHT GAGE

The *optical height gage* is another instrument that applies optics. This instrument combines design and measurement features of gage blocks and a measuring microscope. The optical height gage consists basically of a "stack of gage blocks." These blocks are permanently wrung together. The accuracy of the stack is ±0.00005″ per inch (0.0002mm per 25mm) of height. Measurements are read directly in the eyepiece of the microscope.

Some older height gage models are being retrofitted by their manufacturers to replace the optical feature by an easier reading digital readout. One manufacturer's model, the Digi-Chek® II, is shown in Figure 4–4. This particular height gage includes the digital readout component.

The instrument requires the use of a standard reference bar. The reference bar consists of a master stack of alternating 0.300″ gage block jaws and 0.700″ spacer blocks. The gage blocks are permanently wrung and fastened together to form 1″ increments. The master stack has a special bushing arrangement. The bushing allows the stack to conform to thermal conditions under actual use.

The linear accuracy of the master stack of gage blocks is +0.000004/−0.00000″ per inch of length. The accuracy of a master stack in metric units of measure is +0.00001/−0.0000mm per 25mm of length. The parallelism of the gage surfaces to the base and to one another is 0.000015″ (0.0004mm). The measuring system accuracy is

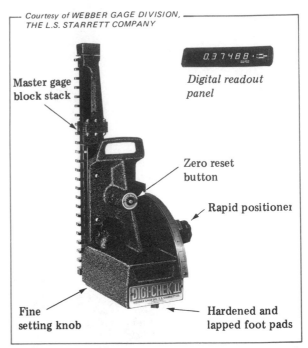

Courtesy of WEBBER GAGE DIVISION, THE L.S. STARRETT COMPANY

Master gage block stack

Digital readout panel

Zero reset button

Rapid positioner

Fine setting knob

Hardened and lapped foot pads

Figure 4–4 Optical Height Gage with Master Gage Block Stack (Standard Reference Bars) and Digital Readout

0.00005″ (0.00125mm). Most digital systems are equipped with a safety power failure flashing display.

INDUSTRIAL MAGNIFIERS

Industrial magnifiers are used in the shop and laboratory.for magnifying lines, surface areas, and design or machining features. Some magnifiers are mounted and provided with an outside source of illumination. They are adjustable. They permit working at a short distance from the workpiece or object. Other magnifiers attach to vernier scales and permit easier and more accurate readings.

Figure 4–5 shows a hand measuring magnifier that has four transparent scales. The magnification of this particular magnifier is 7X. Each transparent scale allows adequate illumination to produce a sharp, flat image. Each of the four scales is contained in a separate mount. The scales are readily interchangeable. The *general-purpose scale* permits measurements in increments of 0.005″, 0.1mm, and 1° and the measurement of 1/16″ to 5/8″ radii and 0.001″, 0.002″,

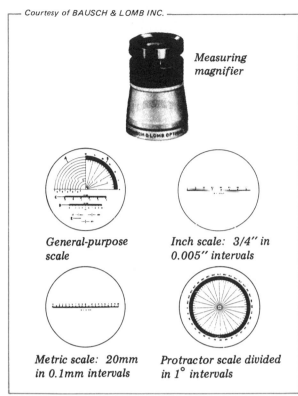

Measuring magnifier

General-purpose scale

Inch scale: 3/4" in 0.005" intervals

Metric scale: 20mm in 0.1mm intervals

Protractor scale divided in 1° intervals

Figure 4–5 Hand Measuring Magnifier equipped with Four Transparent Scales

and 0.003″ line thicknesses. A *separate inch scale* has graduations in 0.005″ intervals up to 3/4″. The *metric scale* is graduated in 0.1mm intervals over 20 mm. The *protractor scale* is divided in single degrees.

Simple designs of magnifiers with single lenses are identified as hand magnifying readers (magnifying glasses).

How to Use a Toolmaker's Measuring Microscope

STEP 1 Mount the specimen in the center of the stage.

STEP 2 Check to see that the light source is adequate.

Note: A vertical illuminator attachment is sometimes used to direct light through the lenses of the objective to focus on the object from above.

STEP 3 Make a rough adjustment of the optical unit. Loosen the clamp screw. Adjust with the locating pin.

STEP 4 Adjust further by turning the knob until the object is in focus.

Measuring Angles

STEP 1 Align the cross lines in the eyepiece with the angular surface to be measured.

STEP 2 Turn the micrometer heads to adjust the workpiece horizontally. The heads move the stage.

STEP 3 Read the graduation on each micrometer head at the start of each setting and at each required measurement.

Note: The graduations on the head permit movements and measurements to an accuracy of 0.0001″ (0.002mm) in either direction.

STEP 4 Adjust the stage by aligning one side of the part to be measured with a cross line.

STEP 5 Take and record the micrometer-head reading.

STEP 6 Move the stage until the side to be measured is in line with the same cross line.

STEP 7 Take this micrometer reading.

STEP 8 Subtract the two readings. The difference (represented by the distance the stage is moved) represents the linear measurement.

Measuring Angles with a Protractor Eyepiece Attachment

Note: The thread angle of a 60° threaded part is to be checked.

STEP 1 Mount the workpiece on the stage. The axis is set parallel to the movement of the table.

STEP 2 Set the angle between the two cross lines at 60° by turning the knurled adjusting ring on the eyepiece.

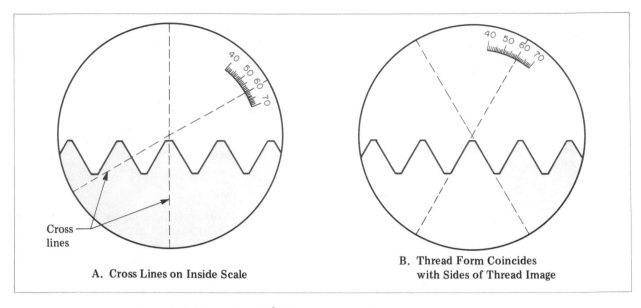

A. Cross Lines on Inside Scale

B. Thread Form Coincides with Sides of Thread Image

Figure 4–6 Measuring a 60° Thread Angle with a Protractor Attachment

STEP 3 Set the eyepiece to one-half the thread angle. The outside scale is used to set the 30° angle.

Note: Since the thread angle is 60°, each slope makes a 30° angle with a line perpendicular to the screw axis (Figure 4–6A).

STEP 4 Turn the micrometer adjustment on the stage to adjust the image.

Note: The thread form must coincide as nearly as possible with the cross lines.

STEP 5 Adjust both cross lines (Figure 4–6A). These lines must coincide with the sides of the thread image (Figure 4–6B).

STEP 6 Read the angle of the thread on the scale in the eyepiece.

Note: When a line bisecting the thread angle is perpendicular to the movement of the stage, the outside scale indicates half the angle between the cross lines.

Note: The outside scale of a 60° thread angle is set to 30°. Any variation in the reading indicates an error in just one-half the thread angle.

Measuring Thread Pitch and Lead

STEP 1 Mount the threaded part so that its axis is parallel to the movement of the stage.

STEP 2 Line up the cross lines with the outline of one thread.

STEP 3 Note the reading on the micrometer head.

STEP 4 Adjust the stage. The cross lines are lined up with the next thread.

STEP 5 Note the micrometer reading. The difference between the two readings is the pitch of a single thread. If the thread is a multiple thread, the difference represents the lead.

PRECISION MEASUREMENT WITH OPTICAL FLATS

CHARACTERISTICS OF OPTICAL FLATS

Optical flats are used for precision checking flatness, parallelism, size, and surface characteristics. The optical flats used for measurement and inspection are made of a high-quality optical quartz. This material has a low coefficient of expansion (approximately 0.32 millionths of an inch for each degree Fahrenheit change). Ultra-low expansion fused silica optical flats are also available.

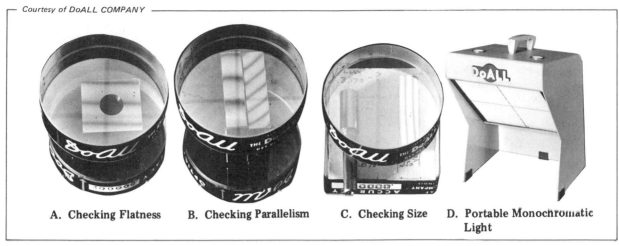

A. Checking Flatness   **B. Checking Parallelism**   **C. Checking Size**   **D. Portable Monochromatic Light**

Figure 4–7 Application of Optical Flats and Portable Monochromatic Light Source

One or two faces may be finished as a measuring surface. A band on the side of the optical flat indicates the measuring surface(s). The accuracy of the flat is marked on the band in millionths of an inch.

Optical flats are manufactured in three grades. *Working flats* are produced with a degree of flatness of 0.000004″ (0.0001mm). No point on a working flat will deviate from any other point by more than 4 microinches (millionths of an inch). *Master flats* are accurate to within 0.000002″ (0.00005mm). *Reference flats*, used primarily under close controlled laboratory conditions, are held to an accuracy of 0.000001″ (0.000025mm).

Optical flats are also furnished with coated surfaces. A thin film of titanium dioxide on the optical flat sharpens the image for greater accuracy.

Commercial optical flats are available in sizes ranging from 1″ to 10″ diameter (25 mm to 250 mm). The thicknesses (depending on diameter) vary from 1/2″ (12mm) to 2″ (50mm). The common sizes of square optical flats are 2″, 3″, and 4″ on a side by 5/8″ and 3/4″ thickness. Metric sizes are 50mm, 75mm, and 100mm by 15mm to 18mm thick.

MONOCHROMATIC LIGHT SOURCE

A monochromatic light source, when used with an optical flat, produces a distinct *fringe pattern*. This pattern, consisting of alternately light and dark bands, shows up clearly to establish the flatness, parallelism, and size (over a limited dimensional range), as well as surface charac-

teristics (Figure 4–7A, B, and C). A portable monochromatic light, or monolight, is adaptable for bench-type applications (Figure 4–7D).

Each color in the solar spectrum has a wavelength that tends to blend with the next color. The wavelengths range from 0.0000157″ (0.0003937mm) for ultraviolet rays to 0.0000275″ (0.0006985mm) for red rays. The light emitted from a monochromatic (rays of one definite wavelength) light is 23.1 microinches (millionths of an inch) or 0.0005867mm.

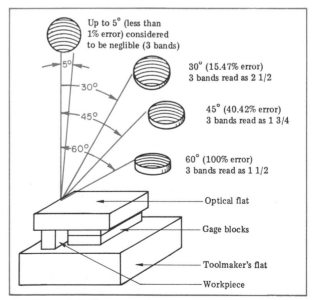

Up to 5° (less than 1% error) considered to be neglible (3 bands)

30° (15.47% error) 3 bands read as 2 1/2

45° (40.42% error) 3 bands read as 1 3/4

60° (100% error) 3 bands read as 1 1/2

Optical flat

Gage blocks

Toolmaker's flat

Workpiece

Figure 4–8 Degrees of Measurement Accuracy as Related to the Viewing Angle

The degree of accuracy in reading the number of fringe bands (measurements in microinches) is subject to parallax errors. For example, Figure 4–8 shows parallax errors for viewing angles of 5°, 30°, 45°, and 60°. Note that the smaller the viewing angle the more accurate the reading of the number of bands in the fringe pattern.

MEASUREMENT OF SURFACE FLATNESS

Thus far, only straight interference bands have been discussed. The straight, parallel, uniform bands relate to a flat surface. Variations in a flat surface produce curved interference bands. Such bands indicate a concave or convex surface, a slight angle, or a radius.

Interference bands that curve slightly at the ends show that the edges of the workpiece are rounded. Interference bands in a series of curved lines that form elliptical circles indicate a high spot on the surface of the workpiece (Figure 4–9A). The pattern may be identified when a slight pressure is applied to the optical flat above the spot. Curvature due to a high spot (hill) points toward the open side of the air wedge. Interference bands as a bent pattern indicate a low spot (valley) in the surface of the workpiece (Figure 4–9B). The curvature points toward the closed side of the air wedge when it is due to a valley.

Table 4–1 Partial Conversion Table for Optical Flat Fringe Bands (Inch and Metric Standard Units of Measure)

Number of Bands	Equivalent Measurement Value		
	Microinches	Inches	Millimeters
0.1	1.2	0.0000012	0.000029
0.2	2.3	0.0000023	0.000059
0.3	3.5	0.0000035	0.000088
. . .	. . .	. . .	. . .
1.0	11.6	0.0000116	0.000294
2.0	23.1	0.0000231	0.000588
3.0	34.7	0.0000347	0.000881
. . .	. . .	. . .	. . .
19.0	219.8	0.0002198	0.005582
20.0	231.3	0.0002313	0.005876

INTERFERENCE BAND CONVERSION TABLE

Part of the full conversion table that appears in the Appendix for converting fringe bands to microinches (or metric equivalents) is reproduced in Table 4–1. Values are given in fractional parts of an inch and of a millimeter. When close approximations are sufficient, one band indicates an accuracy of approximately 10 millionths of an inch (0.00025mm).

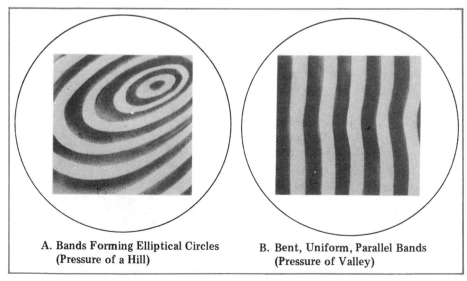

A. Bands Forming Elliptical Circles (Pressure of a Hill)

B. Bent, Uniform, Parallel Bands (Pressure of Valley)

Figure 4–9 Interference Band Patterns Indicating Surface Variations

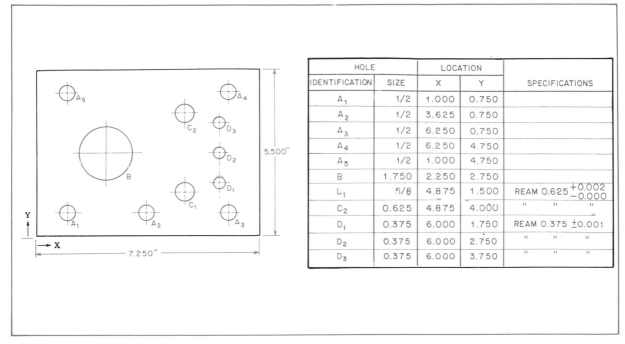

HOLE		LOCATION		
IDENTIFICATION	SIZE	X	Y	SPECIFICATIONS
A_1	1/2	1.000	0.750	
A_2	1/2	3.625	0.750	
A_3	1/2	6.250	0.750	
A_4	1/2	6.250	4.750	
A_5	1/2	1.000	4.750	
B	1.750	2.250	2.750	
C_1	5/8	4.875	1.500	REAM $0.625^{+0.002}_{-0.000}$
C_2	0.625	4.875	4.000	" " "
D_1	0.375	6.000	1.750	REAM 0.375 ± 0.001
D_2	0.375	6.000	2.750	" " "
D_3	0.375	6.000	3.750	" " "

Figure 4–10 Datum Line Drawing with Related Coordinate Chart

Direct measurement with optical flats is restricted to a very narrow range. While optical flats are still used for comparison measurements, some applications are being taken over by high amplification, electronic comparator instruments. These instruments require relatively less skill and are more rapid than optical flats.

TABULAR DIMENSIONING

One technique of dimensioning, known as ordinate dimensioning, was discussed in Unit 3. Another technique is to leave off location and size dimensions and other specifications from drawings and to include the information in chart form. Each size and location dimension on a chart relates to measurements from datums on a drawing. This technique is called *tabular* (or rectangular) *dimensioning*. It prevents cluttering a drawing with dimensions when there are a large number of features. Tabular dimensioning also simplifies the reading of a blueprint or sketch.

Tabular dimensions start at the origin (datum) of the coordinate—that is, each dimension begins at the point where the coordinates intersect. Inch or metric units of measurement may be used.

COORDINATE CHARTS

A *coordinate chart* contains location and size dimensions. Sometimes, a column is added for specifications that relate to a particular feature. Figure 4–10 shows a line drawing with its related coordinate chart. The three columns of the chart —hole, location, and specifications—show the kind of information needed to machine and inspect the part. Each hole size on the drawing is listed in one column. The location dimensions from each of two datums (X) and (Y) are given next. Specifications appear in the last column. The drawing shows the relative position of each hole. The chart provides the dimensions. The worker works from the drawing and coordinate chart to select the cutting tools; determine the machine setups, processes, and tolerances; and check measurements.

Note that the first hole (identified as A_1) is 1/2″ diameter. The centerline of A_1 is 1.000″ from datum X and 0.750″ up from datum Y. Holes C_1 and C_2 are to be reamed to 0.625″ +0.002/0.000. The tolerance on holes D_1, D_2, and D_3 is ±0.001″.

Where a part is fully represented in one view, two datums (like X and Y) are used. However, some parts require two or more views to adequately describe a part or feature. A third datum (Z) is then identified on the drawing. The additional size and location dimensions are given in the coordinate chart in relation to this datum. While the technique of applying tabular dimensioning and coordinate charts to drawings is used for general machining operations, the technique is readily adaptable to numerically controlled setups and processes.

How to Measure a Workpiece With an Optical Flat

STEP 1 Select a master gage block or combination as close to the required size as possible.

STEP 2 Clean the gage block, workpiece, and optical flat.

STEP 3 Place the part (workpiece) and the master gage on a toolmaker's flat or other plane surface.

STEP 4 Measure the width (w) of the gage block (A) and the distance (ℓ) between the reference (contact) points (B).

STEP 5 Place an optical flat so that it is centered over the master gage and the part (C).

STEP 6 Establish the point of contact; in this example, the center line.

STEP 7 Read the number of bands (n) appearing in the width of the master gage. Assume six bands are read.

STEP 8 Use the formula:

$$M = n \times \frac{\ell}{w} \times 11.6 = 6 \times \frac{1\frac{1}{2}}{1} \times 11.6$$

Substitute the known measurements of n, ℓ, and w.

Note: The answer (M = 110.4) is the amount in microinches (or millimeter equivalent) that the part varies from (larger than) the required size.

STEP 9 Disassemble the setup carefully. Clean and properly store all instruments.

Safe Practices in the Use of Industrial Microscopes and Optical Flats

• Normalize the temperature of the measuring instruments, workpiece, and accessory before taking a measurement or carrying on surface inspections. Optical flats have a low conductivity of heat; they normalize at a slower rate than metallic parts.

• Use a fine, soft-bristle brush to wipe over the surfaces of an optical flat or microscope and areas of contact.

• Wipe instrument lenses with a clean, soft cloth or lens tissue. Saturate the cloth with an approved cleansing solution. Any dirt rubbed across a lens will scratch the surface.

• Place a part on the stage of a toolmaker's measuring microscope in order to permit movement over the length or width to be measured.

• Use a precision plane surface, such as a toolmaker's flat, for precision measurements with optical flats and gage blocks.

• Set up the part to be measured for proper viewing. Keep the line of sight as nearly perpendicular to the work surface as possible to reduce parallax in reading.

• Recheck each linear and angular measurement for accuracy.

• Clean and check each gage block and optical flat at the end of a process. Store each instrument in its protective case.

• Take care in adjusting the objective of the instrument so that it does not strike the specimen or part.

TERMS USED IN MEASUREMENT AND INSPECTION WITH OPTICAL INSTRUMENTS

Monochromatic light	A light source that consists of one wavelength (or color) only. A practical ray of light that penetrates an optical flat. Light emitted from a monochromatic lamp having a wavelength of 23.1 micro-inches (0.0005867mm).
Interference	The action of one energy pulse on another when two light rays are brought together.
Fringe bands	Alternate light and dark bands produced by interference. Bands that are one-half wavelength apart.
Optical flat	A quartz, glass, sapphire, or other transparent very nearly perfect plane surface.
Master flats	Flats with high-precision surfaces from which other surfaces may be calibrated.
Coated flat	A coating, such as titanium dioxide, bonded to an optical flat. Permits faster readings by producing sharp, dark fringe patterns.
Parallel separation planes	A set of planes that are parallel to a working surface. Parallel planes that are one-half wavelength (11.6 microinches) apart. Planes that intersect with the part and are seen as fringe bands.
Field	The diameter of the area that may be observed through a magnifier lens (or microscope). A magnified area that is clearly visible when a lens is held at the correct distance from an object.
Object	A part, particle, surface, or feature of a workpiece that is viewed through a magnifier.
Virtual image	The optical counterpart of an object formed by a lens.
Tabular dimensioning	A simplified system of rectangular dimensioning. Dimensioning in relation to one or more datums.
Coordinate chart	A chart containing the identification of features, with size and location dimensions.

SUMMARY

■ Optical flats and microscopes permit measurement and inspection processes to be carried out at a high level of precision.

　■ Optical instruments control light waves to form patterns. These patterns are used to establish a comparison measurement or the accuracy of a surface.

■ The shop, wide field, and toolmaker's measuring microscopes are practical measurement and inspection instruments.

　■ The toolmaker's measuring microscope is especially adapted to precise linear and angular measurements and surface finish inspection.

■ Angular measurements to within an accuracy of five minutes of arc may be read directly on a toolmaker's measuring microscope.
  ■ The optical height gage combines the features of a stack of gage blocks and a measuring microscope.
■ Magnifiers are used in the shop and laboratory for magnifying lines, areas, design features, and surface finish.
  ■ The number of fringe bands and their shape are used in measurement and surface inspection.
■ Optical flats are adaptable for measuring flatness, parallelism, and size, as well as for inspecting the quality and accuracy of a finished surface.
  ■ Working optical flats are accurate to within 4 microinches (0.0001mm); master flats, 2 microinches (0.00005mm); and reference flats, 1 microinch (0.000025mm).
■ Monochromatic light has a single wavelength of 23.1 microinches (0.0005867mm). It is used with optical flats.
  ■ Curved interference bands indicate that a surface is convex or concave or has a slight radius or angle.
■ A coordinate chart contains full information on size and location dimensions and other machining specifications (tabular dimensioning).
  ■ Precision measurement depends on normalizing the temperature of the instruments, workpiece, and work surface.

UNIT 4 REVIEW AND SELF-TEST

OPTICAL FLATS AND MICROSCOPES

1. a. Describe a design feature on a toolmaker's microscope that makes it possible to make measurements to accuracies of 0.001″ (0.02mm) and finer.
   b. Give two applications of a toolmaker's microscope that is equipped with a vernier angle-measurement scale.

2. List three scale mounts that are interchangeable on an industrial measuring magnifier.

3. Interpret what each of the following patterns of interference bands means in relation to a measurement or the accuracy of a machined surface:
   a. Straight, parallel, uniform, and
   b. Straight, parallel, uniform bands that curve slightly at the ends.

4. Give one reason for normalizing the temperature of optical measuring flats, other instruments, and the workpiece in a measurement, layout, or precision inspection setup.

5. a. State two advantages of using tabular dimensioning in place of conventional dimensioning on part drawings that have a great number of design features.
   b. Identify the kind of information normally contained in a coordinate chart.

6. List two precautions to observe when industrial microscopes and optical flats are used.

Measurement: Precision, High Amplification Comparators

Mass production depends on the interchangeability of parts. Machining and inspecting processes for such parts are often carried on independently of the place where the parts are assembled. The parts must function according to specifications and without further fitting. Interchangeability depends on accurate, rapid, and economical methods of measurement. Preceding units dealt with direct measurements taken largely from the scales on precision mechanical and optical instruments.

This unit deals with comparison measurements that require the use of *high amplification comparators*. Comparison measurements play an important function in mass production. The part is compared with a master or standard. The standard represents the basic dimension to be checked.

OBJECTIVES

After satisfactorily completing this unit, you will be able to:

- Understand the features and principles of:
  - High amplification dial indicators for measurements to 0.000020'' (0.0005mm).
  - Reed comparators and scales.
  - Contour measuring projectors.
  - Pneumatic comparator measuring instruments of the flow and pressure types.
  - Electronic comparators, including amplifiers and lever and cartridge heads.
  - Digital electronic indicators with computer output.
- Perform or simulate each of the following processes:
  - Measure with a Mechanical Comparator.
  - Measure with a Reed Comparator.
  - Check a Thread Angle Using an Optical Comparator.
  - Measuring with a Pressure-Type Pneumatic Comparator.
- Use *Terms* which describe the principles, functions, and applications of high-magnification comparators.
- Apply *Safe Practices* when using and storing highly-precise comparators.

MEASUREMENT WITH HIGH AMPLIFICATION COMPARATORS

A comparator is highly sensitive instrument that checks the accuracy of a measurement for any variations from a standard. It incorporates within its design:

- A device for holding a workpiece,
- A master against which the workpiece is checked,
- An amplification unit that permits small variations from basic dimensions to be observed and measured.

GENERAL TYPES OF HIGH AMPLIFICATION COMPARATORS

Six general types of high amplification comparators are available:

- *Mechanical comparators* include high amplification dial indicators and bench comparators.
- *Mechanical-optical comparators* are combination mechanical-optical instruments. A light beam casts a shadow on a magnified scale, and dimensional variations are indicated on the scale.
- *Optical comparators* are projection and reflection comparators. Features are measured by casting a magnified shadow of a part on a screen and then using a chart for comparison measurement.
- *Electrical-electronic comparators* are measuring instruments that have the added feature of power amplification.
- *Pneumatic comparators* require a flow of air between a part and a measuring gage. Pressure and flow changes indicate any variation in dimensional accuracy.
- *Multiple gaging comparators* are a combination of mechanical, electronic, optical, and pneumatic comparators. Multiple gaging comparators are used when a number of dimensions are to be gaged at one time.

Principles underlying each type of comparator, with general applications, are covered in this unit.

HIGH AMPLIFICATION MECHANICAL COMPARATORS

HIGH AMPLIFICATION DIAL INDICATORS

The most commonly used dial indicators are graduated for readings in 0.001″ (0.02mm) and 0.0001″ (0.002mm).

Two mechanical comparators, high amplification dial indicators, are shown in Figure 5–1. The dial indicator pictured in Figure 5–1A has a *discrimination* of 20 microinches. Jeweled bearings are used in the instrument to reduce friction and wear and to ensure *repeatability*—that is, the dial indicator reads (after a number of repeated insertions of a workpiece) to a standard of accuracy of ±one-fifth of a division from an original reading. The dial indicator here is graduated so that each division equals 0.00002″ (0.0005mm); the repeatability is 0.000004″ (0.0001mm).

A precision double dial test indicator, which is particularly suited to machine tool applications,

is shown in Figure 5–1B. This indicator has a balanced 0.004″ (0.01mm) dial (0.002–0–0.002″). The graduations are in 0.00005″ (0.00125mm). The repeatability error of the instrument is ±0.00001″ (0.0002mm). The round steel or carbide contact points are available in three diameters: 0.040″, 0.080″, and 0.120″ (1mm, 2mm, and 3mm).

MECHANICAL COMPARATORS

Mechanical comparators consist of a base, column, adjusting arm, and a gaging head. A mechanical comparator works on the same principles as a dial indicator. It is used for taking

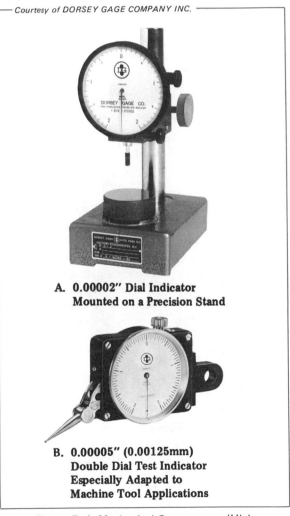

A. 0.00002″ Dial Indicator Mounted on a Precision Stand

B. 0.00005″ (0.00125mm) Double Dial Test Indicator Especially Adapted to Machine Tool Applications

Figure 5–1 Mechanical Comparators (High Amplification Dial Indicators)

accurate linear measurements by comparison with a standard. Any dimensional variation between the part being measured and the standard against which the instrument is set is shown on a magnified scale. The scale is graduated with plus and minus inch or metric units of measurement. The mechanical comparator in Figure 5–1B provides for extremely high amplification.

How to Measure with a Mechanical Comparator

STEP 1 Check the anvil and gage block for nicks or scratches. Clean the instrument and gage block.

STEP 2 Slide the gage block (master) carefully on the anvil.

STEP 3 Lower the gaging head until the contact point just touches this block.

STEP 4 Lock the gaging head in position.

STEP 5 Use the fine adjusting knob to adjust the pointer. Adjust until the scale reading is zero.

STEP 6 Recheck the setting.

STEP 7 Remove the master (standard). Replace it with the part to be measured.

> **Caution:** The size of the part must be within the movement range of the instrument.

STEP 8 Note the reading on the scale. A plus reading indicates an oversize measurement. A minus reading shows the amount the part is undersize.

STEP 9 Remove the part carefully. Replace the gage block in a case. Cover the comparator.

MECHANICAL-OPTICAL COMPARATORS

REED COMPARATOR

A *Reed comparator* (Figure 5–2) combines the features of a mechanical instrument and an optical instrument. It consists of a base, column, and a gaging head and bracket. The main design features of a Reed comparator are shown in Figure 5–2A. A lamp is required in the magnification system. This lamp is housed in the base. The base supports an anvil, a back stop, and a column. The gaging head is moved by a rack and pinion on the column. The locking screw clamps the gaging head in position. The gaging head may also be swung away from the base so that parts that are larger than the working height of the comparator can be measured.

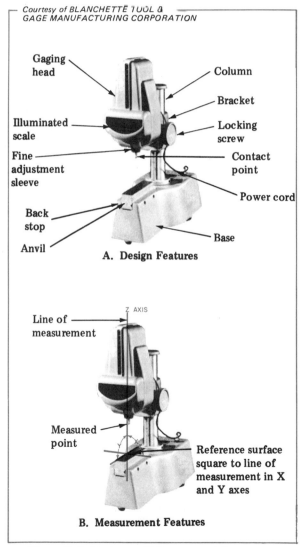

Courtesy of BLANCHETTE TOOL & GAGE MANUFACTURING CORPORATION

A. Design Features

B. Measurement Features

Figure 5–2 Design and Measurement Features of a Reed Comparator

The measurement features are shown in Figure 5–2B. The reference surface for the Z axis is at a right angle to the line of measurement in the X and Y axes. The general range of reed comparators is from 500X to 20,000X.

Reed Comparator Scales. There are two different scales for a simple reed comparator. The note above the inch-standard scale indicates the instrument has a magnification of 1000X (Figure 5–3). Each whole plus or minus numerical division (0.001″) is divided in tenths. The scale is marked 1/10,000 and indicates that each graduation is equal to 0.0001″. The range of the scale is ±2 thousandths (±0.002″), or 0.004″ overall.

The metric scale also has a magnification of 1000X. Each whole plus or minus numerical division represents 0.01mm. Since each numbered division is divided into five equal parts, each graduation equals 0.002mm. The range of the scale used here is from ±0.05mm or 0.1mm overall. The plus and minus range of the sample comparator scales are ±0.002″ (0.05mm). Reed comparators, having a 10,000 to 1 amplification, are capable of measurement accuracies to 10 microinches (0.000010″) or 0.00025mm.

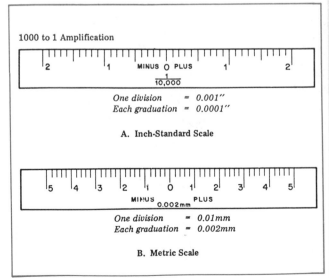

Figure 5–3 Sample Reed Comparator Scales in Customary and Metric Units of Measure

How to Measure with a Reed Comparator

STEP 1   Clean the anvil, contact point, and workpiece.

STEP 2   Unlock the locking screw. Move the gaging head so that the contact point clears the workpiece.

STEP 3   Select a gage block having the required dimension of the workpiece (or gage).

STEP 4   Mount the gage block centrally on the anvil.

STEP 5   Bring the gaging head down carefully until the contact point of the spindle touches the gage block.

STEP 6   Clamp the gaging head to the column. Make further adjustments with the adjusting sleeve. The shadow on the illuminated scale should coincide with the zero on the scale.

STEP 7   Replace the gage block with the part to be measured.

Caution: Slide the part carefully into position under the pointer. The height of the part must fall within the range of the instrument.

STEP 8   Note the reading on the scale. A reading on the plus (+) side indicates the amount the part is oversize. Similarly, a reading on the minus (–) side shows the part is smaller than the standard size.

STEP 9   Remove the workpiece. Turn the electric switch off. Return the gage or standard to a proper storage area. Cover the reed comparator.

Figure 5–4 Application of Optical Comparator (with Digital Readouts) for Contour and Size Measurements

OPTICAL COMPARATORS

Optical measuring instruments have great versatility, range, precision, and accuracy. One of the most popular and widely used optical measuring instruments is the *optical comparator* (Figure 5–4). It is sometimes called a *contour projector*. This projector provides an accurate method of measuring and comparing the contour of irregularly shaped parts. Flat or circular tools, gages, grooves, angles, and radii may be checked for size. Metal, plastic, and soft materials such as rubber may be measured without physical contact. Tolerances within ±0.0005″ (±0.01mm) may be measured accurately.

STAGE AND ACCESSORIES

The stage of an optical comparator may be moved horizontally or vertically. A work table provides a plane surface for accessories and workpieces. The table is provided with a large-diameter micrometer head that permits horizontal movement of the stage to accuracies within 0.0001″ (0.002mm). Vertical travel is obtained with a dial indicator. A work stage that is moved by turning micrometer dials permits the accurate

measurement of linear dimensions. Fixture bases with a small rotary vise or adjustable horizontal and center attachments extend the use of the comparator. Fixtures provide an exact location for a part and a master.

How to Check a Thread Angle by Using an Optical Comparator

STEP 1 Select a lens of sufficient magnification to permit measuring to the required degree of accuracy.

STEP 2 Position the tilting centers on the table.

STEP 3 Set the centers at the helix angle of the thread.

STEP 4 Mount the workpiece between the centers.

STEP 5 Place a thread form chart, if it is a replacement chart, into the screen recess.

Note: A sealed-image overlay chart is placed on top of the screen.

STEP 6 Align the chart to permit accurate vernier angle readings.

STEP 7 Focus the lens with the light turned on. The part is in focus when a clear image is produced.

STEP 8 Center the thread image on the screen. Move the micrometer cross slide stage to position the thread.

STEP 9 Turn the vernier protractor chart to one-half the thread angle.

STEP 10 Adjust the cross slide until the image of one slope of the thread angle coincides with the protractor angle.

STEP 11 Check the other slope of the thread by repeating steps 9 and 10.

STEP 12 Read any variation of the included thread angle directly from the graduated angles on the chart.

CONTOUR MEASURING PROJECTOR

The contour measuring projector is similar in principle and application to the optical comparators previously described. The image is visible in the same position as the operator views the object on the table.

The contour measuring projector has a number of advantages:

- The need to hold the object in special tools or fixtures is eliminated,
- Setup time is reduced,
- The object is placed automatically at a right angle to the beam of light,
- The object may be positioned by hand to permit rapid setup of the image with the lines or outline on the chart.

The table is positioned longitudinally, transversely, and vertically by feed handles. Horizontal measurements are read directly on micrometer dials on each feed screw. Common accessories include center supports and a screw thread fixture. Screens and charts also are used to make linear, angle, radius, and contour measurements.

Today, in addition to establishing the inside size and characteristics of holes, pneumatic comparators are applied in checking outside diameters, concentricity, squareness, parallelism, and other surface conditions.

PNEUMATIC COMPARATORS

Precise comparison measurements may be taken with pneumatic comparators, which utilize a master gage, a stream of air, and a pressure/velocity measuring system. The instruments are known as *pneumatic comparison measuring gages.* They are classified as either air (flow) or pressure-type pneumatic gages. Originally, they were designed to measure the inside characteristics of holes. Measurements were taken by measuring the volume of escaping air from a hole in comparison to a predetermined volume escaping from a gaging head standard. The gaging head was inserted into the hole. Air pressure was applied and a float position was established. Dimensional variation was noted on a scale by a change in the position of the float.

OPERATION OF PNEUMATIC COMPARATORS

The principle of operation of pneumatic comparators is simple. Dependence is placed on the flow of air between the faces of the jets in the gaging head and the workpiece. The clearance between the gaging head and the workpiece controls both the velocity and the pressure of air. The larger the workpiece is than the gaging head, the greater the flow (higher velocity) of air and the lower the back pressure. Conversely, the smaller the clearance between the gaging head and the workpiece, the slower the velocity and the greater the back pressure.

By knowing this principle, it is possible to make a comparison of either the flow of air (velocity) or its pressure. Flow (column-type) pneumatic gages indicate the velocity of air. Pressure-type pneumatic gages are designed to indicate air pressure. Both types of pneumatic gages are operated from either a portable or a plant supply of compressed air.

PNEUMATIC GAGING HEADS

The common forms of gaging heads include plug, ring, and snap gages. A *gaging head* consists of a flow tube and a gaging spindle. The flow tube provides the channel through which the air supply flows and is measured. The gaging spindle consists of a plug with a central air channel. The channel terminates in two or more jets in the side of the spindle. The air flows through these jets. The form of the spindle is determined by the operation to be performed. When the diameter of a hole is to be measured, a spindle with air jets that terminate in annular (around the circumference) grooves is used. When a hole is to be checked for roundness, bell-mouth, or taper, a spindle with air channels that run parallel to the axis of the spindle is used. Two diametrically opposite jets or more may terminate in the channels. The longitudinal (lengthwise) channels permit a point-to-point check. Features of pneumatic gaging heads are illustrated in Figure 5–5.

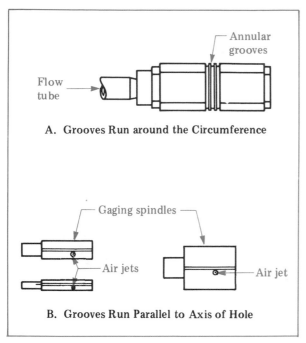

A. Grooves Run around the Circumference

B. Grooves Run Parallel to Axis of Hole

Figure 5-5 Features of Pneumatic Gaging Heads

ADVANTAGES OF PNEUMATIC COMPARATORS

The major advantages of gaging with pneumatic instruments over other measurement instruments are as follows:

- Amplifications range from 1 to 1000 to 1 to 40,000. Pneumatic instruments provide a simple and direct method of high amplification.
- Minute dimensional variations can be controlled within close tolerances.
- Parts may be gaged without contact between the workpiece and the gaging head. Thus, scoring of highly polished surfaces or damage to soft material parts is prevented.
- Less skill is required in using pneumatic gaging instruments than in gaging processes with precision conventional plug, ring, and snap gages.
- The work life of pneumatic gaging heads is longer than for conventional gages.
- Pneumatic gages may be used at a machine or on the bench. The work can be brought to the instrument, or vice versa.

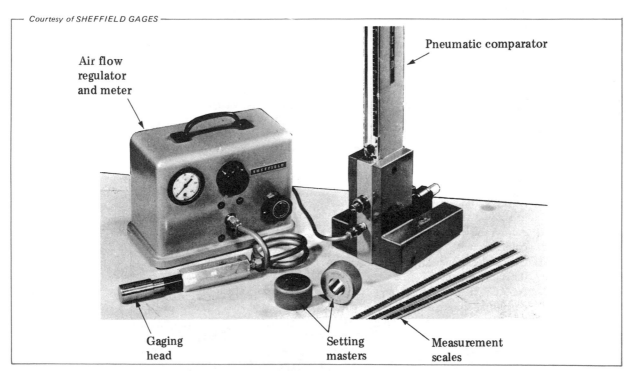

Courtesy of SHEFFIELD GAGES

Figure 5-6 Major Components of a Flow (Column-Type) Pneumatic Comparator

- Multiple diameters may be checked at the same time.
- Pneumatic gages are easier to use than mechanical gages. Holes may be inspected for out-of-roundness, parallelism, concentricity, and other surface irregularities.
- Gaging heads and parts are self-cleaning.

OPERATION OF A FLOW (COLUMN-TYPE) PNEUMATIC COMPARATOR

The flow, or column-type, pneumatic comparator measures air velocity. The major components of such a comparator are shown in Figure 5–6. Air is passed through a filter and a regulator. The air flows through a tapered, transparent tube and reaches the gaging head at about 10 pounds pressure per square inch. The air flow causes a float in the tube to be suspended.

As air flows through a metering gage, it exhausts through the channels of the gaging head and the clearance area between the head and the workpiece.

OPERATION OF A PRESSURE-TYPE PNEUMATIC COMPARATOR

The major features of a pressure-type pneumatic comparator gage system includes a constant air supply which passes through a filter, a pressure regulator, and a master pressure gage. The supply of air is then branched through two channels. The upper branch is identified as the reference channel. The lower branch is identified as the measuring channel. Air escapes from the reference channel to the atmosphere through a zero setting valve. Air from the measuring channel flows through the gage head jets and the part being measured and then into the atomsphere. Note

Courtesy of FEDERAL PRODUCTS CORPORATION

Single setting master

Figure 5–7 Measuring a Diameter with a Pressure-Type Pneumatic Comparator and Gaging Spindle

that the two channels are connected through a differential pressure meter.

SETTING UP THE GAGE SYSTEM FOR MEASUREMENT

The pressure-type pneumatic comparator gage system is set by placing a setting master over the gaging spindle plug. The zero setting valve is adjusted until the gage pointer (needle) reads zero. Any variation between the size of the master gage and the workpiece affects the pressure in the measuring channel. The difference is indicated by the movement of the dial gage pointer.

Table 5–1 Amplification, Discrimination, and Measurement with a
Pneumatic Pressure-Type Balanced System

	Ranges	
Amplification	**Discrimination**	**Measurement**
1,250:1	0.0001″ (0.0025mm)	0.006″ (0.15mm)
2,500:1	0.00005″ (0.00125mm)	0.003″ (0.075mm)
5,000:1	0.00002″ (0.0005mm)	0.0015″ (0.0375mm)
10,000:1	0.00001″ (0.00025mm)	0.0006″ (0.015mm)
20,000:1	0.000005″ (0.000125mm)	0.0003″ (0.0075mm)

The range of amplification of pressure-type pneumatic comparators is from 1250 to 1 to 20,000 to 1. A common pressure-type pneumatic comparator is pictured in Figure 5–7. The data in Table 5–1 indicates the range of amplification, discrimination, and measurement with this type of instrument in a *balanced system*. With increased amplification, a higher level of discrimination and a lower measuring range result.

How to Measure with a Pressure-Type Pneumatic Comparator

STEP 1 Select a pneumatic measuring instrument. The scale must permit measuring to the required degree of accuracy.

STEP 2 Select a setting master of the required dimension.

STEP 3 Mount the gaging head corresponding to the required dimension. The gaging head is attached to the plastic tubing.

STEP 4 Supply a flow of air. Set the pressure gage.

STEP 5 Place the setting master over the gaging spindle plug. Adjust the dial pointer until it indicates zero.

STEP 6 Remove and store the master.

STEP 7 Insert the gaging spindle plug in the hole in the workpiece.

STEP 8 Note any variation from the original zero reading on the dial face.

Note: A clockwise dial pointer movement indicates how much the hole is undersize. A counterclockwise movement indicates an oversize measurement.

STEP 9 Shut off the air supply at the end of the measuring process. Disassemble the gaging spindle plug. Store in an appropriate container.

FACTORS INFLUENCING THE DESIGN OF A GAGING SPINDLE PLUG

Some cautions must be observed with air gages. The quality of surface finish must be considered. Fixed as well as mechanical types of indicating gages check the smallest diameter of a hole or the largest outside diameter. By contrast, pneumatic gages give an average reading between the smallest and largest diameters. Therefore, when air gages are used for holes having a rough surface finish, a special design of gaging spindle is used. This spindle has a body that is smaller than the minimum diameter of the hole. The spindle is fitted with a hardened steel contact blade. The blade either pivots or floats. The action of the contact blade controls the air flow through the jets. The gage then serves as a regular, solid plug gage.

The features of a hole that are to be measured determine the design of plug to use on a spindle. Five different designs are shown in Figure 5–8.

- A single-jet plug is used to check concentricity, location, flatness, squareness, length, and depth.
- A two-jet plug is applied in checking inside diameters and holes that are out-of-round, bell-mouth, or taper.

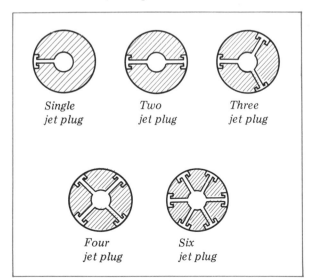

Single jet plug *Two jet plug* *Three jet plug*

Four jet plug *Six jet plug*

Figure 5–8 Cross Sections of Gaging Spindle Plugs with Jets

- A three-jet plug is used for checking triangular out-of-round conditions.

- A four-jet plug provides average diameter readings.

- A six-jet plug produces average measurements for the conditions indicated for two- and three-jet plugs.

ELECTRICAL COMPARATORS

The three major components in an electrical comparator are a gaging head, a power unit, and a microammeter. The microammeter contains a graduated scale. Any movement of the end of the gage head is magnified greatly by the microammeter.

The spindle (lever) of the gage head bears against an armature. The end of the armature is positioned between two electrical induction coils. Current flows between the two coils. When the armature is midway between the two coils, the circuit is balanced. The scale on the instrument reads zero.

Any minute variation in the size of a workpiece causes the spindle to move and change the circuit balance. The amount of movement on the scale is proportional to the movement of the spindle. The magnification factor of the instrument affects the value of each division of the scale. For example, if the magnification of the instrument is 10,000, each division of the scale equals 0.000010″ (10 millionths of an inch) or 0.00025mm (25 hundred-thousandths of a millimeter). The total measurement range on the scale of such an instrument is usually 0.0005″ (0.01mm).

ELECTRONIC COMPARATORS

The versatility of the electronic comparator makes it an ideal instrument for comparison measurements. Standard components may be used in a great number of measurement problems. The absence of mechanical parts makes electronic instruments comparatively rugged as high precision instruments. Also, their sensitivity and speed reduce the time lag during measurements. Electronic comparators are particularly useful, therefore, when measurements of a moving feature are required. Measurements along a strip passing through a rolling mill are a typical example of *dynamic measurement*.

Other considerations for selecting an electronic comparator are as follows:

- Rapidity of operation even at high amplifications,
- Use of the same instrument for multiple amplification ranges,
- Portability of the instruments, which are also self-contained,
- Easily adjusted and understandable controls,
- High instrument sensitivity in all amplification ranges,
- Limited self-checking,
- Instrument accuracy that compares favorably with other instruments.

ELECTRONIC AMPLIFIER AND READOUT

The two major components in an electronic measurement unit are a gage amplifier and one or more gaging heads. The amplifier and a single gaging head serve as a comparator and height gage. In applications where two heads are used, measurements may be taken of differences in size, length, thickness, diameter, flatness, taper, or concentricity. In such applications, the amplifier measures either the difference or the sum of the gaging head outputs.

Electronic gage amplifiers have magnification ranges from ±0.010″ (±0.200mm) to ±0.0001″ (±0.002mm).

Amplifier readouts are provided on either a sensitive meter or a direct digital readout display. The meter version has a precision-calibrated meter dial that is designed for easy readability without parallax error. Both amplifiers contain solid-state circuitry. The result is extreme reliability from variations of temperature, line voltage, and repeatability of measurement.

ELECTRONIC GAGING HEADS

Input into an electronic amplifier is provided by the movement of one or more gaging heads. The two basic types of gaging heads are identified in Figure 5–9 as a *lever (probe) type* and a *cartridge (cylindrical) type*. The gaging heads

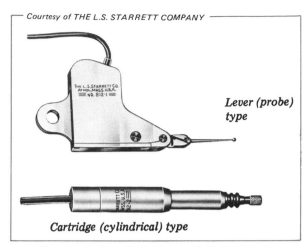

Lever (probe) type

Cartridge (cylindrical) type

Figure 5-9 Two Basic Types of Gaging Heads

are actuated by a contact of 8 to 12 grams of pressure against an object. Three general applications of the lever (probe) type gaging head are illustrated in Figure 5–10.

Lever (Probe) Type Gaging Head. The four major parts in a gaging head are the outer shell, or housing, the transducer, the probe tip, and the lever assembly. The transducer has matched coils and a core and is designed as a compact unit. The transducer is sensitive to fine measurements. Movements of the probe tip are converted into electrical energy in the transducer. These movements are amplified in the electronic amplifier. Direct readouts are indicated by the pointer movement on the amplifier scale.

Cartridge (Cylindrical) Gaging Head. The diameter of the cartridge (cylindrical) gaging head has been standardized. The gaging head may be substituted in holders that accommodate the stem of dial indicators. Thus, the cartridge gag-

ing head may be used interchangeably for many dial indicator applications. The advantage of using the cartridge head is that the contact moves in a frictionless head. The transducer components in the cartridge and lever gaging heads are identical.

DIGITAL ELECTRONIC INDICATORS WITH COMPUTER OUTPUT

The battery-powered electronic indicator with digital output to other interface and computer units is used to perform the following functions:

- Take comparative measurements,
- Provide a clear digital display and accurate analog reading of a measured part feature,
- Flash a light limit signal to identify an "under" or "over" analog reading.

The electronic measurement instrument may be used as a general-purpose comparator. Versatility is increased when the indicator is connected with output to interface with other computer hardware and software. As an example, the model shown in Figure 5–11 has a computer connection that connects with an interface system and communicates measurement readings directly to a hand-held or mainframe computer. Other components like a *statistical analyzer* and an *electronic data converter system* permit decoding and processing measurement data.

ELECTRONIC MEASUREMENT INDICATOR FEATURES

Measurement readings taken with an electronic indicator with computer capability may be transmitted, gathered, stored, and quickly analyzed. The data may then be used to evaluate manufacturing processes, to provide summaries

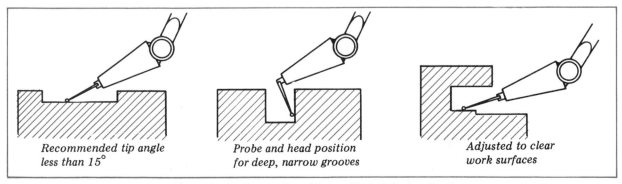

Recommended tip angle less than 15°

Probe and head position for deep, narrow grooves

Adjusted to clear work surfaces

Figure 5-10 Positioning of Lever (Probe) Gaging Head

for *statistical process (quality) control*, and to produce hard copy review printouts.

Important instrument features are shown by callouts in Figure 5–11. Except for the movable measuring spindle, the instrument is totally electronic. The instrument has a *dual display (digital and analog)* for fast, accurate readings. Individual light emitting diode control elements are used with the analog display graduations to show a measurement condition. Flashing *limit signals* provide positive identification of readings as "over" or "under" size.

The *digital readout* indicates the exact deviation of a measurement from zero. The small last place digit on the right gives the value of the reading between graduations.

There are six single-purpose *control switches:* automatic zero, settings of upper and lower limit, count up and count down, and true spindle position.

REMOTE TRANSDUCER APPLICATIONS AND SIZES

A *remote transducer* is often mounted in an existing gaging fixture in place of a regular dial indicator or other gage head. A remote transducer unit includes a measurement *stem* that is located at a distance from the measurement indicator, an extension *cable*, and a *connector* to the indicator head.

Transducers are designed with standard stem lengths ranging from 0.670″ (17mm) to 3.000″ (75mm) and stem diameters of 0.315″ (8mm) and 0.375″ (9.5mm) to fit standard gaging heads and fixtures.

DIGITAL MEASUREMENT RANGES

Digital electronic indicators are available in general measurement ranges from ±0.01000″ to ±0.0400″ and with minimum digital measurement values ranging from 0.00005″ to 0.0001″. The analog range is from ±0.005″ to ±0.040″. The minimum analog graduations are from 0.0001″ to 0.001″.

Metric instruments are available with digital measurement ranges of ±0.199mm and 1.000mm and minimum digital values of 0.001mm; with minimum analog graduations of 0.001mm and 0.01mm, respectively.

Figure 5–11 Electronic Indicator with Digital Output

Safe Practices in the Use and Care of High Amplification Comparators

- Investigate the potential sources of error before making high precision measurements. Check:
  - Direct heat from sunlight,
  - Electric lights or drafts,
  - Heat transfer due to handling,
  - Fastening of the locking nuts for the arm and the gaging head.
- Measure holes that have a rough surface finish with a pneumatic spindle fitted with a hardened contact blade.
- Place the measuring instrument and setup on a vibration-free surface.
- Avoid handling the stand, parts, and gages. Handling may produce uneven temperatures.

TERMS USED WITH HIGH AMPLIFICATION COMPARATORS

Reed-type instrument	A precision measuring instrument. An instrument in which a spring (reed) suspension is substituted for other mechanical systems.
Electronic measuring instruments	Instruments in which amplification is obtained by an electronic system.
Amplifier	A device for increasing (amplifying) electronic signals. When used with different heads, it measures length, thickness, diameter, flatness, taper, and concentricity.
Optical comparator (contour projector)	A comparative measuring instrument. An instrument on which an enlarged object is projected on a screen. A master form against which a projected object is examined for contour conformance and accuracy of measurement.
High amplification comparators	A series of comparative measuring instruments. Instruments in which dimensional variations are multiplied.
Lever (probe) gaging head	A small lever arm mounted to an electronic unit for magnifying small movements of the contact point.
Pneumatic comparison measuring instruments	Instruments that require the flow and recording of the velocity or pressure of air to make comparison measurements. Instruments that translate the flow of air into a measurement in inch and metric standard units.
Transducer	A sensitive device for converting minute movements of a probe (variations from a specific measurement) into electrical energy. This energy is electronically amplified and displayed as a measurement on an indicator scale.
Pressure-type pneumatic comparator (gages)	A pneumatic gaging system in which measurement is based on an air pressure differential.

SUMMARY

- High amplification dial indicators have a discrimination of 20 millionths of an inch (0.0005mm).

   - The reed comparator works on the principle of two spring steel reeds. One reed is attached to a solid block; the other reed, to a movable block. Fine, precise movements are magnified on a pointer.

- Additional magnification on a reed comparator is produced by passing a light beam through an aperture to reflect on a scale. Scale readings within 50 millionths of an inch (0.001mm) are common.

   - Optical comparators depend on the projection of a large image of the object being measured. The light rays against a workpiece are projected through a lens system into a mirror and onto a viewing screen. Shape and size characteristics are compared to an enlarged measurement chart on the screen.

- Pressure-type pneumatic comparators are used to identify variations from specific dimensions based on changes in air pressure. A differential pressure meter is affected by the volume of air escaping between the workpiece and the gaging plug.
    - Plug, ring, and snap gages are three common forms of gaging heads.
- A single jet plug (gage) is used to check concentricity, location, flatness, squareness, length, and depth.
    - Two-jet plugs are needed to check inside diameters of out-of-round, bell-mouth, and taper conditions.
- The electrical comparator is a high magnification instrument. Measurements within 10 millionths of an inch (0.00025mm) are common.
    - Electrical comparators require a power unit, a gaging head, and a milliammeter. Minute variations in current flow are transmitted to a movable pointer on the milliammeter scale.
- Electronic comparators permit the use of the same instrument for measurements at multiple high amplification ranges. The sensitivity and speed of operation permit dynamic measurement of a moving feature.
    - Digital electronic indicators may be interfaced with a statistical analyzer, electronic data converter, and other computer equipment and software programs. Measurement data may be reported on a digital readout or analog display, transmitted, stored, analyzed, and printed.

UNIT 5 REVIEW AND SELF-TEST

1. Name four general high amplification comparison measuring instruments.

2. Compare the accuracy range of a standard, everyday shop dial indicator with the accuracy range of a high amplification dial indicator.

3. List the steps in measuring a 0.9375″ dimension on a flat gage to within +0.000020″/−0.000010″. Use a mechanical comparator.

4. Cite two design features of a reed comparator that differ from similar features of mechanical comparators.

5. Give two advantages of a contour measuring (optical) projector in comparison to a mechanical comparator.

6. State three advantages of using pneumatic gaging instruments over GO and NO–GO gages and other mechanical measuring methods.

7. Give two advantages of a meter readout on an electronic amplifier for precision measurements.

8. Cite three precautions a worker must follow when working with high amplification comparators.

9. Name three functions performed by an electronic indicator with digital output to a computer (electronic) data analyzer.

Surface Texture and Statistical Quality and Process Control

The interchangeability of parts requires surfaces to be machined within specified dimensions, tolerances, and quality of surface texture (finish). Every machined surface has a degree of roughness and other imperfections.

The initial approval of surface quality standards was made by the American Standards Association in 1962. Modifications in the standards continue to accommodate increasingly finer surface finishes, more precise instrumentation, and the capability to more accurately measure, gather, store, and analyze data in order to more efficiently control quality and manufacturing processes.

This *Section* provides a foundation of surface texture characteristics and measuring instruments and interlocks this technology with principles of statistical quality and process control (SPC) and instrumentation.

UNIT 6

Surface Texture: Characteristics and Measurement

OBJECTIVES

After satisfactorily completing this unit, you will be able to:

- Apply *Terms* for specifying and measuring surface texture.
- Interpret functions and principles related to a surface finish analyzer, profilometer, and microscope.
- Use surface texture reference tables and compute arithmetical average (AA) and root mean square (RMS) values.
- Interpret drawings involving surface texture lay symbols and dimensioning.
- Perform each of the following processes:
  - Measure Surface Texture with Comparator Specimens.
  - Measure Surface Finish with an Indicating Instrument.
- Solve the problems assigned from the *Review and Self-Test Items*.

SURFACE TEXTURE TERMS AND SYMBOLS

The terms *surface texture*, *surface finish*, *surface roughness*, and *surface characteristics* are used interchangeably in the shop and throughout this text. However, particular American (ANSI) standard terms, ratings, and symbols are described and illustrated. Use of these terms, ratings, and symbols permits designers, skilled mechanics, and other workers to communicate. Thus, part specifications may be accurately interpreted to produce a given surface quality. Figure 6–1 identifies some common surface texture characteristics and terms.

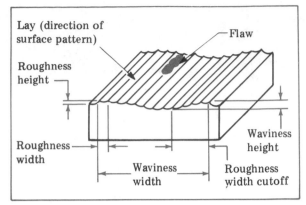

Figure 6–1 Common Surface Texture Characteristics and Terms

Surface Texture. *Surface texture* relates to deviations (from a nominal surface) that form the pattern of the surface. The deviations may be repetitive or random. The deviation ιay result from roughness, waviness, lay, and flaws.

Surface. According to American (ANSI) standards, an object is bounded by a *surface*. This surface separates the object from another object, substance, or space.

Surface Finish. *Surface finish* is indicated by the symbol $\sqrt{}$ and a microinch measurement. For example, a roughness weight of 63 microinches appears on a drawing as $^{63}\!\!\sqrt{}$.

The surface symbol also relates to machining, as follows.

$\sqrt{}$ Surface may be produced by any method.

$\triangledown$ Machining is required (for which a material allowance is provided).

.001$\triangledown$ Material removal allowance (inch or metric millimeter value).

$\oslash$ Material removal is prohibited.

Profile. A *profile* is the contour of a machined surface on a plane that is perpendicular to the surface. The plane may also be at an angle other than 90°, if it is specified. A short section of a surface profile is illustrated in Figure 6–2.

Centerline. The *centerline* is a line that is parallel to the general direction of the profile. Figure 6–2 shows a centerline of an exaggerated

profile. Roughness is measured around the centerline. This line lies within the limits of roughness width cutoff. The centerline in the profile serves as the cutoff line between roughness areas on both sides of the line.

Nominal Surface (Mean Line). A *nominal surface* is considered to be a geometrically perfect surface. Such a surface would result if the peaks were leveled to fill the valleys. The theoretical surface is represented by a *mean line*.

Microinch. A *microinch* is the basic unit of surface roughness measurement. A microinch designates one-millionth of an inch (0.000001″). The numerical value is sometimes followed by the symbol Mμ or μ. Millimeter and micrometer values are used on metric drawings.

Lay. *Lay* is the direction of the predominant surface finish pattern. Lay patterns, symbols, and applications in drawings are described later.

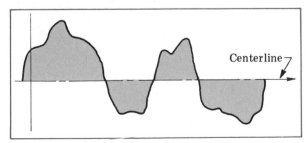

Figure 6–2 Exaggerated Short Section of a Surface Profile

Roughness. The fine irregularities produced by a cutting tool or production process are identified as *roughness (surface roughness).* Roughness includes traverse feed marks and other irregularities.

Roughness Width. The irregularities produced by production processes extend within the limits of the roughness width cutoff (the distance between successive peaks or ridges). *Roughness width* is measured parallel to the nominal surface. The peaks considered constitute the dominant pattern. The maximum roughness width rating, in inches, is placed to the right of the lay symbol. A rating of 0.020″ appears in Figure 6–3A.

Roughness Width Cutoff. A rating in the thousandths of an inch is used for the greatest spacing (width) of surface irregularities that are repetitive. These irregularities are included in the measurement of average roughness height. The *roughness width cutoff* rating appears directly under the horizontal extension of the surface finish symbol. A rating of 0.100″ is given in Figure 6–3B. If no value is shown, a rating of 0.030″ is assumed.

Roughness Height. *Roughness height* is expressed in microinches. The height is measured normal to the centerline. The height is rated as an arithmetical average deviation. The roughness height value is placed to the left of the surface finish symbol leg. If only one value is given—for example, 63 microinches as shown in Figure 6–3C—it indicates the maximum value. When maximum and minimum values for roughness height are given, they indicate the permissible range. In Figure 6–3D the range indicates 63 microinches maximum and 50 microinches minimum.

Waviness. *Waviness* is a surface deviation (defect) consisting of waves. These waves are widely spaced. Waviness may be produced by vibrations, chatter, or machine or work defects. Heat-treating processes and warping strains also produce waviness.

Maximum Waviness Height. The *maximum waviness height* is the peak-to-valley distance of a wave. The maximum waviness height rating is placed above the horizontal extension of the sur-

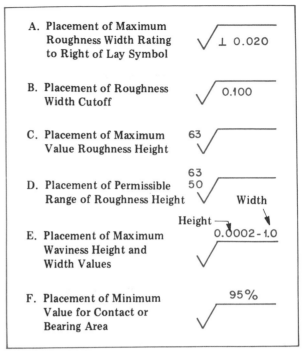

A. Placement of Maximum Roughness Width Rating to Right of Lay Symbol — ⊥ 0.020

B. Placement of Roughness Width Cutoff — 0.100

C. Placement of Maximum Value Roughness Height — 63

D. Placement of Permissible Range of Roughness Height — 63 / 50 — Width

E. Placement of Maximum Waviness Height and Width Values — Height / 0.0002 - 1.0

F. Placement of Minimum Value for Contact or Bearing Area — 95%

Figure 6–3 Representation of Surface Characteristics on Drawings and Specifications

face finish symbol, as shown in Figure 6–3E by the 0.0002″ value.

Waviness Width. *Waviness width* is the spacing of successive wave peaks or valleys and is measured in inches. The value also appears above the surface finish symbol but next to the waviness height, as shown in Figure 6–3E by the 1.0″ value.

Flaws. *Flaws* are surface defects. They may occur in one area of the unit surface. Flaws may also appear at widely varying intervals on the surface. Casting and welding cracks, blow holes, checks, and scratches are common flaws. Normally, the effect of flaws is not included in measurements of roughness height.

Contact or Bearing Area Requirement. The *contact or bearing area* is expressed as a percentage value. This value is placed above the extension line of the surface finish symbol, as shown in Figure 6–3F by the 95%. The value is the minimum contact or bearing area requirement as related to the mating part or reference surface.

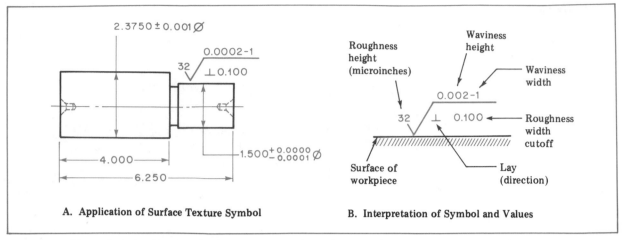

A. Application of Surface Texture Symbol

B. Interpretation of Symbol and Values

Figure 6–4 Application and Interpretation of Lay Symbol and Surface Texture Specifications

REPRESENTATION OF LAY PATTERNS AND SYMBOLS ON DRAWINGS

Six lay symbols are commonly used in drawings and parts' specimens. Tool or surface finish patterns are indicated by perpendicular (⊥), parallel (∥), angular (X), multidirectional (M), circular (C), radial (R), and (P) pitted, non-directional symbols.

The manner in which lay symbols and values are applied on drawings is shown in Figure 6–4A. The interpretation of the symbol and values is given at (B).

COMPUTING AVERAGE SURFACE ROUGHNESS VARIATIONS

COMPUTATION OF ARITHMETICAL AVERAGE (AA)

The *arithmetical average* (AA) of surface roughness is a new unified reading (number) of the average (roughness) variations of a surface above and below a centerline. This average is a measurement of the "hills and valleys" in microinches or metric unit equivalent.

The arithmetical average (AA) is obtained by adding the heights of the surface roughness increments and dividing by the number of intervals. The increments—for example, a, b, c, and so on in Figure 6–5—represent the distance from the centerline to a point on the profile taken at regular intervals. The values in Table 6–1 give the surface roughness measurements above and below the centerline. These values are added. The sum is then divided by the number of intervals. In the example, the arithmetical average (AA) is 8.86 microinches (Table 6–1).

Table 6–1 Roughness Measurements and AA Value (Intervals a through n)

Interval	Roughness Measurement (Microinches)
a	4
b	12
c	18
d	15
e	6
f	5
g	12
h	5
i	8
j	14
k	9
l	3
m	8
n	5
Totals	
14	124

$$AA = \frac{124}{14} = 8.86 \text{ microinches}$$

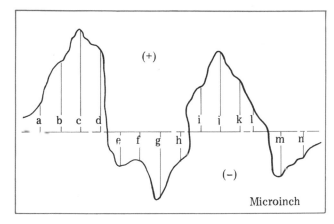

Figure 6–5 Plus (+) and Minus (–) Microinch Measurements Taken at Regular Intervals (Hypothetical Profile)

CONCEPT AND COMPUTATION OF ROOT MEAN SQUARE AVERAGE (RMS)

The concept of the *root mean square average* (RMS) is an older system of roughness measurement than the arithmetical average. Like the AA, the RMS consists of a number. The number is read on the dial of a surface finish analyzer.

The RMS represents the square root of a series of measurements. Each measurement is the distance the actual surface varies from a nominal surface. Like the arithmetical average values, each measurement is taken at equally spaced intervals. Each value is squared and added to the next measurement. The sum of the measurements is divided by the number of intervals (n) used. Finally, the square root of the result is taken. The square root is the root mean square (RMS) value:

$$\text{RMS} = \sqrt{\frac{a^2 + b^2 + c^2 + d^2 \ldots}{n}}$$

Using the microinch intervals (n) and values from the previous example, Table 6–2 gives the squared roughness measurements. The RMS value is computed to be 9.92 microinches.

A roughness measuring instrument that is calibrated for AA values gives an approximately 11% to 12% lower reading than an instrument that is RMS calibrated. Some designs have a selector switch to permit selecting either the

Table 6–2 Squared Roughness Measurements and RMS Value

Interval	Squared Roughness Measurement (Microinches)	
a	a^2 =	16
b	b^2 =	144
c	c^2 =	324
d	d^2 =	225
e	e^2 =	36
f	f^2 =	25
g	g^2 =	144
h	h^2 =	25
i	i^2 =	64
j	j^2 =	196
k	k^2 =	81
l	l^2 =	9
m	m^2 =	64
n	n^2 =	25
	Totals	
14		1,378

$$\text{RMS} = \sqrt{\frac{1,378}{14}} = 9.92 \text{ microinches}$$

RMS or AA scale. Many manufacturers of parts adjust AA readings without changing RMS values found on old drawings.

MEASURING SURFACE FINISHES

A few American standard measurement values of surface roughness height, roughness width cutoff, and waviness height are given in Table 6–3. The full table of surface texture values is included in the Appendix.

FUNCTIONS OF SURFACE COMPARATOR SPECIMENS

Surface comparators (specimens) are produced commercially to measure surface roughness. The comparators have scales that conform precisely with ANSI standards of surface roughness and lay. Surface comparators are generally flat or round. Different scales are designed for gaging common types of machined, ground, and lapped surfaces. The finer microinch-finish

Table 6–3 American Standard Surface Texture Values

Dimensional Measurement	Measurement Values			
	Microinches			
Roughness	1	13	50	200
height	2	16*	63*	250*
	. . .			
	Inches				
Standard roughness width cutoff	0.003	0.010	0.100	0.300	1.000
	. . .				
	Inches			
Waviness	0.00002	0.0001	0.001	0.010
height	0.00003	0.0002	0.002	0.015
	. . .			

*Recommended values

surface comparator specimens are often used with a *surface finish viewer* (Figure 6–6).

SURFACE MEASUREMENT WITH COMPARATOR SPECIMENS

In actual practice, a number of different machined surfaces may be measured by comparator

Courtesy of DoALL COMPANY

Figure 6–6 Surface Finish Viewer and Comparator Specimen for Lay, Process, and Surface Roughness

specimens. Several surfaces are listed in the first column of Table 6–4. General ranges of surface roughness heights and lay appear in the second and third columns respectively. Roughness widths are given in inches in the fourth column for machining processes that produce roughness heights between 2 and 500 microinches. The actual measured values in microinches are given in the fifth column. The letters in the sixth column tell that the measured value was obtained by either a brush analyzer (B) or a profilometer (P).

Surface roughness specimens are available for electrical discharge machined finishes on specific steels such as tool steel (16 to 250 microinches). Surface finish specimens are also provided for grit-and-shot blast surface finishes (32 to 1,000 microinches). Other specimens are available for die-cast and centrifugal, green-sand, and other casting processes.

How to Measure Surface Texture with Comparator Specimens and a Viewer

Direct Inspection

STEP 1 Determine the surface texture specifications from the drawing.

STEP 2 Note the lay pattern, machining process, and the required roughness height and roughness width.

STEP 3 Select a comparator specimen with the same characteristics as the required surface finish.

STEP 4 Compare the surface finishes of the workpiece and the specimen standard.

Inspection with Viewer

STEP 1 Repeat steps 1 through 3 for direct visual inspection.

STEP 2 Slide the surface finish comparator specimen into the viewer.

STEP 3 Place the comparator instrument over the surface. The instrument has a magnifier and built-in illumination.

STEP 4 Turn the eyepiece to bring the machined surface into focus.

STEP 5 Match the surface roughness.

Note: A visual check against a standard establishes the quality of surface finish.

Table 6-4 Machined Surfaces That May Be Measured by Comparator Specimens

Machined Surface	Roughness Height (Microinches)	Lay	Roughness Width (Inches)	Actual Microinch Value	Type* of Instrument
Honed, lapped, or polished	2	Parallel to long dimension of specimen	. . . .	2.4	B
	4		. . . .	3.8	B
	8		. . . .	6.5	B
Ground with periphery of wheel	8	Parallel to long dimension of specimen	. . . .	7.8	B
	16		. . . .	14.	B
	32		. . . .	28.	B
	63		. . . .	64.	B
Ground with flat side of wheel (Blanchard)	16	Angular in both directions	. . . .	17	B
	32		. . . .	28.	B
Shaped or turned	32	Parallel to long dimension of specimen	.002	31.	B
	63		.005	60.	P
	125		.010	130.	P
	250		.020	250.	P
	500		.030	530.	P
Side milled, end milled, or profiled	63	Circular	.010	57.	P
	125		.020	125.	P
	250	Angular in both directions	.100	235.	P
	500		.100	470.	P
Milled with periphery of cutter	63	Parallel to short dimension of specimen	.050	55.	P
	125		.075	115.	P
	250		.125	240.	P
	500		.250	520.	P

Tolerances

Roughness Height (Microinches)	Tolerance (Percent)
2 to 4	+25 –35
8	+20 –30
16	+15 –25
32 and above	+15 –20

*B represents brush analyzer; P, profilometer.

SURFACE ROUGHNESS RELATED TO COMMON PRODUCTION METHODS

Different production methods produce variations in surface texture. Figure 6–7 provides ranges of surface roughness for selected manufacturing processes. A more complete table appears in the Appendix. General and nonconventional machining processes and forming methods are considered.

SURFACE TEXTURE MEASURING INSTRUMENTS

SURFACE FINISH ANALYZER

Variations between the high and low points of the work surface may be measured by a *surface finish analyzer*. The three main components of the instrument are a tracer head, amplifier unit, and a recording dial.

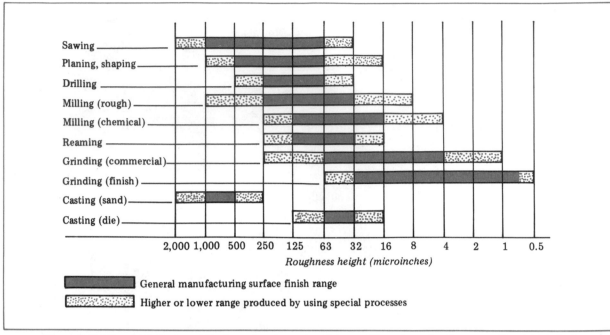

Figure 6–7 Microinch Range of Surface Roughness for Selected Manufacturing Processes (Partial Listing)

Table 6–5 Readings of Machined Surface Profiles
Obtained by Using Two Radii Stylus Sizes

Machining Process	Surface Profiles and Roughness	Stylus Radius (0.0000″)	Average Reading (Microinches)
Shaping		0.0005 0.0001	100–105 110–115
Turning		0.0005 0.0001	45–50 40–44
Grinding		0.0005 0.0001	8–9 10–11
Polishing		0.0005 0.0001	1.4–1.7 1.4–1.7

A diamond stylus is mounted in the tracer head. The tip of the stylus is rounded. The two basic tip radii are 0.0001″ and 0.0005″. The profiles of surface roughness produced by using these two tip radii in four examples of machining processes are presented in Table 6–5. The same wave forms are shown with readings taken using 0.0001″ and 0.0005″ tip radii.

The tracer head is moved across the work surface either manually or mechanically. More precise surface roughness measuring instruments, such as a *brush analyzer*, for example, are moved mechanically. The stylus moves as it follows the surface profile. The minute movements are converted in the tracer head into electrical fluctuations. These fluctuations are magnified in the amplifier. The mechanical movement of the stylus is registered by a needle on the surface recording dial in microinches or metric equivalent values.

PROFILOMETER

The *profilometer* is another surface roughness measuring instrument. It employs a tracing point or stylus to establish the irregularities. These irregularities are greatly magnified and recorded. The profilometer in Figure 6–8 can be set up for manually checking a ground cylindrical surface. The amplimeter pointer shows the roughness on a microinch scale. Direct readings are displayed in microinch or micrometer values on the analog meter as the instrument tracer moves across the work surface.

Courtesy of SHEFFIELD MEASUREMENT DIVISION, WARNER & SWASEY/ A CROSS & TRECKER COMPANY

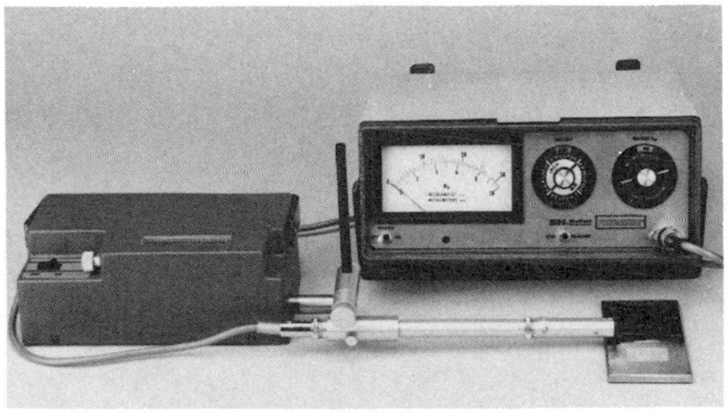

Figure 6–8 Profilometer Surface Texture Measurement System (Microinch and Micrometer Values)

Another type of profilometer produces a recorded profile on a tape, as shown in Figure 6–9. The height of the horizontal line is 25μ (microinches). Waviness, roughness, and waviness/roughness profiles are recorded.

How to Measure Surface Finish with an Indicating Instrument

STEP 1 Read the blueprint. Determine the required surface finish characteristics.

STEP 2 Turn the switch of the surface analyzer on. Allow the instrument to warm up.

STEP 3 Check the machine calibration. Use the 125 microinch test block. Move the stylus on the tracer head over the block. A movement of approximately 1/8″ per second is a good speed.

STEP 4 Adjust (calibrate) the instrument. The pointer should register 125 microinches, the same measurement as the test block.

STEP 5 Use the 0.030 cutoff range for surface roughness of 30 microinches or more. Use the cutoff range of 0.010 for surfaces finer than 30 microinches.

Note: Set the range switch to a high setting if the roughness is not known. Then turn to a finer setting to get the accurate surface finish reading.

STEP 6 Clean the work surfaces to be measured.

STEP 7 Place the tracer over the machined surface. Move the stylus at the rate of 1/8″ per second over the surface.

STEP 8 Read the surface roughness by the movement of the pointer. The reading is in microinches.

Note: On instruments like the surface analyzer, the recording device produces a graph of surface irregularities indicating changes may need to be made in the production tooling or methods of manufacture.

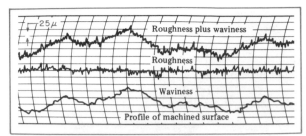

Figure 6-9 Recorded Roughness and Waviness Profiles of a Machined Surface

SURFACE FINISH MICROSCOPE

Surface texture may also be measured with a *surface finish microscope*. This microscope depends on a light wave principle of measurement similar to the principle behind optical flats. The surface to be measured is magnified from 125 to 200 times. The surface finish is then compar·d to a comparator plate of a known standard. The view of one surface is superimposed on the other surface.

The variation indicates the accuracy of the surface finish. Some microscopes have a direct-reading attachment. Variations in surface finish are read on a dial face.

Safe Practices in Using Surface Finish Analyzers and Instruments

- Set the range switch on a surface finish analyzer to a high setting when the surface roughness is not known. After the initial test. a finer setting may be needed to get an accurate surface reading.
- Allow a surface finish analyzer instrument to warm up before taking a measurement.
- Protect the precision stylus point from hitting a surface.
- Check surface finish instruments periodically against a master.

SUMMARY

- Surface texture specifications reflect the ideal surface finish required by the designer.
  - Repetitive and random deviations in surface finish result from roughness, waviness, lay patterns, and flaws.
- Surface roughness refers to the fine irregularities produced by a cutting tool or production process. These irregularities extend within the limits of the roughness width cutoff.
  - Roughness width cutoff is the rating in inches for the greatest spacing of surface irregularities that are repetitive.
- Roughness height is an arithmetical average deviation in microinches. This height is measured normal to the centerline.
  - The common symbols used to show the dominant tool or finishing patterns are: ⊥ (perpendicular), ∥ (parallel), **X** (angular), **M** (multidirectional), **C** (circular), **R** (radial), and **P** (pitted).
- Comparator scales for inspecting surface finish conform precisely with ANSI standards of surface roughness.
  - Comparator scales are marked to show the roughness height, lay pattern, and machining process.

■ Comparator specimens are available for conventional machine processes, electrical discharge machined finishes, casting, and other production techniques.

    ■ The surface finish analyzer consists of a tracer head, amplifier, and recording dial. Readings are obtained using either a 0.0001″ or 0.0005″ stylus.

■ The profilometer also magnifies surface irregularities. The tracer unit may be moved across the work surface either manually or automatically.

    ■ Recording units on surface finish instruments produce charts of waviness and roughness and a combination waviness/roughness graph.

■ The surface finish microscope magnifies a surface finish. The finish is then compared with a comparator specimen or other standard.

    ■ Standard precision instrument safety precautions must be followed in the use of all gages in quality control.

■ Surface finish instruments must be warmed up before measurements are taken. Vibration, dust, and foreign particles affect measurement accuracy.

UNIT 6 REVIEW AND SELF-TEST

1. a. Use a table of surface texture values for general machining processes.
    b. Give the *lay pattern* and the *roughness height in microinches* for the following five machining processes and roughness width values: (1) end milling, 0.005″; (2) straight turning, 0.002″; (3) shaping, 0.001″; (4) cylindrical grinding, 0.0005″; and (5) drilling, 0.005″.

2. a. Describe the functions served by a surface comparator specimen and viewer.
    b. Identify the kind of information the worker obtains by visual and viewer inspection of surface texture measurement.

3. a. Use a table of surface roughness heights for general manufacturing processes.
    b. Give the lower surface finish range in microinches (where special precision manufacturing processes are used) for the following operations: (1) reaming, (2) finish grinding (cylindrical), (3) drilling, and (4) rough milling.

4. a. Cite two main differences between a surface finish analyzer and a comparator specimen.
    b. State one advantage of using a surface finish analyzer over the comparator specimen method.

5. Tell why work surface profiles are tape recorded on a profilometer.

6. State two advantages of using a surface finish microscope in preference to a profilometer for general surface texture applications.

7. Indicate one check that must regularly be made with surface finish measuring instruments.

Statistical Process (Quality) Control (SPC)

The function of statistical process (quality) control is to ensure that the specifications and standards established by design and engineering departments are maintained. Quality control is essential in controlling product quality and to continuous improvements in output. Quality control requires dimensional control, materials testing, inspection of surface texture, assembly run-in periods, and other checks.

This unit deals with measurement instrumentation and processes; classes of fits, allowances, tolerances, and dimensional controls; and principles of statistically gathering, processing, storing, analyzing quality and process control data, and equipment and software components in SPC systems.

OBJECTIVES

After satisfactorily completing this unit, you will be able to:

- Analyze relationships between machine tool and tool setups, manufacturing processes and products, shop floor digital electronic measuring instruments, and interfacing statistical data process and product control (SPC) equipment and systems.
- Carry on parts inspection using different types of gages.
- Understand American National and Unified Form Screw Thread terminology, classification systems, and tolerance symbols for SI metric threads.
- Interpret drawings for surface texture, tolerancing, allowances, and other dimensioning of part features.
- Use a normal curve of distribution, sampling plan tables, simple formulas, and related constants for quality control.
- Relate each new *Term* to quality control principles and techniques, and apply *Safe Practices*.
- Solve the problems assigned from the *Review and Test Items* that follow.

DIMENSIONAL CONTROL USING GAGES

It is more practical and economical in mass production to check parts for size with fixed gages. There is less chance of error when no adjustable parts are involved. Of the many types of gages, six are commonly used. They include the *fixed-size gage*, the *micrometer-feature gage*, a *limit (standard) comparator*, the *indicating comparator*, the *combination gage*, and the *automatic gaging machine*.

DIMENSIONAL TOLERANCES

The permissible variation of a part from its nominal dimension (designated size) is identified by the designer, engineer, and craftsperson. The designer establishes the dimensional limits and surface characteristics of a part and mechanism. The more exacting a dimension, the more difficult and expensive a part is to produce. There is a minute range within which a part will function effectively.

On some dimensions (a shrink fit, for example), only a smaller dimension than the nominal size is permissible. Where a running fit is required, the part may be larger than the nominal size. The amount of variation from the nominal size is called *tolerance*. A tolerance may be *unilateral* (only plus or minus) or *bilateral* (both plus and minus). The amount of tolerance depends on factors such as the function of a part, the kind of material used, the required service life, cost, and production methods and systems.

FIXED-SIZE GAGES

Fixed-size gages are widely used in quality control. Although limited in scope, they are used to check either an internal or external dimension. With use, a plug gage becomes smaller; a ring gage, larger. Therefore, the wear life of fixed-size gages is determined by the number of parts that may be inspected before the gage size exceeds the allowable work tolerance. Constant wear requires that the working gages be checked against a master set of gages. A master gage is easily identified and is used as a dimensional control gage.

In the design of fixed-size gages, a *wear allowance* is provided. For example, if it is established that after 2,000 gagings a 1-1/4" plug gage wears 0.0001" (0.002mm), the plug may be finished within the allowable tolerance of the workpiece to 1.2501". The wear life is thus increased to measure 4,000 workpieces. Wear life may be further increased by chrome plating a worn gage and regrinding and lapping to size. Cemented carbide gages are available for long-run production gaging. Such gages resist abrasive action and wear. Wear life of some carbide gages is increased to almost 100 times longer than wear life of steel gages.

GO AND NO-GO GAGES AND XX, X, Y, AND Z CLASSES OF (GAGE MAKER'S) TOLERANCES

Fixed gages are commercially available in a wide range of sizes for GO and NO-GO measurements. Ring, plug, and thread gages are com-

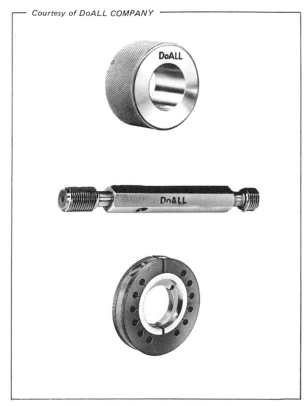

Courtesy of DoALL COMPANY

Figure 7–1 Examples of Ring, Plug, and Adjustable Gages

monly used for such measurements. Some examples of these gages are illustrated in Figure 7–1. These gages are available in inch and metric units of measure and within the same tolerance limits. The accuracy of the gages has been standardized. The tolerances are specified in terms of four classes: XX, X, Y, and Z. Table

Table 7–1 Nominal Sizes and Gage Maker's Dimensional Tolerances

Range of Nominal Sizes (inches)	Gage Maker's Tolerance According to Class			
	XX	X	Y	Z
0.029 to 0.825	0.00002	0.00004	0.00007	0.00010
0.826 to 1.510	0.00003	0.00006	0.00009	0.00012
1.511 to 2.510	0.00004	0.00008	0.00012	0.00016
2.511 to 4.510	0.00005	0.00010	0.00015	0.00020
4.511 to 6.510	0.000065	0.00013	0.00019	0.00025
6.511 to 9.010	0.00008	0.00016	0.00024	0.00032
9.011 to 12.010	0.00010	0.00020	0.00030	0.00040

7–1 gives the range of nominal sizes from 0.029″ to 12″ and the four classes of tolerances. These tolerances are known as *gage maker's tolerances.* The tolerance classes are generally defined as follows:

- *Class XX* refers to master and setup standards that are precision lapped to laboratory tolerances;
- *Class X* refers to working and inspection gages that are precision lapped;
- *Class Y* refers to working and inspection gages that have a good lapped surface finish with slightly increased tolerances;
- *Class Z* refers to working gages for short production runs where the tolerances are wide or commercially finished gages that are ground and polished but are not fully lapped.

Gage manufacturers recommend that the gage used should have an accuracy of one-tenth the tolerance of the dimension the gage is to control. Thus, the gage used for a part having a tolerance of 0.001″ (0.02mm) should have a tolerance of 0.0001″ (0.002mm). The surface finish of the gage should also be finer than the workpiece.

FOUR BASIC GAGE DESIGNS

Most types of plug and ring gages are marked for size and tolerance. There are four basic gage designs. Each of the following designs is adaptable within a range of sizes.

A *reversible cylindrical plug gage* has a pin-type gage design with GO and NO-GO members. Gage maker's tolerances are plus on the GO member and minus on the NO-GO member. This type of gage is primarily used for checking hole sizes and depths, gaging slots, and checking locations, threads, and distances between holes. Pin-type gages are available in steel, plated steel, and carbide. A common range of sizes is from 0.005″ to 1.000″ (0.125mm to 25mm). Each size is produced in the four tolerance classes. Plug members are available in increments of 0.001″ and are provided with collet bushings.

A *trilock plug gage* has a body design that permits GO and NO-GO members to be secured to the ends by means of a fastener (Figure 7–2).

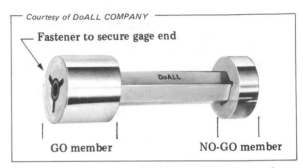

Courtesy of DoALL COMPANY

— Fastener to secure gage end

GO member   NO-GO member

Figure 7–2 Trilock Type of GO and NO–GO Plug Gage

A *taperlock cylindrical plug gage* has plug ends with a taper shank that fits a taper hole in the body (handle) of the gage.

A *progressive plug gage* is a single-end gage. The design permits complete inspection of a hole at one time. The entering portion is the GO gage, which is followed by the NO-GO gage.

While only plug gage and holder designs have been described, many other designs are available. Taper, spline (multiple grooves), flat plug, and special gages are widely used. Figure 7–3 illustrates a GO and NO-GO *stepped gage* with one *plus* and one *minus* gage member.

Snap Gages. Another design of a progressive (combination GO and NO–GO) gage is the *adjustable limit snap gage.* This gage consists of a structurally strong but lightweight and balanced deep-throat C-shape frame. Adjustable gaging pins, buttons, or anvils are brought to size with adjustment screws. They are then secured to and locked in place on the frame with locking screws. A marking disk is stamped. The disk indicates the dimensions and identifies the gage in relation to a production plan.

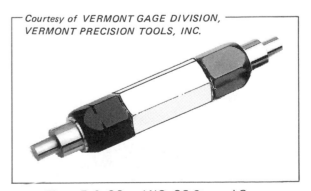

Courtesy of *VERMONT GAGE DIVISION, VERMONT PRECISION TOOLS, INC.*

Figure 7–3 GO and NO–GO Stepped Gage

The GO and NO-GO buttons of an adjustable snap gage are set to gage blocks or other masters. Precision measurements with snap and other gages depend on taking the measurement at the correct angle with respect to the reference planes. Small parts are measured by holding the workpiece with one hand and the gage with the other hand. The gage is then brought to the workpiece and carefully moved along the line of measurement.

TAPER PLUG GAGES

Taper plug gages measure the accuracy of the taper and size. The taper may be one of the standard tapers or a special taper. Figure 7–4 shows an application of a simple solid taper plug gage. The two vertical lines indicate the tolerance and are the GO and NO–GO limits.

ADJUSTABLE THREAD RING GAGES

Adjustable thread ring gages are available for standard metric threaded sizes and pitches and for American National and Unified thread forms. Other gages are available for Acme thread form sizes.

In thread control, *roll-thread gages* are also available. The term indicates that the thread-measuring form is a roll instead of a flat plate.

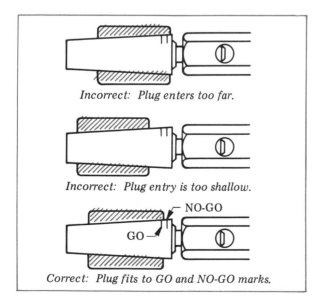

Incorrect: Plug enters too far.

Incorrect: Plug entry is too shallow.

NO-GO

GO

Correct: Plug fits to GO and NO-GO marks.

Figure 7–4 Application of Taper Plug Gage

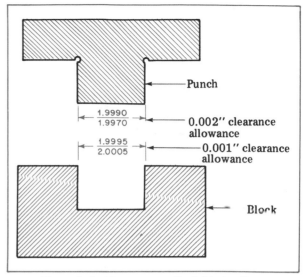

Punch

1.9990
1.9970

0.002″ clearance allowance

1.9995
2.0005

0.001″ clearance allowance

Block

Figure 7–5 Allowance on Each Mating Part for a Clearance Fit

There are even adaptations of roll-thread gages. In some applications, a dial indicator is used to show any variation from the basic size to which the gage is set.

ALLOWANCES FOR CLASSES OF FITS

The term *fit* indicates a range of tightness between two mating parts. This range is expressed on a drawing in terms of the *basic (nominal) size* and the *upper* and *lower (high limit and low limit) dimensions*. Tolerance, to repeat, is the total permissible variation from the basic size. Nominal size is the dimension from which the maximum and minimum permissible dimensions are derived.

CLEARANCE AND INTERFERENCE FITS

Some mating parts must move in an assembly. A *positive clearance* is therefore required. How much clearance to provide depends on the nature of the movement, the size of the parts, material, surface finish, and other design factors. The drawing in Figure 7–5 illustrates limit dimensioning and a positive clearance. There is a 0.002″ (0.05mm) clearance allowance on the punch. The clearance allowance on the block is 0.001″ (0.02mm). The clearance fit has a maximum allowance of 0.0035″ (between 2.0005″ and 1.9970″) and a minimum allowance of 0.0005″ (between 1.9995″ and 1.9990″).

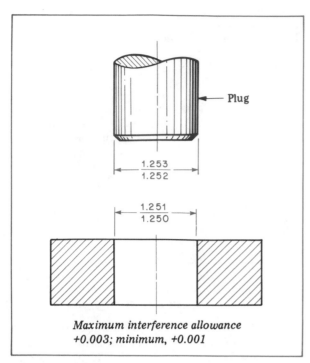

Figure 7-6 Negative Allowance for Interference Fit

ALLOWANCES ON HOLES FOR DIFFERENT CLASSES OF FITS

Standard allowances on holes have been developed for different *classes of fits*. Some allowances relate to tolerances for general purposes. Others relate to *forced, driving, push,* and *running fits*. The term *class* is followed by a particular letter (or combination of letters) that relates to specific design functions.

In the partial Table 7-2, two examples are given of industrial standard allowances and tolerances for reamed holes. A *class A* fit refers to a grade of hole that can be produced by general-purpose reamers. Class A reamed holes are machined to closer dimensional tolerances than *class B* holes. The high and low limits and tolerances are given in this table for a range of diameters from under 1/2″ to 5″. Inch-standard measurements may be converted to equivalent metric-standard units of measure for metric diameters.

The drawing in Figure 7-6 shows a *negative,* or *interference, fit.* The plug is ground to a high limit of 1.253″. The low limit of the mating hole is 1.250″. The size limits of this particular interference fit is 0.001″ minimum and 0.003″ maximum.

The table also shows allowances for *class F* forced fits. A complete table in the Appendix includes allowances for other forced and drive fits, in which the parts must function as a single part. One workpiece that is machined oversize must be forced into another workpiece.

Table 7-2 Allowances and Tolerances on Reamed Holes for General Classes of Fits (Partial Table)

Class of Fit	*Allowances and Tolerances	*Nominal Diameters					
		Up to 1/2″	9/16″ through 1″	1-1/16″ through 2″	2-1/16″ through 3″	3-1/16″ through 4″	4-1/16″ through 5″
A	High limit (+)	0.0002	0.0005	0.0007	0.0010	0.0010	0.0010
	Low limit (–)	0.0002	0.0002	0.0002	0.0005	0.0005	0.0005
	Tolerance	0.0004	0.0007	0.0009	0.0015	0.0015	0.0015
B	High limit (+)	0.0005	0.0007	0.0010	0.0012	0.0015	0.0017
	Low limit (–)	0.0005	0.0005	0.0005	0.0007	0.0007	0.0007
	Tolerance	0.0010	0.0012	0.0015	0.0019	0.0022	0.0024
		Allowances and Tolerances for Forced Fits					
F	High limit (+)	0.0010	0.0020	0.0040	0.0060	0.0080	0.0100
	Low limit (–)	0.0005	0.0015	0.0030	0.0045	0.0060	0.0080
	Tolerance	0.0005	0.0005	0.0010	0.0015	0.0020	0.0020

*These inch-standard measurements may be converted to equivalent metric-standard measurements.

The three grades of running fits are covered in the Appendix Table. Running fits provide for different motion conditions. *Class X* fits provide the greatest allowance for work parts that need to fit easily. *Class Y* fits are used for high speeds and average machine work. *Class Z* fits require a higher degree of precision and are applied in fine tool work and assemblies.

Tables of allowances provide a general guide. The material in the mating parts, the conditions under which a mechanism operates, the kind of work to be done, and a host of other factors must be considered. Usually, the designer indicates the ideal allowances through the dimensions on a drawing. There are situations, however, where the machinist, toolmaker, and inspector must refer to tables and make decisions.

INTERCHANGEABILITY OF AMERICAN NATIONAL AND UNIFIED FORM SCREW THREADS

Interchangeability of Unified and American National Form threads is provided through the standardization of thread form, diameter and pitch combinations, and size limits. The general characteristics of the internal and external Unified Form threads are shown in Figure 7–7.

The form is theoretical. Actually, neither British nor American industry has made a complete changeover. The 60° thread angle, diameter and pitch relationships, and limits are accepted. However, the British continue to machine threads with rounded roots and crests. There are production advantages and extended tool wear life in producing a rounded root. American industry continues with flat roots and crests. Fortunately, the manufacturing limits agreed upon permit parts to be mechanically interchangeable.

IMPORTANCE OF ACCURATE THREAD MEASUREMENT

With increasingly higher physical demands on machines, it has become necessary to produce stronger and more dependable fastenings. Maximum thread strength and dependability are determined by the following conditions:

- *Proper thread lead.* Testing shows that a thread with an error in lead is capable of sustaining only a fraction of the load of a perfectly formed thread;

- *Correct tooth form.* Out-of-round and taper threads wear rapidly, come loose in service, and often fail as a fastener.

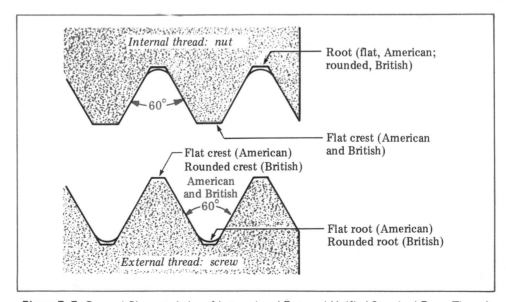

Figure 7–7 General Characteristics of Internal and External Unified Standard Form Threads

THE BASIC SIZE (CONTROLLING POINT) IN THREAD MEASUREMENT

Allowances for classes of fit for screw threads are designated differently from diameter sizes. All thread measurements must be derived from a controlling point. This point is the theoretically correct size, or *basic size*. The basic size relates to a full form thread. Thus, for each thread, there is a *basic major diameter*, *basic minor diameter*, and *basic pitch diameter*.

In actual practice, it is impossible to form a screw thread to its precise basic sizes. Limits have been established within which dimensional accuracy and form may be held. The deviations from basic sizes are identified as the *minimum* and *maximum limits*. A product may be accepted when its size limits are within the minimum limit and the maximum limit. Like dimensions for holes, the difference between the upper and lower limits is the tolerance. In general practice, tolerance for an external thread is *below* the basic size. The tolerance for an internal thread is *above* the basic size.

IMPACT OF TAP DRILL SIZES AND THREAD HEIGHT ON STRENGTH

There are a number of advantages for producing a partially threaded hole. Production efficiency, added tap wear life, reduced tap breakage, and decreased costs are four main advantages. The major requirement of a tap drill is to produce a hole size that permits tapping within specified limits. Too large a tap drill produces a thread that may be too shallow. Too small a tap drill requires excessive cutting force with possible tap breakage. Many production difficulties are caused by the use of tap drills that are too small.

After years of testing in practical applications, the 100% depth thread has proven to be only 5% stronger than a thread cut to 75% depth. Thus, the 75% depth thread is recommended for average conditions. The 50% depth thread is generally unsatisfactory in terms of strength requirements. This depth of thread may be stripped easily. Manufacturers recommend that for the vast majority of tapped hole requirements, a minor diameter tapped 55% to 75% of full depth is adequate.

CHARACTERISTICS OF THE UNIFIED SCREW THREAD SYSTEM

Just a few years ago, *serial taps* were used for cutting precise internal screw threads. Today, these taps have been replaced with *ground thread taps*. Such taps are identified by the letter *H* followed by a numeral. For external threads, taps are identified by the letter *L* and a numeral are used. The letters and numbers are a method of designating limits. The letters indicate the range of allowances above (H) or below (L) the *basic pitch diameter*. The numbers correspond to *class of fit*. The series of H1 through H10 is used for taps through 1″ inclusive. H4, H5, and

Table 7-3 Class of Thread, Tap Drill, and Thread Height Specifications for Tapping 1/4-20 UNC and NC Threads

Tap Suggested		Tap Drills					Recommended Hole Size Limits for Various Lengths of Engagement					Theoretical Hole Sizes to Give Various Percentages of Threads	
		Nearest Available Commercial Drills											
Class of Thread	Type Tap	Nominal Size	Decimal Equivalent	Probable Oversize Will Cut	Probable Hole Size	% Thread	Diameters	1B and 2B		3B		% Height of Thread	Required Hole Size
								Max.	Min.	Max.	Min.		
1B	Cut	9	.1960	.0038	.1998	77	To 1/3 dia.	.202	.196	.2013	.1960	83 1/3	.1959
2B	H5	8	.1990	.0038	.2028	73	1/3 to 2/3	.204	.199	.2040	.1986	75	.2012
3B	H3	7	.2010	.0038	.2048	70	2/3 to 1 1/2	.207	.202	.2067	.2013	70	.2046
2	H3	13/64	.2031	.0038	.2069	66						65	.2078
3	H2	6	.2040	.0038	.2078	65	1 1/2 to 3	.210	.204	.2094	.2040	60	.2111

H6 ground taps are widely used for sizes over 1″ and through 1 1/2″ and for 6TPI and finer pitches. The limits start at the basic pitch diameter.

The kind of background information the worker needs to know to supplement drawing specifications is given in Tables 7–3, 7–4, 7–5 and 7–6. A ground thread tap size of 1/4–20 UNC and NC is used throughout as an example. In Table 7–3, the suggested type of tap is given in the first column. As the table shows, an H5 ground thread tap will produce a class 2B fit. The standard (cut) tap will produce a class 1B fit.

Tap drill sizes and the probable hole size for general thread depths (77% to 65%) appear in the second column. The percent of thread depth is also affected by the length of thread engagement. Thus, maximum and minimum hole size limits are shown for lengths up to 3 times the thread diameter. These limits appear in the third column. The fourth column of Table 7–3 gives recommended hole sizes for thread depths varying from 83 1/3% to 60%.

Additional information for internal threads is provided in Table 7–4. Maximum and minimum limits are given for minor and pitch diameters for class 1B, 2B, 3B, and 2 and 3 fits. Thread tolerances are also provided. Table 7–5 covers similar measurements for external threads.

Table 7–6 provides major and pitch diameter thread limits for ground thread taps H1, H2, H3,

Table 7–4 Limits and Tolerances for Minor and Pitch Diameters of 1/4–20 UNC and NC Internal Threads

Class of Thread	*Minor Diameter (Basic 0.1850)			Pitch Diameter (Basic 0.2175)		
	Max.	Min.	Tol.	Max.	Min.	Tol.
1B	.2070	.1960	.0110	.2248	.2175	.0073
2B	.2070	.1960	.0110	.2224	.2175	.0049
3B	.2067	.1960	.0107	.2211	.2175	.0036
2	.2060	.1959	.0101	.2211	.2175	.0036
3	.2060	.1959	.0101	.2201	.2175	.0026

*Extreme limits

and H5 and for a cut thread tap. Note that the thread limit tolerance on the major diameter is 0.0010″ for a ground thread and 0.0025″ for a cut thread. The pitch diameter thread tolerance

Table 7–5 Limits and Tolerances for Minor and Pitch Diameters of 1/4–20 UNC and NC External Threads

Class of Thread	Minor Diameter (Basic 0.2500)			Pitch Diameter (Basic 0.2175)		
	Max.	Min.	Tol.	Max.	Min.	Tol.
1A	.2489	.2367	.0122	.2164	.2108	.0056
2A	.2489	.2408	.0081	.2164	.2127	.0037
3A	.2500	.2419	.0081	.2175	.2147	.0028
2	.2500	.2428	.0072	.2175	.2139	.0036
3	.2500	.2428	.0072	.2175	.2149	.0026

Table 7–6 Major and Pitch Diameter Thread Limits for Ground Threads and Cut Thread Taps (1/4–20 UNC and NC)

	Major Diameter (Basic 0.2500)		Pitch Diameter (Basic 0.2175)						
	Ground Thread	Cut Thread	H1	H2	H3	H4	H5	H6	Cut Thread
Max.	.2550	.2557	.2180	.2185	.2190	—	.2200	—	.2200
Min.	.2540	.2532	.2175	.2180	.2185	—	.2195	—	.2180
Tol.	.0010	.0025	.0005	.0005	.0005	—	.0005	—	.0020

limits with the H1, H2, H3, and H5 ground thread taps are all 0.0005″. The pitch diameter tolerance of the cut thread tap is 4 times greater (0.0020″).

H AND L PITCH DIAMETER TAP LIMITS

Another specification of ground thread taps refers to the dimensional factors of H and L. H dimensions are *above* the basic pitch diameter. L dimensions are *below* the basic pitch diameter. The two letters are followed by numbers—for example, H1 and H2 or L1, L2, and so on. The H and L dimensions indicate a range of allowances above or below the basic pitch diameter. Individually, each H or L designation indicates a single thread depth tolerance (Figure 7–7).

The use of H and L designating limits for a ground thread tap and the tap limit sizes for a 1/4–20 UNC and NC thread are shown in Figure 7–8. The basic pitch diameter in this thread example is 0.2175″. The H1 ground thread tap will produce a thread having a pitch diameter of 0.2185″. That is, the tap cuts to a tolerance of 0.001″ (or 0.0005″ tolerance for the single thread depth).

The right side of Figure 7–8 relates the H range of class of fit. Note the use of the designa-tions 1B, 2B, 3B and 2 and 3. The "B" following a thread notation refers to an inside thread. A drawing marked 1/4–20 UNC–3B provides thread specifications that include the class of fit. The 3B for this particular size thread (and according to the figure) has a tolerance of 0.0036″. The allowable range of acceptance of a 1/4–20 UNC thread for a class 3B fit for a tapped hole is from the basic pitch diameter of 0.2175″ (minimum) to 0.2211″. If within this range the maximum pitch diameter is 0.2190″, an H3 tap will produce this class of fit.

BASIC ISO METRIC THREAD DESIGNATIONS AND TOLERANCE SYMBOLS

ISO metric threads are designated by the letter *M*. This letter is followed by the nominal size in millimeters. The size is separated from the pitch size by the sign ✕. M16✕1.5 is an example of the format. This format is used on a drawing for all ISO metric thread series except the coarse pitch series. Coarse pitch ISO metric threads are designated in a simpler form. Only the letter (M) and the nominal size in millimeters are given, for example, M16. The practice of using this form is common outside the United States. Internally, United States industry includes the pitch in order to avoid confusion.

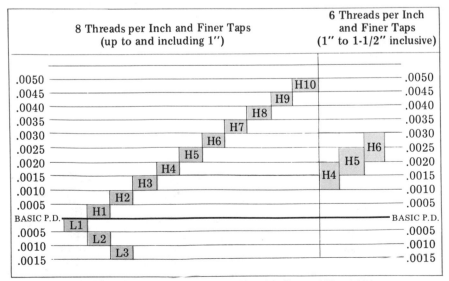

Figure 7–7 Method of Designating H and L Ground Tap Limits

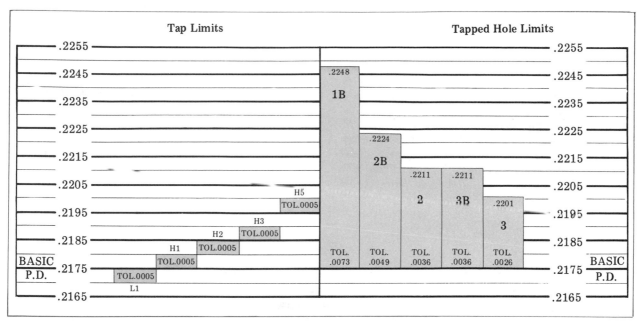

Figure 7–8 Application of Tap Limits to Class of Fit for a 1/4–20 UNC and NC Thread

TOLERANCE SYMBOLS FOR ISO METRIC THREADS

Letters and numbers are used with ISO metric threads to identify the *position* and *amount of thread tolerance*. Letters designate the position relative to basic diameters. The lowercase letters *g* and *h* are used for external metric threads. The capital letters *G* and *H* apply to internal metric threads. The position of the tolerance establishes the clearance (allowance) between internal and external threads.

A *tolerance symbol* is a combination of the number indicating the amount of tolerance and the tolerance position letter. The tolerance symbol indicates the actual maximum and minimum limits for internal or external threads. The first number and letter combination usually is the pitch diameter tolerance symbol. The second number and letter combination is the tolerance symbol for the crest diameter.

Figure 7–9 shows how a drawing is dimensioned for an external ISO metric thread. The tolerance symbols 5g and 6g are used together. For the M16–1.5 thread, the 5g indicates that the pitch diameter tolerance is grade 5 and the thread is an external one. The 6g refers to the crest diameter tolerance of grade 6 on the same external thread.

A single combination symbol is used when the crest and pitch diameter tolerances are the same. For example, an ISO metric thread may be marked: M16–1.5 6g. This screw thread has an outside diameter of 16mm and a pitch of 1.5mm. The thread is cut to a pitch diameter and crest diameter tolerance of grade 6.

IDENTIFYING THE AMOUNT OF TOLERANCE

The amount of tolerance permitted on either internal or external threads is defined by a *grade number*. This number precedes the letter that

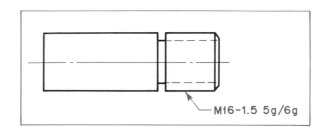

Figure 7–9 Application of Pitch and Crest Diameter Tolerance Symbols on an Outside ISO Metric Thread

designates the thread type. The smaller the grade number, the smaller the tolerance. For example, an external thread marked with a grade 4 tolerance is machined to closer tolerances than a thread dimensioned with a grade 6 tolerance. A grade 8 tolerance is larger than a grade 6. Grade 6 is a general-purpose grade of screw thread. This grade is recommended for internal and external threads wherever possible. Tables of ISO metric screw thread standards provide additional specifications and limits. Table 7–7 summarizes ISO thread designations that are most commonly used on drawings.

QUALITY CONTROL METHODS AND PLANS

The control of the quality of a process is agreed upon by the parts designer and industrial management. *Quality control* is the means of ensuring that a standard of acceptance (the product quality) of machined parts is maintained. Further, the inspection of the parts is kept at an economical level. Quality control requires a mathematical approach based on a normal frequency-distribution curve, sampling plans, and control charts.

Table 7–7 Common Applications of Basic ISO Metric Thread Designations

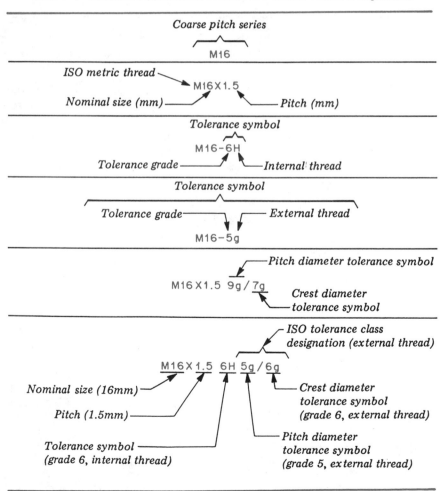

NORMAL FREQUENCY-DISTRIBUTION CURVE FOUNDATION

Tests carried on over many years indicate that measurement variations occur during manufacturing processes. When they are plotted on a graph, the normal size variations follow a curved pattern. The general shape of the pattern is shown in Figure 7–10A. A limited number of the parts are larger or smaller than the basic dimension. The curve formed is known as the *normal frequency-distribution curve.*

This curve is divided into six zones or divisions (Figure 7–10B and C). Each zone is mathematically equal in width. The average on the curve is represented by the centerline. Note that there are three zones on each side of the centerline. The Greek letter sigma (σ) is used with a number to indicate the range of measurements.

For example, the 50 machined lathe parts in Figure 7–10A were measured by micrometer. The diameters varied from 0.9372″ to 0.9378″. Note that the greatest number of parts fall along the centerline (the basic diameter of 0.9375″). The next largest numbers are 0.9376″ and 0.9374″—either +0.0001″ or –0.0001″. These parts are said to fall within plus or minus one sigma ($\pm 1\sigma$).

Statistically, 34% of the workpieces fall within this first sigma. Thus, the diameters of 68% of the parts lie between 0.9374″ and 0.9376″. At $\pm 2\sigma$, the diameters of 95 1/2% of the parts fall between 0.9373″ and 0.9377″. At $\pm 3\sigma$, 99 3/4% of the parts measure between 0.9372″ and 0.9378″. The $\pm 3\sigma$ limit is the *natural tolerance limit.*

REFINING THE MANUFACTURING PROCESSES

Excessive process spread in relation to design tolerance indicates that the manufacturing processes need to be refined. The products fall outside the tolerance requirements. More precise tooling is needed. The machined dimensions must be controlled to produce all parts within the specified tolerance. Only accidental deviations need to be inspected. Where the products fall outside the tolerance, the designer sometimes corrects the problem by redesigning the part or mechanism. The design tolerance is then brought within the limitations of the machine and setup.

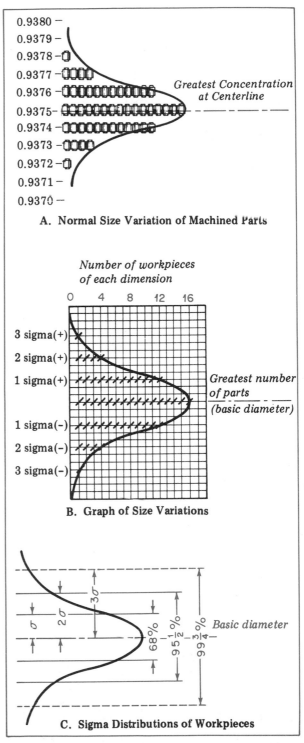

A. Normal Size Variation of Machined Parts

B. Graph of Size Variations

C. Sigma Distributions of Workpieces

Figure 7–10 Graphic Representation of Normal Size Variations of 50 Machined Workpieces

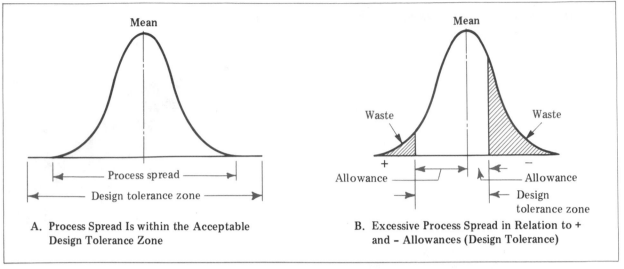

A. Process Spread Is within the Acceptable Design Tolerance Zone

B. Excessive Process Spread in Relation to + and − Allowances (Design Tolerance)

Figure 7–11 Acceptable and Unacceptable Process Standards

The conditions shown in Figure 7–10 are based on the fact that the mean (centerline) of the frequency curve of distribution is correct. Many times the mean is on one side of the tolerance. The *mean* represents the point (or line on a graph) where there are an equal number of parts above and below the basic dimension. Figure 7–11B shows a common condition in manufacturing. The mean leans toward one side of the design tolerance zone. The parts produced by the machine tool and setup thus require a greater amount of inspection time. The waste product requires a redesign of the part or changes in manufacturing processes.

SAMPLING PLANS USED IN QUALITY CONTROL

Unless there is automatic inspection, it is costly and not practical to carry on a 100% inspection. The term *sampling* is used in regard to quantities (batches) of parts that have been machined, cleaned, and are ready for inspection. The plan for sampling is referred to as a *sampling plan*. Such plans are based on the premise that a definite number of defective parts is allowable. This number is identified as the *acceptable quality level (AQL)*.

The AQL is usually based on previous trial runs on machined parts. For example, the AQL for a particular part that is machined on a bar chucking machine may indicate that 0.8% to 1.4% are defective. This small percent of defective workpieces is allowable.

The same batch may have an *average outgoing quality limit (AOQL)* of 0.9% to 1.6% defective. This AOQL means that the batch of parts leaving the inspection station will average no more than the fixed percent of defects. The batches (lots) meet the sampling requirements. Any lot that fails to meet the sampling requirements is re-inspected. The defective parts are sorted out before they move out of inspection.

There are several different types of sampling plans. The *single-sampling*, *double-sampling*, and *sequential-sampling* plans are most common.

SINGLE-SAMPLING PLAN

Sampling plans for the part and lot sizes are shown in Table 7–8. The first column lists the three most common types of sampling plans. If a single-sampling plan is used, a random number of machined parts is required for inspection. In this example, the single-sampling size is 55, as shown in the third column. As a result of inspection, if there are 3 or less parts defective, the lot is accepted. If 4 parts in the lot are defective, the lot is rejected. The acceptance and rejection numbers appear in the sampling plan in the fifth and sixth columns.

DOUBLE-SAMPLING PLAN

Rejecting the lot means that each part in the sample is inspected. The defective parts are replaced. In a double-sampling plan, Table 7–8 shows that a first double-sampling size of 36 is required. Note that if there is more than 1 defective part within the range between the acceptance number (1) and the rejection number (5), a second screening must take place. In the second screening, 72 additional parts are needed. The acceptance number is now 4; rejection, 5. The lot is accepted if there are 4 or less defective parts. If there are 5, the lot is rejected. The lot is subjected to further screening. In this double-sampling plan, the total number of parts checked is 108 when a second sample is needed.

SEQUENTIAL-SAMPLING PLAN

The representative sequential-sampling plan in Table 7–8 for the 800 to 1,300 lot parts lists seven samples. The first sample size consists of 15 parts. If 0 or 1 part is defective, a second sample must be inspected. If 2 or more parts in the first sample are defective, the lot is rejected. If the number of defectives in the second sample is between 0 and 3, a third sample of 15 is used. The process continues with additional samples of 15 as long as the combined samples are within the acceptance and rejection numbers given in the table.

SELECTION OF SAMPLING PLAN

Each sampling plan serves a different function. Major factors affecting the choice of plan to use are cost and additional inspection time. Single-sampling plans are used on complicated precision parts that require close dimensional and surface finish accuracy. Sequential sampling is used with homogenous, large lots and provides a quick and comparatively inexpensive check. Any lots that fail to meet sampling requirements are reinspected and sorted (screened). The defective parts are replaced with acceptable parts.

QUALITY CONTROL CHARTS

Control charts show graphically the processes undergoing inspection. There are four common types of control charts:

- c charts are used to plot the number of defects in one workpiece;

- p charts show the percent of defective parts in a sample;

- $\overline{X}$ charts provide a graph of the variations in the averages of the samples

- $\overline{R}$ charts show variations in the range of samples.

Table 7–8 Example of Sampling Plans for Inspecting Lot Sizes Ranging from 800 to 1,300 Parts (1.1 to 2.1 AQL)

Type of Plan	Sample and Sequence	Size of Sample	Combined Samples		
			Cumulative Size of Sample	Lot Acceptance Number	Lot Rejection Number
Single-Sampling	First	55	55	3	4
Double-Sampling	First	36	36	1	5
	Second	72	108	4	5
Sequential-Sampling	First	15	15	*	2
	Second	15	30	0	3
	Third	15	45	1	4
	Fourth	15	60	3	5
	Fifth	15	75	3	5
	Sixth	15	90	3	5
	Seventh	15	105	4	5

*Requires inspection of two samples to permit acceptance

Figure 7–12 is a combination $\overline{X}$ and $\overline{R}$ chart. The sample averages are recorded at one-half hour intervals. Usually the size of the sample is 4 or 5 parts. The average dimensions are plotted on the chart. The range (the difference between the largest and smallest dimensions) is also plotted as a bar (Figure 7–12).

The grand average ($\overline{\overline{X}}$) is computed after all the samples are plotted. The $\overline{\overline{X}}$ is marked and distinguished as a colored, dashed line. The range of variations in tolerance ($\overline{R}$) is averaged in a similar manner. The average is drawn as a solid, colored, distinguishing line.

COMPUTING THE UPPER AND LOWER AVERAGES CONTROL LIMITS

The upper and lower averages control limits are computed by using the following formulas:

$$\text{upper control limit (UCL)} = \overline{\overline{X}} + (A_2\overline{R})$$

$$\text{lower control limit (LCL)} = \overline{\overline{X}} - (A_2\overline{R})$$

The symbol A_2 refers to a constant that is based on the sample size. A simple listing of control-chart constants is given in Table 7–9.

Example: Assume a sample size consists of 4 parts. The average range of variation in tolerance is 0.0004". The grand average is 0.9376". Both the upper and lower control limits are required.

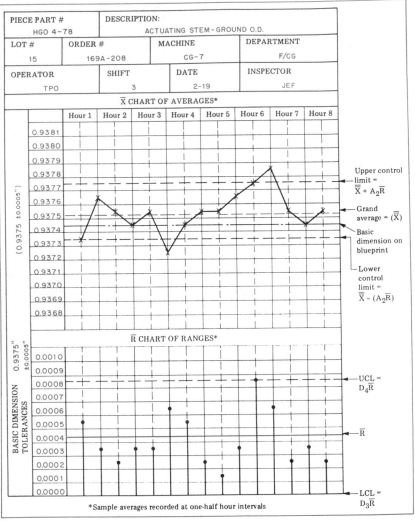

Figure 7–12 Average and Range Chart with Inspection Data and Control Limits for a Manufactured Part

Table 7–9 Averages and Ranges Control-Chart Constants (Small Samples)

Number of Inspections in Sample	Factors for Control Limits (Averages)			Factor for Central Line	Factors for Control Limits (Ranges)			
	A	A₁	A₂	d₂	D₁	D₂	D₃	D₄
2	2.1210	3.7590	1.8800	1.1280	0	3.6860	0	3.2680
3	1.7320	2.3940	1.0230	1.6932	0	4.3581	0	2.5740
4	1.5000	1.8800	0.7289	2.0590	0	4.6980	0	2.2820
5	1.3420	1.5959	0.5770	2.3260	0	4.9180	0	2.1140

$\overline{\overline{X}}$ is given as 0.9376″, and $\overline{R}$ is given as 0.0004″. The A_2 constant for 4 parts in a sample (from Table 7–9) is 0.7289. By substituting these values in the formula:

$$UCL = \overline{\overline{X}} + (A_2\overline{R})$$

The upper control limit may be found as follows:

$$\begin{aligned}UCL &= 0.9376 + (0.7289 \times 0.0004) \\ &= 0.9376 + 0.0003 \\ &= 0.9379″\end{aligned}$$

The lower control limit is found by subtracting $(A_2\overline{R})$:

$$\begin{aligned}LCL &= 0.9376 - 0.0003 \\ &= 0.9373″\end{aligned}$$

COMPUTING THE RANGE CONTROL LIMITS

The upper and lower range control limits are also found by simple formulas and the table of constants (Table 7–9). The D_4 constant is used for the upper range control limit. The D_3 constant applies to the lower limit. From the previous example, $D_4 = 2.2820$ and $\overline{R} = 0.0004″$. The formula for the upper range control limit is:

$$UCL\ range = D_4 \times \overline{R}$$

Therefore,

$$\begin{aligned}UCL\ range &= 2.2820 \times 0.0004 \\ &= 0.0009″\end{aligned}$$

The lower range control limit is found by using the following formula:

$$LCL\ range = D_3 \times \overline{R}$$

Control limits are established after a number of random samples are taken. The number of samples depends on the degree of accuracy to which the part is to be held, material in the workpiece, type of machine and cutting tools, and part size and complexity. The cost of inspecting too many parts in a sample may be excessive. Too few parts in a sample may give an untrue condition.

APPLICATIONS OF CONTROL CHARTS

Control charts are an indicator of both positive and negative conditions occurring in inspection. The designer, engineer, manager, and craftsperson refer to control charts. The samplings of machined parts indicate the nature and extent of inspection required. The results also show how closely one or more processes must be controlled for the product to meet acceptable standards.

A great number of rejects may call for the part to be redesigned. The tolerance limits may require different tooling. The process may need to be done on another machine or according to another method. For example, if a part being turned cannot be held within a tolerance limit, the diameter may need to be ground to size. Sometimes, the control-chart data suggest that the material specifications be reexamined.

Control charts are helpful in providing experimental information. Comparisons of machining setups and speeding-up techniques may be made. Charts permit study and decision making as to whether or not machining or inspection time may be reduced.

INTEGRATED STATISTICAL QUALITY AND PROCESS CONTROL (SPC)

An integrated system for statistical quality and process control (SPC) includes at least the following four major pieces of equipment, called *links*.

- *Electronic digital gages* and *measuring* and *inspection instruments.*
- *Electronic data transfer* units.
- *Electronic data collection* and *processing* units.
- *Data analysis* and *documentation software.*

With this combination of digital electronic and software SPC units, it is functional and effective to take measurement and make inspection tests, to transmit statistical data, to make on-the-spot analysis of the data in terms of making decisions affecting the manufacture of specific parts, and to produce either a digital readout or a printout.

Courtesy of ELCOMETER INC.

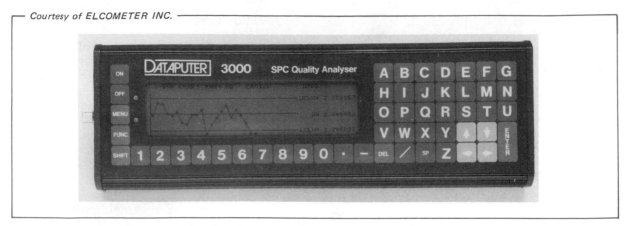

Figure 7–13 Instrument for Providing SPC Data by Fully Graphic Screen

Other more sophisticated SPC systems include computer capability for *networking*. Networking includes sophisticated *memory* (*archiving* and increased storage capability), greater capacity to carry on advanced statistical processes, and the combining of output from many substations to form a totally-integrated factory automated manufacturing system.

A shop-floor automatic digital electronic instrument that provides SPC data by fully graphic screen is illustrated in Figure 7–13.

SPECIAL FEATURES OF DIGITAL ELECTRONIC INSTRUMENTS

SPC precision inspection and measuring instruments and devices are designed to interface (be used with) on-the-spot *data collecting, transfer, processing, printing,* and *data storage equipment (hardware)* and *software systems* for handling the data.

A few examples are presented of precision instruments that are used singly or in multiple dimension and surface texture applications (using fixture holding and positioning devices).

- *Digital electronic indicators.* Errors generally resulting from counting graduations and revolutions or plus-minus deviations are eliminated.
- *Digital electronic micrometers.* This instrument provides direct or comparative measurements between parts samples and a standard. Electronic micrometers display GO (if within tolerance) or NO–GO

measurements in relation to a preset nominal dimension and upper and lower tolerance limits. Plus (+) and minus (−) dimensional values can be displayed.

- *Digital electronic column gages.* This type of gage is used in combination with a *remote transducer* (for fixture gaging). Measurements are shown as a *digital readout* and as an *analog display*.

Digital electronic column gages may be used singly for "stand-alone" applications or a number of column gages may be "ganged" together with each column gage displaying the reading of one or two transducers.

Digital electronic column gages are designed with a number of single-purpose switches to control the following functions.

- Automatic zero.
- Upper and lower limit settings.
- Count up and count down.
- True spindle position.

Other design features include a colored backlight display for analog bar graph and digital LCD readouts, and limit signals to identify *good, over,* and *under* readings. The digital output of the instrument permits displayed measurement/inspection values to be communicated from on-the-spot models to a variety of compatible statistical quality and process control units and to computers on networking (mainframe) systems.

- *Digital electronic height gage.* This instrument may be used as a "stand-alone" height gage or it may be interfaced into an SPC system. Measurements in decimal inch or metric millimeter values are quickly converted. A *floating zero* permits comparative measurements and absolute dimensions to be taken from any reference point. A measurement reading is held in memory until it is read.

- *Digital electronic caliper.* This direct reading caliper (Figure 7–14A) produces decimal inch and equivalent millimeter inside and outside measurement values to a fourth decimal place accuracy. Provision is made for instant true inch/metric measurement conversion.

  The instrument is designed for tolerance calssifications. There are light (LCD) segments that show in color whether a dimension is *within tolerance* (green), *out-of-tolerance* (red), or *reworkable* (yellow). These conditions are illustrated in Figure 7–14B.

  When taking comparative measurements, zero can be set at any point. Variations from the set zero point can be read from the display. The minus sign (–) indicates a below standard (negative) value.

  The digital electronic caliper is designed as an essential unit for SPC systems. There is direct transmission of measurement/inspection data from the instrument to other statistical gathering, processing, and interfacing SPC equipment.

FEATURES OF SPC SYSTEM EQUIPMENT (UNITS)

DATA PROCESSOR/PRINTER

This instrument produces instant instrument to tape SPC printouts. The model illustrated in Figure 7–15 has the following features.

- Capability to *store measurements* of up to five features per part. Other models or combinations permit a greater number of part features.
- Generate $\overline{X}$-*bar* and $\overline{R}$-*charts* based on recorded data and basic or batch statistics for selected features of all subgroups.
- Direct keyboard entry of part measurements and limits.

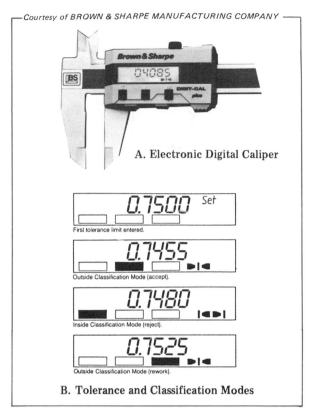

A. Electronic Digital Caliper

First tolerance limit entered.

Outside Classification Mode (accept).

Inside Classification Mode (reject).

Outside Classification Mode (rework).

B. Tolerance and Classification Modes

Figure 7–14 SPC Features of an Electronic Digital Caliper

- Capability to collect a maximum of 9999 measurements, 99 subgroups, and 20 parts per subgroup.
- Calculate minimum/maximum part sizes.
- *Enter upper* and *lower specifications limits* and enter the limits of each feature for comparison with measurements as they are taken.
- *Calculate* and *store control limits* in memory.
- *Change the mode* from inch to metric measurements and the external mode to an internal measurement.
- Generate a *histogram* plot.

DATA BANK, DATA RECORDERS, REPORT PROCESSORS

These SPC units are available as shop-floor computers for collecting, storing, and analyzing a great number of measurements that are accepted from compatible digital electronic instru-

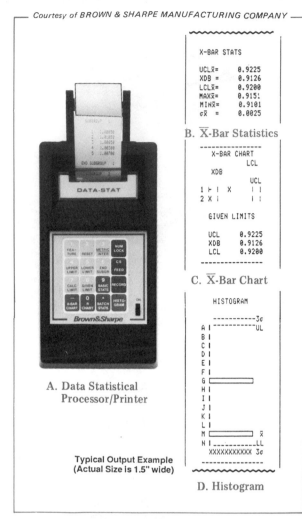

A. Data Statistical
Processor/Printer

Typical Output Example
(Actual Size is 1.5" wide)

```
X-BAR STATS

UCLX̄=    0.9225
XDB =    0.9126
LCLX̄=    0.9200
MAXX̄=    0.9151
MINX̄=    0.9101
σX̄  =    0.0025
```

B. X̄-Bar Statistics

```
X-BAR CHART
                 LCL
     XDB
                 UCL
1 ⊢ I   X     I I
2 X I         I I

  GIVEN LIMITS

  UCL    0.9225
  XDB    0.9126
  LCL    0.9200
```

C. X̄-Bar Chart

```
HISTOGRAM

  ----------- 3σ
A I ----------- UL
B I
C I
D I
E I
F I
G ⊏======⊐
H I
I I
J I
K I
L I
M ⊏==========⊐        x̄
N I ----------- LL
  XXXXXXXXXXX 3σ
```

D. Histogram

Figure 7–15 Design Features of a Data Statistical Processor/Printer with Examples of Graphic Output

ments. The data generated is sent directly to data statistical and printer units that have compatible SPC software.

DATA SWITCHBACK

This equipment provides for a series of electronic gages, fixturing devices, and other data output measurement/inspection tools to function in one SPC setup with one output to a data bank or data statistical unit. Additional measurement input to the system is accommodated by changing the instrument setting on a data switchback unit.

COMPATIBLE SOFTWARE

Many SPC systems are compatible with only units produced by a single manufacturer. Other SPC systems have the capability to interface precision electronic measuring/inspection instruments and quality and process control data with pre-programmed software standards produced by a number of SCP equipment manufacturers.

Compatible software is available to generate charts, variable data graphs, frequency histograms, X̄-bar, R̄ range, sigma control charts, and other reports essential in manufacturing to quality and process control.

Safe Practices in SPC Measurement and Inspection

- Pay particular attention to the abrasive action of metals on the dimensional accuracy of a gage. Cast iron, cast aluminum, and some of the plastic materials have a higher abrasive action than steels, brass, and bronze.

- Take care when gaging close tolerance holes in soft metals to prevent seizing and loading. Excessive wear may be produced on the gage and an inaccurate measurement may result. A special gage lubricant is produced to coat the gage.

- Burr, degrease, and thoroughly clean machined parts. Wash all parts in a lot that are to be used for quality control.

- Inspect gages frequently for size and scratches. Check working gages against a master gage.

- Return a worn gage for reprocessing or scrapping. The wear on plug and ring gages begins at the end. Wear gradually advances along the length of a gage.

- Clean and examine all measuring/inspection tools and SPC data equipment regularly for operating condition and accuracy.

TERMS USED IN STATISTICAL QUALITY AND PROCESS CONTROL

Statistical, Quality and Process control (SPC)	Techniques and measurement systems used in manufacturing to ensure conformity with standards. Dimensional and materials control obtained through sampling plans. Maintenance of quality and the improvement of product resulting from continuous inspection.
Dimensional tolerance	The permissible variation of a workpiece from its nominal dimension. (This variation may be the result of unilateral or bilateral tolerances.)
Classes of tolerances (holes)	Four classes of (gage maker's) tolerances: XX, for laboratory standards; X, for inspection gages; Y, for working gages that are lapped; and Z, for ground working gages.
Fit	A range of tightness between two mating parts. Dimensioned on a drawing by the basic size and the upper and lower limits.
Interference fit	A negative fit. The machining of a part to a dimension that requires the mating parts to be forced together.
Classes of fit (holes)	Standards accepted by manufacturers, designers, and engineers for mating parts. Specifications generally used by the craftsperson for sliding parts, forced fits, average machine fits, fine tool and work assemblies, and others.
Controlling point in thread measurement	The theoretically correct (basic) size from which all thread measurements are derived.
H and L pitch diameter limits	A letter specification for ground thread taps. (The letter is followed by a number corresponding to a class of fit.) A range of allowances above (H) or below (L) the basic pitch diameter.
Tolerance symbols (ISO metric threads)	A number and letter system to identify the position and amount of thread tolerance. The maximum and minimum tolerance limits for internal (G and H) and external (g and h) threads.
Control charts	Types of charts that provide technical information about the parts undergoing inspection. $\overline{X}$, $\overline{R}$, c, and p charts. Charts that show the number of defective parts in a sample or variations in the averages or range of samples.
Normal frequency-distribution curve Acceptable product Sampling plan Digital electronic inspection tools and measuring instruments (SPC)	A profile with equal distribution on both sides of the mean (centerline). A symmetrical line graph with six equally spaced zones. Workpieces that fall within the required tolerance. A scheme for inspecting a given number of parts in a batch. Precision gages and measuring tools and instruments designed for accurate digital electronic readout of dimensional, tolerancing, and other surface texture measurements. SPC instruments with capability to interface measurement and quality data to another component in an SPC system.
Data base, data transfer, data collection, data processing (SPC)	Terms used to identify compatible units that are part of a statistical quality and process control system. SPC equipment serving a special function as, for example: data transfer, data gathering, data storage, data analyzing, data printing, etc.

SUMMARY

- Statistical quality and process control involves dimensional measurements, materials, assembly, and other inspections.

  - Fixed gages are used in mass production. These gages are set or calibrated against masters or other standards. Fixed gages are commercially available.

- The four tolerance classes are: XX, X, Y, and Z. Fixed gages may be used to check parts ranging in precision from lapped surfaces (class XX) to commercially finished gages that are ground but not lapped (class Z).

  - The upper and lower limits of a basic size indicate the required range of tightness between two mating parts.

- Engineering tables provide standards for allowances for different classes of fits. Some relate to standard tolerances for general purposes. Others give high and low limits for forced fits, driving fits, push fits, and running fits.

  - Ground thread taps are designed with dimensional factors H and L. H refers to dimensions above the basic pitch diameter; L, below this diameter.

- Numbers that follow H and L designations relate to class of fit.

  - Letters and numbers are used with ISO metric threads to indicate the position and amount of thread tolerance. Lowercase g and h are used for external threads. Capital G and H apply to internal threads.

- The distribution of acceptable manufactured parts generally falls within a normal frequency-distribution curve.

  - Products that fall outside the tolerance requirements indicate a need to refine the manufacturing processes.

- The most common sampling plans in quality control include single-sampling, double-sampling, and sequential-sampling plans.

  - Control charts in quality control present graphically what is happening in inspection. c charts reflect the number of defects in one workpiece. p charts provide the percent of defective parts in the sample. $\overline{X}$ charts show the variations in the averages of the samples. $\overline{R}$ charts indicate variations in the range of the samples.

- Upper and lower averages control limits are computed by formula:

$$UCL = \overline{\overline{X}} + (A_2\overline{R})$$
$$LCL = \overline{\overline{X}} - (A_2\overline{R})$$

  - The upper and lower range control limits require the use of constants for D_4 and D_3 limits:

$$UCL \ (range) = D_4 \times \overline{R}$$
$$LCL \ (range) = D_3 \times \overline{R}$$

■ Digital electronic micrometers, indicators, calipers, height gages, column gages, and other precision instruments permit fast, accurate work station checks of dimensional accuracy and surface texture characteristics. The use of transducers makes it possible to instantly transmit data for purposes of controlling product quality and manufacturing processes.

■ SPC systems provide linkages that permit on-the-floor measurement and inspection data to be transmitted through other processing instruments where the data are gathered, analyzed, reported out as visual charts and printed graphs and reports, and stored in memory.

UNIT 7 REVIEW AND SELF-TEST

1. Identify four different kinds of testing that are necessary to quality control.

2. State the importance of *nominal size* to dimensions of mating features.

3. a. Refer to a table of classes of fits and gage maker's dimensional tolerances.
   b. Determine the gage maker's tolerance for parts (1) through (4) according to the class of fit given for each part: (1) 25.4mm, class XX; (2) 1.125″, class X; (3) 203.2mm, class Y; and (4) 10.375″, class Z.

4. Explain why interchangeable Unified and American National Form threads will also interchange with SI metric threads having corresponding pitches.

5. Give three corrective steps a machine operator takes when a quality control inspection shows a machined part falls outside the tolerance.

6. a. Use a sequential-sampling plan for the inspection of 900 parts at an acceptable quality level of 1.5.
   b. Indicate (1) the size of the next sample and (2) the cumulative size of sample if there are 4 or more rejects in the fourth sample.

7. a. Tell how information from a control chart may affect the part design and/or the manufacturing processes.
   b. Indicate how control charts are valuable in time-motion studies.

8. a. Refer to an ANSI table of standard running and sliding fits.
   b. Determine (1) the standard tolerance limit for a 1″ nominal diameter shaft and mating hole and (2) the minimum and maximum clearance for the shaft and mating hole. Note that the parts are assembled to an RC_1 fit, which indicates minimum running play.
   c. Give the maximum and minimum clearance between the shaft and the hole.

9. Explain briefly the meaning of *compatibility* as applied to measurement and inspection data processing functions performed by using a statistical quality and process control system.

10. List six processes that can be performed by employing a SPC system.

11. Record four different types of precision measuring and inspection instruments that are used in a SPC system.

PART 3 Turning Machines: Advanced Technology and Processes

Advanced Engine Lathe Work

The construction of smaller and more compact (microminiaturized) mechanisms requires applications of fine screw threads. Therefore, the American Standard Unified Miniature Screw Thread (UNM) is considered in this unit. Tabular information for this 60° thread is related to the Unified (inch-standard) and ISO (metric-standard) thread form and systems.

The square and the 29° thread form are also important thread forms. Three series in the 29° included angle forms are considered: the general-purpose American Standard Acme, Stub Acme threads, and the worm thread (Brown & Sharpe). Characteristics, design features, and the cutting of these threads on an engine lathe are covered in the unit. Production thread-making processes and the technology and cutting of multiple-start threads are also considered.

UNIT 8

Thread-Cutting Technology and Processes

OBJECTIVES

After satisfactorily completing this unit, you will be able to:

- Describe basic methods of producing screw threads.
- Relate design features, sizes, characteristics, and tables relating to American Standard Unified Miniature, Stub Acme, General-Purpose Acme, and Square Threads to machine setups and the cutting of these threads.
- Understand functions of multiple-start threads and cutting methods.
- Apply formulas for following and lead helix angles to the grinding of single-point thread cutting tools.
- Interpret shop drawings with designations of miniature, multiple-start, square, and Acme threads.
- Perform the following processes.
  - Setting Up and Cutting a Square Thread on a Lathe.
  - Cutting a Right-Hand Acme Thread on a Lathe.
- Follow recommended *Safe Practices* and correctly apply new *Terms* for advanced thread cutting.
- Solve the assigned problems from the *Review and Self-Test Items*.

PRODUCTION METHODS OF MAKING THREADS

The seven basic methods of producing screw threads are as follows:

- *Casting*,
- *Rolling*,
- *Chasing*,
- *Die and tap cutting*,
- *Milling*,
- *Grinding*,
- *Broaching*.

CASTING METHODS

The most widely used casting methods are: sand casting, die casting, permanent-mold casting, plastic-mold casting, and shell-mold casting. The old sand-casting method produces an extremely rough thread.

Die Casting and Permanent Mold Casting. Threads produced by die casting and permanent mold casting have a high degree of accuracy and a good surface finish. The threads (internal) are cast in parts that are usually fastened together and are not disassembled.

The disadvantage of die cast and permanent-mold cast threads is in the low melting point alloy that is used. The parts are comparatively soft and have limited durability if reused. Many die cast parts are designed with steel and other inserts. These inserts are either cast in place or threaded into a hole. The inserts overcome the problem of rapid wear that results when a soft, die cast metal is used.

Plastic Mold Casting. Plastic mold casting is used on plastic materials only. Metal inserts (aluminum, brass, and steel) may be cast in place. Where a plastic material is strong enough to hold a fastener without stripping, it is economical to tap the threads.

ROLLING METHOD

Rolling threads is the fastest and most economical method of producing external screw threads. The rolling and flowing (*plastic deformation*) action reduces mechanical problems such as tension, fatigue, and shear. Also, no chips are produced because no metal is removed.

Furthermore, the method produces a *burnished* thread surface. Surface finish accuracies as close as 4 microinches (AA value) are finer than the accuracies produced by any other thread-forming method.

Cold-rolling threads requires the use of thread-roll dies. The dies are made of high-speed steels, high-carbon–high-chrome steels, and silicon carbides. The dies do not require sharpening because the threads are formed by *upsetting*, which is similar to the raising of diamond or straight-line surfaces by knurling.

Rolled threads are formed by displacing metal using flat or round dies. Those dies are shaped in the exact form of the finished thread. A sliding motion of the dies burnishes and work hardens the threads. No burrs are left. The accuracy of the thread lead, pitch diameter, and thread angle is maintained over longer runs with cold rolling than with any multiple-point cutting tool process.

MILLING METHOD

A milled thread is formed by a revolving milling cutter. The shape of the cutter conforms to the required thread form. Both internal and external threads may be milled. This method produces a more accurate thread than a thread produced by using taps and dies. Several advantages of the milling process are as follows:

- The formation of a coarse-pitch or a long thread is particularly suited to this process;
- Lead screws may be milled to close tolerances under fast production conditions;
- The full thread depth may be cut in one pass;
- A simple cutter may be used to mill more than one thread size.

GRINDING METHOD

Hardened parts are threaded by grinding. Grinding is the most precise method of generating a screw thread. Pitch diameters may be held to an accuracy of ±0.0001″ per 1″ (±0.002mm per 25.4mm). The accuracy of lead may be ground within 0.0003″ in 20″ (0.01mm in 508mm).

Both internal and external threads may be ground. Single or multiple-form grinding wheels are used. The threads may be completely ground to depth from solid stock or finish ground. Some advantages of grinding are as follows:

- Distortion resulting from heat treating may be eliminated. The threads may be ground from solid stock after hardening;
- Parts that may be distorted by milling can have the threads ground to depth without distortion;
- Hardening and stress cracks from preformed threads are eliminated;
- Grinding threads to close dimensional and form accuracies is practical.

The work speed for general thread grinding is from 3 to 10 inches per minute (ipm) or 75mm to 250mm per minute (mm/min). During each revolution of the work, the work is moved past the grinding wheel a distance equal to the thread pitch.

Thread Grinding Wheels. Resinoid-bond wheels are used when a fine edge must be maintained and where a limited degree of accuracy is required. These wheels operate at 9,000 to 10,000 sfpm (2,750 to 3,050 m/min). Vitrified-bond wheels are more rigid than resinoid-bond wheels. Vitrified-bond wheels are used for extreme accuracy. The recommended speed range is from 7,500 to 9,500 sfpm (2,300 to 2,900 m/min).

Threads are ground in carbide and hard alloys by using diamond wheels. These wheels are of rubber or plastic bond. Diamond chips are set in the bond. Threads may also be produced by centerless grinding. Product examples of such grinding include taps, worms, lead screws, thread gages, and hobs.

BROACHING (SCRU-BROACHING)

Presently, broaching has limited applications to internal threads. The method, often referred to as *scru-broaching*, is used principally in the automotive field. Workpieces such as steering-gear ball nuts, lead screws, and rotating ball-type automobile assemblies may be scru-broached.

An opened positioning fixture and tooling setup for a broaching machine are shown in

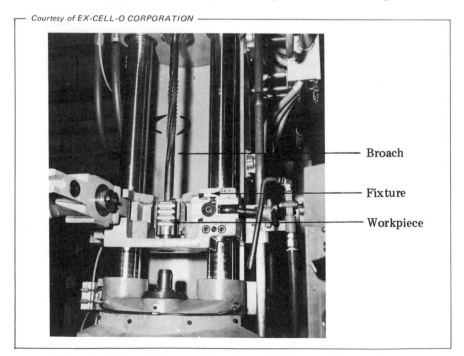

Broach

Fixture

Workpiece

Figure 8–1 Opened Fixture, Broach, and Scru-Broach Machine Setup for Broaching an Internal Thread

Figure 8–1. The broach has a pilot end and spiral-formed teeth on the body. The broach is guided by a lead screw. As the broach turns, the workpiece and fixture are drawn up to cut the thread.

AMERICAN STANDARD UNIFIED MINIATURE SCREW THREADS

SIZES IN THE UNM SERIES

The *American Standard Unified Miniature Screw Thread Series* (UNM) is an addition to the standard coarse and fine thread series of the American National and the Unified thread series. The UNM thread series provides general-purpose fastening screws for applications on instruments and microminiature (exceedingly small, precise) mechanisms. UNM threads are also known as *Unified Miniature Screw Threads*. The 60° UNM basic thread form and series range from 0.0118″ to 0.0551″ (0.30mm to 1.40mm) in diameter.

Fourteen sizes and pitches have been endorsed as the foundation for the Unified standard. These sizes and pitches coincide with a corresponding range endorsed by the International Organization for Standardization (ISO). The UNM screw thread form is compatible with the basic profiles of both the Unified (inch-standard) and ISO (metric-standard) systems. Threads in either series are interchangeable.

Characteristics of the UNM Thread Form. The 60° UNM thread profile is identical to the 60° basic thread form in the Unified and American National systems. However, the height and depth of engagement equals 0.52 × pitch, which is used instead of 0.5413 × pitch to permit more precise agreement between metric and inch dimensions.

There is only one thread class in the UNM screw thread standard. This class establishes a zero allowance. The screw threads are identified on drawings by the major (outside) diameter in millimeters and the letters UNM. The designation indicates that the threads are in the Unified Miniature Screw Thread Series. For example, a designation of 1.20 UNM on a drawing gives the major diameter as 1.200mm (0.0472″). The thread is in the UNM thread series. From tables, the metric pitch is established as 0.250mm (or the equivalent 102 threads per inch).

UNM Thread Form Tables. Tables are available that give the *limiting diameters* according to size and tolerances for internal and external threads. The limiting diameters correspond to the major, pitch, and minor screw thread diameters. Table 8–1 provides an example.

The minimum flat on a thread-cutting tool equals 0.136 × pitch. The thread height at the minimum flat is 0.64 × pitch. The thread

Table 8–1 Sample of Size and Tolerance Limits of Unified Miniature Screw Threads

Size Designation*	Metric Pitch (mm)	Metric-Standard External Threads (mm)						Lead Angle at Basic Pitch Diameter	
		Major Diameter		Pitch Diameter		Minor Diameter			
		Maximum[a]	Minimum	Maximum[a]	Minimum	Maximum[b]	Minimum[c]	Deg.	Min.
0.30 UNM	0.080	0.300	0.284	0.248	0.234	0.204	0.183	5	52

Size Designation*	Threads per Inch	Inch-Standard Internal Threads (0.001″)							
		Minor Diameter		Pitch Diameter		Major Diameter			
		Minimum[a]	Maximum	Minimum[a]	Maximum	Minimum[d]	Maximum[c]	Deg.	Min.
0.30 UNM	318	0.0085	0.0100	0.0098	0.0104	0.0120	0.0129	5	52

*Boldface denotes preferred series
[a]Basic dimension
[b]Use minimum minor diameter of internal thread for mechanical gaging.

[c]Reference only. The thread tool form is relied on for this limit.
[d]Reference only. Use the maximum major diameter of the external thread for gaging.

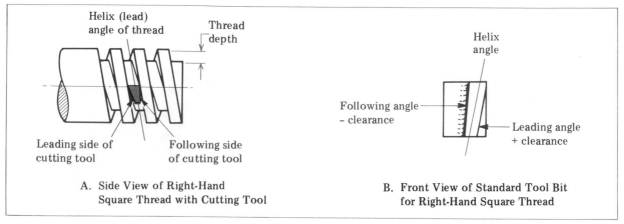

Figure 8–2 Leading and Following Side Angles of a Square-Threading Tool

flat and depth may be either computed or read from UNM tables of design data.

TECHNOLOGY AND CUTTING PROCESSES FOR SQUARE THREADS

The *square thread* is still widely used, although it is not as common as it was years ago. Square threads transmit great force. They usually have a coarser pitch than threads that are cut to a 60° form. Unfortunately, it is difficult and often impossible to compensate for wear on square threads. Cutting tools for square threads must, therefore, be ground with clearance for the leading and following sides.

Figure 8–2A and B shows the helix (lead) angle of a thread and the side angles of a tool bit for cutting a right-hand square thread. The helix angle of the thread varies. The angle depends on the thread lead and the thread diameter. The helix angle is needed to calculate the leading and following side angles of the cutting tool. A clearance of 1° is required on the leading and following sides of the cutting tool, as shown in Figure 8–3.

CALCULATING THE LEADING SIDE ANGLE

Since a right triangle is formed, the tangent of the leading side angle equals the lead (l) of the thread divided by the circumference of the minor diameter (D_2). Expressed as a simple formula:

$$\tan \text{ leading side angle} = \frac{\text{lead of thread (l)}}{\text{circumference of minor diameter } (D_2)}$$

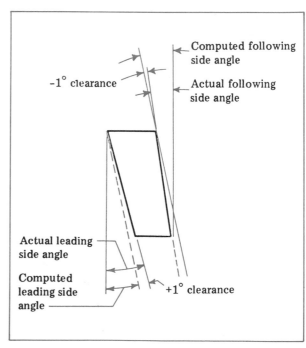

Figure 8–3 Clearance Allowances for Leading and Following Side Angles of a Square-Threading Tool Bit

Example: A 1 1/4″–4 square thread is to be cut with a single-point cutting tool. Calculate the leading side angle.

Note: The lead of a single depth is equal to 1 divided by the number of threads per inch. In this example, the lead is 0.250″ and the minor diameter is 1.000″:

$$\text{tan leading side angle} = \frac{0.250}{1.000 \times \pi}$$

$$= 0.0796$$

A table of natural trigonometric functions is used to establish the helix angle that corresponds to the value for the tangent of the leading side angle. Thus, 0.0796 = 4°33′. By adding 1° for clearance, the leading side of the cutting tool is ground at an angle of 5°33′.

CALCULATING THE FOLLOWING SIDE ANGLE

The tangent of the following side angle is found in a similar manner. The lead (l) of the thread is divided by the circumference of the major diameter (D_1). In the preceding example:

$$\text{tan following side angle} = \frac{\text{lead of thread (l)}}{\substack{\text{circumference} \\ \text{of major} \\ \text{diameter } (D_1)}}$$

$$= \frac{0.250}{1.250 \times \pi}$$

$$= 0.0637$$

The natural trigonometric function value of 0.0637 as a tangent is 3°39′. The side clearance of 1° is subtracted. Thus, the tool bit must be ground to 2°39′ on the following side. In practice, the craftsperson grinds the tool bit to 2°40′ for ease in reading and checking the ground angle.

How to Cut a Square Thread

STEP 1   Calculate the width of the tool bit.

Note: For roughing out coarse-pitch threads, grind the tool point width from 0.010″ to 0.015″ (0.2mm to 0.4mm) smaller than the thread groove width. The square cutting tool point is ground 0.002″ to 0.003″ (0.05mm to 0.08mm) larger for finish threading.

STEP 2   Position the quick-change gears for the required threads per inch or metric pitch in millimeters.

STEP 3   Set the compound rest at 0°.

STEP 4   Set up the workpiece in a chuck, between centers, or in a fixture or other work-holding device.

STEP 5   Turn a groove to the minor diameter at the end of the threaded section, if possible.

STEP 6   Set the threading tool square with the work axis.

STEP 7   Start the lathe. Set the cross feed graduated collar at zero.

Note: Some workers feed the cutting tool by using the compound rest.

STEP 8   Proceed to cut the thread to depth. Feed the tool from 0.005″ to 0.010″ (0.1mm to 0.2mm) for each roughing cut.

Note: Use feeds of approximately 0.002″ to 0.003″ (0.05mm to 0.08mm) for each finishing cut.

STEADY REST AND FOLLOWER REST FUNCTIONS

The cutting (chasing) of long square, Acme, and other thread forms often requires a support device for the workpiece. The *steady rest* and the *follower rest* are two lathe accessories that serve to support workpieces and to prevent springing due to cutting forces.

The steady rest has three jaws that adjust to accommodate workpieces of varying diameters. The top section is hinged so the steady rest may be opened to nest the workpiece. When locked securely to the machine bed, the steady rest provides an additional work support.

By contrast, the follower rest has two jaws that are positioned to be in contact with the workpiece (above and in back of the area where cutting forces are exerted). The follower rest is secured to the saddle, moves along with the cutting tool, and provides support against the cutting forces.

One or more steady rests are often used to support heavy parts in combination with a follower rest that serves to offset tool cutting forces.

TECHNOLOGY AND MACHINING PROCESSES FOR ACME THREADS

All Acme screw threads have a 29° included angle. Although not as strong, the Acme thread is replacing the square thread.

GENERAL-PURPOSE (NATIONAL) ACME THREAD FORM

The *general-purpose Acme screw thread* (National Acme Thread Form) consists of a series of diameters and accompanying pitches. Table 8-2 shows some recommended sizes, pitches, and basic minor diameters for all thread classes (of fits).

Basic Thread Classes. Three classes of general-purpose Acme threads are: 2G, 3G, and 4G. Each class has clearance on all diameters to provide free movement. The class of mating external and internal threads should be the same for

Table 8-2 Recommended Threads per Inch and Basic Minor Diameters for General-Purpose (National) Acme Thread Series (1/4″ to 1 1/2″)

Size	Number of Threads per Inch	Basic Minor Diameter
1/4	16	0.1875
5/16	14	0.2411
3/8	12	0.2917
1/2	10	0.4000
5/8	8	0.5000
3/4	6	0.5833
7/8	6	0.7083
1	5	0.8000
1 1/4	5	1.0500

general-purpose screw thread applications. Class 2G is used for such assemblies. Classes 3G and 4G provide for finer fits with less backlash or end play.

Basic Dimensions. A number of general-purpose Acme screw thread tables are found in trade handbooks. These tables provide basic dimensions; formulas; limiting diameters (minimum and maximum major, minor, and pitch diameters for 2G, 3G, and 4G thread classes); and pitch diameter allowances and tolerances. The pitch diameter allowance on external threads may be calculated by multiplying the square root of the outside diameter by 0.008 for class 2G, 0.006 for class 3G, and 0.004 for class 4G.

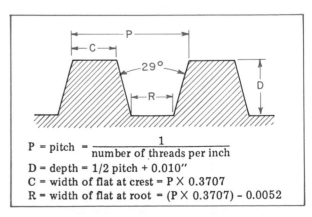

$$P = \text{pitch} = \frac{1}{\text{number of threads per inch}}$$
D = depth = 1/2 pitch + 0.010″
C = width of flat at crest = P × 0.3707
R = width of flat at root = (P × 0.3707) - 0.0052

Figure 8-4 Profile and Basic Formulas for the General-Purpose Acme Thread Form

Table 8–3 Basic Formulas and Dimensions of Two Selected American Standard Stub Acme Screw Threads

Design Features →	Threads per Inch	Pitch	Height of Thread (Basic)	Total Height of Thread	Thread Thickness (Basic)	Width of Flat	
						Crest of Internal Thread (Basic)	Root of Internal Thread
Formulas →	N	P = 1/N	0.3P	0.3P + 1/2 allowance*	P/2	0.4224P	0.4224P − (0.259 × allowance)*
Basic dimensions (rounded to four decimal places) →	16	.0625	.0188	.0238	.0313	.0264	.0238
	14	.0714	.0214	.0264	.0357	.0302	.0276

*The allowance for 10 or more threads per inch is 0.010"; for less than 10 threads per inch, 0.020".

The thread profile drawing in Figure 8–4 shows the formulas used by the worker to calculate the basic dimensions of the general-purpose Acme thread form.

General Thread Clearance. A clearance of 0.010" (0.2mm) is generally provided at the crest and root of the mating threads for pitches that are finer than 10 threads per inch (or 2.5mm pitch). The hole diameter on an internal Acme thread, for coarser pitches than 10 threads per inch, is 0.020" (0.5mm) larger than the minor diameter of the screw. The outside diameter of an Acme tap is correspondingly larger than the major (outside) diameter of the screw.

AMERICAN STANDARD STUB ACME SCREW THREADS

The *Stub Acme screw thread* has a 29° included angle with a flat crest and root. The thread is used for unusual applications where a coarse-pitch, shallow-depth thread is needed. While the formula for pitch is the same as for general-purpose Acme threads, the formulas for thread height, tooth thickness, basic width of flat at crest and root, and the limiting diameters are different. Table 8–3 provides formulas for the basic dimensions of two selected Stub Acme threads with sizes of 16 and 14 threads per inch.

29° WORM THREAD (BROWN & SHARPE)

The *worm thread* (Brown & Sharpe) is also a 29° form thread. Worm threads are generally combined with worm gears to provide mechanical movement for transmitting uniform angular motion rather than power. Such applications permit the design of a deeper thread form than on general-purpose Acme threads. The tooth thickness at the crest and root are correspondingly smaller. The 29° worm thread form and simple formulas for the basic dimensions are shown in Figure 8–5.

MULTIPLE-START THREADS

Multiple-start threads are used to increase the rate and distance of travel along a screw thread per revolution. The lead of a multiple-start thread is increased without increasing the depth to which a thread is cut. The three common multiple-start threads are known as *double*, *triple*, and *quadruple threads*. The relationship between the pitch and leads of double and triple multiple-start threads and a regular single-pitch screw thread are shown in Figure 8–6.

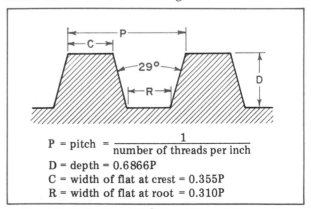

$$P = \text{pitch} = \frac{1}{\text{number of threads per inch}}$$
$$D = \text{depth} = 0.6866P$$
$$C = \text{width of flat at crest} = 0.355P$$
$$R = \text{width of flat at root} = 0.310P$$

Figure 8–5 Profile of 29° Worm Thread (Brown & Sharpe) and Formulas for Basic Dimensions

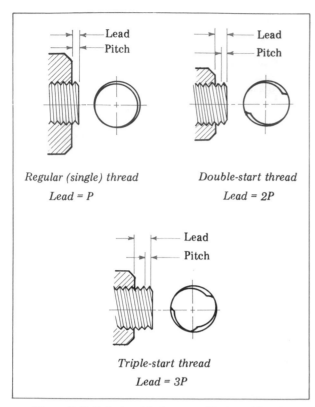

Regular (single) thread
Lead = P

Double-start thread
Lead = 2P

Triple-start thread
Lead = 3P

Figure 8–6 Relationship between Pitch and Lead
for Single-, Double-, and Triple-Start Threads

METHODS OF CUTTING
MULTIPLE-START THREADS

The positioning of the single-point chasing (cutting) tool, the setting of the quick-change gears, cutting speeds, depths of cuts and the rough and finish cutting of multiple-start threads are the same as for single-pitch screw threads. However, there are two major differences:

- The quick-change gears must be set to cut the thread to the required *lead;*
- The workpiece must be advanced a fractional part of one revolution for a multiple thread. For example, a double thread has two starting places located exactly 180° from each other.

If a thread-chasing dial is used, the cutting tool may be positioned at the next entry position after each cut. Thus, on a double thread, each thread is cut alternately to the same depth. Another method of positioning for the next start is to advance the cutting tool by using the compound rest to advance the tool a distance equal to the pitch.

How to Cut a Right-Hand Acme
(29° Form) Thread on a Lathe

STEP 1   Grind the tool bit to fit the 29° Acme thread gage. The slot used should correspond with the thread pitch.

STEP 2   Set the quick-change gears to cut the required number of threads per inch or metric thread pitch.

STEP 3   Position the compound rest at 14 1/2° to the right.

STEP 4   Locate the thread-cutting tool at center height and position it with a 29° Acme thread gage at an angle of 90° with the workpiece axis (Figure 8–7).

STEP 5   Feed the cutting tool to thread depth (minor diameter). Turn the end of the workpiece to this minor diameter for about 1/16″ (1.5mm) long.

STEP 6   Back the cutting tool out to position it at thread cutting depth for the first roughing cut.

STEP 7   Feed the tool in at the 14 1/2° angle. Take a series of roughing and finishing cuts.

Note: Coarse-pitch Acme threads may be roughed out with a square-point cutting tool.

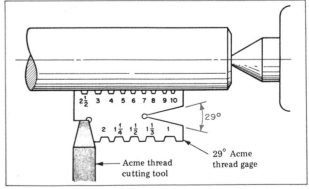

Figure 8–7 Positioning Acme Threading Tool
Square with Workpiece

Safe Practices in Advanced Thread-Cutting Applications

• Operate vitrified-bond thread-grinding wheels within a 7,500 to 9,500 sfpm (2,300 to 2,900 m/min) range. Resinoid-bond wheels may be operated between 9,000 to 10,000 sfpm (2,750 to 3,050 m/min).

• Check the leading and following side angles on single-point threading tools to permit the cutting edges to cut without interference from rubbing against the sides of the threads.

• Machine the end of a workpiece to the major (outside) or minor (inside) thread diameter for about 1/32″ (0.8mm) to provide a checkpoint for cutting to thread depth. Burrs are thus prevented from projecting beyond the face of the workpiece. The small turned area permits mating threaded parts to engage easily.

• Cut a groove at the end of the thread, if possible, to help in backing out the cutting tool quickly at the end of the cut. Sometimes, the threading tool is permitted to end in the groove, while the chasing mechanism and carriage movement are disengaged. The cutting tool is then backed out.

• Use a 29° Acme screw thread instead of a square thread when less force is required and compensation must be made for wear.

• Use a multiple-start screw thread rather than a comparatively coarse and deep thread for applications where the thread depth reduces the strength of the workpieces beyond an allowable limit.

• Remove the thread-cutting tool at the end of each cut. Reset it at the next depth to start each new cut. Check the cutting-tool depth before taking each cut.

• Back the cutting tool out and away from the workpiece before positioning it for the next successive multiple-start thread.

• Remove the fine, wedge-shaped burr at the starting and end threads, particularly if they are square or Acme threads.

TERMS USED IN ADVANCED THREAD CUTTING AND PRODUCTION METHODS

Casting method of making threads	Production of threads by flowing molten metal into a preformed mold. Producing screw threads by using conventional casting processes such as sand casting, die and permanent-mold casting, and plastic molding.
Rolling threads	Displacing metal by forcing a part to be threaded through flat or round thread dies.
Milled thread	A screw thread that is generated on a milling machine. A thread form produced as a revolving workpiece is advanced past the cutting teeth of a formed milling cutter.
Ground threads	Precision threads formed by a preformed grinding wheel. Threads produced by grinding, usually on a thread-grinding machine.
Broaching (scru-broaching)	Forming a screw thread by using a multiple-tooth cutting tool. The teeth are spiral-formed on a body that has a pilot end. The broach is turned by a feed screw that draws the workpiece into the cutting tool.

Unified Miniature Screw Thread Series	An American Standard series of 60° form threads that are compatible with the Unified inch-standard and the ISO metric-standard screw thread systems. Threads designated on drawings as UNM.
Leading and following side angles	Thread angles resulting from the slope of a thread form around the periphery of the workpiece. The leading side refers to the thread side of the thread-cutting tool that advances with the thread-cutting process. The following side relates to the opposite thread-cutting tool side.
National Acme thread form	A 29° thread form with a flat crest and root. A series of standard thread pitches for specific diameters.
General-purpose Acme thread classes	Three classes of fit designated as 2G, 3G, and 4G. Thread classes for general-purpose applications of Acme threads ranging from coarse (2G) to finer, precise fits (4G).
Stub Acme screw thread	A modified 29° form, shallow-depth thread. Variations in thread height, tooth thickness, basic flat at crest and root; with corresponding changes in the limiting screw thread diameters.
Limiting diameters 29° worm thread (Brown & Sharpe)	Major, pitch, and minor diameters applied to screw threads. A standard series Acme thread form, generally used in a gear train with a worm gear to transmit uniform motion.

SUMMARY

- Screw threads are manufactured by methods such as casting, rolling, chasing, die and tap cutting, grinding, and broaching.
    - The flow of metal in rolling threads produces a high-quality and dimensionally accurate thread form and size.

- Resinoid-bond wheels are used for general-purpose thread grinding. Vitrified bonds are better suited in producing extremely accurate thread forms.
    - A spiral-formed broach may be used for rough and finish threading.

- Deep, coarse, long screw threads are often milled with a form cutter on the milling machine.
    - The 14 thread sizes in the UNM series range from 0.30mm to 1.40mm. The basic thread form and thread height permit agreement between metric and inch-standard threads in the series.

- Multiple-start threads are cut to the depth of a single thread. When the thread-chasing dial method is used, it is practical to cut each thread by engaging each thread at its fractional position.

- The leading and following thread angles are computed by using a tangent formula and locating the angle in a table of trigonometric functions. One degree is added to the lead angle. One degree is subtracted for the following angle. The one degree provides clearance for the cutting tool.
- Square threads are usually cut 0.002″ to 0.003″ (0.05mm to 0.08mm) wider than thread width to provide clearance between the mating threads.
  - General-purpose Acme screw threads consist of a pitch-diameter series of 29° form threads. The series is standard and has three thread (fit) classes: 2G, 3G, and 4G.
- The 29° worm thread is used generally as the driver to turn a worm wheel in order to transmit uniform motion. Deeper thread flanks are required, while the crests and roots are thinner than the general-purpose form.
  - The American Standard Stub Acme screw thread is adapted to parts that require a coarse-pitch, shallow-depth thread of the 29° form.
- The general safety precautions for mounting and holding the workpiece, positioning the cutting tool, backing the tool out to take successive cuts, and using a cutting fluid must be followed.

UNIT 8 REVIEW AND SELF-TEST

1. a. Describe two production methods for manufacturing screw threads, other than chasing on an engine lathe.
   b. Cite one advantage of each production method in comparison to the chasing of threads.

2. a. Identify the general purpose for which the American Standard Unified Miniature Screw Thread (UNM) Series was designed.
   b. Describe the basic thread form and size range of the UNM series.
   c. Tell why a UNM screw thread is interchangeable with a same pitch ISO miniature screw thread.

3. Interpret the meaning of the following dimension found on a drawing of a fast-operating clamping screw for a drill jig:

   ### 1.500 – 0.25 PITCH, 0.75 LEAD
   ### ACME, CLASS 2G, LH

4. a. Refer to appropriate Acme screw thread tables.
   b. Give the dimensions needed for both machining and inspecting three parts that are threaded to conform to general-purpose (National) screw thread series standards. The outside diameters of the three parts are 0.500″, 1.000″, and 2.000″.

5. Explain the difference in setting up an engine lathe to chase a multiple-start thread as compared to cutting a single-lead regular screw thread.

6. State what precautions must be taken to avoid the following screw-thread cutting problems:
   a. Thread groove rubs against the following side of a square-thread cutting tool.
   b. Burrs project beyond the threaded-end face of the workpiece.

Production Turning Machines: Technology and Processes

Production chucker and bar turning machines vary in design and size from hand-fed to sophisticated computer numerically controlled models. This *Section* deals with general-purpose turret-tooled turning machines; major components, tooling setups, work and cutter-holding devices, and the planning of machine setups.

This unit provides a brief historical review of machine tool developments; characteristics and components of bar and chucker turning machines and turret lathes; turret, spindle, and cross slide tooling; and basic cutter mounting and bar turner preparation. *Unit 10* covers fundamental internal and external production machining processes. *Unit 11* deals with the planning of turret turning tooling. The Section concludes *(Unit 12)* with technology and processes for automatic screw machines and multi-purpose, multi-axis machines.

UNIT 9

Bar and Chucker Turning Machines; Turret Lathes

OBJECTIVES

After satisfactorily completing this unit, you will be able to:

- Appreciate the fact that highly-automated, modern bar and chucker turning machines evolved as a result of continuous machine tool and tooling design improvements.
- Describe the functions of bar and chucker turning machines and turret lathes; fixed-center turret, cross-sliding turret, plain and compound cross slide turrets; and production turning centers.
- Identify major components; attachments such as thread-chasing and taper turning, toolholder and other accessories.
- Establish the relationship of single, multiple, combination, and successive cuts in planning machine setups.
- Describe headstock spindle tooling, including extended nose, positive stop, and expanding collets; different types of standard chucks; and applications of faceplates and angle plates.
- Interpret the functions of slide tools (including box and roller back box); interchangeable turning, knurling, recessing, drilling, reaming, counterboring, and cutting-off tools and holders; and other floating toolholders.
- Identify how to set up releasing tap and die holders and self-opening die heads.
- Perform each of the following processes.
  - Mount Cutters on Center.
  - Set Up a Bar Turner.
- Follow recommended *Safe Practices* for turret lathes and production turning machines and use new *Terms* correctly.

EVOLUTION OF MODERN BAR AND CHUCKING MACHINES

EARLY TURRET LATHE DEVELOPMENTS (1845–1945)

Although the age of interchangeable manufacturing was ushered in around 1800, mass production developed very slowly. It took about 45 years before a new type of metal turning lathe was built that would accelerate the mass production movement. The eight-station turret lathe of Stephen Fitch revolutionized turning processes. The addition of a turret to the iron lathe of the day made it possible to hold and position a number of cutting tools on the same machine. All of the cutting tools mounted in a turret could be used to machine part after part without changing the setting.

One of the early turret lathes was developed by Reed Warner and Ambrose Swasey. Both men were machinists who had worked in the Pratt and Whitney shops in Hartford, Connecticut. They moved to Cleveland, Ohio, to establish a plant for the manufacture of machine tools. The Warner & Swasey Company has served the machine tool industry in making production machines from 1880 until now.

An early development of the original 1880 turret lathe is shown in Figure 9–1. While simple, the features of the 1880 machine are the same as bar and chucker (turret) turning machines produced today. The 1880 model had a step-cone pulley. The spindle was driven from an overhead countershaft by a flat leather belt. The three cones and a belt reversing shifter permitted three forward and three reverse speeds. The cross slide permitted front and back mounting positions of cutting tools. The round turret was mounted on a ram. Each tool in the turret was positioned and secured by a hand clamp. Each cutting tool was fed by a hand lever. This hand lever controlled the tool movement of the ram. The coolant dripped onto the workpiece from the container on the back of the machine. A chip pan under the bed collected the chips and permitted the coolant to be strained into a reservoir that extended below the chip pan.

The standard turret lathe of the early 1900s is pictured in Figure 9–2. This lathe was built rigidly to withstand increased cutting forces.

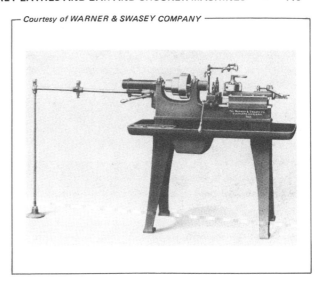

Courtesy of WARNER & SWASEY COMPANY

Figure 9–1 1890-Model Turret Lathe

These forces were due to increased speeds, feeds, and the harder materials that were being machined. Speed changes were made by shifting a clutch lever. Power feed was added to the turret unit. By 1912, a universal cross slide was incorporated. Thus, more tool setups could be mounted and positioned on the cross slide.

During World War I, a 12-speed geared-head turret lathe was in production. This lathe was equipped with a universal cross slide with square

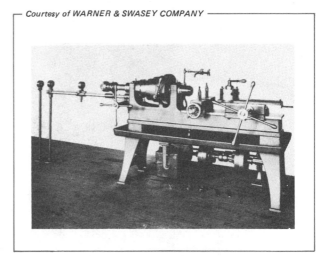

Courtesy of WARNER & SWASEY COMPANY

Figure 9–2 Geared-Head and Power-Fed Turret Lathe of World War I Era

turret and power feeds for the hexagon turret and cross slide. A rapid traverse mechanism permitted fast movement of heavy hexagon turrets. A coolant system was added. Coolant flow at a constant rate became automatic. Independent adjustable stops permitted accurate control of linear dimensions. A typical 1 1/2″ bar machine of 1914 had a three-horsepower (3HP) drive and a maximum spindle speed of 450 RPM.

A. TURRET TURNING MACHINES: COMPONENTS AND ACCESSORIES

As new, harder-to-machine metals were produced, new alloys for cutting tools were developed. In turn, tough alloy steels were used to develop more powerful machines. By the end of the World War II era the versatility and productivity of turret lathes (bar and chucking machines) increased significantly.

AUTOMATED BAR AND CHUCKING MACHINES

A modern *fully automatic chucker and bar machine* is shown in Figure 9–3. This machine may be programmed directly. Trip blocks operate microswitches, which control the functions of the carriage, cross slide, vertical slide, and threading head, as well as speed and feed changes. The reverse feed is also independent of the forward feed so that rapid return may be programmed. The machine is capable of being programmed for the following operations:

• Straight turning,
• Taper turning,
• Facing,
• Forming,
• Chamfering (beveling),
• Single-point threading,
• Die-head threading,
• Cutting off,
• Drilling,
• Straight or taper boring,
• Reaming,
• Recessing,
• Tapping.

Courtesy of HARDINGE BROTHERS INC.

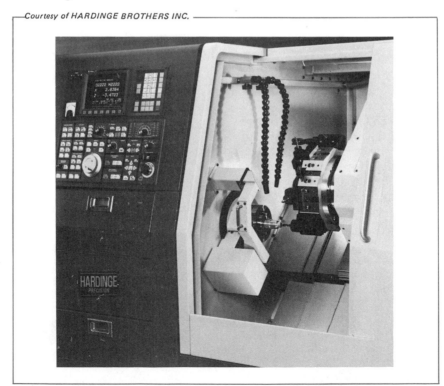

Figure 9–3 Fully Automatic CNC Slant Bed Chucker and Bar Machine

A. Boring and Inside Diameter Turning Tools on Inner Turret Row

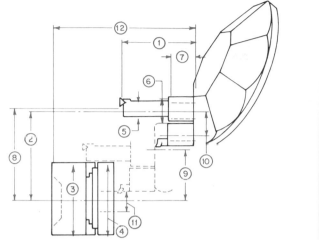

B. Workpiece and Chuck Clearance for Boring Bar and Right Angle Facing Tool on the Same Station

Figure 9–4 Multi-Tool 16-Station Slant Turret and Chucking Setups on Universal (NC) Turning System

There are 34 spindle speeds on this machine. The range is from 125 to 3,000 RPM. Four speeds within this range are available throughout the machining cycle. Tolerances within 0.0001″ (0.002mm) are obtainable on the cross slide by using a "tenth" (0.002mm) dial indicator. Similarly, a carriage dial indicator reading in increments of 0.001″ (0.02mm) is used for dimensions of length. This chucker and bar machine is a small model, having one 2 HP spindle drive motor and another 2 HP motor for the hydraulic pump.

TURNING CENTERS (UNIVERSAL TURNING SYSTEMS)

Another modern adaptation of the turret lathe is a group of machine tools identified as *turning centers* or *universal turning systems*. These machine tools are usually numerically controlled. Part of a universal numerically controlled (NC) turning system is illustrated in Figure 9–4A. A multi-tool single turret design and chucking setups are shown. The turret slants 35°. This feature places the tooling for outside diameter operations at a plane perpendicular to the plane of operation. Thus, maximum turning force and interference-free turning are provided.

Note that the inner row of eight tools on the turret are for inside diameter operations, as shown in Figure 9–4B. The boring and inside diameter turning tools are positioned in this inner row of the turret. This single turret accommodates 16 tools. There are three spindle speed ranges: (1) 65 to 2,000 RPM, (2) 80 to 2,500 RPM, and (3) 90 to 3,000 RPM. This model heavy-duty turning machine can accommodate bar stock up to 2 1/2″ (64mm) diameter. It has a chucking capacity up to 12″ (305mm). Boring bars up to 2″ diameter, with a maximum length of 9 1/2″ for bar and holder, may be mounted in the turret. This particular model requires a 40 HP motor to drive the spindle.

CLASSIFICATION OF BAR AND CHUCKING (TURRET) TURNING MACHINES

Multiple tooling is provided on a turret lathe (turning machine) by replacing the usual lathe tailstock spindle and assembly with a hexagonal turret. Also, the regular tool post on a compound rest is replaced by a square turret to hold four separate cutting tools. A square turret may also be positioned at the back of the cross slide. Thus, 16 different cutting tools may be set up at one time.

Horizontal turret lathes (turning machines) may be classified into *bar machines* and *chucking machines*. Bar machines are designed for feeding bar stock through a spindle or for machining castings or forgings that may be held in collets. Chucking machines are used for workpieces that are held and driven by a chuck or work-holding fixture. Chucking machines are used for machining regular and irregular-shaped castings, forgings, and large-size cut bar stock.

RAM AND SADDLE TYPE TURNING MACHINES

Bar and chucking machines may be of the *ram type* or the *saddle type*. On the ram type, the base on which a ram moves is clamped securely on the bed of the bar and chucker turning machine or turret lathe. The turret is mounted on the slide or ram. The ram movement toward or away from the spindle is controlled by hand or power feed. The ram type is adapted for bar work and chuck work where it is possible to

control the overhang of the ram. The main features of an older model bar machine with a ram slide are identified in Figure 9–5.

On the saddle type of turret lathe, the turret is mounted on a saddle. The saddle with its apron and gearbox move back and forth on the machine ways. The saddle type has a longer stroke than the ram type and a more rigid turret mounting. The tool overhang remains constant as the whole carriage and turret setup advances as a unit. The side-hung carriage makes it suitable for heavier chuck work and long turning and boring cuts.

Preselector and Cross Slide Features of the Ram-Type Turret Lathe. There is a direct-reading speed preselector on many geared-head models. It permits shifting to a preselected speed by setting the preselector dial. The preselector shown in Figure 9–6 reads directly in surface speed, RPM, and work diameter.

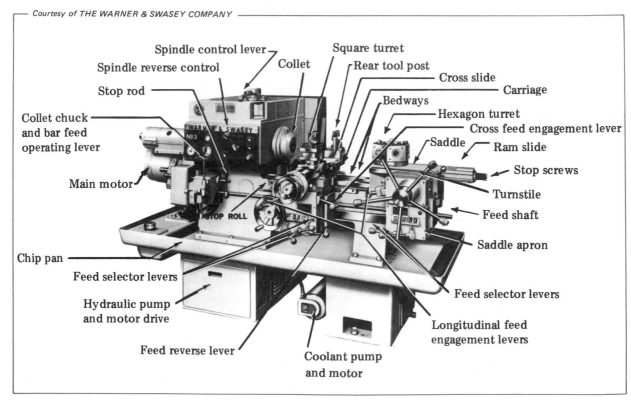

Courtesy of THE WARNER & SWASEY COMPANY

Figure 9–5 Main Features of Older Model, Manual Feed/Control, Ram-Type Turret Lathe

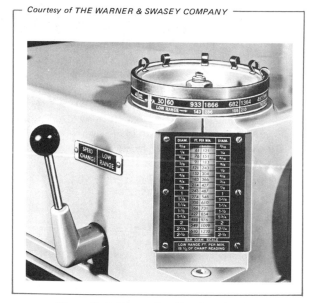

Figure 9–6 Direct Reading Spindle Speed Preselector for Dial Settings

The cross slide may be *plain* or *universal.* The plain cross slide is adjusted along the bed and locked at a required position. The universal cross slide is mounted on a carriage. This cross slide permits movement in two directions to cut parallel (longitudinally) and at right angles, crosswise to the bed ways (Figure 9–7).

Classification of Saddle-Type Turret Lathes. Saddle-type turret lathes may be of the *fixed-center turret*, *cross-sliding turret*, or *compound cross slide* type.

The *fixed-center turret* is fixed to align with the spindle axis. The turret moves parallel to this axis. The *cross-sliding turret* (Figure 9–7) may be fed by power or hand.

The *compound cross slide* (Figure 9–8) has a compound rest that may be set directly at any angle. This feature permits cutting bevels and steeper tapers than it is possible to machine with a taper attachment.

MAJOR COMPONENTS OF THE TURRET LATHE

HEADSTOCK

Automatic turret lathes have an infinite number of variable spindle speeds and may be programmed to 3,000 RPM. The spindle on a turret lathe is designed to run clockwise or counterclockwise. Spindle brakes are available

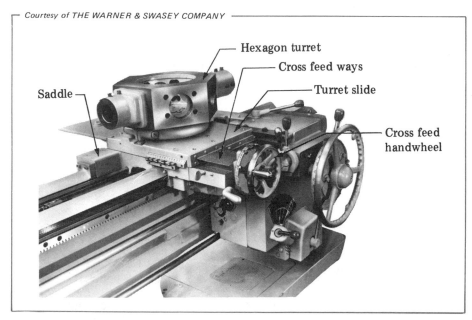

Figure 9–7 Cross-Sliding Turret with Movements Parallel and at Right Angles to Spindle Axis

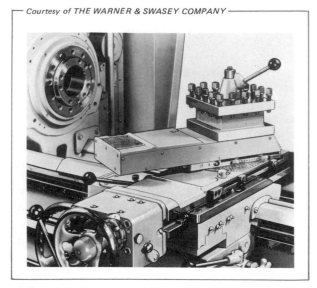

Figure 9–8 Compound Cross Slide

as a machine and personal safety precaution.

A quickly adjusted *stop rod* is located under the headstock. Individual feed-stop screws control each tool position to within 0.001″ (0.02mm).

CARRIAGE OR CROSS SLIDE

The two basic types of cross slides are: (1) *side-hung* and (2) *reach-over*. The *side-hung cross slide* provides maximum swing capacity over the cross slide.

The *reach-over cross slide* is also called a *bridge cross slide*. This cross slide is supported on both bed ways and has the advantage over the side-hung type because it is possible to add a tool post or cross slide turret in the rear position.

Cross Slide Position Stops. The cross slide may be positioned for different depths of cut by adjustable stops. The set of stops on the front end controls the depth of cut of the front tools. The rear stops control the depth of the rear turret tools. The cross slide travel is controlled by dogs, which engage the stops.

Carriage Feeds. The carriage may be fed longitudinally or transversely, by hand or power.

The power feed may be rapid traverse or regular. Trip dogs are used to move the feed lever to the off position. The feed may be reversed to provide tool movement in either direction. Handwheels are provided for cross slide and longitudinal movement of the main turret.

A second set of feed stops, at the left side of the carriage, controls the longitudinal movement of all carriage tools. The feed stops are part of a *stop roll*. This stop roll has a center stop that permits adjusting all of the stops the same amount, without disturbing the individual stop screws.

MAIN AND SECONDARY TURRETS

Main turrets generally have six, eight, or twelve stations. Some main turrets are designed with inside and an outside row. Such turrets accommodate two sets of tools: one set of eight tools for internal operations; another set of eight for external operations. Some main turrets are mounted with the ram or saddle in a horizontal plane. Other turrets are mounted vertically on the ram or are designed with the bed and turrets inclined for easier access and machine tool set-ups. *Secondary turrets* on the cross slide may be mounted horizontally or vertically.

All turrets are designed to receive adapters and to position each tool along the central axis. Each tool is advanced to a required position by setting an adjustment stop screw. This screw is located in the turret slide. The turret is automatically indexed to the next position by action of the handwheel.

The tool at each turret station is set up to perform a specific work process. After the turret is indexed, a power feed may be engaged. This feed is automatic and is disengaged by stops that are set to a definite length of travel for each particular tool. Turret positioning on bar and chucking machines is done automatically. Thus, turret indexing by hand is eliminated.

Cutting fluids are usually supplied through the center of the turret. The coolant is directed on the individual cutting tool when it is involved in a cutting process. Cutting tools on the cross slide receive a supply of coolant from a source other than the main turret.

THE FEED TRAIN AND RAPID TRAVERSE

Power to operate the cross slide and turret is transmitted through the *feed train*. The feed train consists of the gearing at the headstock end (lead screw gearbox), the feed shaft, the cross slide carriage apron or gearbox, and the main turret apron or gearbox. Feeds are first selected by setting the possible range in the head-end gearbox. Cross feeds and main turret feeds are set by shifting the levers on the aprons.

The main turret on saddle-type turret lathes usually travels a considerable distance up to and away from the workpiece. A *power rapid traverse* is designed in such lathes to facilitate positioning and removing cutting tools.

BASIC MACHINE ATTACHMENTS

THREAD-CHASING ATTACHMENTS

The two basic kinds of threading attachments are: (1) *leader and follower* and (2) *independent lead screw*. These attachments are used for chasing threads with single-point thread-cutting tools or for accurately leading-on taps and die heads from the main turret. The *leading-on attachment* attached to the main turret on a ram-type turret lathe feeds the turret at the correct pitch for the thread being cut. An *automatic knock-off mechanism*, operated by the turret stop screws, provides a depth control for threading cuts to shoulders or in blind holes.

Lead Screw Attachment for Ram-Type Turret Lathe. The cross slide is actuated for thread cutting by opening and closing half nuts with a lever located on the side of the cross slide apron. The length of travel is controlled by the automatic knock-off mechanism. This mechanism may be used for right- and left-hand threads.

Lead Screw Attachment for Saddle-Type Turret Lathe. The main units required in thread cutting on a saddle-type turret lathe include the *selective quick-change gear box* and the *lead screw gear box*. Movement for thread cutting is set by the *independent lead screw selector lever*. Threads are cut using either the *carriage* or the *saddle half-nut control lever*.

Figure 9–9 Roller Rest Taper Turner

TAPER TURNING ATTACHMENTS

Tapers are turned on bar and chucker turning machines and turret lathes by three general methods. Tapers may be turned by using (1) a formed tool, (2) a roller rest taper turner, or (3) a taper attachment.

A *roller rest taper turner* (Figure 9–9) may be used to produce long, accurate tapers on bar

Figure 9–10 Taper Attachment for Ram-Type Turret Lathe

jobs. The taper turner can be set quickly for size by adjusting a graduated dial setting. The angle of taper is controlled by the *taper guide bar*. As the cutting tool advances toward the headstock, the cutter and rolls in the roller rest taper turner recede to produce the required taper.

A *taper attachment* (Figure 9–10) provides a third general method of turning an internal or external taper or chasing threads. When turning a taper, the attachment is clamped to the bed. The cross slide or the cross sliding main turret moves along the bed. The *pivoted guide plate* of the taper attachment is set at the angle of the required taper. The guide plate guides the cross slide and the mounted cutter along the taper angle. This movement produces the required taper.

CNC CHUCKER AND BAR TURNING MACHINES

The manual adjustment of feed and speed controls and the operation of handwheels and levers that controlled turret tool positions and movements are now superceded on high production turning machines by computer numerical control input.

Figure 9–11 shows two major control panels. The machine control panel at (B) controls speeds, feeds, cutting direction, coolant flow, and other machine functions.

The numeric CNC panel at (C) features color graphics and tool path simulation and CNC controls for programming work processes for machining complex workpieces. Programming may be done directly from a part print or by other standard programming for a complete automatic cycle. Cycling includes the opening and closing of step chucks and collets; operation of cross slide, carriage, and all stations of the turret, and (where applied to the machine) automatic threading, taper turning, and vertical slide attachments.

The multi-tooled interchangeable turret top plate shown at (D) is another machine feature. The machine model displayed has a 1 5/8″ diameter round, square and hexagon (across flats) bar through-collet capacity, a 5″ chucking (gripping) capacity, and a 14″ swing. Spindle speeds are infinitely variable up to 5,000 RPM. Cross slide and carriage feeds are also infinitely variable within the feed range.

Courtesy of HARDINGE BROTHERS INC.

A. Machine Spindle Area

B. Machine Control Panel

C. Numeric CNC Control

D. Multiple-Tooled Interchangeable Turret Top Plate

Figure 9–11 Chucker and Bar Turning Machine: Controls and Tooling Features

B. SPINDLE, CROSS SLIDE, AND TURRET TOOLING

KINDS OF MACHINE CUTS

The four kinds of cuts that may be taken on bar and chucker turning machines and turret lathes are: (1) *single*, (2) *multiple*, (3) *combined*, and (4) *successive*. A *single cut* involves one cutting tool that performs one operation at one time. *Multiple cuts* are two or more cuts taken at one time from one turret station. *Combined cuts* are cuts taken by tools mounted on both the cross slide and hexagon turret at the same time. *Successive cuts* relate to cuts that are made from successive faces of the hexagon turret in consecutive order.

As a result of being able to use different cutters, work-holding and positioning devices, combinations of movements of the cross slide and hexagon turret, and required speeds and feeds for each process, bar and chucking turning machines and turret lathes have the following advantages over the engine lathe.

- Combined and multiple cuts permit several operations to be performed at the same time.
- Tooling setup time is reduced; each tool is set to positive stops; and each cut is identical over the production run.
- Precision machining to close dimensional and surface finish tolerances is possible under production conditions.
- Permanent multi-tooling setups on the machine conserve downtime; combined mul-

tiple, and successive cuts are adapted to quantity production requiring common tooling.
- A single turret turning machine has the tooling capacity to mass-produce parts that would otherwise require a number of engine lathes as second, third, and so on operation machines.

HEADSTOCK SPINDLE TOOLING

COLLET CHUCKS

Standard special alloy steel collets have a spring-tempered body. These collets are manufactured in round and other regular shapes in fractional, metric, decimal, letter, and number sizes. Special-shaped collets are available for holding odd-shaped parts and extruded stock.

Extended-nose collets have a soft face and pilot hole. These collets may be machined to accommodate a special size or odd shape. The collet may be drilled, bored, or stepped out to the exact required size. The extended nose permits deep counterbore and tool clearance for extended work processes.

Positive-stop collets (Figure 9–12) have all the design features of the standard collet plus a precision-threaded section at the back end of the collet bore. The positive-stop collet may be fitted with a *solid positive* stop, an *ejector* stop, or a *long* stop. All stops are threaded into a positive-stop collet. Each stop pin is adjustable to the desired part length.

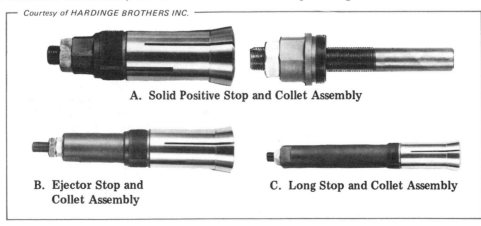

Courtesy of HARDINGE BROTHERS INC.

A. Solid Positive Stop and Collet Assembly

B. Ejector Stop and Collet Assembly

C. Long Stop and Collet Assembly

Figure 9–12 Positive-Stop Collets

Dead-length collets and *dead-length step chucks* permit shoulders and faces to be machined to exact length regardless of variations in the outside diameter. An adjustable solid stop is threaded into an inner collet. The inner collet is spring-loaded against the spindle face so that no lateral movement occurs.

Expanding collets are valuable for close-tolerance machining because there is no lateral movement of the master expanding collet or pads. There are three main parts of a master expanding collet: (1) A master collet that is hardened and ground. (2) Pads that are soft. The shoulder on the machinable pad is faced to locate the work. Repetitive parts are thus faced to the same length. Turning the pad on the machine spindle ensures exact concentricity. (3) A limit ring is used to hold the pads at nominal size while they are turned to a required outside diameter. The pad segments are marked to permit replacement in the same position. The pads may be remachined to hold subsequent parts of smaller diameter.

Precision expanding collets are used for chucking on internal surfaces. The expanding collet

assembly holds the workpiece in a previously machined bore. This assembly permits the part to be machined with concentric and square shoulders, faces, and diameters in relation to the bore.

A complete expanding collet assembly and a cutaway section showing the main parts are featured in Figure 9–13. The work-locating stop is faced and bored in position in the assembly to assure an absolutely square locating face. The spindle collar has four adjusting screws. These screws are for concentricity adjustment. The three cap screws are for mounting the work-locating stop. The draw collet controls the movement of the draw plug and the expanding collet. Expanding collets are available in standard round form in fractional, decimal, and metric sizes.

STEP CHUCKS AND CLOSERS

Step chucks and *step chuck closers* are designed for accurately holding large-diameter work. The step chuck and closer shown in Figure 9–14 accommodates workpieces up to 6″ (152mm) diameter. Parts such as castings, molding, stampings, and tubing may be held

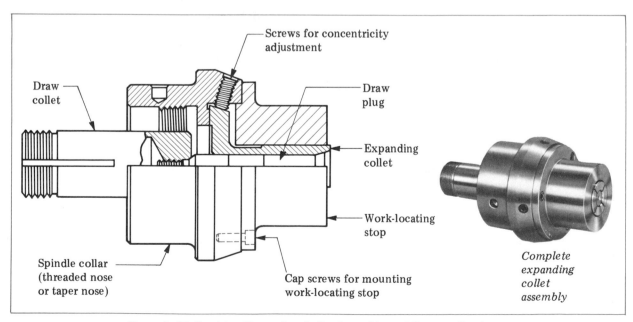

Figure 9–13 Precision Expanding Collet Assembly and Cutaway Section

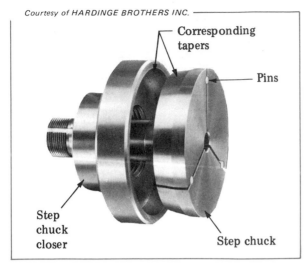

Courtesy of HARDINGE BROTHERS INC.

Corresponding tapers

Pins

Step chuck closer

Step chuck

Figure 9-14 Step Chuck and Step Chuck Closer

rigidly and accurately, without crushing or distortion.

CHUCKS

Standard chucks include independent-jaw chucks, universal (scroll) chucks, and combination independent-jaw and universal chucks. The step design of the jaws permits their use over a wide range of diameters.

One main difference between bar and chucker turning machine and turret lathe and engine lathe chucks is in the use of standard reversible-top soft jaws. These jaws are used for inside and outside chucking operations on certain production jobs.

Chuck Jaw Serrations and Jaw Types. Standard chuck jaws are *serrated*. Grooves are cut at 60° to a medium depth of 0.078″ (2mm) and pitch of 1/8″ (3mm). A limited-depth (fine) serrated jaw is used for finer gripping requirements.

Rocking jaws are used on castings, forgings, weldments, and other rough surfaces. Surface roughness and irregularities in the gripping area require the use of rocking jaws. The rocking jaw is undercut so that the clamping forces grip the part at the ends only.

Wide or wrap-around jaws are often used on second operation work. Thin-wall, fragile parts require wide or wrap-around jaws. These jaws distribute the gripping forces and minimize distortion.

It is important that the radius of the chuck jaws be ground to within ±0.001″ (±0.02mm) of the workpiece diameter to ensure that the gripping forces around a fragile part are distributed evenly over the entire chucking area.

SPECIAL FIXTURES AND FACEPLATES

Special fixtures are often used under the following conditions:

- A difficult-to-machine job must be held very rigidly to withstand heavy or extremely accurate cuts;
- The use of a fixture is more economical on a high-production job;
- The part may be loaded and unloaded faster and easier than by conventional chucking.

Fixture designs vary considerably. Some fixtures are mounted on standard chucks and complement the chuck jaws. Other fixtures are used for second and third operations to nest workpieces on a previously machined surface. Indexing is used on some fixtures for the multiple positioning of a workpiece.

Faceplate Applications. The use of a T-slotted faceplate for holding a fixture is a common shop practice. Angle plates that fasten directly to the T-slots of the faceplate are used.

Angle fixtures are used when a surface to be machined is in an exact relation to a previously machined flat surface.

Indexing fixtures permit the movement of workpieces that require similar machining operations in two or more locations. A common holding problem in turret lathe work is encountered with workpieces that require identical turning, boring, facing, chamfering, and threading operations on two or more bosses.

Fixture plates are available commercially. They are round plates with a flange for direct application to the headstock spindle. Fixture plates may be machined to become a fixture and for mounting special-purpose chucks or other fixtures.

UNIVERSAL BAR EQUIPMENT FOR PERMANENT SETUPS ON TURRET LATHES

A permanent setup of universal bar equipment for a six-station turret generally includes two

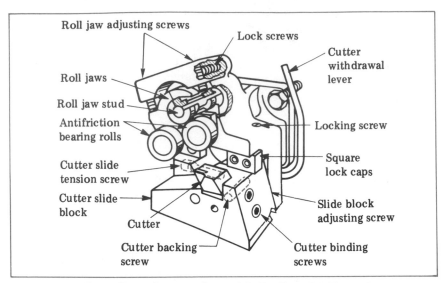

Figure 9–15 Principal Parts of Roller-Type Bar Turner

short, flanged tool holders (one of which holds a revolving stock stop), a large, flanged tool holder, a combination end face and turner, and two single cutter turners. Other cutting tools and accessories are: a starting drill and drill chuck, adjustable knee tool, clutch tap and die holder, floating tool holder, die head, combination stock stop and starting drill, center drilling tool, and a revolving center. With this combination of holders and cutting tools in the hexagon turret, it is possible to position a workpiece, face, turn, chamfer, undercut (groove), center, drill, bore, ream, and thread. Other tooling may be provided on the square turret and rear tool post to provide for additional single, multiple, or combined cuts. External straight and taper turning, facing, shoulder turning, thread chasing, and cutting off processes are added.

BAR TURNER FUNCTIONS AND OPERATIONS

The *bar turner* (Figure 9–15) is the most widely used cutting-tool holder and work-support device for bar work. Each bar turner consists of a rigid cutter holder and self-contained rolls or flat carbide shoes that support the cutting action and serve as a steady rest. The rolls or shoes also support the workpiece to produce a concentric, turned surface.

The rolls are adjustable to accommodate a wide range of finished diameters. During the cutting action, the cutting forces produce a constant vibrationless relationship between the workpiece and the cutting edge. The rolls are usually set just slightly beyond the turned radius (nose of the cutter). In this position, the cutting forces that keep the turned surface against the rolls burnish the diameter. The effect is a finished diameter that is free of tool marks.

A shank-type cutter of high-speed steel, brazed-tip carbide, or insert-tip carbide is used. The cutter is set on end and is ground to cut in this position.

STARTING BAR TURNERS PROPERLY

Supporting the Workpiece. Bar turners are used when a bar extends far enough beyond a collet chuck or chuck to require support. Support may be provided by chamfering or center drilling the end and using a center support.

After a starting cut is made by the tool on the square turret, the bar turner is positioned to turn the entire outside diameter for the required length.

Adjusting the Cutter Slide Block. The bar holder is designed to accommodate variations in cutter thickness. A cutter slot, set at an angle in the block, permits gripping the cutter and automatically establishing front and side clearances.

A pictorial view of a roller-type bar turner (Figure 9–15) shows the adjustments of the

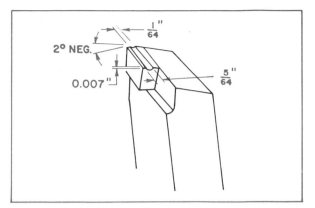

Figure 9–16 Specifications of Ground Chip Groove in Brazed-Tip Carbide Bar Turner Cutter

slide block. The slide block adjusting screw permits adjustment of the cutter for size. The spring, screw, and shoe provide tension against the adjusting screw. The screw is turned clockwise to feed the cutter toward the work.

Compensation must be made for variations in cutter thickness. An adjustable backing screw repositions the cutter (after it is reground or changed) in relation to the rolls.

At the end of a cut, the cutter withdrawal lever is turned to retract the cutter.

The cutter slide block is adjusted to increase the tension when turning large diameters. The adjustment is made by tightening the cutter slide tension screw. Similarly, the slide block is adjusted to the extreme forward position for small work where less tension is required.

Selecting Cutters and Materials. The three recommended bar turner cutter materials are cobalt high-speed steel, brazed-tip carbides, and insert-tip carbides. The insert-tip carbides are recommended for long production jobs where cutters must be changed during the run. Chip breakers are used on insert-tip cutting tools.

Maintaining Chip Control. The shape, angles, and dimensions given in Figure 9–16 are recommended for brazed-tip carbide bar turner cutters. In the case of insert-tip carbide cutters, separate chip breakers are used. The chip breaker is clamped on top of the insert tip.

SLIDE TOOLS FOR TURRETS

A *box tool* (Figure 9–17) is designed for right- or left-hand general turning work on soft metals such as brass and aluminum. It is adaptable for light finishing cuts where a fine-quality surface finish and close tolerances are to be maintained. Carbide back rests are used to extend wear life. These V-rests may be single or double. Box tools are made for right- or left-hand turning with one or two cutters. Some box tools are arranged to hold a centering tool or center drill in the shank.

A *balance turning tool* (Figure 9–18) consists of blade blocks, a blade block retainer, and fine adjusting screws. These parts are interchangeable with equivalent-sized *roller back box tools*. One-piece setting gages are available for adjusting each blade. The blades may be set for equal depth or for roughing and finish turning

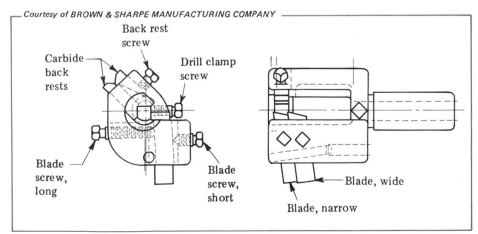

Courtesy of BROWN & SHARPE MANUFACTURING COMPANY

Figure 9–17 Right-Hand (Two-Blade) Box Tool with Carbide V-Rest

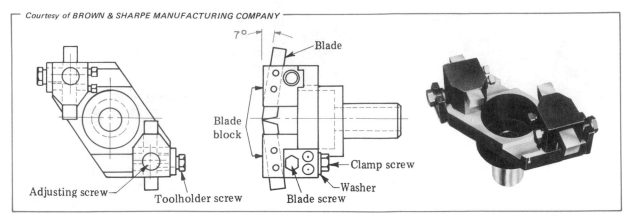

Courtesy of BROWN & SHARPE MANUFACTURING COMPANY

7°

Blade

Blade
block

Clamp screw

Washer

Adjusting screw

Toolholder screw

Blade screw

Figure 9–18 Principle Parts of a Balance Turning Tool

cuts. The equal-depth setting permits machining fine finishes to close tolerances with increased tool life. Setup time is decreased and repeatability of settings is ensured.

Slide tools provide a rugged, movable slide that permits the turning of tapers, turning behind shoulders, turning irregular shapes or contours, necking, forming, or cutting off. The cutting tool remains on center during the full length of slide travel. The three common interchangeable heads for a slide tool body include

Courtesy of TRW GEOMETRIC TOOL DIVISION

Figure 9–19 Self-Opening Stationary Die Head (Outside Trip Type)

a turning head, a recessing tool head, and a knurling head.

An *adjustable toolholder* accommodates drilling, reaming, counterboring, chamfering, and similar tools. The arm extending from the side of an adjustable holder holds a square-face chamfering tool bit at an angle of 45°. Round chamfers (rounded corners) may be produced by using a chamfering tool bit ground to the required radius. Drills may be held by a bushing in an adjustable head. The adjustable toolholder permits setting center drills, drills, reamers, and other end-cutting tools at the exact center of the workpiece.

A *self-aligning floating reamer holder* floats freely on antifriction bearings. These bearings provide alignment when the reamer enters a hole. After being adjusted to an approximate alignment with the work, the reamer aligns itself with the hole that is being reamed. This alignment eliminates bell-mouthed and egg-shaped holes.

An *adjustable stub collet holder* permits tool and collet changes without realigning the holder. Stub collets are used. They provide a range of 0.015″ (0.4mm): +0.005″ (0.12mm) to –0.010″ (0.25mm) for variations in tool diameters. Stub collet holders eliminate the need for special bushings for fractional, decimal, letter, number, and metric sizes of tool shanks.

A *standard drill chuck* has wide application in universal bar tooling and other turret lathe setups.

A *releasing "acorn" die holder* is used for cutting standard external threads. A *releasing*

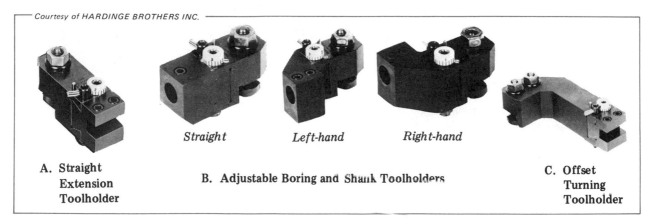

A. Straight
   Extension
   Toolholder

Straight   Left-hand   Right-hand

B. Adjustable Boring and Shank Toolholders

C. Offset
   Turning
   Toolholder

Courtesy of HARDINGE BROTHERS INC.

Figure 9–20 Cutting-Tool Holders for Turret Tooling

tap holder is used for tapping internal threads.

A *self-opening stationary die head* (Figure 9–19) is used to cut threads on machines where the die head does not rotate. The chasers (thread form cutters) are adjustable to a required pitch diameter. Die heads are available with an inside or outside trip. The length of thread on an inside trip model is accurately controlled.

CUTTING-TOOL HOLDERS

While the designs of toolholders used on multiple station turrets differ, the functions are similar. The general design features of one manufacturer's cutting-tool holders are shown in Figure 9–20. A *straight extension* toolholder (Figure 9–20A), *boring* and *shank* toolholders (Figure 9–20B), and an *offset turning* toolholder (Figure 9–20C) for turning large-diameter work are displayed. These holders are used for boring tools, square tool bits, and shank-type tooling.

One unique feature is the fine adjustment provided by the graduated collar. The graduations are in increments of 0.0002″ (0.005mm). Each fifth graduation reads in 0.001″ (0.025mm). Each graduation indicates the amount of change on the diameter of the workpiece. A movement between two graduated lines shows the diameter

is changed by 0.0002″ (0.005mm). The cutter is locked in position when the required diameter is reached.

A *slide tool* is used for turret turning and boring operations. The fine pitch adjusting screw permits dial settings of 0.0002″ (0.005mm). Each fifth dial graduation is marked to represent 0.001″ (0.025mm).

CUTTING-OFF TOOLS

Cutting-off processes are generally performed by mounting the cutoff tool in a holder on the square turret or rear tool post. On some bar and chucker turning machines and turret lathes, the cutting-off process is done from a vertical cutoff slide. In all instances, the cutter must be on center.

One grinding technique for increasing tool life on insert carbide cutoff tools that are used for parting solid stock is to grind the face to a 3/4″ (19mm) radius. The edges are ground at a 45° angle chamfer. The radius form stabilizes the cutter. A smoother, flatter surface results. When tubing is to be parted, a slight lead angle of 3° to 5° is ground on carbide inserts. The lead angle reduces the fine burr normally produced on the workpiece.

How to Mount Cutters on Center

Square Turret or Rear Tool Post Cutters

STEP 1 Place the end of a steel rule (scale) on the top face of the cross slide.

STEP 2 Hold the steel rule vertically.

STEP 3 Read the cutting-tool edge (tip) height. Adjust until the scale reading and cutting-tool point coincide with the spindle axis height.

Note: Two alternate methods may be used. (1) The center height may be established by positioning the cutting edge on line with a hexagon turret center. (2) The cutter may be positioned by using a right-angle gage. The cutter is adjusted for height until the cutting edge coincides with the center height as marked on the gage.

Multiple Turning Head Cutters

STEP 1 Grind the cutting tool for the required inside or outside cut to be taken.

STEP 2 Secure the cutter in the overhead cutting-tool holder.

STEP 3 Rotate the shank of the cutter holder until the grooves on the shank and turning head face coincide.

Bar Turner

STEP 1 Mount a turning cutter in the square turret.

STEP 2 Turn a section to within 0.001" (0.02mm) of the desired diameter for about 1/2" (12mm).

STEP 3 Swing the roll jaws out of position.

STEP 4 Set the bar turner cutter above center.

STEP 5 Advance the bar turner with the spindle in a brake position.

STEP 6 Bring the cutter against the turned portion by adjusting the cutter slide.

STEP 7 *Rub* a light shine mark on the turned diameter.

STEP 8 Set the cutting edge of the cutter at the center of the shine mark.

Slide Tool

STEP 1 Insert a cutter in the slide tool.

STEP 2 Check the cutting edge alignment. Hold a steel rule (scale) to the cutting-tool edge.

STEP 3 Adjust the cutter until the scale cuts the centerline of the slide tool and its center.

How to Start a Bar Turner

The square end remaining after a part is cut off must be chamfered or turned to a specific diameter to start a bar turner properly. Otherwise, the end of the roll turned diameter will have rough surface indentations resulting from uneven starting. Four general methods are used to start a bar turner.

Method 1: Chamfering Tool on the Hexagon Turret

STEP 1 Feed the bar to length.

STEP 2 Mount an end working (chamfering) tool or pointing tool in the hexagon turret. (The setup is shown in Figure 9–21.)

STEP 3   Chamfer the end of the bar.

> Note: Chamfer to a slightly smaller diameter than the required size.

STEP 4   Index the turret to the next station on which the bar turner is mounted. Adjust the bar turner to accommodate the workpiece and to take a required cut.

Method 2: Turning the End of the Workpiece

STEP 1   Extend the workpiece a short length from the collet.

STEP 2   Use a turning cutter in the square turret. Turn a small area to the required size.

STEP 3   Feed the stock to full length.

STEP 4   Turn the diameter using a bar turner to support the workpiece as the cut is taken.

Method 3: Chamfering Tool in the Square Turret

STEP 1   Cut off the workpiece.

STEP 2   Chamfer the bar stock by using a chamfering (bevel) cutter.

> Note: The bar stock is often cut off and chamfered by using a combination cutter.

STEP 3   Feed the stock to the required length.

STEP 4   Set up the bar turner rolls and cutter to machine to the specified diameter.

Method 4: Center Drilling

STEP 1   Feed the bar stock to the required length.

STEP 2   Center drill the end.

> Note: Use the center drilling tool for cold finished bars. On hot rolled stock, use a center drill mounted in a drill chuck.

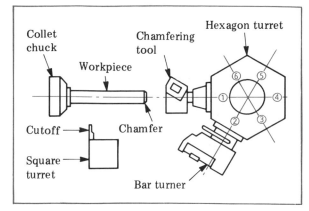

Figure 9–21 Use of Chamfering Tool on Hexagon Turret

A. Safe Practices in the Care and Maintenance of Bar and Chucker (Turret) Turning Machines

- Shut down the bar and chucker turning machine or turret lathe before checking all automatic oiling systems. Make sure that there is an adequate supply of oil and that the forced feed systems are working.
- Go through a complete cycle at reduced speeds to establish that all tools and stops are positioned correctly. Check the machined part for dimensional accuracy and quality of surface finish.
- Examine the overhang of cutting tools and follower rolls. Reduce the lengths to which they extend in order to provide maximum tool support without interfering with the cutting process.
- Check the aligning hole on the turret at each station before inserting or sliding a positioning or cutting tool into the turret.
- Check the cutting fluid systems for the main turret, secondary turret, square tool post, and back position toolholder. The nozzles must be capable of delivering a required flow of coolant to the place where it properly serves to cool and aid in the cutting process.
- Test the cutting fluid with a manufacturer's gage to establish the strength and condition of the solution.

B. Safe Practices in Setting Up Spindle, Cross Slide, and Turret Tooling

- Shut down the bar and chucker turning machine or turret lathe. Clean the spindle bore, collets, and other spindle accessories before mounting. Use a wire hand brush to remove foreign particles from threads. An air hose should not be used. The air pressure forces dirt and grit into bearing surfaces.
- Check all seating areas on holders to be sure each cutting tool rests solidly.
- Select cutters, holders, and square, hexagon, octagon, etc. turret accessories that provide maximum rigidity for holding, driving, and machining.
- Select chuck jaws so that two jaws grip an area that is less than one-half the diameter of the workpiece. The two jaws thus tend to center the workpiece against the third jaw.

- Regrind the chip groove on a bar turner cutter so that chips are formed and removed properly.
- Check for possible interference between the workpiece, tooling setups, and the indexing of each turret face.
- Tighten the cutter and binding screws on a bar turner so that the cutting edge does not move below center.
- Make sure each stop roll is adjusted to trip the power longitudinal and transverse feeds at the place where the workpiece is machined to the specified dimension.
- Use rocking jaws on rough surfaces (castings, forgings, weldments, and so on). Such jaws compensate for surface inaccuracies and imperfections.
- Withdraw the cutters and stop the bar and chucker turning machine or turret lathe if machine vibration or tool chatter occurs.

A. TERMS USED WITH BAR AND CHUCKER TURNING MACHINES; TURRET LATHES

Chucker (chucking) machine	A turning machine where forgings, castings, large and irregular workpieces are held in a chuck or a fixture.
Automatic bar and chucker (chucking) machine	A turret lathe or bar and chucker turning machine that may be programmed so that trip blocks actuate microswitches. These switches control the functions of the carriage, cross slide, vertical slide, threading head, speed and feed changes, and forward and reverse directions.
CNC turning center	A numerically controlled bar or chucker turning machine or universal turret lathe with a greater range of speeds and feeds and number of tool stations than a standard bar or chucking machine or combination.
Cross-sliding turret	A main turret that may be moved crosswise (transversely) from the axis of the turning machine.
Universal cross slide	A cross slide that may be moved, by hand or power, lengthwise or crosswise in relation to the bed.
Stop roll	A device that contains the individual stops for depth of cut. When designed with a center stop, all individual stops may be adjusted without resetting the individual stops.
Leader and follower threading-chasing attachment	A mechansim in which a half nut (follower) engages a lead screw (leader) to feed the cross slide at the required thread pitch.
Leading-on attachment	An attachment used for either thread chasing with a single-point threading tool or for leading-on die heads and taps from the main turret.

B. TERMS USED WITH SPINDLE, CROSS SLIDE, AND TURRET TOOLING

Headstock spindle tooling	Primarily work-holding devices for bar and chuck work. Collets, step chucks, regular chucks, and faceplates are mounted in the bar and chucker machine or turret lathe spindle.
Extended-nose collet	A machinable collet that extends the normal length of a collet to permit deeper counterboring and tool clearance.
Dead-length collet	A collet design that permits inserting an adjustable solid stop. The collet is spring-loaded so that no lateral movement occurs.
Step chuck	A soft-face chuck that may be bored to size on the bar and chucker machine or on the turret lathe. A collet chuck bored to hold large-diameter (short-length) parts.
Rocking jaws	A combination of recessed jaws that grip in alternate locations along the length of irregular or rough-surfaced bars, castings, or welded parts.
Jaw serrations	Indentations cut across the gripping surface of each chuck jaw. Medium or fine pitch and depth cuts that improve the gripping surfaces of chuck jaws.
Faceplate fixture	A work-holding device mounted on the faceplate of a bar and chucker machine or turret lathe. A device for positioning a workpiece accurately and holding it securely during machining processes.
Universal bar equipment (ram- or saddle-type turret lathe)	A combination of standard cross slide and multiple-station turret tools. A setup of tools established by manufacturers after years of extensive experience. A tooling setup that may be left permanently on turret lathes and bar and chucker machines for short-run production jobs requiring basic internal and/or external cuts.
Bar turner	A hexagon turret accessory that is used widely to control concentricity during an external turning process. A cutting tool assembly consisting of adjustable guide rolls that serve as a follower rest and a turning tool during an external turning process.
Roller rest taper turner	An attachment for producing long, accurate tapers on bar work. As the cutting tool advances, the cutter and the rolls, actuated by the guide bar, recede to produce the required taper.
Roller back box tool	A toolholder design in which a number of different cutter block slides may be interchanged. The cutter in the setup is always positioned at the diametral setting.
Adjustable toolholder	A toolholder for drills, reamers, counterbores, and chamfering and similar end-cutting tools. A holder that permits setting each cutting tool on the spindle axis. Also, a toolholder with a graduated collar.
Self-opening holders	Usually, toolholders for die heads that trip at the end of a required length and release thread chasers.

——————————————————— SUMMARY ———————————————————

- A main turret fed by a ram or saddle, a square or hexagon secondary turret on the cross slide, the back position on the cross slide for other cutting tools, and multi-speed geared headstocks or motorized spindles are distinguishing features of turret lathes and bar and chucker machines.

  - Turning centers combine additional tooling stations to increase the number of machine processes to eliminate two or more machines.

- There are multiple speed ranges up to 3,000 RPM on the small precision toolroom bar and chucking machines. Heavier-duty machines, with bar stock capacities up to 2 1/2" (64mm) diameter and chucking up to 12" (305mm), operate at speed ranges up to 3,000 RPM.

  - Bar and chucking machines are designed with 6, 8, or 16 tooling turret stations. The turret may be placed flat, vertically, or at an angle.

- Turret lathe and bar and chucker machine tooling are considered in relation to the headstock, front and rear stations of the cross slide, and the hexagon turret.

  - Collets provide an easy, economical, and accurate method of feeding and holding bar stock manually or automatically.

- Solid positive stop, ejector stop, and long stop parts collets are used to control the accurate positioning and holding of workpieces.

  - Step chucks permit shoulders and faces to be turned to exact lengths on parts where the outside diameter may vary. Step chucks are usually used for workpieces of larger diameter than workpieces that may be held in standard collets.

- The rigid cutter holder on the bar turner supports the cutting action against the end of the cutting tool. The tool design and supporting rolls produce a vibrationless cut.

  - Turning tools are generally of cobalt high-speed steel, brazed-tip carbides, and insert-tip carbides.

- Slide tools for turrets permit the fast changing and accurate replacement of cutting tools. Slide tool bodies accommodate cutting tools for all basic internal and external cuts.

  - Tapers may be turned by using a formed tool, a taper attachment (similar in construction and operation to an attachment for an engine lathe), or a roller rest taper turner. The taper guide bar controls the simultaneous movement of the support rolls and the cutter.

- External threads may be cut with a die head. The self-opening stationary head is tripped at thread depth. The chaser inserts are automatically backed away from the thread.

  - Cutting-off tools are generally mounted in the cross slide square turret or back tool post. Some spindles are fitted for a permanently mounted cutting-off head.

UNIT 9 REVIEW AND SELF-TEST

A. TURRET TURNING MACHINES: COMPONENTS AND ACCESSORIES

1. a. Describe a bar machine and a chucking machine in terms of holding and feeding workpieces.
    b. Distinguish between a ram-type and a saddle-type bar and chucking machine.
    c. Indicate the difference between a fixed-center turret and a cross-sliding turret on a saddle-type turret lathe.

2. Cite three functions of a secondary turret on a cross slide.

3. Describe briefly the location and operation of a leading-on thread-chasing attachment.

4. Describe how a taper is produced using a roller rest taper turner.

5. Cite three possible problems that may develop with improper care and maintenance of an indexing turret.

6. State three important safety checks the bar or chucker turning machine or turret lathe operator makes before starting a production run.

B. SPINDLE, CROSS SLIDE, AND TURRET TOOLING

1. Describe the difference between combined cuts and successive cuts as related to bar and turning machine (turret lathe) work.

2. Cite two advantages of an extended-nose collet as compared to a conventional spring collet.

3. a. State the purpose of positive-stop collets.
    b. Name three general types of stops for positive-stop collets.
    c. Explain a common design feature for positive-stop collets.

4. a. Give applications for the use of rocking jaws on a standard independent-jaw chuck on a turret lathe.
    b. Tell what function wrap-around jaws serve.

5. State two conditions under which special fixtures and other faceplate setups are more practical for bar and chucker machine and turret lathe operations than are standard chucks or step chucks and closers.

6. a. Describe how cutter wear life is increased by features that are incorporated in the design of a bar turner.
    b. Name two cutter design features for controlling chips in carbide-tip or brazed-tip insert applications.

7. a. Name two slide tools for turret lathe work.
    b. Describe the function of each slide tool.

8. State why air pressure should be avoided in cleaning the spindle bore, work-holding accessories, and cutting tools on a turret lathe.

9. Explain the functions of a stop roll in relation to the cross slide carriage.

10. Identify three possible causes of turning a finished diameter undersize when a bar turner is used.

11. List two practices to consider when taking external and internal cuts at the same time.

12. State three corrective steps to take to avoid tool chatter problems.

13. List three safety checks the bar and chucker turning machine or turret lathe operator makes with respect to a tooling circle setup.

Basic External and Internal Machining Processes

Basic external cuts refer to straight, taper, and bevel turning; facing; grooving; thread cutting; cutting off; and knurling (which, technically, is not a cutting process).

Basic internal cuts relate to center drilling and drilling; straight, angle, and taper boring; counterboring; countersinking; reaming; undercutting (recessing); and threading either chased or tapped threads.

This unit covers tooling and setups for basic external and internal cuts using bar and chucker turning machines and turret lathes. Cutting speeds and feeds are established and sequencing combinations of machine tooling and turning processes are treated in detail.

OBJECTIVES

After satisfactorily completing this unit, you will be able to:

- Use cutting speed, feed, and depth of cut rates from machinability tables.
- Straight turn using the square turret, bar turner, and multiple turning head.
- Set up and adjust a bar turner for size and finish.
- Set up and turn a taper using a taper attachment or second method.
- Perform facing, grooving, radius forming, chamfering, and cutting off processes, and thread with a die head or single-point cutting tool (or simulate these external and the following internal processes).
- Perform the following internal processes: drilling (with start and core and spade drills for deep-hole drilling); straight and taper boring and reaming; recessing, back facing, tapping with collapsing taps, and threading with single-point thread cutting tools.
- Arrange for sequencing external and/or internal cuts.
- Follow *Safe Practices* for external and internal bar and chucker turning machine and turret lathe processes and correctly use new *Terms*.

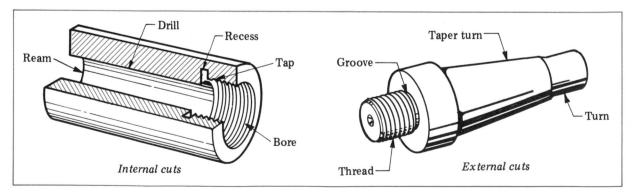

Figure 10–1 Basic Internal and External Cuts on Bar and Chucker Turning Machines and Turret Lathes

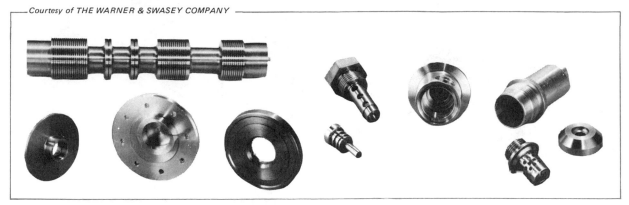

Figure 10–2 Typical Bar Work Machined on Bar and Chucker Turning Machines and Turret Lathes

Basic bar and chucker turning machine and turret lathe processes are illustrated by line drawing (Figure 10–1). Examples of typical bar machine work are shown by actual photos (Figure 10–2). Representative parts requiring chucker tooling and machining processes are illustrated in Figure 10–3.

BASIC EXTERNAL CUTS

TURNING CUTS

Straight turning is generally performed by one or a combination of three different methods: (1) side turning from the square turret; (2) overhead turning from the hexagon or octagon turret; and (3) long-bar turning using a bar turner.

Side Turning from the Square Turret. In side turning, the cutting tool is mounted in the square turret. Turning cuts may be taken simultaneously with processes that are carried on from turret tooling stations. Roughing and finish turning cuts may be taken by positioning the cutting tools at successive square turret stations.

Straight Turning with a Bar Turner. Bar turners are capable of machining a fine surface to tolerances within 0.001″ (0.02mm). Bar turners, equipped with either brazed-tip or insert-type carbide cutters, are used to perform most of the heavy metal turning processes.

Overhead Turning from the Hexagon Turret. The multiple turning head on the hexagon turret provides rigid tool support that is essential in

heavy metal-removing processes. Cutter holders are mounted in the multiple turning head (Figure 10–4A). A reversible plain cutter holder (Figure 10–4B) is used for rough turning cuts. A reversible adjustable cutter holder (Figure 10–4C) has a built-in micrometer feature that permits a cutter to be set to size accurately and quickly.

Taper Turning. Internal or external tapers may be machined by using a taper attachment. The cutting tool is mounted in the square turret on ram-type and fixed-center saddle-type turret lathes. The taper attachment is set at the required taper angle and locked to the bed. The cutter travel is controlled by the taper attachment guide plate (slide block).

FACING CUTS

Square Turret Facing. The square turret is used for most facing cuts. A single-point facing

Figure 10–3 Samples of Cast and Forged Flanges Requiring Chucker Tooling

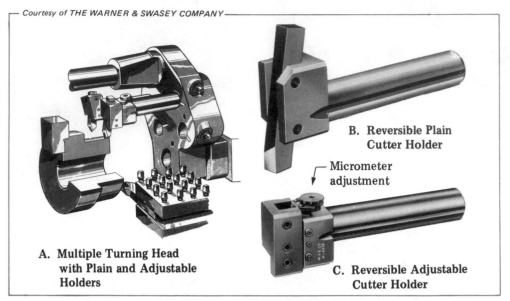

Courtesy of THE WARNER & SWASEY COMPANY

B. Reversible Plain Cutter Holder

Micrometer adjustment

A. Multiple Turning Head with Plain and Adjustable Holders

C. Reversible Adjustable Cutter Holder

Figure 10–4 Overhead Turning from Hexagon Turret with Multiple Turning Head

cutter is generally mounted in one station of the square turret. The cross slide carriage is positioned along the bed ways at the required length. It is clamped in position for facing cuts. Workpieces may also be faced by combining the facing cut with drilling, boring, or other processes.

Hexagon Turret Facing. Short facing cuts are often taken with tools mounted on the turret in a *quick-acting slide tool.* The slide tool is actuated by a hand-operated lever. This lever feeds the mounted cutter at right angles to the spindle axis.

Facing across large-diameter workpieces is performed on cross-sliding turret machines. The cutter is held on the turret in a through-slot cutter holder, boring bar, or tool-base cutter block.

End Facing. An *end former* or *combination turner and end former* (Figure 10–5) is used to support the end of a workpiece that would otherwise spring under the cutting action. The rolls are set ahead of the cutter to support the workpiece.

FORMING AND CUTOFF CUTS

Forming Cuts. A fast method of producing a finished shape and diameter is by *forming.* The

four basic forming cuts are: (1) necking (cutting a recess close to a shoulder), (2) chamfering a beveled surface, (3) radius forming, and (4) grooving. Single forming cuts are usually taken with the cutter mounted in the square turret. A *necking cutter block* mounted on the rear of the cross slide is used when several necking or grooving cuts are required.

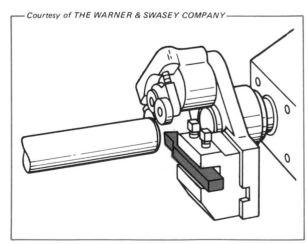

Courtesy of THE WARNER & SWASEY COMPANY

Figure 10–5 Application of Combination Turner and End Former to Facing

The hexagon turret may also be used for forming cuts. A bar and cutter mounted in a quick-acting slide tool may be positioned and fed to cut a groove or chamfer. An internal facing cutter is usually held in a boring bar or toolbase cutter block on cross slide turret machines.

Cutoff Cuts. Cutting-off operations are performed by mounting the cutoff-tool holder in the square turret or in the rear cutter block. High-speed steel and carbide cutoff blades, mounted in a holder, permit the rapid changing of cutters as they wear. Maximum rigidity in the setup ensures fast, smooth cutoff.

THREADING

Threading with Die Heads. A continuous, even force must be applied when tapping so that the turret moves with the die head. For most threads, the die head may be led on by hand—that is, the turnstile is turned manually in starting and following the thread.

Long, accurate threading requires that a positive lead be provided for uniformly feeding the die head across the workpiece. A leading-on attachment or a thread-chasing attachment set at the required thread pitch, is generally used.

Precision Threading with Single-Point Cutters. Single-point thread cutting is used to produce threads that must be concentric with other diameters and meet precise tolerances.

Some bar and chucker machines and turret lathes are equipped with a thread-chasing attachment. This attachment is fitted to the cross slide and feeds the thread cutter in the square turret. Threads may also be cut from the hexagon turret when it is fitted with a thread-chasing attachment.

CUTTING SPEEDS (sfpm) AND FEEDS (ipr) FOR BAR TURNERS

Table 10–1 gives one manufacturer's recommended starting cutting speeds and feeds for bar turner processes on ram-type turret lathes. The use of these cutting speeds and feeds provides for maximum machining production rates, quality of surface finish, and dimensional accuracy (concentricity and diametral setting).

All cutting speed and feed tables provide a baseline of values that serve to guide the turning machine operator. Compensation may need to be made for factors such as chip control, rate of tool wear, machine horsepower, specifications, and other conditions governing all machine tools and setups.

Table 10–1 Starting Points for Cutting Speeds and Feeds for Bar Turner Processes on Turret Lathes

	Metals to Be Machined				
	C-1070	A-4340	C-1045	C-1020	B-1112
	Cutting Speeds (sfpm)				
High-speed steel cutter	50	70	90	120	150
Premium Carbide cutter	220	245	320	420	490
	Feeds				
Depth of cut	1/8″	1/4″	3/8″	1/2″	5/8″
Feeds (ipr) (0.001″ per revolution)	0.030	0.018	0.012	0.0075	0.0045

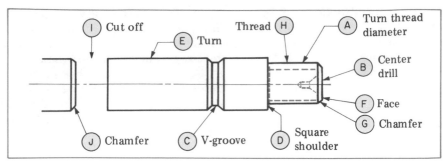

Figure 10-6 Sequence of External Cuts for a Threaded Shaft

How to Sequence External Cuts

Example: Set up a ram-type turret lathe to mass-produce the threaded shaft shown in Figure 10-6.

Note: The thread is to be die cut. The die head is to be led on by hand.

STEP 1 Feed the stock to length against a revolving bar stop.

Note: It is assumed that the bar end is already chamfered.

STEP 2 Position (index) the bar turner that is mounted on the hexagon turret.

STEP 3 Turn the thread diameter (A).

STEP 4 Index the center drilling tool on the turret. Center drill the end of the bar (B).

STEP 5 Index to position the revolving center. Support the workpiece with the center.

STEP 6 Form the V-groove with a cutter mounted in the square turret (C).

STEP 7 Position the shoulder-turning cutter on the square turret. Square the shoulder at the end of the thread (D).

STEP 8 Turn the outside diameter by using a turning tool mounted in the square turret (E).

STEP 9 Remove the center. Position the end former or combination turner and end former on the turret.

Note: The tools have rolls that hold the workpiece securely against the cutter.

STEP 10 Face (F) and chamfer (G) the end of the threaded shaft.

STEP 11 Turn the hexagon turret to the self-opening die head position.

STEP 12 Check the thread length trip stop. It should be adjusted so that the die head opens for the chaser insert to clear the workpiece just before the die head reaches the shoulder. Cut the thread (H).

STEP 13 Feed the cutoff tool mounted on the rear tool post to cut the workpiece to length (I).

STEP 14 Position the chamfering tool in the square turret. Chamfer the end of the next workpiece (J).

The tooling setup and sequencing of cuts required to machine the threaded shaft are illustrated in Figure 10-7.

How to Set Up a Bar Turner

STEP 1 Extend the bar stock about 3" from the collet.

STEP 2 Turn the diameter 0.001" (0.02mm) under the required size for 1/2".

STEP 3 Position the bar turner cutter slightly above center. Loosen the roll jaw adjusting screws until the rolls clear the turned diameter.

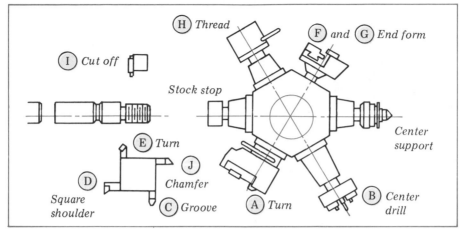

Figure 10–7 Tooling Setup and Sequence of Cuts for Machining the Threaded Shaft

STEP 4 Move the bar turner so that the cutter is located over the area of the turned surface. Check to see that the withdrawal lever is in the closed position.

STEP 5 Adjust the cutter slide block until the cutter just grazes the turned diameter.

Note: The spindle must be in the brake position and the cutter on center.

STEP 6 Adjust the cutter slide block 0.0015″ away from the turned diameter.

STEP 7 Clear the cutter from the turned surface by retracting the cross slide until the cutter touches the turned shoulder.

STEP 8 Loosen the roll jaw stud. Adjust the rolls longitudinally. The roll radius is positioned behind the cutter radius.

STEP 9 Start the spindle. Turn the roll adjusting screws until the rolls lightly contact the workpiece and roll. Clamp the rolls securely.

STEP 10 Adjust the coolant nozzle and start the flow.

STEP 11 Advance the cutter to the cutting position. Take a trial cut.

STEP 12 Return the bar turner to its original position. Stop the spindle.

STEP 13 Check the accuracy of the turned diameter and the quality of the surface finish.

How to Adjust a Bar Turner for Size and Finish

Correcting an Oversized, Finish Turned Diameter

STEP 1 Check the position of the cutter and securely tighten the binder screws. Adjust the cutter to the centerline.

STEP 2 Check the back roll. A loose roll does not provide a firm support; consequently, the workpiece may spring away from the cut.

STEP 3 Tighten the roll if it is loose. Inspect the cutting edge for wear. Adjust the cutter slide to compensate for wear.

Correcting an Undersized, Finish Turned Diameter

STEP 1 Check the cutting edge of the cutter for buildup. Remove the buildup by stoning. Readjust the cutter, if needed.

STEP 2 Check the front roll for adjustment and looseness. These conditions permit the work to vibrate and cause the cutter to chip, flake, or wear rapidly. Correct the problem by tightening the front roll. Both rolls are then checked for proper contact.

STEP 3 Check the rolls for tightness. Tightness produces an overburnished, undersized diameter. Readjust the tightness of the rolls.

Correcting Variable Diameter Sizes

STEP 1 Check the looseness of the back roll. Looseness produces a tapered diameter, particularly on long turned parts. The condition is corrected by resetting the back roll with the bar turner positioned near the collet.

STEP 2 Check the spring tension on the cutter slide block. If there is any play due to improper tension, the turned diameter may vary over its length. Tighten the spring adjusting screw at the rear of the slide block to increase the tension.

Correcting Surface Irregularities

STEP 1 Check the rolls for proper setting. A series of undersized grooves along the turned surface indicates that the rolls are improperly set.

STEP 2 Check for alignment, looseness, and wear on the rolls. Any one of these conditions can cause the rolls to dig into a finished surface to produce a rough, irregular surface.

STEP 3 Check the position of the rolls in relation to the nose radius of the cutter. Advance the rolls if they are set too far behind the cutter.

STEP 4 Decrease the feed if the cutter dips in at intervals. Check the cutting speed. Adjust, if needed.

STEP 5 Check the shape and condition of the chip groove. Regrind for the chips to form properly and to be removed without being forced between the rolls and the finished diameter.

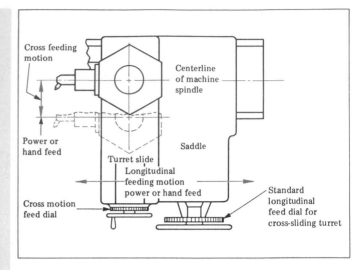

Figure 10–8 Longitudinal and Cross Feeding Movements of the Cross-Sliding Hexagon Turret

ADDED MACHINING CAPABILITY WITH THE CROSS-SLIDING TURRET

The turret of the cross-sliding hexagon turret may be fed longitudinally and transversely (cross feed). Movement may be by hand or power feed. The positioning of the cutting tool for an internal or external cut with cross feed or longitudinal motion is controlled by micrometer dial settings. The cross feeding and longitudinal feeding motions and positioning dials are shown in Figure 10–8.

With a cross-sliding turret, all internal operations are normally performed with the boring bar as the primary tool. The boring bar is mounted on the hexagon turret. The square turret is used for all external cuts like turning, grooving, facing, chamfering, and so on. The cross-sliding turret is ideal for short-run lots. A great range of diameters may be machined by merely setting dial clip positions for each new diameter. Cuts may be taken from both the cross-sliding hexagon and the square turret at the same time.

Added machining capability is provided by a built-in positive stop. This stop permits the cross-sliding turret to operate also as a fixed-center turret. Tooling is then aligned with the centerline of the turret lathe or bar and chucker turning machine.

BASIC INTERNAL PROCESSES

DRILLING

Parts are drilled to produce holes to specifications, to center stock, and to remove material from existing holes. Drilling cuts require different drills. Parts may be center drilled or spotted with a *start drill.* Holes may be produced from the solid by using a *twist drill.* Holes that are larger than 2" (50mm) diameter are drilled with a *spade drill.* Previously drilled, cored, or pierced holes may be enlarged with a *core drill.* Deep-hole drilling sometimes involves a *coolant-feeding drill.*

Core drilling requires a three- or four-fluted cutter. The core drill is used to enlarge previously drilled, cast, or pierced holes. Such holes must first be chamfered. Long holes are often started by boring to the core drill diameter for about 1/2" (12mm).

Spade drilling is a practical method of drilling holes that are 2" (50mm) or larger in diameter. The face of the cutter is ground to a cutting angle similar to the angle of a twist drill face. A spade drill has two cutting edges.

Deep-hole drilling relates to holes that are four or more times deeper than the drill diameter. Most deep-hole drilling is done with drills designed so that a cutting fluid may be pumped to the cutting edges.

BORING CUTS

Boring cuts are used to turn concentric, straight holes to precise dimensional limits. Boring bars are designed for internal rough and finish turning, facing, counterboring, undercutting, chamfering, bevel and taper turning, and thread chasing. Boring bars hold square high-speed steel tool bits, brazed-carbide or carbide-tip cutters, ceramic inserts, and chip breakers.

Straight Boring. *Turret boring* on a ram-type or fixed-center saddle-type turret lathe requires the cutter to be mounted in a slide tool. The boring bar and cutter are thus permitted to be adjusted to the bore size. The slide tool set-up permits micrometer adjustment of the cutter to establish the depth of cut and bore diameter. Once the cutter is set to depth, the slide is clamped securely and provides the rigidity necessary to take roughing cuts or to finish bore a hole.

Cross-sliding hexagon turret boring does not require the use of a slide tool. Adjustments for bore diameter are made by moving the cross slide directly.

Square turret boring is used for limited operations. These include the boring of short holes, taking lighter cuts, and using slower feeds. The solid, forged cutter designed for square turret use is generally less rigid than the heavier, solid boring bars.

Taper Boring. The square turret is also adapted for *taper boring.* The taper attachment is set at the required taper and is secured to the bed. The bore size is established by moving the cross slide handwheel. For accurate taper boring the cutter is fed *away* from the spindle. The cut is in the direction of the largest taper diameter.

Large taper boring operations are usually performed on a cross-sliding hexagon turret machine which permits the use of heavier, rigid boring tools adapted to heavy cuts (Figure 10–9).

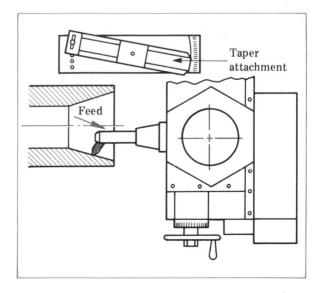

Figure 10–9 Taper Attachment Setup for Steep-Taper Boring on Cross-Sliding Hexagon Turret

REAMING CUTS

Reaming usually follows boring. Boring permits machining to a specific size and leaving the proper amount of material to remove in the finish reaming process.

Solid and *expandable reamers* are used. These reamers are held in a *floating reamer holder* which is mounted on the turret and is designed to float. Thus, the reamer is able to align itself with a previously drilled or bored hole.

Taper reamers are used to finish a taper bore or to produce an internal taper directly. A pair of roughing and finish cutting taper reamers is used, especially when steep tapers are to be cut.

Adjustable floating-blade reamers are designed for reaming large-diameter holes. There are two adjustable cutting blades in a floating blade reamer. This reamer is mounted in a flanged toolholder. The two reamer blades are aligned in a previously formed hole by the floating action of the cutter blades.

INTERNAL RECESSING (UNDERCUTTING), FACING, AND BACK-FACING CUTS

Relief areas for threading, grinding, shouldering, and taper surfaces must often be produced as internal cuts. A combination of longitudinal and cross slide movements is required. The cutting tools are brought to position inside the workpiece. The tools are then fed by cross feeding to undercut (internally recess) or produce an internally faced surface.

TAPPING AND SINGLE-POINT THREAD CUTTING

Sufficient force must be exerted to start the tap in a tap hole. Once started, the tap must be self-feeding. A tap that is forced during cutting, or *dragged* while it is being withdrawn, will produce defective threads. The turret and holder must be fed at a rate equal to the thread pitch.

Tapping with Solid and Collapsing Taps. *Solid taps* are generally used for thread sizes up to 1 1/2″ (38mm). These taps are held on the hexagon turret in a releasing tap holder. The forward motion of the turret and tap is stopped by the setting on the turret stop. The tap is withdrawn by reversing the direction of spindle rotation.

Collapsing taps are used for thread sizes larger than 1 1/2″ (38mm). These taps are designed to collapse. The cutting edges pull away from the workpiece at the end of the cutting stroke. Collapsing taps may be mounted directly on a hexagon turret or in a flanged toolholder. One advantage of a collapsing tap over a solid tap is that the cutter is withdrawn without reversing the spindle direction.

Single-Point Thread Cutting. The single-point threading tool is held in a boring bar that is mounted in a hexagon turret slide tool. A flanged holder is used for mounting the threading tool on cross-sliding turrets. The cutter may also be held in a thread-chasing cutter holder that is mounted on the square turret of the cross slide. Multiple passes are required until the cutter is fed to thread depth.

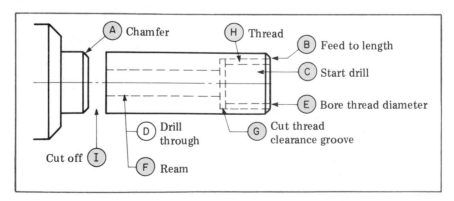

Figure 10–10 Sequence of Work Processes for Machining the Threaded Sleeve

How to Sequence Internal Cuts

Example: Set up the sequence of internal cuts to mass-produce a quantity of the Threaded Sleeve shown in Figure 10–10.

Note: The chamfered end and outside diameter are turned as external cuts from the square turret.

STEP 1   Chamfer the workpiece (A).

STEP 2   Feed to the required length against a combination stock stop (B).

STEP 3   Start drill the end (C).

STEP 4   Position the reamer size drill. Drill the hole at least 1/8″ (3mm) deeper than the length of the threaded sleeve (D).

STEP 5   Bore the thread diameter (E).

Note: A slide tool mounting for the boring bar permits size adjustments.

STEP 6   Ream the drilled hole (F).

Note: A fluted reamer is generally mounted in a floating reamer holder to ensure alignment with the drilled hole.

STEP 7   Cut the thread clearance groove by feeding the grooving cutter to thread depth (G). A quick-acting slide tool is used for the cutter and boring bar mounting.

STEP 8   Position the tap and releasing tap holder. Turn the turnstile to feed the tap to the required depth (H).

Note: Nonstandard and special fit threads are generally chased. The single-point thread cutter is fed by engaging the lead screw in connection with either the cross slide square turret or the ram turret.

Note: Remove the tap be reversing the spindle direction.

STEP 9   Cut off the workpiece to length (I).

Note: Remove the outside diameter burr during the cutting-off process. Burr the remaining fine-burred edge from around the reamed hole.

The positioning of the standard tools for the sequence of cuts is diagrammed in Figure 10–11.

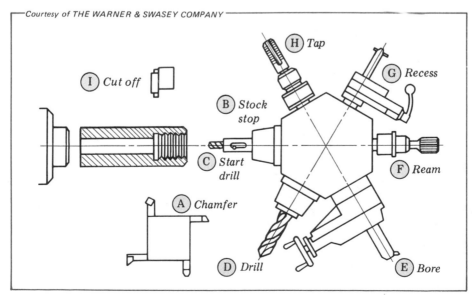

Courtesy of THE WARNER & SWASEY COMPANY

Figure 10–11 Tooling Setup for Producing the Threaded Sleeve

APPLICATIONS OF CUTTING SPEED AND FEED TABLES TO EXTERNAL AND INTERNAL CUTS

The bar and chucker turning machine or turret lathe operator must be able to establish appropriate cutting speeds and feeds. Reference is made to manufacturers' suggested starting points, which provide a baseline of cutting speed and feed values. These values are modified as one of the first considerations in machinability. Metals that are commonly machined on the turning machines are grouped into *machinability classes* as listed in Table 10–2. Information on machinability class is used with other cutting speed tables. For example, Table 10–3 is a partial table (showing machinability classes A and B) of cutting speeds for a few basic external cuts on turret turning machines.

Table 10–4 is a partial table of data for starting feed rates (ipr) for external cuts on turret turning machines. Complete and additional cutting speed and feed tables are included in the Appendix. Four factors must be considered in determining the final feed for roughing cuts:

- The rigidity of the cutter and the tooling strength,
- The rigidity of the work-holding device and the nature and design of the workpiece,
- The condition and capacity of the machine,
- End thrust.

Table 10–2 Machinability Class of Commonly Machined Metals

Machinability Class	Commonly Machined Metals
A	B-1112 C-1118 Bronze (phosphor 64)
B	C-1010 and 1015 (tubing) C-1019, C-1020, C-1137, C-1141 Brass (Naval 73) Bronze (Tobin) Cast iron (soft) Stainless (#416)
C	C-1040, C-1045, C-1050 C-4140, A-4150, A-4340, A-4615 Bronze (aluminum 68) Cast iron (hard) Malleable iron Stainless (#302 and #304)
D	E-3310, E-4160 Cast steel Stainless (#440C)
E	C-1070, C-1090 E-52100 Monel
F	Aluminum Brass (free-machining)

Table 10–3 Suggested Starting Points for Cutting Speeds on Turret Turning Machines for Basic External Cuts (Partial Table)

Machinability Class	Nature of Cut	Turning — Overhead and Cross Slide Cutter Material Carbide	Turning — Overhead and Cross Slide Cutter Material High-Speed Steel	Turning — Roller Turners Carbide	Facing Carbide	Forming High-Speed Steel	Forming Carbide	Cutting-Off High-Speed Steel	Cutting-Off Carbide
A	Roughing	490	150	490	490	150	490	120	500
A	Finishing	560			560				
B	Roughing	420	120	420	420	120	420	100	500
B	Finishing	490			490				

Table 10–4 Starting Feed Rates on Turret Turning Machines: External Cuts (Partial Table)

Depth of Cut	Turning			Facing	Forming		Cutting Off	
	Overhead and Cross Slide		Roller Turners				High-Speed Steel	Carbide Insert
	Hexagon Turret	Square Turret			Steel	Cast Iron		
	Feed per Revolution of Spindle							
							Stock to 1″ (25.4mm) diameter 0.0045″ (0.1mm)	Machinability Class: A, B, F: 0.003″ (0.1mm) C, D: 0.004″ (0.1mm) E: 0.005″ (0.1mm)
1/16″	0.030″	0.019″		0.017″				
1.5mm	0.8mm	0.5mm	0.030″	0.4mm				
1/8″	0.018″	0.019″	(0.8mm)	0.017″	0.004″	0.006″		
3mm	0.5mm	0.5mm		0.4mm	(0.1mm)	(0.2mm)		
1/4″	0.012″	0.012″	0.018″	0.010″				
6mm	0.3mm	0.3mm	0.5mm	0.3mm				

Note: Millimeter values are rounded to nearest 0.1mm.

Coarse feed ranges are used for roughing cuts. High cutting speeds and fine feeds are used for high-quality, accurate cuts to produce precision surface finishes.

Table 10–5 provides technical information relating to threading on bar and chucker turning machines and turret lathes with high-speed taps or chasers. Cutting speeds are listed for a range of threads from 3 to over 25 threads per inch (8.5mm to less than 1 mm metric pitches).

Table 10–5 Recommended Starting Cutting Speeds for Threading with High Speed Steel Taps or Chasers on Bar and Chucker Turning Machines and Turret Lathes (Partial Table)

Material	Threads per Inch (tpi)			
	3 tc 7 1/2	8 to 15	16 to 24	25 and up
	Equivalent Metric Thread Pitch (mm)			
	8.5 to 3.5	3.2 to 1.5	1.5 to 1	1 and less
	Starting Cutting Speeds in sfpm*			
Aluminum	50	100	150	200
Bakelite	50	100	150	200
Brass				
Bar stock and castings	50	100	150	200
Forgings	25	40	50	80
Iron				
Cast	25	40	50	80
Malleable	20	30	40	50
Wrought	15	20	25	30
Steel				
Carbon 1010–1035	20	30	40	50
Carbon 1040–1095	15	20	25	30
Chrome	8	10	15	20
Stainless	8	10	15	20

*Use 75% of the cutting speed for cutting taper pipe threads.

THE PLANNED TOOL CIRCLE

Production efficiency requires that operator effort and operating time be reduced, where practical. Efficient machining may be accomplished by positioning the cutting tools and setups so that they fall within a *tool circle*. The saddle should also be positioned so that each cutting tool travels the shortest possible distance before machining begins.

In a tool circle, all positioning, support, and cutting tools extend an equal distance from the hexagon turret. Figure 10–12 shows a production setup where all of the tools and holders fall within a tool circle. As each position of the turret is indexed, the longitudinal travel to the end of the workpiece is approximately the same. The operator is able to set up uniform operation timing motions to index and position each tool to the cut. The longitudinal turret movement stroke and the turnstile movement are kept short.

Note: The distance between the saddle and the end of the workpiece must be great enough for the binder clamp to lock the hexagon turret securely.

Note: Each tool on the turret must be checked for interference. When indexed, each tool must clear the square turret and rear tool post tools.

Figure 10–12 Planned Production Tooling and Setups That Fall within a Tool Circle

PERMANENT SETUPS WITH CHUCKING TOOLING

Castings, forgings, large-diameter rolls, collars, gear blanks, and flanged parts may be machined with chucking tooling. Machining usually involves combined and multiple cuts. As many processes as possible are completed in chucking in order to ensure concentricity and squareness

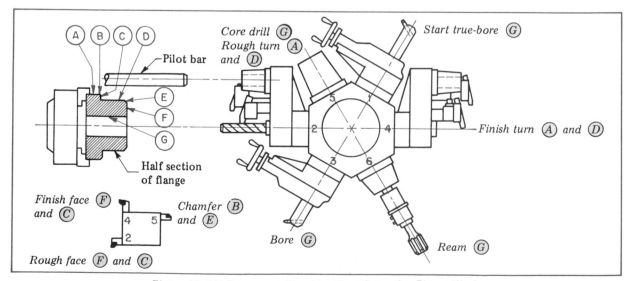

Figure 10–13 Permanent Chucking Tool Setup for Flange Work

between internal and external surfaces. Rough castings and forgings require three distinct cuts:

- A first cut for heavy stock removal,
- A second cut for truing up,
- A third cut for dimensional accuracy.

REPRESENTATIVE CHUCKING TOOLING (FLANGES)

Cast or forged flanges are considered as a representative sample of a group of chucking jobs. Each part may be chucked on a bar and chucker turning machine or a turret lathe and machined on the inside and outside.

A permanent setup for machining flanges with chucking tooling is shown in Figure 10–13. The representative flange is shown as a half section. The flange is held in a chuck. The surfaces to be machined are lettered (A) through (G). The square turret and hexagon turret tooling corresponding with each process are also lettered (A) through (G). The sequence of each cut is identified by a numeral.

TOOL CHATTER PROBLEMS

Tool chatter produces a high-pitched, shrieking, whistling sound. Most tool chatter problems are caused by excessive tool overhang. The tool setup is not rigid enough to withstand the cutting force and speed. Corrective steps for tool chatter problems are as follows:

- Keep the overhang of the toolholder and cutter to a minimum;
- Use the largest possible size of cutting tool, boring bar, or shank-type cutter holder;
- Grip workpieces close to the chuck or spindle nose;
- Use a revolving center to support heavy cuts;
- Tighten and recheck all clamping screws in cutting-tool holders;
- Position cutters on center;
- Replace a dull cutter with one especially sharpened for a specific job and keep the cutter sharp;
- Reduce the feed rate;
- Eliminate *dwell* so that a cutting tool does not remain in contact at the same machined surface position after the process is completed.

Safe Practices for Basic External and Internal Machining Processes

- Use manufacturers' tables of recommended cutting speeds and feeds. Adjust the recommended values based on the specifications of the piece part, the accuracy and quality of surface texture, rate of tool wear, chip control, and machine capability.
- Check the working clearance between the cross slide and turret tooling.
- Mount bar turner rolls on a finished diameter ahead of the cutter to produce a precise, concentric diameter.
- Mount bar turner rolls behind the cutter to support the cutting force and burnish the turned surface.
- Use a multiple turning head on the hexagon turret to provide rigid tool support for heavy cuts.
- Use a leading-on attachment to cut fine, precise threads.
- Chamfer or bore the end of a cored or pierced hole for a short distance to center guide the lips of a drill.
- Use a spade drill for drilling holes over 2″ diameter to minimize the overhang of the drill.
- Use the heaviest possible boring bar with the shortest overhang for heavy, long boring cuts.
- Bore an internal taper on the finish cut with the direction of feed toward the large diameter.
- Set the turret stop about 1/8″ (3mm) less than thread depth to control threading to a shoulder or other fixed point of a workpiece.
- Check the composition of the cutting fluid and the rate of coolant flow at each station. Position splash guards to contain the coolant and chips.
- Keep the work area around the machine dry and free of chips.

TERMS USED WITH EXTERNAL AND INTERNAL TOOLING, AND MACHINING SETUPS

Sequencing of cuts	Planning the tooling for each internal and/or external cut and the position of each tool and accessory on the turret or cross slide.
Quick-acting tool slide	A tool-holding device that makes it possible to adjust a cutter. Once set, the cutter may be fed quickly to depth by a quick-acting lever on the tool slide.
Combination turner and end former	A tooling device consisting of work-supporting rolls and a cutter that may be fed to end face a workpiece.
Leading-on attachment	A turning machine attachment that may be set to feed a cutter at a rate equal to the pitch of the required screw thread. An attachment that permits the feeding and withdrawal of a threading setup at the pitch rate of the thread to be cut.
Indexing (turret lathe)	Moving a hexagon, square, or other multistation turret to the next position (face). Positioning the tooling for each successive process in relation to the axis of the workpiece.
Adjustable floating-blade reamer	A two-blade reamer for reaming large-diameter holes. Floating blades that align with a formed hole.
Internal recessing	The cutting (undercutting) of a groove inside a workpiece. Usually, a clearance area at thread depth or a groove at a shoulder.
Core drilling	The use of three- and four-fluted drills to drill a cored, pierced, or previously drilled hole to size.
Collapsing tap	A tap design generally used for sizes over 1 1/2″ diameter. A tap with threaded inserts that move away (collapse) from a thread. A feature that permits a tap to be withdrawn without reversing the spindle direction of rotation.
Stop roll	A device that permits settings for each turret face (station). A multiple setting device for controlling longitudinal and transverse movement of tooling on the cross slide and hexagon turret.
Tool circle	Positioning the tooling on a turret to uniform starting positions.
Buildup (cutter)	The addition of the material being cut on and around the cutting edge of a cutter. Excess material, fused or adhering to the cutter, that results from tremendous force and heat in the cutting area.
Chucking tooling setups	The combination of tooling and positioning cutters and accessories at each turret position to perform required machining processes.
True bore	An initial cut or cuts on a cast or rough surfaced hole for the purpose of providing a concentric straight starting hole for subsequent operations.
Dwell	Leaving a cutter in contact with a workpiece as it continues to turn and no cutting takes place.

SUMMARY

- The basic internal cuts in bar and chucker turning and turret lathe work include center drilling and drilling; straight, angle, and taper turning; counterboring and countersinking; reaming; undercutting (recessing); and threading.
  - The basic external cuts cover straight and taper turning, bevel cutting and chamfering, facing, grooving (for turning), thread cutting, and cutting off.
- Facing cuts are taken with a cutter mounted and fed from a square turret, a slide tool mounting, or a fixed position on a cross-sliding turret machine.
  - Cutting-off cuts are generally made with a cutoff tool positioned in either the square turret or the rear tool post of the cross slide.
- Straight turning may be done with the cutter mounted in a square turret on the cross slide or the hexagon turret. Bar turners are used for light turning processes; multiple turning heads, for heavy-duty overhead turning.
  - The heaviest boring bar or forged boring tool should be mounted as short as possible. Boring may be done from a cross slide or hexagon turret.
- Internal taper finish boring cuts are taken feeding toward the large diameter.
  - Drilling processes usually require center drilling and/or start drilling. Cored holes are core drilled. Large-diameter holes (approximately 2″ diameter and over) are spade drilled.
- Threads are machine tapped with solid taps for sizes up to 1 1/2″. Collapsing taps are used for larger diameter threads.
  - Cast, forged, or flame cut holes are true bored to the diameter of a core drill. Boring ensures proper drill alignment and concentricity.
- Holes are reamed with solid, taper, or adjustable floating-blade reamers.
  - Internal recessing, grooving, and back-facing cuts require tool positioning inside the workpiece. The cutters are fed by cross slide movement. On a fixed-center turret, the tool is mounted in and fed from a quick-acting slide tool.
- After setup, the complete cycle of cuts should be taken carefully and checked for correct cutting action, dimensional accuracy, surface finish, and safe clearance in the work area.
  - Machinability tables help to establish the size, shape, and design of the cutter(s) and cutter holder(s); cutting speeds and feeds; the type and size of machine tool; and the composition of the cutting fluid, if required.
- Machine stops are used to control the movement of the cross slide and hexagon turret.
  - A bar turner is adjusted to correct turning oversized, undersized, and variable-sized diameters and surface irregularities.
- Permanent tooling and setups are common shop practices for workpieces that may be grouped according to similar design features and chucking processes.

■ Machine vibration and tool chatter are conditions that affect surface finish and accurate dimensional measurement control. These conditions may be identified by sound during machining or by checking with instruments. Balance of the workpiece; trueness and condition of the spindle; overhang of the setup, tooling, or workpiece; cutter position and sharpness; and feed rate are some checkpoints.

UNIT 10 REVIEW AND SELF-TEST

1. Explain how a burnished, quality surface finish is produced by using a bar turner.

2. Describe how large-diameter workpieces may be faced on machines equipped with a cross-sliding turret.

3. a. List four basic forming cuts for bar and chucker turning machine work.
   b. Indicate two general locations for cutting-off tools. State why such locations are functional and desirable.

4. Use machinability and cutting speeds and feeds tables for bar and chucker turning or turret lathe operations. Determine the starting cutting speeds in sfpm for the following three jobs:
   a. Rough facing class A machinability parts with a carbide-tipped cutter.
   b. Cutting off class B machinability parts with a high-speed steel cutter.
   c. Rough cutting class B machinability parts with a high-speed steel cutter by using a bar turner.

5. Determine the starting feed rates for machining on a bar and chucker turning machine or turret lathe for each of the following processes:
   a. Turning a workpiece, using a roller turner setup, for a 3mm depth of cut;
   b. Forming steel parts to a 1/4″ depth of cut;
   c. Cutting off a 1″ diameter workpiece with a high-speed steel cutter.

6. Tell what function is served by each of the following drills: (a) start drill, (b) spade drill, and (c) core drill.

7. Describe two turret setups that permit a boring bar and cutter to be adjusted for each successive cut when straight boring.

8. Explain the difference between the cutting of an internal, square recess and back facing a square shoulder on a bar and chucker turning machine.

9. Give two advantages of using collapsing taps in preference to solid taps for certain applications on a turret lathe.

10. Identify three characteristics for a planned tool circle in production machining.

11. State three safety precautions the bar and chucker turning machine or turret lathe operator should take to prevent personal injury or damage to the machine, setup, or workpiece.

Single- and Multiple-Spindle Automatic Screw Machines

The automatic screw machine is an adaptation of the engine lathe, and the turret lathe. The screw machine is used principally for semi-automatic and fully automatic, high-volume production of identical parts. The three basic types of screw machines are: (1) plain, (2) automatic single-spindle, and (3) automatic multiple-spindle. Major components, accessories, and typical tooling setups, including cam features, are described. The descriptions build upon engine and turret lathe technology and machining processes. Cutting speeds, feeds, and cutting fluids are related to general screw machine processes.

Attention is paid to common problems relating to: feeding, stock stops, basic internal machining processes, turret tooling, and cross slide tooling. Causes of these problems and corrective actions are identified.

Bar and chucker turning machines and turning/milling machining centers are also adaptable to the mass production of similar products. Computer numerical control (CNC) applications to automatic screw machines and 4- and 6-axis high production mill/turn centers are covered later in the text.

OBJECTIVES

After satisfactorily completing this unit, you will be able to:

- Classify single- and multiple-spindle automatic screw machines and describe essential components.
- Select tooling for automatic screw machine production.
- Interpret functions and the design and manufacture of cams.
- Identify appropriate cutting fluids and the qualities of new cutting oils.
- Understand accessories and setups for second machine operations for automatic screw machines.
- Diagnose screw machine troubleshooting problems from examples of defective workpieces or damaged tools. Identify corrective steps related to problems in stocking and rod feeding, stock stop adjustments, hole forming and knurling, turret tooling, and turret and cross slide.
- Follow *Safe Practices* and precautions in screw machine operation and correctly use new *Terms*.

CLASSIFICATIONS OF SCREW MACHINES

Originally, the screw machine developed from the engine lathe for the economical manufacture of interchangeable screws and bolts. In addition to the feeding of regular bar and wire stock, today's machines permit forgings, castings, and irregular-shaped parts to be fed by hand or automatically by *magazine*.

Like turret lathes and bar and chucker turning machines, the tooling for screw machines is mounted and operated from front and rear stations on the cross slide and turret. Some machines are equipped with upper front and rear slides that are located in front of the spindle nose in a vertical plane and at an angle. The upper slides provide additional tool positions for rough and finish forming, knurling, rolling, and other processes.

153

Attachments provide added capability to cross drill, cross tap, and mill flats and slots. (Cross drilling and cross tapping refer to the drilling and tapping of holes at a right angle to the axis of the turned part.)

PLAIN SCREW MACHINE

Screw machines are designed to perform a single machining operation or several operations at one time or in a rapid sequence. The main slide, turret, and other slides of the *plain screw machine* are operated manually. By contrast, these components and the feeding of bar stock on the semiautomatic (wire feed) screw machine are operator controlled.

AUTOMATIC SCREW MACHINE

The *automatic screw machine* is fully automatic in operation. Once the cutting tools are set up on the cross slide and turret and the controlling cams are positioned, the stock is automatically advanced and clamped in the chucking device. The turret and tools are automatically indexed in sequence.

Single-Spindle Automatic Screw Machine. The *single-spindle* automatic screw machine is a general-purpose machine in which a single piece of stock is fed through the spindle at one time. The workpiece is machined in the sequence in which the tools in the indexing turret and on the cross slide are set up.

Usually, round, hexagon, square, and regular-shaped bar stock is processed on the single-spindle automatic screw machine. The bar stock is securely held in a spring collet chuck. The chuck is actuated by a friction clutch inserted between two pulleys on the spindle. The spindle drive allows the direction of rotation to be changed (reversed).

Multiple-Spindle Automatic Screw Machines. The *multiple-spindle* automatic screw machine has four, five, six, eight, or more work-rotating spindles. Additional production capacity is possible due to the increased number of operations that can be carried on at the same time on one machine.

MAJOR COMPONENTS OF AUTOMATIC SCREW MACHINES

A Brown & Sharpe ultramatic screw machine (Figure 11–1) is an example. However, all models include cams, levers, clutches, trip dogs, and collet or chucking devices. Bar stock in 10, 12, and 20 foot (250mm, 300mm, and 500mm) lengths is automatically advanced.

SPINDLE DRIVE MECHANISM

A constant-speed motor drives the speed and ratio change gears in a speed case. In turn, the gears drive the spindle through chains and sprockets. A small, friction-brake motor powers the drive shaft, chuck, and feed cam drive at the end of the machine.

SPINDLE

The spindle is mounted in antifriction bearings. The capacity for the Brown & Sharpe ultramatic screw machine (Figure 11–1) is from 3/4″ (19mm) to 1 5/8″ (41mm) diameter. Large-capacity #3 machines have either a 2″ (50mm) or 2 3/4″ (70mm) capacity. Two-speed and four-speed screw machines are available. There are 18 spindle speeds in the two-speed range from 20 RPM to 5,018 RPM for the 3/4″ (19mm) capacity machine. The #2 machine, 1 5/8″ (41mm) size, has 17 speeds in the two-speed range from 14 RPM to 2,906 RPM. High speeds and low speeds are produced in a clockwise and counterclockwise direction.

FEED CYCLE CHANGE GEARS

Feed cycle change gears (Figure 11–2) control the rate of production. The gears are installed at the right end of the machine. The safety cover provides for the automatic disengagement of the drive shaft.

The gears that are used are determined by the time (in seconds) required to machine one piece. The feed change gear plate on the machine shows the position of the driver and driven gears on the stud. Gear combinations are given for a range of time (in seconds) to machine one piece from 1.6 to 1,000. The right column on the gear plate shows the gross production per hour from 2,250 (1.6 seconds per piece) to 3.6 parts (1,000 seconds).

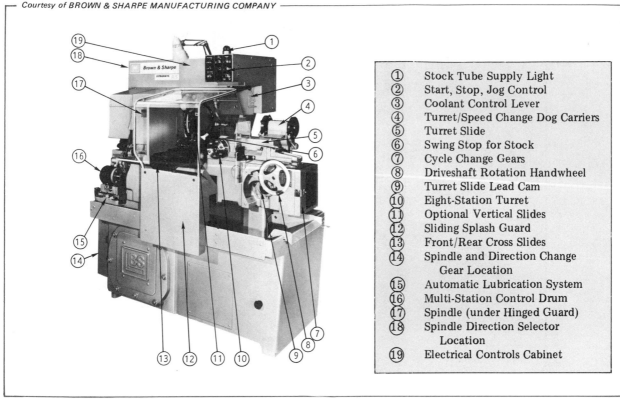

①	Stock Tube Supply Light
②	Start, Stop, Jog Control
③	Coolant Control Lever
④	Turret/Speed Change Dog Carriers
⑤	Turret Slide
⑥	Swing Stop for Stock
⑦	Cycle Change Gears
⑧	Driveshaft Rotation Handwheel
⑨	Turret Slide Lead Cam
⑩	Eight-Station Turret
⑪	Optional Vertical Slides
⑫	Sliding Splash Guard
⑬	Front/Rear Cross Slides
⑭	Spindle and Direction Change Gear Location
⑮	Automatic Lubrication System
⑯	Multi-Station Control Drum
⑰	Spindle (under Hinged Guard)
⑱	Spindle Direction Selector Location
⑲	Electrical Controls Cabinet

Figure 11-1 Major Features of an Ultramatic Screw Machine

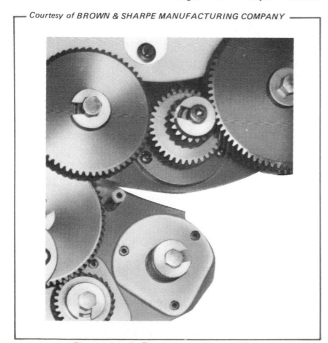

Figure 11-2 Feed Cycle Change Gears

TRIP LEVERS, CAMSHAFTS, DOGS, AND DOG CARRIERS

The chuck clutch trip lever controls the chuck clutch. The chuck clutch operates mechanisms through gearing to the chuck camshaft. A chuck and feed cam on this shaft operates a chuck fork (Figure 11–3). The fork causes the collet to open and then to close. The stock is automatically advanced and positioned by an adjustable feed slide. Adjustable dogs on dog carriers control the opening of the chuck before the stock is advanced during the stock feeding cycle.

An automatic operating mechanism controls such operational movements as indexing the turret and changing the spindle speeds. One camshaft carries the control drum, upper slide cams, chuck dog carrier, and the cross slide cams. A second camshaft carries the lead cam for the turret slide and the rapid pull-out cam.

Dog shaft carriers control the turret dog carrier (for timing the turret indexes) and the reversing dog carrier (for spindle clutch and spindle

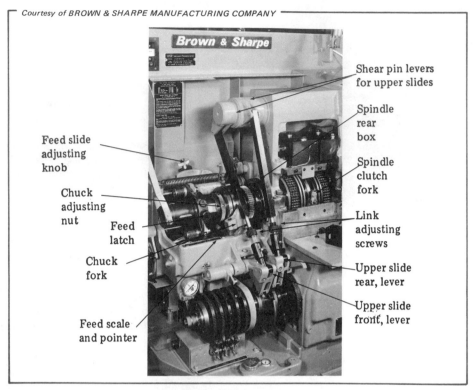

Figure 11–3 Spindle and Feed and Upper Slide Adjustments

speed changes). Adjustable trip dogs on the turret dog carrier are set to move the trip lever for single or double indexing. Single indexing is used to index each position of the hexagon turret. Double indexing is used to index every other tool position.

UPPER FRONT AND REAR TOOL SLIDES

Upper slides provide additional capacity for turning and cutoff processes. Figure 11–4 illustrates an upper front slide equipped with a forming tool post. This slide accommodates right- and left-hand turning tools. The upper rear slide has two cutoff toolholders. These are adapted for right- and left-hand "T" type cutting-off blades.

TURRET SLIDE

The turret slide may be moved toward or away from the spindle. Movement is controlled by the relationship between the lead lever, withdrawal cam, and the rack of the turret slide. The turret slide is provided with a positive stop

screw. This screw is used to maintain extreme accuracy of depth to close dimensional tolerances.

TOOLING FOR AUTOMATIC SCREW MACHINES

Standard tool bits are used with many holders. The cutters may be altered slightly to be converted for direct use. Tools for machining external surfaces are mounted on the turret. These tools include balance turning tools, box tools, knee tools, plain and adjustable hollow mills, and swing tools. End-of-work machining tools include centering and facing tools, pointing tools, and pointing toolholders for circular tools. Internal surfaces are cut with tools such as drills, reamers, counterbores, and recess cutters. These tools may be held in rigid or floating holders.

Internal threads are cut with solid and adjustable taps. External threads are cut with self-opening die heads or thread rolls mounted in cross slide toolholders. Knurling tools and top and side knurl holders are held on the cross slide.

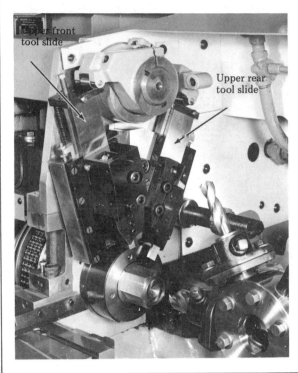

Figure 11–4 Upper Front and Rear Tool Slides

Other knurling tools and adjustable knurl holders are mounted on the turret. Circular cutting-off and forming tools are used in the regular cross slide tool post or an adjustable tool post. Thin, straight-blade cutting-off tools and square tools are supported on the cross slide.

The *turret-mounted tools* include: angular cutting-off tools, support and auxiliary tools, back rest for chuck, fixed and adjustable guides for operating spring tools, and spindle brakes.

GUIDELINES FOR AUTOMATIC SCREW MACHINE TOOL DESIGN AND SELECTION

Maximum efficiency in the operation of an automatic screw machine requires the exercise of judgment in tool selection and design. The operator should consider the following guidelines.

- *Straight rough-turning operations.* The balance turning tool is an all-around cutting tool. Plain hollow mills are used for fixed-size turn-

ing. Adjustable hollow mills permit turning within a range of diameters.
- *Straight finish-turning operations.* Turning box tools are adapted to finish turning.
- *Removing scale on hot-rolled bar stock.* The special knee tool is recommended for bar stock that does not have the external finish of screw-machined bar stock.
- *Straight turning long, slender workpieces.* A pointing tool of the box type steadily and rigidly supports the work in a sleeve ahead of the cutting tool. This setup prevents the work from deflecting under the force of the cut.
- *Taper and form turning a long, shallow, and continuous curve.* The swing tool is used for taper turning. It sometimes replaces a circular forming tool for other forming processes.
- *Regular form turning and straight turning behind shoulders.* Circular forming tools are recommended when the length of cut is limited in relation to the diameter of the workpiece. Circular forming tools are mounted on the cross slide for regular form turning and for straight turning behind shoulders.
- *Centering, drilling, reaming, and counterboring.* The particular tool for each process is held in a floating toolholder. This holder permits easy adjustment so that the centerlines of the cutting tool and of the workpiece are aligned.
- *Rounding off, chamfering, and pointing.* These processes are performed with a circular tool. This tool is held in a pointing toolholder. In pointing, the workpiece must be rigid so that it will not spring under the cutting forces.

OPERATION OF FULLY AUTOMATIC SCREW MACHINES

Bars are loaded in each of the hollow spindles on a fully automatic screw machine. Operations at each spindle are carried on simultaneously and in sequence. The end-of-work machining tools (except for the die head) are all fed as a unit by movement of the tool slide. The die head is fed independently. Similarly, the tools on the cross slide are fed separately by the drum cams. Figure 11–5 shows five of the six spindles on an automatic screw machine.

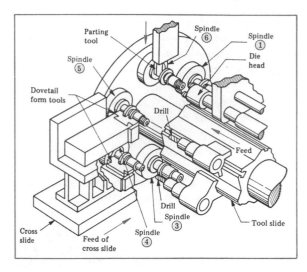

Figure 11–5 Schematic of Partial Setup on a Six-Spindle Automatic Screw Machine

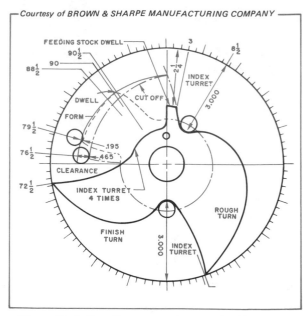

Figure 11–6 Cam Features and Design for Machining a Workpiece on an Automatic Screw Machine

At spindle ④, the dovetail form tool is used to rough form the head while the hole is partially drilled. Simultaneously, the previously drilled hole is drilled to depth and size, and the outside diameter is finish turned in spindle ⑤. Interestingly, at spindle ⑥, a separate feed is set for the die head. After threading, the cutting-off tool is fed from a vertical cross slide to part the workpiece. All of the end-of-work machining tools are fed as a unit by the movement of the tool slide.

Each spindle rotates as a single-spindle machine. Each spindle has a spring collet chuck. Cutting tools such as drills, reamers, counterbores, and taps are mounted in front of each spindle. Turning, forming, and cutting-off tools are secured on the cross slides. When an operation is finished, each spindle turns to the next position. The spindles are rotated until the cycle is completed.

FUNCTIONS AND DESIGN OF CAMS

Cams control the timing and position of all cutting tools, and the action of work-holding devices. Cams impart motion to the turret and the front, rear, and vertical cross slides. This motion controls the cuts. Cams are designed to *rise*, *fall*, or *dwell*—that is, they cause the cutting tools to advance, be withdrawn, or remain stationary.

The cam contour is also designed so that it revolves with the camshaft to enable it to return the tool slide to an inactive position. The operating cam is used to bring the turret to its operating position. Cam designs for cross slides and the turret must provide for starting and stopping each operation at the correct place and time.

An example of three superimposed cams is shown in Figure 11–6. The cams are used for rough turning, finish turning, and cutting-off operations. This particular cam is used to automatically machine a part on a Brown & Sharpe automatic screw machine. Note that the cam perimeter is divided into 100 parts. The speed of the camshaft determines the number of seconds represented by each division.

CUTTING SPEEDS AND FEEDS

Cutting speeds and feeds depend on the following:

- Material and shape of the workpiece.
- Type and size of screw machine.
- Material, type, and number of cutting tools.
- Design features of the toolholders.
- Depth of cut.
- Kind of cutting fluid.
- Required quality of surface texture.

CUTTING FLUIDS

As with all other machining processes and machine tools, a cutting fluid serves as a coolant and a lubricant. In addition, the cutting fluid must contain properties that permit it to separate rapidly from the chips.

Straight mineral oils are usually used as a blending medium for general-purpose screw machine operations. The mineral oil is mixed with a base cutting oil that possesses the necessary lubricating qualities.

Mineral lard oils are used on parts that require a higher quality surface finish than can be obtained by using straight mineral oils. Mineral lard oils have noncorrosive properties. These properties make mineral lard oils especially suitable for machining copper and copper alloys.

Sulphur cutting oils and *sulphurized oils* are recommended to increase tool life. Sulphur increases the cooling and lubricating characteristics of oils. Sulphurized oils wet surfaces faster than regular cutting oils—that is, sulphurized oils penetrate into remote places. Such oils have excellent film strength and help prevent chips from welding on the tool point.

Mineral oils are adapted for light machining operations on some steels and brass. Mineral oils are used for tapping and threading some nonferrous metals and for other difficult work.

RECENTLY DEVELOPED CUTTING OILS

The major claims for new, light-colored, transparent, and odorless cutting oils are as follows:

- The transparency of recently developed oils makes the machined part clearly visible to the machine operator during machining;
- The cutting oils have maximum lubricity, which results in reduced heat and tool wear;
- Antiweld qualities minimize buildup at the cutting edge of the tool for a considerable distance from the cutting edge;
- Increased extreme-pressure properties reduce the frictional heat developed by the rubbing chips on the extreme-pressure cutting areas;
- Exceptional cooling characteristics permit heat that is generated by the rubbing of the chips and workpiece to dissipate rapidly;
- The oils are corrosive and are recommended for all kinds of machining operations on all grades of steel. Noncorrosive oils are used for nonferrous metals.

Major oil producers have also developed special sulphurized and other cutting oils. These oils are all compounded to be noncorrosive and nonstaining. In addition to the previously listed claims, such cutting oils are claimed to produce equally satisfactory results on ferrous and non ferrous metals.

AUTOMATIC SCREW MACHINE ATTACHMENTS

Second machine operations are often eliminated by adding special attachments to the machine being used. Attachments are available for slotting

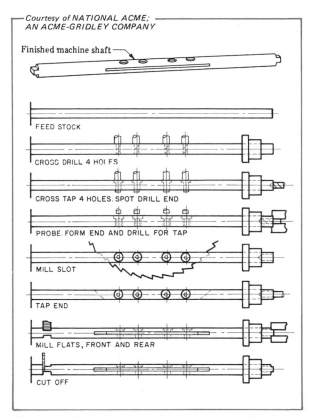

Figure 11-7 Applications of Attachments to Eliminate Second Machine Operations on a Shaft (Produced on an Eight-Spindle Automatic Screw Machine)

screws, burring, tapping nuts, rear-end threading, cross drilling (at right angles to the longitudinal machine axis), slotting, and machining flats. An example of a machine shaft produced on an eight-spindle automatic screw machine is shown in Figure 11–7. The drilling and tapping of the four holes and the milling of the center slot and the two slots on the end are performed on the same screw machine. Attachments are used for these operations. The production rate for this particular drilled, slotted, and threaded shaft is 200 parts per hour.

Features of and special functions performed by four common screw machine and chucking machine attachments follow. There are still other attachments that are designed for end-of-workpiece screw slotting and slabbing (parallel) cuts; spindle stopping, locating, and indexing; automatic second-operation loading; tool length setting; and additional operations that otherwise might require a second machine.

LONGITUDINAL TURNING ATTACHMENT

This device permits movements both longitudinally and "in-feed". Diameters may be turned behind a shoulder to a total tolerance of 0.0005″ (0.013mm). The slide on the attachment may be swiveled in two directions to turn tapers. Turning operations may be overlapped with turret tools.

SINGLE-POINT THREADING ATTACHMENT

This attachment (secured to the front cross slide) is designed for external and internal, right- and left-hand threading operations where a high degree of dimensional and concentricity accuracy is required. Single-point threading is used for thread sizes that exceed the capacity of a die head or thread rolls and when unusual and multi-start thread forms are to be produced.

Other turret, rear cross slide, and upper slide tooling may be operated simultaneously. The single-point threading attachment is gear driven from the machine spindle. A clutch (actuated by *dogs* on the *control drum*) starts and stops the attachment.

The movement of the threading tool into the work is controlled by the *front slide cam*. At the end of each "pass" (cut), the tool is withdrawn and the cross slide is retracted. Thread depth accuracy is controlled by a *positive cross slide stop*.

MILLING ATTACHMENTS

A *rear cross slide milling attachment* is designed for milling slots, forms, and flat areas at a right angle to a workpiece. A *brake* is used to hold the machine spindle stationary during the milling process.

A *turret milling attachment* is used for slotting, parallel milling, and form milling at the end of a workpiece (when the machine spindle is stationary). The cutter spindle assembly and the attachment are mounted and secured in any turret station that is provided with a motor drive. One or more cutters may be mounted on an attachment spindle. The milling cutters may be rotated in a horizontal or vertical plane.

OFFSET DRILLING ATTACHMENTS

Off-center holes may be drilled in the end of a workpiece with an *offset drilling attachment*. One or more of these attachments may be located in as many of the turret stations. Off-set holes are drilled with the machine spindle held in a stationary position.

The drill spindle is ball bearing mounted. Offset drilling attachments (such as the model shown in Figure 11–8) can be adjusted from "on center" to 13/32″ (10mm) "off center" on either side of the centerline. The drill capacity on the #00 screw machine is 3/32″ (2.4mm) diameter. A larger size offset drilling attachment is available for drill sizes up to 3/16″ (4.8mm) diameter for use on #2 and #3 screw machines and chucking machines.

Drill speed ranges to 4175 RPM for the #2 and #3 machines and to 4350 RPM for the smaller #00 machine are standard. However, there are slower and faster speed ranges available.

Cross Drilling Attachment. This attachment is designed for hole drilling at a right angle to the axis of the workpiece. The drill spindle is motor driven. The cross drilling attachment is positioned "on center" and is mounted on the back cross slide.

ADAPTATION OF ATTACHMENTS TO CHUCKING MACHINES

Features and functions of screw machine attachments and accessories are adaptable for automatic chucking machines. These machines are used for round and irregular-shaped castings, stampings, forgings, powdered metal formed, cold headed (requiring "over-the-head" chucking), and other parts that are not suitable for conventional screw machines.

The spindle is stopped for hopper or magazine loading. Standard and specially formed chuck jaws are usually air operated. Machined parts are ejected automatically.

Courtesy of BROWN & SHARPE MANUFACTURING COMPANY

Figure 11–8 Offset Turret Drilling Attachment

AUTOMATIC SCREW MACHINE TROUBLESHOOTING

Examples of automatic screw machine troubleshooting problems, causes, and corrective action are provided in Table 11–1. Some screw machine manufacturers furnish complete troubleshooting information for rod feeding and stocking; hole forming, turning, knurling, and other processes; turret adjustments, and accessories (Chart 1, Appendix).

Table 11–1 Examples of Automatic Screw Machine Problems, Causes, and Corrective Action (Partial Table)

Problem	Possible Cause	Corrective Action
Stock Stop Adjustment Variation in length	—Projection left because the cutting-off tool does not remove all of the teat.	—Reset the cutting-off tool on center.
	—Feed-out period on the lead cam is too short.	—Replace with the correct cam.
	—Turret cutting tools push the stock back into the collet.	—Sharpen each cutting tool. Check for correct feed.
	—Worn roll and pin on lead cam lever.	—Remove and replace the worn parts.
	—Chuck or feed finger tension too loose (or worn).	—Adjust the chuck to the required tension or replace a worn feed finger.
Turret Tooling *Knee Tools* Cutting with a taper	—Bar too small to use a knee tool; length of cut too long.	—Replace with a box tool, balance turning tool, or hollow mill.
	—Too much force required for the cut.	—Grind the correct clearance on the tool bit.
	—Tool not on center; too much feed; spindle speed too fast.	—Set tool on center or a few thousandths above; correct the feed and/or speed.

Safe Practices in Setting Up and Operating Screw Machines

- Use noncorrosive oils on machine tools where there is a possibility of leakage into the bearing system.
- Place enclosing guards around chucking and machining areas before operating a screw machine to prevent personal injury that can be caused by tool breakage, flying chips, or the workpiece being thrown out of a chuck.
- Check the proper placement of each tool and how securely each is held. In addition to personal injury, work damage may result from a tool working loose or being forced out of a holder.
- Shut down the machine if any tooling changes or setup adjustments are to be made.
- Safeguard bar stock that projects beyond the spindle. Guard rails are placed to cover the extended length in order to shield workers from a revolving workpiece. Note that sections of piping through which bar stock may be fed should be installed to prevent long bars from whipping around.
- Use short sheet metal or plastic screens to intercept and direct the cutting fluid to the chip pan, screens, and oil reservoir.
- Install a fine mesh, nonskid floor mat for worker protection against any oil that may be thrown from the screw machine.
- Change the sawdust or oil-absorbent compound around a screw machine as common practice. (Many departments have adopted nonslip, nonflammable floors for areas surrounding screw machines.)
- Wash hands and face thoroughly at regular intervals. Avoid contact with the cutting fluid as much as possible to prevent skin disorders that are caused by certain cutting fluids.
- Change work clothes frequently. Otherwise, they may become saturated with the cutting fluid.

TERMS USED IN SCREW MACHINE TECHNOLOGY AND PROCESSES

Magazine feeding	A device for holding, positioning, and automatically feeding precut, preformed, and/or premachined parts for chucking in a screw machine.
Single-spindle automatic screw machine	A general-purpose screw machine in which bar stock is automatically fed and machined. A one-spindle machine for multiple turning and cutting processes.
Multiple-spindle automatic screw machine	Four, six, eight, or more work-rotating spindles on one machine. An automatic screw machine that permits operations to be performed simultaneously on the workpiece on each spindle.
Feed cycle change gears	A gear train that permits gear combinations to be set up to control the rate of production. Gear combinations that control the time (in seconds) to machine one workpiece.
Chuck clutch trip lever	A lever that controls the automatic feeding and clamping of stock in a collet-type chuck.
Dog shaft carriers	Carriers that provide control of the turret indexes and spindle clutch and speed changes.
End-of-work machining tools	Cutting tools used for facing, center drilling, reaming, countersinking, and counterboring. Cutting tools that machine or are fed into a workpiece from the end.

Cam rise, fall, and dwell	Motion and rest imparted by a cam to operate the turret and front, rear, and vertical cross slides. Cam design features for rough, finish, form turning, and cutting-off operations.
Wet surfaces faster (sulphur cutting fluids)	Properties of sulphur cutting fluids that permit the coolant to penetrate into hard-to-reach places during cutting.
Maximum lubricity	A quality of a cutting fluid to maintain its lubricating properties under severe pressure and temperature machining conditions.
Second machine operations	Additional operations performed on a machine using special attachments. Operations normally performed on a second machine that are possible on a screw machine by adding attachments.
Troubleshooting	Recognizing a machine or tooling setup malfunction, diagnosing the problems and possible causes, and taking corrective steps.
Stocking and feeding	Mechanism and processes of opening and closing a chuck and automatically advancing stock on an automatic screw machine.

SUMMARY

- The three basic types of screw machines are: (1) plain, (2) automatic single-spindle, and (3) automatic multiple-spindle.
    - Screw machine tooling is mounted on a turret, on front and rear stations of the cross slide on front and rear tool slides, and on additional attachments.
- Screw machine spindles are usually mounted in antifriction, preloaded bearings. Spindle speeds range from 20 RPM to over 5,000 RPM.
    - Screw machine production rates are controlled by feed cycle change gears. The gear combinations permit varying the machining time (in seconds).
- Multiple-spindle automatic screw machines are highly productive. Additional operations may be performed, depending on the number of spindles. Usually, one operation is performed simultaneously for each spindle.
    - Cams control the advance, withdrawal, and stationary positions of cutting tools on the cross slide and turret.
- Chucks and collets are opened and closed after a workpiece is advanced and positioned (by an adjustable feed slide) by a chuck clutch trip lever.
    - Cam carrier shafts carry the control drum, slide cam, chuck dog carrier, cross slide cams, turret slide lead cam, and the rapid pull-out cam.
- External machining cuts are made with cutting tools held in box tools, balance turning tools, swing tools, slide tools, and knee tools. Fixed and floating holders, top and side knurl holders, and adapters are used.
    - The screw machine operator usually troubleshoots problems in five general categories: (1) stocking and rod feeding; (2) stock stop adjustment; (3) basic hole forming processes such as drilling, reaming, counterboring, threading, and knurling; (4) turret tooling problems with knee, box, and balanced turning tools and with the use of hollow mills; and (5) turret and cross slide.

- Cam designs are based on the nature of a workpiece, machine processes, spindle RPM for each operation, RPM required to machine one part, necessary tool travel, and feeds.
  - Speeds and feeds for screw machine processes are governed by the same factors as for similar cutting processes on other machine tools.
- Straight mineral oils, mineral lard oils, and sulphurized oils are recommended for general-purpose processes, for machining tough noncorrosive metals, and for deep penetration.
  - New, light-colored, transparent, and odorless cutting oils provide maximum lubricity, dissipate heat rapidly, maintain properties under severe pressure and temperature conditions, and provide the same cutting and lubricating properties on ferrous and nonferrous metals.
- Special attachments permit second machine operations such as slotting, tapping, rear-end threading, cross drilling, and machining flats.
  - Personal safety relates to the use of guards and enclosures and protective goggles, checking tool setups, protection against revolving overhanging bars, and hygiene in removing cutting fluid from the skin and clothing.

UNIT 11 REVIEW AND SELF-TEST

1. List three attachments that permit additional second machine operations to be performed on a screw machine.

2. Describe briefly the function of (a) feed cycle change gears, (b) the chuck clutch trip lever, and (c) the dog shaft carriers of an automatic screw machine.

3. a. Name five standard tools for automatic screw machines.
   b. Give the major process for which each tool is designed.

4. Indicate five workpiece specifications (in addition to dimensions) that are required for the designer to lay out screw machine cams.

5. State six conditions that control screw machine speeds and feeds.

6. a. Identify four properties of new screw machine cutting oils that make them different from standard mineral oils.
   b. Describe the importance of each property in relation to either tool wear life, quality of surface finish, heat removed, or machining.

7. Record the corrective steps to follow in troubleshooting the following screw machine problems:
   a. Stock stop adjustment: Scored end of workpiece; stock stop center hole too large.
   b. Turret tooling: (1) Knee tool cutting with a taper; tool not on center. (2) Second cutter of balanced turner back of center; shoulder steps produced.

8. Describe three safety precautions an operator should observe before starting a screw machine.

PART 4 Milling Machines: Advanced Technology and Processes

Vertical Milling Machines

This Unit deals with precision boring heads, tooling and machine setups, and the machining of accurately bored holes; principles and applications of baseline, ordinate, and tabular dimensioned drawings; and the use of the vertical shaper attachment for hole forming operations.

Hole Forming, Boring and Shaping

OBJECTIVES

After satisfactorily completing this unit, you will be able to:

- Apply standard work-holding and positioning tools and devices for drilling, reaming, boring, counterboring, countersinking, tapping, and performing shaper operations on the vertical milling machine.
- Identify design features of the offset boring head.
- Determine cuttings speeds and feeds for hole-forming and shaping processes.
- Understand causes and treatment of tool deflection problems.
- Establish shaper head and cutter stroke settings of a shaper head.
- Perform right angle, simple angle, and compound angle work processes.
  - Machining Holes (conventional vertical head).
  - Boring Holes with an Offset Boring Head.
  - Simulate Shaping Cuts Using the Vertical Shaper Attachment.
- Solve hole-forming problems and setups which depend on baseline, ordinate, and tabular dimensioning.
- Follow each recommended tool, machine attachment, and personal *Safe Practice* and correctly use each related *Term*.

Design features of the vertical milling machine permit positioning a workpiece precisely by the longitudinal and cross slide movements of the table. These movements fall along the X axis and the Y axis. The vertical movement of the knee makes it possible to set and machine a workpiece according to dimensions on the Z axis. The graduated micrometer collars on the cross slide, table, and knee provide a quick, accurate, precision method of locating cutting tools and a workpiece. The micrometer collars permit direct measurements to accuracies within 0.001″ (0.02mm). Vernier, bar, and indicator attachments extend the accuracies to 0.0001″.

The vertical milling machine is uniquely adapted to *hole-forming processes* such as drilling, reaming, countersinking, counterboring and tapping with basic cutting tools, and to *boring*.

Still other processes may be performed on the vertical milling machine by using standard attachments and accessories. The dividing head, rotary table, sine table, and angle plates are commonly used in toolrooms and custom jobbing shops. The shaper head attachment extends the machine capability to include *shaping processes* that otherwise would have to be performed on a second machine.

Hole and other surface locations and dimensions for milling processes are often specified by ordinate, tabular, and baseline dimensioning. Essential layout and dimensional information as given in relation to the X, Y, and Z axes (Figure 12–1) is covered later in the unit.

WORK- AND TOOL-HOLDING DEVICES FOR BASIC HOLE-FORMING PROCESSES

In drilling, reaming, counterboring, countersinking, and tapping processes, the same construction and sizes of drills, machine reamers, counterbores, countersinks, and machine taps that are applied to drilling machines, horizontal milling machines, lathes, and screw machines are used on a vertical milling machine. These fixed-diameter, multiple-edge cutting tools may be held in a chuck, adapter, or collet. Taps are turned for threading and removing from a workpiece by a tap adapter or a reversing tapping attachment.

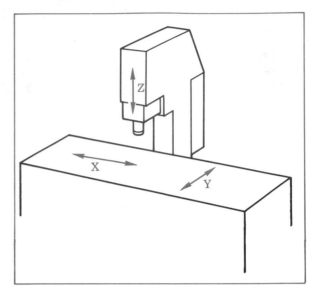

Figure 12–1 Three Basic Axes of Vertical Milling and Numerically Controlled Machines and Machining Centers

In basic hole-forming processes, four methods are commonly used to position and hold workpieces on a vertical milling machine. These methods require the use of:

- A standard plain, swivel, or universal vise;

- Straps, parallels, step and V-blocks, T-bolts, and clamps to secure workpieces in an aligned position;

- Fixtures for nesting, positioning, and holding parts;

- Stops that fit snugly into the table T-slots against which a workpiece is aligned and held parallel to the column face.

As in horizontal milling processes, a dial indicator is used to accurately align the solid jaw of a vise. A dial indicator may also be used to align the edge of a workpiece precisely to be parallel or at a right angle to the table axis. In alignment applications, the dial indicator is fastened to the vertical spindle to indicate a machined locating surface.

FACTORS AFFECTING HOLE-FORMING PROCESSES

In vertical milling, the material of which a hole-forming cutting tool is constructed, its cutting speed for a particular process and workpiece, feed rates, and cutting action are all influenced by the same factors that apply to drill press, horizontal miller, and lathe work. However, the following additional factors must be considered:

• The vertical head may be positioned at a single or compound angle. Holes may be formed with the spindle axis (and the axis of the cutting tool) set to cut at a required angle.

• The table may be moved longitudinally or transversely to any precise dimension. Hole locations to tolerances of 0.0001″ (0.002mm) are possible by using measuring rods and a tenth dial indicator.

CUTTING SPEEDS/FEEDS FOR BORING TOOLS

Cutting speeds of high-speed steel, carbide-tipped, and solid carbide boring tools require consideration of the same factors that apply to cutting speeds for other similar machining processes. In addition to the usual rigidity requirements, if an offset boring head produces chatter and excess vibration, the spindle RPM may need to be reduced.

Boring is often used to correct eccentric, roughly formed, bell-mouth holes that may not be parallel with a central axis. The depth of the first cut or cuts, therefore, varies around the periphery of a previously formed hole (Figure 12–2). This variable depth continues until the bored hole is concentric with the spindle axis. The cutting action for the first few cuts may require a reduction in cutting speed.

A boring tool may be hand-fed and/or power-fed by the quill. The coarser quill feeds of 0.003″ and 0.006″ (0.1mm and 0.15mm) per revolution are used for general roughing cuts. Finish cuts are taken at the finest quill feed of 0.0015″ (0.04mm) per revolution. At the end of a finish cut, some experienced operators reverse the feed direction. The boring tool is then fed back up

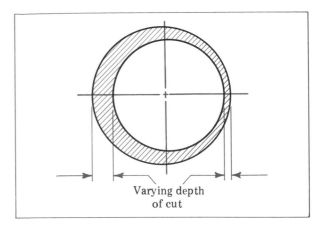

Figure 12–2 Varying Depth of Cut Produced by Eccentrically Formed Hole

through the workpiece without resetting the cutting-tool position. In general practice, the boring tool is usually withdrawn from the hole at the end of the cut by stopping the spindle and moving the cutting edge clear of the finished surface. Otherwise, a fine, spiral groove may be cut into the work surface as the spindle turns.

SELECTION OF BORING TOOLS

Operator judgment is required in the selection of boring tools. Selection of the shortest and heaviest boring bar or boring tool should be consistent with the bore diameter and hole depth, the kind of material being bored, the nature and quantity of the cutting fluid, and the process itself.

CONSIDERATION OF TOOL DEFLECTION

The deflection of the cutting tool is an important consideration in establishing the depth of cut, cutting speed, feed, and dimensional and surface finish accuracy. Tool deflection may be caused by any one or combination of the following factors:

• Inaccurately ground or worn and dulled cutting edge.
• Too deep a cut for the body size of the cutter.

- Too heavy a feed.
- Too high a cutting speed for the boring process.
- Improperly supported and secured workpiece.
- Incorrect cutting fluid.
- Improperly positioned or unlocked tool slide.
- Interrupted cutting, as in the case of boring a radius.

BORING WITH THE OFFSET BORING HEAD

The *offset boring head* is used to bore a hole to any standard or other precise size. Boring produces straight and concentric holes about a fixed center. The quality of the surface texture may be controlled by the form and sharpness of the boring tool, depth of cut, and feed rate. Boring holes that are larger than 1″ in diameter is more economical than maintaining a whole inventory of drills, machine reamers, and other cutting tools to cover a wide range of sizes.

BORING SHOULDERED SURFACES

A combination boring and facing head is used to form a shoulder as in counterboring. Holes are bored through a workpiece or to a specific depth. Holes are counterbored by boring the side walls to depth. The shoulder area may then be squared by feeding the end-cutting edge of the boring tool toward the center of the hole. A combination boring and facing head is also used to machine a stepped circular area (Figure 12–3). The same cutter is used to bore the hole to size and to machine the flat base.

MACHINING A RADIUS

The boring head is also commonly used in applications where an accurate radius is to be machined into a surface. In such applications, the workpiece is strapped to the table or held in a fixture. The centerline for the radius is located at the axis of the boring head and spindle. The table is also positioned. The center of the hole to be bored is centered as close as practical.

A scrap plate is clamped opposite to the workpiece. Its edge is the same distance from the center (boring head axis) as the edge of the workpiece. A trial cut is taken with the offset boring head for about 1/8″ (3mm) deep. The diameter is measured and the cutting tool is adjusted for depth of cut. One or more roughing and finish cuts may be needed to produce the required radius.

THE OFFSET BORING HEAD AND CUTTING TOOLS

Offset boring heads are available in a number of different models and sizes. Adjustments for machining diameters in increments of 0.001″ (0.02mm) are made rapidly by moving the cutting tool laterally in the holder. The amount of movement is read directly on a micrometer dial on the boring head. Microboring heads permit diametral adjustments within 0.0001″ (0.002mm).

FEATURES OF THE OFFSET BORING HEAD

The major construction features of an offset boring head are illustrated in Figure 12–3. The micrometer dial is graduated into 50 divisions that permit adjustments to 0.0005″ (0.01mm). There is a positive lock when the tool slide is set at a required size.

The boring head set includes three *boring bars.* Two of these bars permit the use of standard-size, round boring tools. The third bar holds the different sizes of solid-shank boring bars in a set. The two *reducer bushings* accommodate two different sizes of boring tools. Like the boring bars, the reducer bushings are held securely in a particular cutting-tool location in the tool slide by a setscrew.

SPECIFICATIONS OF SELECTED BORING HEAD FEATURES

The boring head in Figure 12–3 is furnished with shanks that fit R–8 and #30, #40, and #50 NS spindles. The diameters that may be bored range from 1/4″ to 6″ (6mm to 150mm) for the R–8 size. The largest #50 NS shank boring head accommodates hole sizes from 5/16″ to 11 1/2″

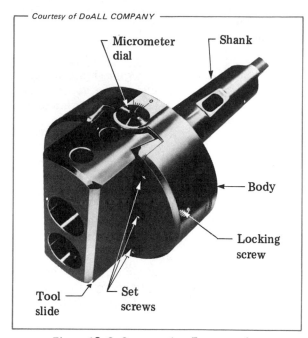

Micrometer dial

Shank

Body

Locking screw

Tool slide

Set screws

Figure 12–3 Construction Features of an Offset Boring Head

(8mm to 292mm). The total travel of the tool slide varies for each head. The N–8 size has a travel of 31/32" (24mm). The #50 NS shank boring head has a travel of 1 11/16" (42.8mm).

The R–8 and the #30 NS shank boring heads hold 1/2" (12.5mm) diameter boring bars. The #40 NS and #50 NS heads accommodate 5/8" (15.8mm) and 3/4" (19mm) diameter bars, respectively.

How to Machine Holes by Using a Conventional Vertical Milling Machine Head

STEP 1 Select an appropriate cutter holder, adapter, collet, or tapping unit. Secure the tool-holding device in the vertical spindle.

STEP 2 Mount and secure the cutting tool.

STEP 3 Determine the correct cutting fluid, if needed. Set the nozzle to direct the quantity and flow of cutting fluid.

STEP 4 Position the cutting tool at the center of the hole. Lock the knee, cross slide, and

table. Set each of the micrometer collars at zero.

Note: Ordinate, tabular, and baseline dimensioned drawings give each dimension from a zero position on the X, Y, and/or Z axes. The operator usually sets the table, cross slide, and knee micrometer collars at zero. All table movements to position the cutting tool relate to the zero settings.

STEP 5 Proceed to carry on the hole-forming processes by following the steps that are used on a drilling machine, lathe, or screw machine.

How to Bore Holes with An Offset Boring Head

Boring a Predrilled or Formed Hole Vertically (0°)

STEP 1 Align the vertical spindle head at 0°.

STEP 2 Select an offset boring head to accommodate the size and type of tool bit needed for the job. Check for and remove any burrs. Mount and secure the boring head in the spindle.

STEP 3 Mount the boring tool so that the cutting face is aligned with the center line of the head. Tighten the cutter setscrew.

STEP 4 Align the spindle (and boring head) axis with the zero coordinates on the X and Y axes. Set the cross slide (saddle) and table feed micrometer collars at zero.

STEP 5 Move the table horizontally and transversely to the required center line dimension.

STEP 6 Adjust the boring tool for a trial cut. Take the cut for about 1/16" (1mm to 2mm) deep.

Note: The cut should clean up to more than half the diameter to permit taking an accurate diametral measurement.

STEP 7 Set the micrometer dial on the boring head at zero. Turn the micrometer dial adjusting screw on the boring head to feed the cutter for a roughing cut. Lock the tool slide on the boring head.

STEP 8 Bring the cutter into contact with the workpiece. Continue feeding by engaging the power quill feed.

Note: The proper cutting fluid and volume must be used to flow the chips out and away from the area being bored.

STEP 9 Return the boring tool to the starting position. Stop the machine. Remeasure the bored diameter.

STEP 10 Set the boring tool for subsequent roughing and/or finish cuts.

Note: The feed rate is decreased to obtain a higher-quality surface finish. Some operators use the down-feed and, at the end of the cut, change to an up-feed without changing the finish cut position of the boring tool.

Note: The cutting edge must be moved clear of the workpiece if it is brought up and out of the bored hole on the final cut.

Drilling and Boring a Hole

STEP 1 Select a drill that is smaller than the diameter of the hole to be bored.

STEP 2 Mount the drill in a tool-holding device. Secure it to the spindle.

STEP 3 Set the spindle RPM and feed rate for drilling the first hole. Position the drill at the center of the hole. Drill.

STEP 4 Remove the drill and adapter. Replace with an offset boring head and boring tool.

STEP 5 Proceed to bore the hole as for conventional boring.

Boring a Hole at a Simple or Compound Angle

STEP 1 Position and lock the vertical milling machine head at the required angle in the plane that is parallel to the column face. Read the angle setting on the graduated quadrant of the head.

STEP 2 Proceed to bore the hole to size.

Boring a Radius or Stepped Circular Area (Figure 12–4)

STEP 1 Insert a centering rod in a boring head bushing or a chuck. Move the table longitudinally and transversely until the spindle axis is aligned at the layed out center line at the end (Y datum) of the workpiece.

STEP 2 Move the table longitudinally on the center line. The distance moved along the X axis should center the boring head at the center line of the radius to be bored.

STEP 3 Insert a tool bit, with correctly ground rake and clearance angles, in a boring bar. Mount the tool bit so that the cutting edge is aligned with the center axis of the boring head tool slide block.

STEP 4 Move the boring cutter until it grazes the end of the workpiece. Set the micrometer adjusting screw at zero.

STEP 5 Feed the tool outward for a trial cut. Take the cut for approximately 1/16″ (1mm to 2mm) deep.

STEP 6 Stop the spindle. Measure the diameter.

STEP 7 Increase the depth of cut, if possible. Use a coarse feed for a roughing cut.

Note: The depth of cut (amount the boring tool is moved on the tool slide) is established by the micrometer dial reading.

STEP 8 Reset the cutting tool for subsequent roughing and finish cuts.

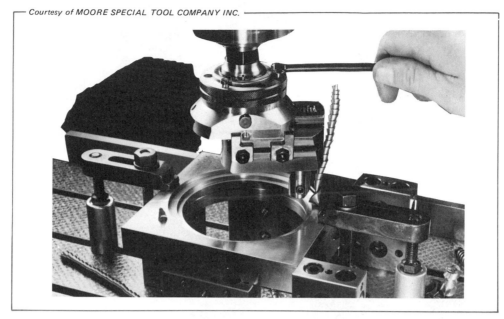

Figure 12–4 Use of Boring and Facing Head to Machine a Stepped Circular Area

Boring a Counterbored Area or a Flat Shouldered Surface

STEP 1 Mount a boring and facing head on the quill so that the boring tool is driven by the spindle.

STEP 2 Select a boring tool that is ground to produce the specified square, round, or angular corner for the shoulder.

STEP 3 Take a series of cuts to bore the outside diameter to depth and to remove the excess material within the area to form the flat face.

Note: The cutting tool is fed downward to depth by the quill feed handwheel. The flat shouldered surface is cut by turning the boring and facing head handwheel to feed the boring tool inward.

STEP 4 Remove the chips. Clean and burr the workpiece. Measure the diameter and depth. Check the surface finish against a comparator specimen.

SHAPING ON A VERTICAL MILLING MACHINE

The vertical shaper head attachment is used to cut internal keyways, slots, and splines. The sides of openings and other intricate forms may be shaped on jigs, fixtures, punches, dies, and other workpieces that are to be precisely machined. Gear teeth and racks may be cut with the shaper head.

Some heads are adapted to blind-hole machining processes. The shaper head (Figure 12–5) often replaces a second machine—for example, a broaching machine for broaching.

The cutting tools are formed for side and end cutting, with corresponding rake angles and clearance. The tools are mounted in the tool head. A cutting tool is positioned by the table and cross slide handwheels.

Vertical positioning of a workpiece is controlled by elevating or lowering the knee. The length of stroke is adjusted directly on the shaper head. The attachment may be set for shaping at a 0° vertical angle or at any other

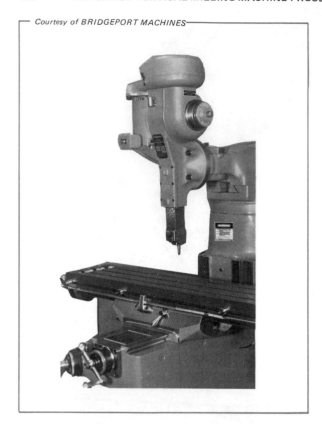

Figure 12–5 Vertical Shaper Head Attachment

single or compound angle. The settings are made directly. The angle is read on the graduated shaper head and/or the ram, depending on the job specifications.

Surfaces that require shaping in a circular direction are positioned, mounted, and secured on a rotary table or by using a dividing head. The accessories permit workpieces to be indexed or turned in relation to other vertical milling processes.

The shaper head (in general) has a stroke capacity of 0″ to 4″ (0mm to 102mm). The length of stroke can be dialed in increments of 1/8″ (3mm). The head is available with a slow speed and a standard speed range (strokes per minute). The slow speed range is 35, 50, 75, 100, 145, and 210 strokes per minute. The standard speed range is 70, 100, 145, 205, 295, and 420 strokes per minute.

How to Use a Shaper Attachment on a Vertical Milling Machine

STEP 1 Replace the vertical spindle head on the ram with a self-contained shaper head.

Caution: Place a wooden pad on the table under the vertical spindle. Secure assistance in handling both the vertical head and the shaper head.

STEP 2 Position and lock the shaper head at 0° or whatever simple or compound angle is required.

STEP 3 Select the size and style of cutting tool to accommodate the cuts to be taken. Mount the cutting tool.

Note: Shaper tool sets are commercially available to machine square, angular, and round forms and combinations of these forms as corners and flat or irregular surfaces. Additional tools are forged for special applications.

STEP 4 Determine the cutting speed (number of strokes per minute). Set the shaping head dial at this cutting speed.

STEP 5 Position the cutting tool a short distance up from the top face of the workpiece to provide clearance for taking measurements. (The operator may also easily see the layout lines and cutting action.)

STEP 6 Set the length of the ram stroke.

Note: The distance must permit the cutter to clear the workpiece at the top and bottom end of each stroke. In the case of cuts to a specified depth, the ram is positioned to bottom out at a specific depth.

STEP 7 Locate the cutting edge of the cutter at the depth for the first cut.

STEP 8 Take a trial cut for a short distance. Measure the workpiece. Adjust the cut, if needed.

Note: At this step, the operator must determine whether any adjustments are needed. These adjustments relate to depth of cut, feed rate, and cutting speed.

Note: The heaviest possible cutting tool is used for efficient cutting with the least amount of deflection.

STEP 9 Reset the depth of cut and take successive cuts to rough out the required shape.

Note: Large areas are usually roughed out by ending each successive cut near a corner in a series of steps. When the surface is roughly shaped to depth, the excess corner material is cut away.

STEP 10 Replace the roughing cutter with a cutter ground for finish cuts.
STEP 11 Relocate the cutting edge at the roughed-out edge. Feed to depth.
STEP 12 Take a trial cut. Measure the surface. Adjust the cutter, as needed. Take the finish cut.
STEP 13 Reset the machine for other shaping or additional vertical milling operations that may be required. Then, burr the workpiece.
STEP 14 Follow standard machine and tool cleanup and storage practices.

Shaping and Indexing Radial Surfaces

STEP 1 Mount a rotary table centrally on the vertical milling machine table.
STEP 2 Position the workpiece on the rotary table. The location must permit feeding and revolving the workpiece at a specified radius and angle setting.

Note: The workpiece should be positioned in relation to the zero degree reference graduation setting on the rotary table.

STEP 3 Position the cutter in the shaper-tool holder to permit shaping the required form.
STEP 4 Feed the cutter to depth for a first cut.

Note: A formed cutter is usually used for spline cutting. The cutter is fed to depth by taking a series of cuts in the path of the first cut. The workpiece is then indexed to the next spline, positioned with the rotary table handwheel. The regular table movement and shaping are continued.

Note: The cutter is fed along the required radial path by feeding the rotary table handwheel for each cutting stroke.

STEP 5 Continue to take a series of roughing cuts. Also, shape away any stepped areas that may be formed at the corners.
STEP 6 Replace the roughing cut cutter with a finely ground finish shaping tool.
STEP 7 Reposition the cutter. Feed to the depth of the finish cut. Measure and adjust, if required.
STEP 8 Turn the rotary table handwheel to feed the workpiece along the required radial path.
STEP 9 Remove and burr the workpiece if no additional machining operations are to be performed on the machine.
STEP 10 Leave the machine clean. Return the rotary table and all tools to proper storage places.

BASELINE, ORDINATE, AND TABULAR DIMENSIONING

BASELINE DIMENSIONING

Baseline dimensioning is used on part drawings that require precision layout and/or positioning on a machine. All measurements are related to one or more common finished surfaces. These surfaces are called *baselines*. When successive measurements are taken from a single baseline, dimensional errors are not cumulative.

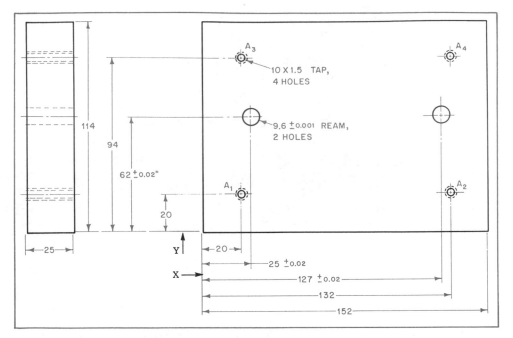

Figure 12-6 Drawing Showing Simple Application of Baseline Dimensioning

Baseline dimensioned drawings simplify the positioning of a workpiece from corresponding X, Y, and Z machine axes (Figure 12–6). The micrometer collars on the horizontal and cross slide feed screws, on the table and knee, are used to position a workpiece for successive cuts from baseline dimensions.

The drawing in Figure 12–6 shows a simple application of baseline dimensioning. Here, the two baselines are at a right angle to each other. Baseline dimensioning is often applied with the baseline as a center line. Measurements are then taken from a fixed plane position. Baseline dimensioning may also be applied in laying out an irregular shape (Figure 12–7).

DATUMS

A *datum* is a feature from which dimensions are referenced. A datum may be a plane, line, or other exact reference point. Generally, shop and laboratory drawings have two datums. The datums are identified as *zero coordinates*. All mea-

surements are taken from mutually related zero coordinates (datum planes). For example, in baseline dimensioning, measurements are taken from X and Y datums.

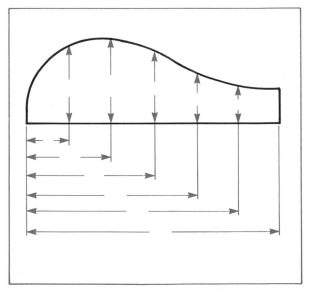

Figure 12-7 Baseline Used in Dimensioning Irregular Form

ORDINATE DIMENSIONING

Ordinate dimensioning differs from baseline dimensioning in that dimension lines and arrowheads are not used. Figure 12–8 is an example of ordinate (rectangular datum) dimensioning. All dimensions are related to the zero coordinates (datums) along the X and Y axes. Each dimension appears at the end of the extension line of a feature to which it refers. A feature is located by the intersection of two datum dimensions. For example, the center of one A hole in Figure 12–8 is 0.750″ from the X axis and 0.625″ up from the Y axis.

Ordinate dimensioning, like baseline dimensioning, permits greater reading and machining accuracy than does a conventionally dimensioned drawing. Dimensional errors are controlled and not multiplied.

TABULAR DIMENSIONING

Tabular dimensioning is another form of simplified, accurate dimensioning. All dimensions are related to datums and originate at zero coordinates. Tabular dimensioning is used for milling machine, jig borer, numerically controlled, machining center, and other machining processes. Tabular dimensioning is recommended when a great number of common features make a con-

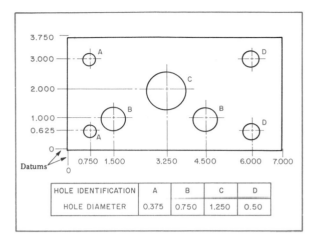

Figure 12–8 Example of Ordinate Dimensioning

HOLE IDENTIFICATION	A	B	C	D
HOLE DIAMETER	0.375	0.750	1.250	0.50

ventionally dimensioned drawing difficult to read. Figure 12–9 is an example of tabular dimensioning.

The dimensions of each feature on a tabular dimensioned drawing are given in a table, or *coordinate chart* (Figure 12–9). The dimensions of each hole are given as they originate at the X and Y datums. The table also provides sufficient information about a feature to permit accurate machining. Tabular dimensions may be given in inch and/or metric units of measure.

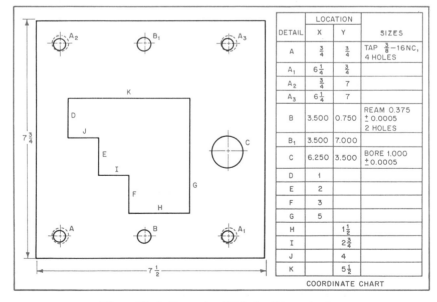

DETAIL	LOCATION		SIZES
	X	Y	
A	$\frac{3}{4}$	$\frac{3}{4}$	TAP $\frac{3}{8}$–16NC, 4 HOLES
A$_1$	$6\frac{1}{4}$	$\frac{3}{4}$	
A$_2$	$\frac{3}{4}$	7	
A$_3$	$6\frac{1}{4}$	7	
B	3.500	0.750	REAM 0.375 ± 0.0005 2 HOLES
B$_1$	3.500	7.000	
C	6.250	3.500	BORE 1.000 ±0.0005
D	1		
E	2		
F	3		
G	5		
H		$1\frac{1}{2}$	
I		$2\frac{3}{4}$	
J		4	
K		$5\frac{1}{2}$	

COORDINATE CHART

Figure 12–9 Example of Tabular Dimensioning

Safe Practices for Using Boring and Shaper Tools and Attachments

- Start with the recommended starting cutting speed and feed. Adjust the speed and feed rates, as the cutting action may require, after all factors are considered.
- Check the workpiece for effects on surface texture and dimensional accuracy caused by any eccentric forces of the boring head. Reduce the spindle RPM if vibration or chatter occurs.
- Set the depth of cut for a boring tool or shaper tool so that the cutting forces will not deflect the cutter and cause an incorrect work surface. The feed may also need to be reduced.
- Rotate the boring tool and head slowly and carefully when the cutting edge is set to take the first cut in forming a radius.
- Move the cutting tool away from the finish bored surface before it is withdrawn from a workpiece.
- Place a wooden pad over the table surface to protect it from possible damage when removing or replacing a vertical spindle head with a shaper head. Two persons should be involved in removing any cumbersome and heavy attachment, accessory, or workpiece.
- Set the cutting stroke length to permit the shaping tool to clear the workpiece at the end of the cut. Adequate clearance must be provided in the starting position to take measurements safely.
- Stop the machine spindle before any measurements are taken. The workpiece must be clean and burred.

TERMS USED IN HOLE FORMING, BORING, AND SHAPING

Vertical axis (0°) boring	Positioning the vertical milling machine head in a zero position (at a perfect right angle to the table) and boring.
Offset boring head	A tool-holding device with a shank that fits the machine spindle. (Cutting tools are held in boring bars, reducer bushings, or directly in a tool slide. The cutter is positioned for boring by moving the tool slide with a micrometer adjusting screw).
Boring head set	An offset boring head with different sizes of boring bars, two reducer bushings, and wrenches.
Feed increments (boring head)	The amount a tool may be fed for a cut. The distance in 0.001″ or 0.0001″ (0.02mm and 0.002mm) that represents the movement of the slide between two graduations on the micrometer dial of a boring head.
Cleaning up the surface (boring)	Machining a previously formed surface. Removing tool marks or other surface imperfections while correcting inaccuracies such as eccentricity, bell mouth, and so on.
Stepped circular area	A radial area that is bored below another surface, usually a horizontal plane.
Shaping splines	The process of shaping a series of grooves that are equally spaced around a circumference.

Datum   A precise point from which all dimensions and measurements are referenced.

Ordinate dimensioning   Dimensions placed at the end of an extension line without using dimension lines or arrowheads. Each dimension relates to zero coordinates along one or more axes.

Zero coordinate   A starting point from which all dimensions along a specific axis originate.

Coordinate chart   A table accompanying a part drawing. A table that gives dimensions to locate each feature with respect to zero coordinates.

SUMMARY

- Cuts may be taken at any simple or compound angle from 0° (right angle to the milling machine table) to a horizontal plane angle (180°).

  - The longitudinal and cross slide table and the vertical knee micrometer collars or programmed readouts are used in positioning hole-forming and shaping processes.

- The combination boring and facing head extends the range of machining operations. The boring process may be followed by feeding a cutting tool across the face to form a shoulder as in counterboring or producing a stepped surface.

  - Boring head shanks are commercially available to fit R-8, #30 NS, #40 NS, and #50 NS spindles.

- The same factors governing the selection and use of cutting tools for lathe work, drilling machines, and horizontal milling machine processes apply to boring tools. Consideration is given to rake, relief, and clearance angles and the shape of the cutting edge.

  - Quill feeds of 0.003″ and 0.006″ (0.08mm and 0.15mm) are used for roughing cuts. The 0.0015″ (0.04mm) quill feed is usually used for finish cuts.

- The following causes of tool deflection must be considered and corrected: worn or incorrectly ground cutter; too high a cutting speed and/or too coarse a feed and/or too deep a cut; poorly supported or secured workpiece; incorrect quality and quantity of cutting fluid, and interrupted cut and/or movement of the tool slide.

  - The vertical spindle head accessory is used with cutting tools that require rotary and/or planetary motion, as in the case of a facing process.

- The shaper head attachment permits second machine processes to be performed on the vertical miller. Through- or partial-depth surfaces may be shaped vertically or at any simple or compound angle.

  - The ram on the shaper head may be adjusted for position in relation to the workpiece. Cutting strokes may be varied from 0″ to 4″ (0mm to 102mm) on toolroom models.

■ The shaper head is adapted to perform operations that often require special broaches and to cut gear teeth, racks, and other intricate shapes.

- ■ Shaper heads are designed to operate with any one of six speeds within a slow speed range of 35 to 210 strokes per minute. High speed ranges of from 70 to 420 strokes per minute are also available.

■ Baseline, ordinate, and tabular dimensioned drawings are used to accurately lay out a part, indicate the amount of table movement, and take measurements. Dimensions are related to datums and X, Y, or Z axes.

- ■ Radial surfaces may be shaped or holes may be positioned and formed by using a rotary table or a dividing head.

■ Standard sine bars and sine plates and other work-positioning accessories and tools are applied to hole-forming, boring, and shaping processes.

UNIT 12 REVIEW AND SELF-TEST

1. a. Identify three basic hole-forming processes that are common to vertical milling machine work.
    b. Indicate the cutting tool that is used for each of the three processes.
    c. Name the tool-holding device for each cutting tool.

2. Explain how the T-slot grooves on a vertical milling machine table may be used to position a rectangular workpiece for work processes that are parallel to the table axis.

3. State three advantages of using a vertical milling machine instead of using a drilling machine for drilling, reaming, counterboring, countersinking, and tapping.

4. a. Name three design features of an offset boring head.
    b. Explain the function served by each feature.

5. a. Secure the specifications of an offset boring head.
    b. Furnish technical information about the following design features: (1) offset boring head manufacturer, (2) shank style and size, (3) draw-in bar thread specifications, (4) boring diameter range, (5) boring bar set lengths, (6) diameter of boring bar, and (7) total tool slide travel.

6. State two practices to observe to ensure maximum rigidity in setting up a boring head and boring tool.

7. Give (a) two possible causes of deflection in a boring tool and (b) the corrective steps that the operator may take.

8. State two unique applications of the shaper head that save second machine setups or special tooling.

9. Tell how ordinate dimensioned drawings are used by the vertical milling machine operator to position the table for drilling and reaming.

10. State two tool or machine safety factors to consider to ensure dimensional accuracy when shaping is performed.

SECTION TWO

Advanced Horizontal Milling Machine Processes

The actual machining of cams and gears involves a knowledge of design features, the interpretation of drawings and specifications, and the use of formulas to establish machining dimensions. This section provides information on methods of forming and representing spur, helical, bevel, and worm gears, and cams. Actual jobbing shop setups for the milling machine, dividing head, and other accessories; cutter selection; and step-by-step milling procedures and measurement practices are covered in detail.

UNIT 13

Gear Production, Design Features and Computations

OBJECTIVES

After satisfactorily completing this unit, you will be able to:

- Describe gear broaching, shaving, casting, hot and cold rolling, extruding, stamping, powder metallurgy, and other conventional gear milling production methods.
- Understand precision fini-shear, burnishing, and form grinding methods of finishing gear teeth.
- Calculate common gear tooth measurements and use a gear tooth vernier or best wire size system for measuring, using handbook formulas and tables.
- Perform each of the following processes.
  - Calculating Bevel Gear Dimensions.
  - Calculating Large- and Small-End Gear Tooth Dimensions.
  - Selecting Bevel Gear Cutters.
  - Determining the Amount of Gear Tooth Taper Offset.
- Interpret drawings and sketches for spur, rack, pinion, helical, bevel, and worm gearing for design features and applying ANSI and SI metric gearing standards.
- Correctly use production, design feature, and computational *Terms.*

EARLY DEVELOPMENTS AND GEAR-CUTTING METHODS

Gears have been used for light-duty purposes, such as clock mechanisms and simple machines, for hundreds of years. Starting with the Indus-trial Revolution, gears and gear trains for heavy-duty purposes became necessary because machines in such fields as textiles, printing, and simple manufacturing required positive and uniform driving mechanisms.

FORM MILLING

The first *gear-cutting machines*, dating back to 1800, were used principally to cut gears for simple, low-power drives. Further applications of gears in machine design and the construction of gear-cutting machines accelerated during the 1850s due to increased manufacturing demands for gears capable of transmitting greatly increased power loads at faster speeds.

Specialized gear-cutting machines were developed. Most of these machines require a formed milling cutter. The milled form of the tooth and the space between teeth permitted mating teeth to fit. The flank of a driver gear tooth form rolled against the flank of a corresponding driven gear tooth produces motion.

GEAR HOBBING

By 1900, *hobbing machines* were produced for highly specialized production gear machining. Hobbing machines are still used today, particularly for extremely large gears of 15 or more feet (4.51m) in diameter. They are constructed with horizontal and vertical spindles.

The gear hobbing process is the fastest of the generating processes. Several teeth are cut at the same time. The cutter is continuously meshed with the gear blank. This process contrasts with the conventional jobbing shop method of milling one tooth, disengaging the cutter and workpiece, indexing, and then cutting the next tooth.

The hobbing cutter (hob) resembles a worm thread (Figure 13-1). Cutting edges are formed on the cutter by gashes made parallel to the axis of the bore. The tooth form is the same as a gear rack. A gear tooth is generated when the hobbing (worm) cutter turns one revolution while the gear moves one space. The tooth is formed by feeding the cutter to tooth depth (Figure 13-2).

Spur, helical, and herringbone gears; splines; gear sockets; and other symmetrical shapes may be hobbed. Spur gears are hobbed by setting the cutter with its teeth parallel to the gear blank axis. Helical teeth are cut by setting the hob

Courtesy of BARBER-COLMAN COMPANY; MACHINE AND TOOLS DIVISION

Figure 13-1 Gear Hobbing Cutter Features

axis at the angle required to produce the helix. The number of teeth cut on a gear is controlled by the gear ratio between the hob and the gear blank.

GEAR SHAPING

Gear shapers were first produced around the turn of the twentieth century. They use a cutter formed like a mating gear but relieved to produce cutting edges. The cutter requires a reciprocating

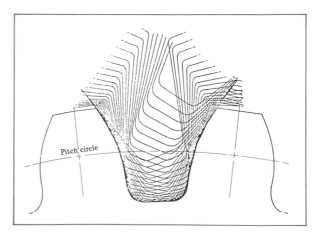

Pitch circle

Figure 13-2 Gear Tooth Generating Action of a Hob

movement. As the cutting progresses and the tool is fed to the required depth, the cutter and gear blank slowly rotate together. A gear is thus produced in one rotation of the blank.

At the end of each cutting stroke, the work rapidly clears the cutter to prevent rubbing. The work is returned to position for the next cutting stroke. Cutters are set to reciprocate from 100 to 2,000 strokes per minute for fine-tooth gears.

Gear shaping is practical for cutting symmetrical spur, helical, and herringbone gears; internal and cluster gears; racks; elliptical gears; and other special shapes.

The increasing use of *hypoid gears* created a need to produce quiet, accurately meshing gear teeth. A *hypoid gear generator* was developed to meet this need. Hypoid gears are a modification of bevel gears. Spiral teeth are formed on the bevel surface. The axes of the gears are offset for the pinion (small gear) and gear. Hypoid gears are widely used in automotive differentials and other mechanisms where motion is transmitted by a driving shaft at a right angle and offset from the driven shaft.

OTHER METHODS OF FORMING GEARS

BROACHING AND GEAR SHAVING

Broaching is a fast, economical method of producing individual and nonclustered gears. Broaching is a practical method for machining an accurate straight-tooth or helical-tooth gear form having a pressure angle of less than $21°$.

Gear shaving is a technique for producing gear teeth on soft metals, mild steels, and other materials up to a 30 Rockwell C hardness. Surface finishes in the $32R_C$ to $16R_C$ range are possible. Gear shaving requires the use of a rotating cutter that may be positioned at an angle to the axis of a workpiece for diagonal shaving. The teeth are formed as the shaving cutter is fed to depth into a free-rotating blank (Figure 13–3).

The cutter revolves at high speed and reciprocates across the entire surface of the gear blank. Gear shaving produces teeth that are accurate in terms of shape, dimensions, and surface texture.

SHAVING GEAR

FREE-ROTATING GEAR BLANK

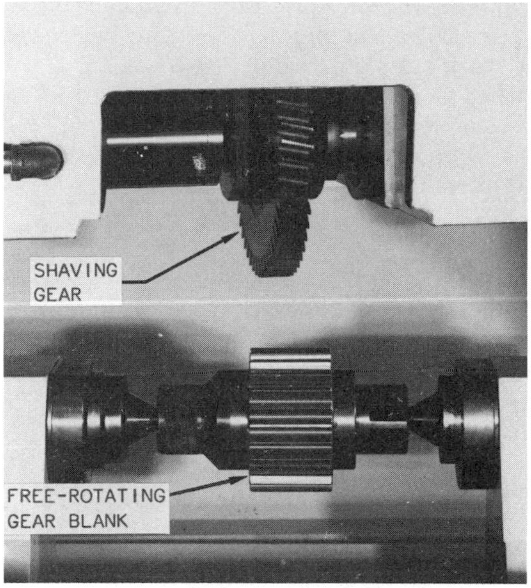

Figure 13–3 Gear Shaving a Spur Gear

CASTING

The *sand casting* method is still used for large gears where the teeth are later machined. Other casting methods include permanent-mold, shell-mold, and plastic-mold casting. Many gears are *die cast* or *lost-wax cast* and require limited, if any, machining.

HOT AND COLD ROLLING

Hot rolling refers to the production of gear teeth from a solid, heated bar of steel. Gear-shaped rollers are impressed into a revolving heated bar. The force between the rollers and heated material causes the metal to flow and conform to the gear tooth shape and size.

Cold rolling is becoming widely used for the high-volume production of spur and helical gears. Material is saved in producing each gear. Second and third machining operations are eliminated from most gear requirements. Cold rolling produces a smooth, mirror-finish surface texture.

A high-volume cold rolling gear machine is illustrated in Figure 13–4. Hardened and precision-ground rolling dies are used on this machine to roll (form) helical pinions and other helical gears. The work parts are loaded into a magazine. They are then picked up and prelocated by a work arbor. The gear teeth are formed by applying a tremendous force (as in hot rolling) for the rolling dies to impress the tooth form in the blank.

Extruding

Extruding is limited to brass, aluminum, and other soft materials. In some cases, bars are formed to gear tooth shape by forcing (extruding) the material through forming dies. Each gear is then cut off to width. Second and other operations are required on extruded parts. For example, the hole must be pierced. The teeth are often finish machined to precise limits.

Stamping

Gears for watches, small-gear mechanisms, and many business and household devices are mass-produced by *blanking (stamping)*. Fine blanking permits the stamping of gear parts up to 1/2" (12.7mm) thickness within close tolerances and to high-quality surface finish requirements.

Powder Metallurgy

Powder metallurgy is a method that produces gears in three major steps:

- *Mixing*, in which selected metals and alloys in powdered form are mixed in specific proportions;
- *Compressing*, in which the mixture in a particular size of gear-forming die is subjected to extreme force;
- *Sintering*, in which the shaped gear is brought to high temperatures and the metal particles are sintered (fused) together to form a gear of a required form and size.

Large gears produced by this method require the *preform* (gear as it comes from a powder metallurgy press) to be machined to a precise size and surface finish.

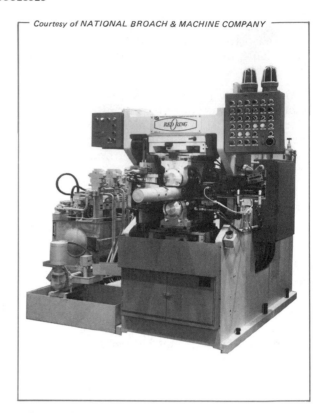

Figure 13–4 High-Volume, Double-Die Cold Rolling Gear Forming Machine

GENERATING MACHINES FOR BEVEL GEARS

GENERATING STRAIGHT-TOOTH BEVEL GEARS

Gear tooth generators are built to machine straight-tooth bevel gears that range in size from 0.025" to 36" (0.64mm to 914mm) in diameter. Two reciprocating cutting tools are mounted on the generating machine cradle. The cutting tools form the top and bottom sides of a tooth respectively. The cradle and gear blank roll upward together.

At the top of the upward movement, a tooth is completely generated and the cutting tools are automatically withdrawn. The machine indexes while the cradle moves down to the starting position for the next tooth.

GENERATING SPIRAL-TOOTH BEVEL GEARS

Spiral bevel gear generators produce spiral teeth. Tooth profiles for spiral bevel gears (with intersecting axes) and *hypoid gears* (with *nonintersecting axes*) may be generated on the same machine. The tooth profile and spiral shape are produced by combining straight-line and rotary motions between the gear blank and the cutter.

GEAR FINISHING MACHINES AND PROCESSES

More accurate gear tooth shapes and precise surface finishes are produced by such second operations as shearing, shaving, and burnishing. Heat-treated gears are finished by grinding, lapping, and honing.

SHEARING

Fini-shear is a new process for accurately producing and finishing gears. It uses a carbide cutter resembling a master gear. The cutter is rotated with and reciprocates against a gear blank. The cutting tool is brought into contact with the work at an angle to the work axis. The shearing, cutting action produces a high-quality surface finish.

GEAR BURNISHING

Gear burnishing is a cold-working process. The machine rolls a gear in contact with three hardened burnishing gears. Burnishing produces slightly work-hardened, smooth gear teeth.

FINISHING GEAR TEETH ON HARDENED GEARS

Two common gear tooth finishing processes for hardened gears include *grinding* and *lapping*. Gear teeth may be form ground or generated. Form grinding is similar to form milling. The grinding wheel is formed by diamond dressing. The diamond holders are guided by templates that conform to the gear tooth form. The formed grinding wheel reciprocates parallel to the work axis to produce accurately formed teeth with a high-quality surface texture.

Gear teeth on hardened gears may also be finish ground on a generating-type grinder. The tooth flanks are ground by the flat face of the wheel. The gear tooth is rolled past the revolving grinding wheel.

GEAR TOOTH MEASUREMENT AND INSPECTION

Gear tooth measurements and tests relate to:

- Concentricity (runout),
- Tooth shape,
- Tooth size,
- Surface texture,
- Lead (on helical gears),
- Tooth angles (on bevel gears).

TOOTH INSPECTION WITH AN OPTICAL COMPARATOR

In gear tooth inspection, as in other uses of an optical comparator, a transparent drawing of the tooth is placed on the screen. The drawing scale conforms to the magnification power of the lens. The enlarged shadow of the gear tooth outline that is reflected on the screen is compared directly with the drawing. The nature and extent of error in tooth shape, size, and pitch are visually established.

SPECIAL TESTING FIXTURES

Special fixtures employing a master gear are used for testing concentricity (runout). The testing relates to out-of-trueness on spur gears (parallelism of the teeth and gear axes), angular runout on bevel gears, and the lead of helical and worm gears.

GEAR TOOTH VERNIER CALIPER MEASUREMENT

Gear teeth must be precisely formed and dimensionally accurate if they are to mesh properly. One technique for measuring gear teeth is to use a vernier caliper. The accuracy of the caliper measurement is influenced by the accuracy of the outside diameter of the gear.

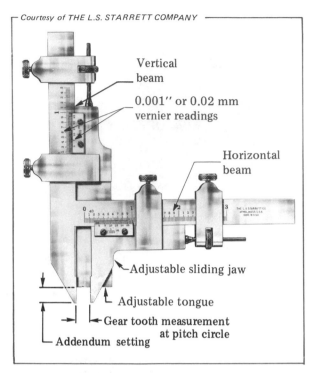

Vertical beam

0.001″ or 0.02 mm vernier readings

Horizontal beam

Adjustable sliding jaw

Adjustable tongue

Gear tooth measurement at pitch circle

Addendum setting

Figure 13–5 Gear Tooth Vernier Caliper Features and Application

The *gear tooth vernier caliper*, illustrated in Figure 13–5, is designed to measure (1) the chordal thickness of a gear tooth at the pitch circle, and (2) the addendum distance from the top of the tooth to the pitch circle.

There are two beams on the caliper. The horizontal beam is used for measuring the chordal thickness and has an adjustable sliding jaw. The vertical beam has an adjustable tongue. The jaw and tongue are adjusted independently. Different sizes of gear teeth may be accurately measured when proper allowance is made for any outside diameter variations of the gear blank.

The jaws of the vernier caliper are hardened and ground. The measuring surfaces are lapped. Vernier calipers are also provided with tungsten carbide measuring surfaces.

The vernier caliper is graduated in thousandths of an inch (0.001″). It is available in two sizes for measuring gear teeth from 20 to 1 diametral pitch. Metric gear tooth vernier calipers are graduated to fiftieths of a millimeter (0.02mm).

The comparable gear tooth range in the English diametral pitch unit system is from 1.25 to 25 millimeter modules.

TWO-WIRE METHOD OF MEASURING GEAR TEETH

A micrometer measurement over the outside diameter of two wires is another technique for checking the accuracy of gear teeth. Two cylindrical wires (pins) of a specified diameter are placed in diametrically opposite tooth spaces for an even number of teeth or for an uneven number of teeth.

Standard wire (pin) diameters are specified in handbook tables according to the diametral pitch of the gear. The *diametral pitch* of a gear designed according to ANSI standards equals the number of teeth in each inch of pitch diameter. *Pitch diameter* represents the diameter of the pitch circle at which the tooth thickness (chordal distance) is measured.

Van Keuren Wire Diameters. Selected Van Keuren wire diameters for external and internal gears are listed in Table 13–1. The Van Keuren standard wire sizes range from 0.8640″ to

Table 13–1 Selected Van Keuren Wire Diameters for External and Internal Gears

External Gears		Internal Gears
Wire Diameter = $\dfrac{1.728}{\text{Diametral Pitch}}$		Wire Diameter = $\dfrac{1.440}{\text{Diametral Pitch}}$
Diametral Pitch	Wire Diameter	Wire Diameter
2	0.8640	0.7200
2 1/2	0.6912	0.5760
3	0.5760	0.4800
4	0.4320	0.3600
64	0.0270	0.0225
72	0.0240	0.0200
80	0.0216	0.0180

Table 13-2 Sample Dimensions for Checking External Measurement over Wires

Even Number of Teeth	Standard Pressure Angle			
	14 1/2°	20°	25°	30°
	Dimension over Wires*			
14	16.3746	16.3683	16.3846	16.4169
16	18.3877	18.3768	18.3908	18.4217
18	20.3989	20.3840	20.3959	20.4256

*The table dimensions are for 1 diametral pitch gear teeth using Van Keuren standard wire sizes.

0.0216″ diameter for external gears having diametral pitches from 2 through 80, respectively. The wire diameters for internal gears between this same range of diametral pitches is from 0.7200″ to 0.0180″.

Tables of Measurements over Wires. Once the standard Van Keuren wire size is determined, information on the measurement over wires is found in other tables. Handbook tables provide measurements for 1 diametral pitch, external and internal teeth, even and odd numbers of teeth in a gear, and a range from 6 through 500 teeth (Table 13-2). Values are given for four different pressure angles: 14 1/2°, 20°, 25°, and 30°.

$$\frac{\text{measurement}}{\text{over wires (M)}} = \frac{\text{table value for pressure angle}}{\text{diametral pitch of the gear}}$$

For example, the overall micrometer measurement for a 16-tooth, 25° pressure angle, 10 diametral pitch requires the use of the measurement formula and wire table. The table value for 16 teeth, 25° pressure angle equals 18.3908.

$$M = \frac{18.3908}{10} = 1.8391''$$

Allowance for Play between Teeth. The measurement over wires indicates the overall dimension when the pitch diameter is correct and no allowance is made for play (backlash) between mating teeth.

Handbook tables are available on standard backlash allowances for external and internal spur gears.

SPUR GEAR DRAWINGS (REPRESENTATION, DIMENSIONING, AND TOOTH DATA)

Drawings of gears conform to ANSI standards. The drawings contain such information as material and heat treatment (where required), quality of surface texture, gear tooth data, and manufacturer's identifying markings, and finished dimensions. Common spur gear terms and measurements are illustrated in Figure 13-6.

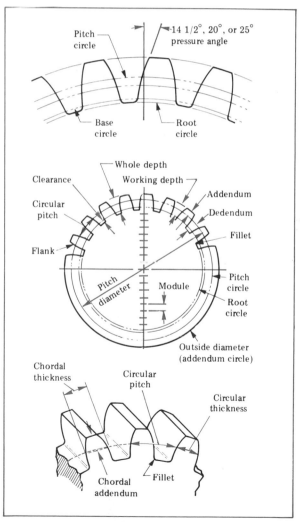

Figure 13-6 Identification of Common Spur Gear Terms and Measurements

The view or views used depend on design features. Gears, pinions, or worms that are machined on a shaft are represented by a view that is parallel with the axis. This view is the preferred representation.

Gears with holes and hubs are generally drawn as a section view. The section cuts through the axis. The section may be partial or full. Usually just one view is used. When a drawing needs to be clarified, a section (front) view and a few teeth may be drawn, as shown in Figure 13–7.

Other dimensions relating to the gear teeth are included as *gear tooth data.* Sometimes, *reference dimensions* are included for purposes of computing other needed values.

GEARING TERMINOLOGY

ANSI and Gear Manufacturer's Association definitions are applied to gear terms.

The *pitch circle* is an imaginary circle that represents the point at which the teeth on two mating gears are tangent. A gear size is indicated by the diameter of the pitch circle. Most gear definitions are computed in relation to the pitch circle.

The *pitch diameter* is the diameter of the pitch circle.

The *addendum* is a vertical distance that ex-tends from the pitch circle to the addendum circle. The adjustable tongue on the vertical beam of a gear tooth vernier caliper is set to the addendum. At this depth, the gear tooth thickness may be accurately measured at the pitch circle.

The *addendum circle* is the outside (adden-dum) diameter.

The *dedendum* is the vertical distance that extends from the pitch circle to the root circle. The dedendum accommodates the addendum of the mating tooth. It also provides additional clearance between the outside diameter of the mating gear and the root circle.

The *root circle (diameter)* is located at the root (bottom) of the gear teeth.

The *base circle:* is the circle from which the tooth profile is generated. The diameter of the base circle is found by drawing a line at a 14 1/2°, 20°, or 25° angle through the pitch point. The angle depends on the pressure angle used in the gear design.

The *pressure angle* is the angle that estab-lishes the tooth shape. It represents the angle at which the forces from the tooth of one gear are transmitted to the mating tooth on another gear.

While many 14 1/2° involute and composite system full-depth teeth and 20° involute stub teeth are still used and tables for these teeth are included in handbooks, these tooth forms are no longer recommended. Revised American Na-

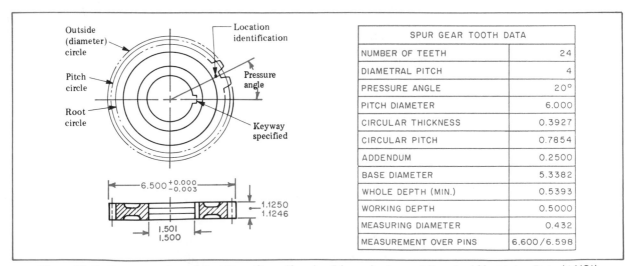

SPUR GEAR TOOTH DATA	
NUMBER OF TEETH	24
DIAMETRAL PITCH	4
PRESSURE ANGLE	20°
PITCH DIAMETER	6.000
CIRCULAR THICKNESS	0.3927
CIRCULAR PITCH	0.7854
ADDENDUM	0.2500
BASE DIAMETER	5.3382
WHOLE DEPTH (MIN.)	0.5393
WORKING DEPTH	0.5000
MEASURING DIAMETER	0.432
MEASUREMENT OVER PINS	6.600/6.598

Figure 13–7 Representation of Cast Spur Gear, Showing Dimensioning and Required Machining Data (ANSI)

tional Standard (ANSI B 6.1) tables provide information on 20° and 25° involute spur gear forms. The gear tooth forms are identical except for the different pressure angles and the minimum allowable number of teeth.

The ANSI 25° standard form provides even greater strength and permits the machining of gears with fewer than 18 teeth. The 25° form produces lower contact compressive strength and greater gear tooth surface durability.

The *circular pitch* is the circular distance that extends along the pitch circle from a point on one tooth to the corresponding point on the next tooth.

The *circular thickness* is the thickness of a gear tooth as measured along the arc of the pitch circle.

The *chordal thickness* is the tooth thickness as measured along a chord of the pitch circle. The measurement is taken by the movable leg on the horizontal beam of the gear tooth vernier caliper.

The *chordal addendum* extends from the outside diameter to the chord at the chordal thickness. This dimension differs from the addendum.

The *whole depth* indicates the overall height of a gear tooth or the depth to which each tooth is cut.

The *working depth* is the depth to which one tooth extends into the gear tooth space on a mating gear or rack.

The *clearance* is the difference between the working depth and the whole tooth depth.

MODULE SYSTEMS OF GEARING

The *English module* represents the *ratio* between pitch diameter in inches and the number of teeth. That is, *diametral pitch* equals the number of teeth per inch of pitch diameter. A 10 diametral pitch gear, for example, has 10 teeth for each inch of pitch diameter.

The *SI Metric module* is an *actual dimension:*

$$\text{SI Metric module} = \frac{\text{pitch diameter (mm)}}{\text{number of teeth in gear}}$$

For example, a 25-tooth gear with a pitch diameter of 75mm has a module of 3—that is, there are 3mm of pitch diameter for each tooth.

Most of the gearing produced in the United States is designed around the *diametral pitch system.* This system provides a series of standard gear tooth sizes for each diametral pitch. For example, in a 10 diametral pitch series of gears, a 40-tooth gear has a pitch diameter of 4″; a 42-tooth gear, 4.2″; a 44-tooth gear, 4.4″; and so on.

METRIC MODULE EQUIVALENT

A metric module equivalent of a diametral pitch may be found by dividing 25.4 by the diametral pitch. Where necessary, the answer is rounded off to the closest module in the standard

Table 13–3 SI Metric Spur Gear Rules and Formulas for Required Features of 20° Full-Depth Involute Tooth Form (Partial Table)

Required Feature	Rule	Formula*
Module (M)	Divide the pitch diameter by the number of teeth	$M = \dfrac{D}{N}$
	Divide the outside diameter by the number of teeth plus 2	$M = \dfrac{O}{N + 2}$
Addendum (A)	Multiply the module by 1.000	$A = M \times 1.000$
Clearance (S)	Multiply the module by 0.250	$S = M \times 0.250$
Whole depth (W^2)	Multiply the module by 2.250	$W^2 = M \times 2.250$
Tooth thickness (T)	Multiply the module by 1.5708	$T = M \times 1.5708$

*Dimensions are in millimeters.

Table 13-4 ANSI Spur Gear Rules and Formulas for Required Features of 20° and 25° Full-Depth Involute Tooth Forms (Partial Table)

Required Feature	Rule	Formula*
Diametral pitch (P_d)	Divide the number of teeth by the pitch diameter.	$P_d = \dfrac{N}{D}$
	Add 2 to the number of teeth and divide by the outside diameter.	$P_d = \dfrac{N + 2}{O}$
	Divide 3.1416 by the circular pitch.	$P_d = \dfrac{3.1416}{P_c}$
Whole depth of tooth (W^2)	Divide 2.250 by the diametral pitch.	$W^2 = \dfrac{2.250}{P_d}$
Thickness of tooth (T)	Divide 1.5708 by the diametral pitch.	$T = \dfrac{1.5708}{P_d}$
Center distance (C)	Add the pitch diameters and divide the sum by 2.	$C = \dfrac{D^1 + D^2}{2}$
	Divide one-half the sum of the number of teeth in both gears by the diametral pitch.	$C = \dfrac{1/2\,(N^1 + N^2)}{P_d}$

*Dimensions are in inches.

series. For example, if the metric module equivalent of a 10 diametral pitch is required, the nearest standard module to 2.54 is 2.5.

Table 13-3 is a partial listing of the SI Metric spur gear rules and formulas for required features of a 20° full-depth involute tooth form. Table 13-4 is a partial listing of the ANSI spur gear rules and formulas for required features of 20° and 25° full-depth involute tooth forms. The complete table is included in the Appendix.

CHARACTERISTICS OF THE INVOLUTE CURVE

The term *involute curve* identifies the almost exclusive profile used for gear teeth. An involute curve is generated as the path of a point. The point, theoretically held taut on a straight line, scribes an involute curve as it unwinds from the periphery of the base circle.

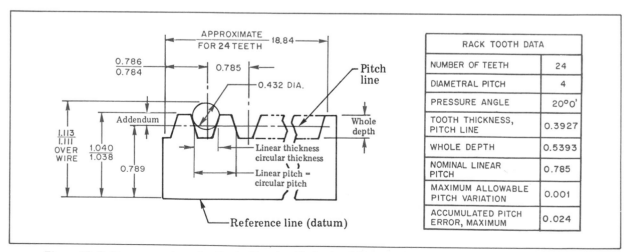

Figure 13-8 Representation of a Rack, Showing Methods of Dimensioning and Required Machining Data

Some of the characteristics of an involute curve are as follows:

- The form of the involute curve depends on the base circle.
- The action of two mating involute gears provides uniform motion.
- The rate of motion is established by the diameters of the base circles of the driver and driven gears.
- A straight line of action is produced by the contact between intermeshing teeth on a driver and a driven gear.
- The pressure angle represents the angle between the line of action and a line perpendicular to the common centerline of the mating gears.

RACK AND PINION GEARS

A *rack* is a straight bar with gear teeth cut on one side. The profile and design features of the teeth on a rack are identical with the mating spur gear *pinion* teeth (Figure 13–8).

Some different terms are used with rack and pinion gears because all circular dimensions on the rack become linear. The *lineal pitch*, as shown in Figure 13–8, corresponds with circular pitch:

$$\text{lineal pitch} = \frac{3.1416}{\text{diametral pitch}}$$

The *lineal thickness* is equal to the circular thickness of a spur gear.

REPRESENTING AND MACHINING A RACK

A rack is usually represented and dimensioned on a drawing as shown in Figure 13–8. Tooth data provide additional information for machining and inspecting the gear teeth.

Racks are machined in jobbing, maintenance, and small shops on a milling machine. Short rack lengths may be held on parallels in a vise or strapped to the table.

In conventional milling, the workpiece and formed gear tooth cutter axes are parallel. The cutter is accurately moved by the cross slide an amount equal to the circular pitch of the gear. Each tooth is cut by feeding the rack longitudinally past the cutter.

The formed gear tooth cutter may be mounted in a rack-cutting attachment. The table is moved a distance equal to the lineal (circular) pitch to position each gear tooth. An indexing attachment is usually used to position (index) the workpiece to the required lineal pitch. The teeth may also be indexed by using the graduated dial on the table feed screw.

A straight-tooth rack is cut by mounting the gear cutter axis parallel with the gear axis. The teeth on a helical rack require the workpiece to be set at the gear helix angle in relation to the cutter axis.

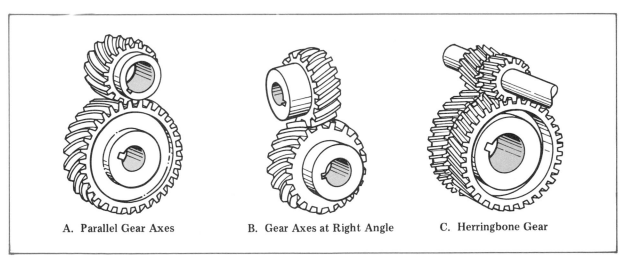

A. Parallel Gear Axes   B. Gear Axes at Right Angle   C. Herringbone Gear

Figure 13–9 Common Applications of Helical Gearing

HELICAL GEARING

Helical gears transmit rotary motion between two shafts. The shafts may have parallel axes that lie in the same plane or the axes may be at any angle up to 90° to each other (Figure 13–9).

Helical gear teeth permit continuous, smooth, quiet operation as compared to spur gear teeth. More than one helical gear tooth is engaged at one time. Thus, the gear strength is increased for teeth of equal size and pitch.

The *herringbone (double-helical) gear* used on parallel shaft movements combines the features of a right- and left-hand helical gear in one gear. Since the end thrusts are equalized, no end thrust bearings are required.

CALCULATING HELICAL GEAR DIMENSIONS

The basic features, symbols, rules, and formulas for helical gear calculations and measurements are found in handbook tables and manufacturer's technical data sheets (Table 13–5).

NORMAL CIRCULAR AND DIAMETRAL PITCHES

The terms for helical gears are similar to the terms used for spur gears. Two terms, *normal circular pitch* and *normal diametral pitch*, require further discussion. The normal diametral pitch represents the diametral pitch of the helical gear tooth cutter.

Table 13–5 Helical Gearing Features, Symbols, Rules, and Formulas (Partial Table)

Feature	Symbol	Feature	Symbol	Feature	Symbol
Addendum	A	Circular pitch	P_c	Tooth helix angle	
Center distance	C	Normal circular pitch	P_{nc}	Cosine	cos a
Cutter number	C#	Diametral pitch	P_d	Cotangent	cot a
Helix angle of teeth	H	Normal diametral pitch	P_n	Tangent	tan a
Lead of tooth helix	L	Pitch diameter (D)		Normal tooth thickness	
Number of teeth	N	Gear 1	D_1	at pitch line	T_n
Outside diameter	O	Gear 2	D_2	Whole depth of tooth	W

Required Feature	Rule	Formula
Pitch diameter (D)	Divide the number of teeth by the product of the normal diametral pitch and the cosine of the tooth helix angle.	$D = \dfrac{N}{P_n \times \cos a}$
Center distance (C)	Add the pitch diameters of the two gears and divide by 2.	$C = \dfrac{D_1 + D_2}{2}$
Number of teeth for selecting the formed cutter (C#)	Divide the number of teeth by the cube of the cosine of the tooth helix angle.	$C_\# = \dfrac{N}{(\cos a)^3}$
Helix angle of the gear teeth*	Divide the normal circular pitch by the circular pitch.	$\cos a = \dfrac{P_{nc}}{P_c}$
	Divide the diametral pitch by the normal diametral pitch.	$\cos a = \dfrac{P_d}{P_n}$
	Divide the number of teeth by the product of the normal diametral pitch and the pitch diameter.	$\cos a = \dfrac{N}{P_n \times D}$
	*Each quotient equals a natural trigonometric cosine function, which is converted into the helix angle of the gear teeth (H).	

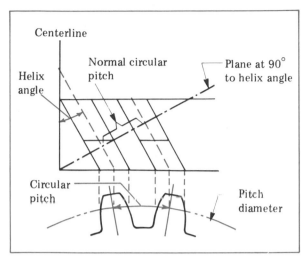

Figure 13–10 Relationship of Normal Circular Pitch of a Helical Gear to Circular Pitch

The normal circular pitch of a helical gear is the distance between a reference point on one gear tooth and the corresponding reference point on the next tooth. The measurement is taken at the pitch circle on a reference plane that is at a right angle to the tooth helix angle (Figure 13–10).

SELECTING A HELICAL GEAR CUTTER

The formed gear cutter used on a helical gear is the same type as is used on a spur gear. However, the cutter size differs since the diametral pitch of the spur gear changes to be the normal diametral pitch of the helical gear. This change in cutter size involves the number of teeth to be cut and the gear tooth helix angle.

The number of teeth for which a cutter is to be selected ($C_\#$) is determined by dividing the number of gear teeth (N) by the cube of the cosine of the helix angle. Expressed as a formula,

$$C_\# = \frac{N}{(\cos a)^3}$$

Example: Determine the cutter number to use to mill a 32-tooth helical gear having a helix angle of 30° by substituting the values in the formula,

$$C_\# = \frac{32}{(0.86603)^3} = 49.26$$

A reference is made to a handbook table of gear milling cutters for different numbers of gear teeth. From such a table, a #3 gear milling cutter

is selected because the 49 teeth fall within the 35 to 54 tooth range of this cutter.

A simplified technique of selecting a cutter for helical gear milling is to use a table or a chart such as the one shown in Figure 13–11. The cutter number to use is determined by drawing a straight line from the number of teeth in the helical gear to be cut (through the wide band of cutter numbers) to the helix angle. For the previous example, the straight line would extend from 32 teeth to the 30° helix angle. The line, drawn in Figure 13–11, cuts through the cutter #3.

The chart shows the number of teeth in the range for each cutter size from 12 teeth through 135. Where greater accuracy of tooth shape is required, an intermediate half-number series of cutters is available. These cutters are listed in manufacturers' catalogs and handbooks.

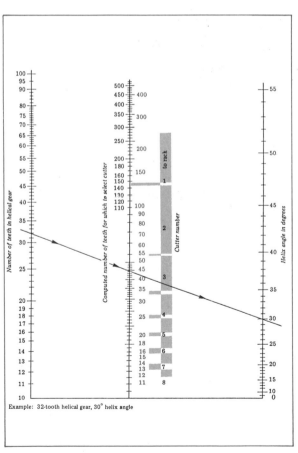

Figure 13–11 Cutter Selection Chart for Helical Gear Cutting

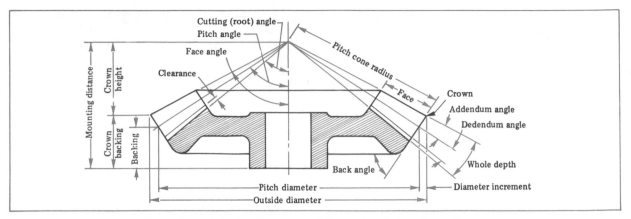

Figure 13-12 Design Features and Terms Applied To Bevel Gears

BEVEL GEARING

DESIGN FEATURES

Bevel gears are used to transmit motion and power between two shafts whose axes intersect. Most bevel gear axes intersect at 90°. The axes may also be at any angle. The design features of a bevel gear and the accompanying terms appear in Figure 13-12.

The teeth on bevel gears conform to standard 14 1/2°, 20°, and 25° pressure angles. The tooth form follows the same involute form as a spur gear. However, the teeth taper toward the apex of the cone. Two right-angle (90° axes) bevel gears of the same module, pressure angle, and number of teeth are called *miter gears.* The small gear in a bevel gear combination is identified as a *pinion.*

WORKING DRAWINGS OF BEVEL GEARS

A one-view, cross-sectional drawing is usually used to represent a bevel gear. A second view is included when the gear has spokes. The relationship of a bevel gear and pinion combination is shown in a single-view drawing of the two gears. Gear teeth are generally not drawn, except on an assembly drawing. The dimensions for machining a bevel gear blank are included on the gear drawing. Cutting dimensions and data appear in a table to supplement the information contained on the drawing. Figure 13-13 is the representation of a bevel gear showing methods of dimensioning with required machining data.

COMPUTING BEVEL GEAR DIMENSIONS

The formulas for computing spur gear dimensions are also used for the following bevel gear features:

- Addendum,
- Dedendum,
- Whole tooth depth,
- Module,
- Diametral pitch,
- Circular pitch,
- Chordal thickness,
- Circular thickness.

Dimensions for such bevel gear features as: *pitch cone distance*, *diameter increment; addendum, dedendum, face,* and *cutting angles; face width;* and the *whole depth* and *tooth thickness* at the *small end of the tooth* are calculated. Formulas and other technical data are provided by gear tooth manufacturers and are found in handbooks. Formulas are also given for determining the precise and the approximate amount a *gear blank* is *offset* from the center line when bevel gear teeth are cut on a milling machine.

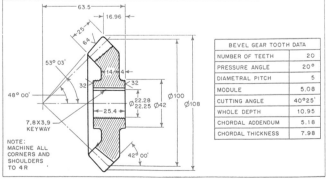

BEVEL GEAR TOOTH DATA	
NUMBER OF TEETH	20
PRESSURE ANGLE	20°
DIAMETRAL PITCH	5
MODULE	5.08
CUTTING ANGLE	40°25'
WHOLE DEPTH	10.95
CHORDAL ADDENDUM	5.18
CHORDAL THICKNESS	7.98

Figure 13-13 Representation of a Bevel Gear, Showing Methods of Dimensioning and Required Machining Data

How to Calculate Bevel Gear Dimensions

Example: Calculate the dimensions for machining a pair of 10 diametral pitch miter bevel gears and for measuring the gear teeth. Each miter gear has 24 teeth.

STEP 1 Calculate the addendum (A):

$$A = \frac{1}{\text{diametral pitch}} = \frac{1}{10}$$

$$= 0.1000'' \text{ addendum}$$

STEP 2 Calculate the dedendum (D):

$$D = \frac{1.157}{\text{diametral pitch}} = \frac{1.157}{10}$$

$$= 0.1157'' \text{ dedendum}$$

STEP 3 Calculate the pitch diameter (D_p):

$$D_p = \frac{\text{number of teeth}}{\text{diametral pitch}} = \frac{24}{10}$$

$$= 2.0000'' \text{ pitch diameter}$$

STEP 4 Calculate the pitch cone distance (P_{cd}):

$$P_{cd} = \frac{\text{diametral pitch}}{2 \times \text{sine pitch cone angle of } 45°}$$

$$= \frac{2.0000}{2 \times (0.707)}$$

$$= 1.4144'' \text{ pitch cone distance}$$

STEP 5 Calculate the diameter increment (D_i):

$$D_i = 2A \times \text{cosine pitch cone angle}$$

$$= (2 \times 0.1000) \times 0.707$$

$$= 0.1414'' \text{ diameter increment}$$

STEP 6 Calculate the addendum angle ($\angle A$):

$$\text{tangent addendum angle} = \frac{\text{addendum}}{\text{pitch cone distance}}$$

$$= \frac{0.1000}{1.4144}$$

$$= 0.0707$$

Use a natural trigonometric table to establish the angle. The rounded addendum angle equals $4°3'$.

STEP 7 Calculate the dedendum angle ($\angle D$):

$$\text{tangent dedendum angle} = \frac{\text{dedendum}}{\text{pitch cone distance}}$$

$$= \frac{0.1157}{1.4144}$$

$$= 0.0818$$

Again, use a natural trigonometric table to establish the angle. The rounded dedendum angle equals $4°41'$.

STEP 8 Calculate the face angle ($F\angle$):

$$F\angle = \text{pitch cone angle} + \text{addendum angle}$$

$$= 45° + 4°3' = 49°3'$$

STEP 9 Calculate the cutting (root) angle ($C\angle$):

$$C\angle = \text{pitch cone angle} - \text{dedendum angle}$$

$$= 45° - 4°41' = 40°19'$$

STEP 10 Calculate the face width (F_w):

$$F_w = \frac{\text{pitch cone radius}}{3}$$

$$= \frac{1.4144}{3} = 0.4713''$$

How to Calculate Large- and Small-End Gear Tooth Dimensions

Large End

STEP 1 Calculate the whole depth.

$$\text{whole depth} = \frac{2.157}{\text{diametral pitch}}$$

$$= \frac{2.157}{10} = 0.2157''$$

STEP 2 Calculate the tooth thickness at the pitch line.

$$\text{tooth thickness} = \frac{1.571}{\text{diametral pitch}}$$

$$= \frac{1.571}{10} = 0.1571''$$

Small End

Note: The size of the small end is approximately two-thirds of the large-end measurement.

STEP 3 Calculate the whole depth.

$$\frac{\text{whole}}{\text{depth}} = \frac{\text{whole depth at large}}{\text{end} \times 2/3}$$

$$= 0.2157 \times 2/3 = 0.1438''$$

STEP 4 Calculate the tooth thickness at the pitch line.

$$\frac{\text{tooth}}{\text{thickness}} = \frac{\text{tooth thickness at}}{\text{large end} \times 2/3}$$

$$= 0.1571 \times 2/3 = 0.1047''$$

How to Select the Bevel Gear Cutter

STEP 1 Calculate the number of teeth (N):

$$\frac{\text{number of teeth}}{\text{for cutter selection}} = \frac{\substack{\text{number of teeth} \\ \text{in gear}}}{\substack{\text{cosine of pitch} \\ \text{cone angle}}}$$

$$= \frac{32}{0.707} = 45.26$$

Note: The nearest whole number of teeth is used: N = 45 teeth.

STEP 2 Refer to a handbook table of gear milling cutter sizes.

STEP 3 Select the number of the gear cutter that includes 45 teeth in its cutting range.

Note: A #3 formed involute gear milling cutter is selected. This cutter has a tooth cutting range of 35 to 54 teeth.

The milling of bevel gear teeth with a formed rotary cutter requires two cuts to be taken for each tooth. The gear blank is slightly offset from the center line. After the first cut is completed on all teeth, the table is moved in the opposite direction an equal amount beyond the center line.

How to Compute the Amount of Gear Tooth Taper Offset

Precision Method of Computing

STEP 1 Compute the ratio to be used in determining the gear tooth cutter factor:

$$\text{ratio} = \frac{\text{pitch cone radius}}{\text{face width}}$$

$$= \frac{1.4144}{0.4713} = \frac{3}{1} \text{ or } 3{:}1$$

Note: The 3:1 ratio of the pitch cone radius to width of face is used to determine the gear tooth cutter factor for the #3 cutter. This factor is found in handbook tables for obtaining offset for milling bevel gears.

Calculate the amount of table offset:

$$\frac{\text{table}}{\text{offset}} = \left[\frac{\substack{\text{thickness of gear cutter} \\ \text{at pitch line at small} \\ \text{end of tooth}}}{2} \right]$$

$$- \left[\frac{\text{gear tooth cutter factor}}{\text{diametral pitch}} \right]$$

$$= \left[\frac{0.1047}{2} \right] - \left[\frac{0.266}{10} \right] = 0.0258''$$

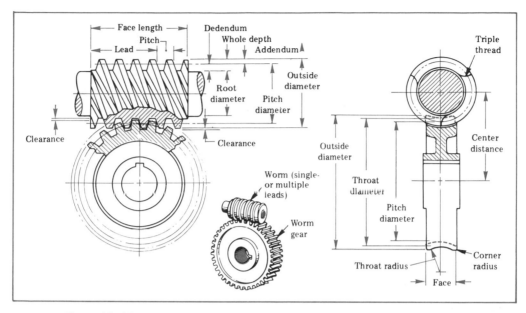

Figure 13-14 Worm and Worm Gear Features, Common Terms, and Dimensions

WORM GEARING TERMINOLOGY

The following terms and formulas are specific to worm gearing (Figure 13-14).

The *axial pitch* is the distance that extends between corresponding reference points on adjacent threads in a worm. The axial pitch of a worm is equal to the circular pitch of the mating worm gear.

The *axial advance (lead)* is the distance the worm thread advances in one revolution. A single thread advances a distance equal to the thread pitch; a double thread, twice the pitch; and so on.

The *lead angle* is formed by the tangent to the helix of the thread at the pitch diameter and a plane perpendicular to the worm axis. By formula,

$$\text{Lead Angle (tangent to Thread Helix Angle)} = \frac{\text{lead}}{\pi \times \text{Pitch Diameter of Worm}}$$

The *throat diameter of a worm gear* is the overall measurement of the worm at the bottom of the tooth arc. The throat diameter equals the pitch diameter plus twice the addendum.

REPRESENTATION OF WORMS AND WORM GEARS

A one-view partial or half-section view of a worm gear and a single view of the worm are used to represent a worm gearing combination. ANSI standards for gear and worm dimensions are indicated in Figure 13-15. The thread and tooth data needed for machining accompanies the drawings. Complete rules and formulas for calculating dimensions, ratios, and other design features are contained in trade handbooks.

GEAR MATERIALS

Gears are made of many different materials. Large and small gears used for light- to medium-duty applications are generally made of cast iron. Where one gear is subject to great force or revolves a greater number of times in relation to the mating gear (as in the case of a pinion and bevel gear combination), two different materials are used. Steel and cast iron, steel and bronze, and non-metallic material combinations are common.

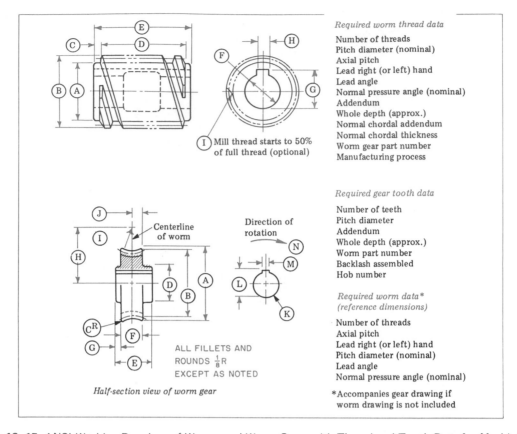

Figure 13-15 ANSI Working Drawings of Worms and Worm Gears with Thread and Tooth Data for Machining

The three categories of materials used on gears are *ferrous*, *nonferrous*, and *nonmetallic*. Cast iron and steel are widely used. Low-carbon steels are used if gears are left soft or are to be case hardened. Higher-carbon steels are used for gears that are to be hardened, tempered, and ground.

The nonferrous materials are generally applied to light-duty operations or where corrosion resistance is required. Bronze; die cast metals; and aluminum, copper, and magnesium base metals are the most common.

Nonmetallic gears that are corrosion resistant to air and chemicals are used for quiet, high-speed mechanisms. Such gears are often meshed with metallic gears. Gears are made of plastics (where temperature is controlled), thermoplastics (for example, nylon), and resin-impregnated canvas (for example, formica).

MACHINING WORMS AND WORM GEARS

Gear hobbing machines are used in the production milling of worms, particularly worms with multiple threads. During hobbing, all (multiple) threads are machined simultaneously instead of by taking separate cuts and indexing for each thread.

Worm thread generating machines are also used. A helical, spur gear type of cutter generates a single or multiple thread as it turns in mesh with the workpiece.

Worms may also be milled or cut on a lathe. A setup similar to helical milling is used when milling the worm thread (Figure 13-16). A thread milling cutter is mounted on a universal milling attachment. The axes of the cutter and attachment are set at the required helix angle of the worm and for the thread direction (right- or left-hand). The gear train ratio between the table

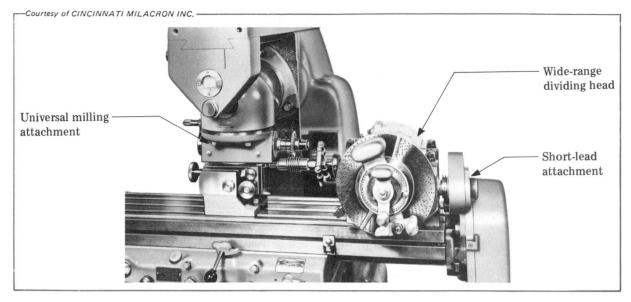

Wide-range
dividing head

Universal milling
attachment

Short-lead
attachment

Figure 13-16 Worm Thread Milling Setup Using Universal Milling and Short-Lead Attachments and a Wide-Range Dividing Head

feed screw and the worm of the dividing head rotates the worm blank as it advances to produce the required lead. Double- and triple-lead threads require the worm to be positioned at the next thread starting location by indexing after each pitch thread is milled to depth.

Hardened worm threads are first roughed out by milling with a formed cutter. They are then precision ground. The disk-shaped cutter has straight edges; the grinding wheel, straight sides. The straight-sided grinding wheel produces a thread having convex sides.

Worm gears are generally hobbed on a gear hobbing machine. The teeth may also be gashed or hobbed on a milling machine.

SUMMARY

- Gear teeth may be machine milled, hobbed, shaped, broached, and shaved.
  - Gears may be hot rolled, cold rolled, extruded, stamped, and formed by powder metallurgy processes.
- Straight- and spiral-tooth bevel gears may be produced on a gear generator. The helical tooth profile is generated by a combination of straight-line and rotary motion between the cutter and gear blank.
  - Gear tooth shapes are precisely finished by shearing, shaving, and burnishing processes. Hardened gear teeth are ground to form and lapped or honed.
- Most terms for spur gears are applicable to helical, bevel, and worm gears.
  - Gear tooth measurements are usually taken mechanically by vernier caliper or micrometer; tooth profile accuracy, on an optical comparator.
- Tables are used to establish Van Keuren wire sizes and measurements for specified numbers of gear teeth and pressure angles.
  - Gears are represented in drawings according to ANSI standards.

- The three common pressure angles used in gear design are 14 1/2°, 20°, and 25°. The latest 20° and 25° angle designs provide greater load-carrying qualities and smoother, quieter, operation.
    - The English module gear tooth system provides a ratio between the pitch diameter in inches and the number of teeth.
- The SI Metric module is an actual dimension obtained by dividing the pitch diameter (mm) by the number of teeth in the gear.
    - Standard ANSI terms, formulas, and rules for spur, helical, bevel, and worm gears are contained in handbook tables.
- Bevel gear cutting on a milling machine requires two cuts. Each cut is slightly offset beyond the center line.
    - The teeth in a worm gear resemble a coarse-pitch thread with a single, double, or triple lead.
- Gear materials depend on conditions of use and design features such as number of teeth and gear size. The three basic materials groups are: ferrous (high- and low-carbon steels and cast iron); nonferrous (die cast metals, bronze, brass, aluminum); and nonmetallic (nylon, formica, and other plastics).

UNIT 13 REVIEW AND SELF-TEST

1. a. Describe gear hobbing and gear shaping.
   b. List four different gear forms that may be cut on a gear shaper.

2. List five basic categories of gear manufacturing methods.

3. Give two distinguishing design features of a bevel gear generating machine as contrasted to a conventional milling machine.

4. List three important measurements or tests a craftsperson makes in milling a spur gear.

5. Describe briefly the shop practice of measuring the thickness of standard spur gear teeth with a gear tooth vernier caliper.

6. Calculate the pitch diameter, outside diameter, and whole depth of tooth for the following 20° full-depth spur gear tooth forms: (a) 4 diametral pitch, 48 teeth; (b) 16 diametral pitch, 89 teeth; (c) 6 module SI Metric, 24 teeth; and (d) 1 module SI Metric, 127 teeth.

7. State the difference between normal circular pitch and normal diametral pitch as applied to helical gearing.

8. a. Use a table for cutter selection for helical gear cutting.
   b. Determine the cutter number to use for cutting the following helical gears: (1) 20 teeth, 22 1/2° helix angle; (2) 25 teeth, 30° helix angle; (3) 36 teeth, 45° helix angle; and (4) 50 teeth, 30° helix angle.

9. Determine the following dimensions for a pair of 36-teeth, 10 diametral pitch, miter bevel gears: (a) pitch diameter, (b) pitch cone angle, (c) cutting (root) angle, (d) tooth thickness at large end, and (e) tooth thickness at small end.

10. State two design features of worm gearing that differ from bevel gearing features.

Machine Setups and Practices for Gear Milling

The four basic types of gears that are machined in a custom shop include spur (and rack), helical, bevel, and worm gear forms. They are usually cut on a horizontal milling machine. An involute form cutter is used to mill the tooth profile. Tooth spaces are indexed with a dividing head.

Typical machine setups and gear-cutting practices are described in this unit. Consideration is given to cutting gear teeth in the English module (diametral pitch) and SI Metric module systems and for pressure angles of 14 1/2° and of 20° and 25° (newer standard form angles).

OBJECTIVES

After satisfactorily completing this unit, you will be able to:

- Establish the range of sizes, characteristics, and factors to consider in selecting involute gear cutters.
- Understand the functions of gear racks, rack milling, and rack milling attachments.
- Apply design features and setup requirements to machine spur, helical, and bevel gears.
- Setup to machine worm threads and worm gears.
- Perform or simulate each of the following processes.
  - Milling a Spur Gear.
  - Milling a Rack.
  - Milling a Helical Gear.
  - Milling Miter Bevel Gears.
  - Milling a Worm Thread and Worm Gear.
- Apply standard gear formulas, technical data, and dimensions from handbook tables.
- Follow *Safe Practices* relating to machine setups, workpiece, and personal safety and correctly use appropriate *Terms*.

INVOLUTE GEAR CUTTER CHARACTERISTICS

Standard involute gear cutter sizes range from 1 to 48 diametral pitch. Figure 14–1 shows examples of gear tooth sizes from 4 to 16 diametral pitch (P_d). Special cutters are available for teeth that are smaller in size than 48 diametral pitch.

A *set* of gear cutters is available for each diametral pitch. Each set consists of eight cutters (#1 through #8). Sometimes, these eight *full-size* cutters are combined with seven *half-size* cutters (#1 1/2 through #7 1/2). The gear tooth range of full- and half-size standard involute form gear cutters is given in Table 14–1.

A gear cutter is selected according to (1) the diametral pitch and (2) the number of teeth in the gear. The gear cutters in a set are made so that each numbered cutter has a slightly different shape. Shape variations ensure that the gear teeth in a pair of gears having the same diametral pitch will mesh properly. For example, the tooth shape on a pinion must be machined with a greater curvature to permit accurate, smooth matching with the teeth on a larger mating gear.

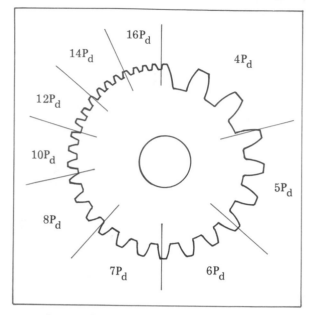

Figure 14-1 Examples of Comparative Gear Tooth Sizes from 4 to 16 Diametral Pitch (P_d)

There is a gradual change in the cutter tooth profile from the #1 cutter, which has slightly curved sides, to the #8 cutter (to cut 12 to 13 teeth), which has a greater tooth curvature.

How to Mill a Spur Gear

STEP 1   Mount the dividing head so that the cutting action takes place with the table positioned centrally.

STEP 2   Position the involute form gear cutter on the arbor as close to the column as the setup will permit.

STEP 3   Mount the gear blank and mandrel between centers. Tighten the footstock center and lock.

STEP 4   Align the cutter teeth with the vertical center line of the gear blank.

STEP 5   Lock the cross slide.

STEP 6   Check to see that the cutting teeth clear the circumference and face of the gear.

STEP 7   Select and mount the required index plate. Set the index pin.

STEP 8   Position the spindle speed and cutting feed dials and the coolant nozzle.

STEP 9   Set the table trip dogs.

STEP 10   Start the coolant flow and the spindle. Raise the table until the cutter teeth just graze the gear circumference of the gear blank.

Table 14-1 Range of Teeth Cut by Full- and Half-Size Standard Involute Form Gear Cutters

Cutter Number		Range of Teeth
Full Size	Half Size	
1		135 to rack
	1 1/2	80 to 134
2		55 to 134
	2 1/2	42 to 54
3		35 to 54
	3 1/2	30 to 34
4		26 to 34
	4 1/2	23 to 25
5		21 to 25
	5 1/2	19 and 20
6		17 to 20
	6 1/2	15 and 16
7		14 to 16
	7 1/2	13
8		12 to 13

STEP 11 Set the graduated knee feed collar at zero.

STEP 12 Move the cutter clear of the gear blank. Then, raise the table to three-fourths of the tooth depth.

STEP 13 Rough out the first tooth.

STEP 14 Repeat the indexing and cutting steps for the remaining teeth.

STEP 15 Loosen the knee clamp. Raise the table to the working gear tooth depth. Lock the knee clamp.

STEP 16 Reduce the cutting feed and/or increase the cutting speed to produce the desired surface finish.

STEP 17 Finish cut the first tooth. Stop the spindle. Return the cutter to starting position.

STEP 18 Continue to index and finish cut the remaining teeth.

> Note: Sometimes, the table is set to machine within 0.002″ (0.05mm) of finish size. Adjustments may then be made, after measurement with a gear tooth vernier micrometer caliper, to machine the gear teeth within specified limits of accuracy.

RACK MILLING

A *gear rack* is a form of spur gear having all teeth in one plane. As for standard gear tooth measurements, measurements on a gear rack are taken from the *straight pitch line*. This pitch line for English module gears is located one addendum (1/diametral pitch) below the top face of the tooth.

An important consideration in rack tooth milling is the pitch. The pitch is the linear distance the workpiece is moved to mill each tooth. This *linear pitch* must be equal to the *circular pitch* with which the rack teeth are to mesh.

Teeth on short racks may be indexed by moving the cross slide the distance equal to the circular pitch of the gear. The teeth on racks longer than the movement of the cross slide are positioned by moving the table longitudinally. The circular pitch distance is read directly on the graduated collar of the table lead screw.

RACK MILLING/INDEXING ATTACHMENTS

Milling cutters are often held, positioned, and driven by a rack milling and indexing attachment (Figure 14–2).

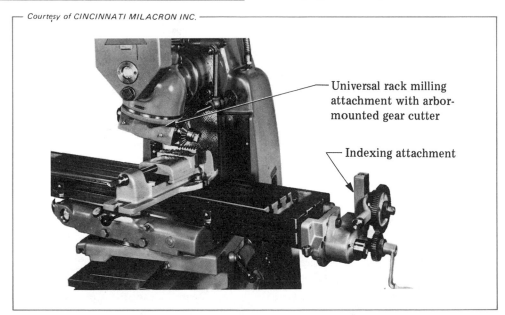

Courtesy of CINCINNATI MILACRON INC.

— Universal rack milling attachment with arbor-mounted gear cutter

— Indexing attachment

Figure 14–2 Horizontal Milling Machine Setup with Rack Milling and Indexing Attachment for Cutting Teeth on a Rack

The *rack milling attachment* is driven by the spindle. The attachment permits setups for milling straight teeth or teeth at an angle. The cutter is held at 90° to the position used in conventional milling (Figure 14–2).

The *rack indexing attachment* ensures accuracy in positioning the workpiece for successive teeth to be milled. A standard rack indexing attachment may be used for indexing as follows:

- From 4 to 32 diametral pitches;
- From 1/8″ to 3/4″ in increments of 1/16″;
- For table movements of 0.1429″, 0.1667″, 0.2000″, 0.2857″, 0.1333″, and 0.4000″.

How to Mill a Rack

STEP 1 Use handbook tables or formulas to calculate the whole depth of the tooth and the linear pitch.

STEP 2 Determine the change gears and the indexing setup if the rack length requires the use of a rack milling attachment and/or a rack indexing attachment.

STEP 3 Mount and secure the change gears and the indexing setup.

STEP 4 Determine the spindle speed and cutting feed. Adjust the machine speed and feed accordingly.

STEP 5 Position the rack milling attachment at 0° or any other required angle.

STEP 6 Mount and secure the selected form milling cutter.

STEP 7 Align and secure the workpiece.

STEP 8 Position the work so that the center line of the first tooth to be milled is aligned with the center line of the cutter.

STEP 9 Start the coolant flow and spindle. Raise the table slowly until the cutter just grazes the work surface. Move the cutter clear of the workpiece.

STEP 10 Set the graduated collar on the knee at zero. Raise the workpiece the distance of the whole depth of the tooth.

Note: When the size and depth of the tooth require considerable material to be milled, the teeth are usually rough cut. Sufficient stock is left for a finish cut and a finer feed. The finish cut and feed depend on the required surface texture and accuracy of the gear tooth profile.

STEP 11 Feed by hand. Engage the power feed when it is established that the setup and cutter action are correct.

Caution: Lock the table or cross slide (depending on how a cutter is fed) before a tooth is milled. Release for indexing.

STEP 12 Stop the spindle. Return the cutter to starting position.

STEP 13 Index (or use the reading on the graduated collar) for the next tooth.

STEP 14 Start the spindle and machine the tooth.

Note: Some craftspersons take a trial cut for a short distance to form one tooth. The spindle is then stopped. The partial tooth is burred and cleaned. The chordal thickness is checked with a gear tooth vernier caliper. Adjustments are made in the working tooth depth to compensate for any error.

STEP 15 Continue steps 13 and 14 until all teeth are machined. The table or cross slide is locked before machining a tooth.

STEP 16 Stop the machine and coolant flow. Clean all accessories, the machined rack, and the machine.

STEP 17 Remove the burrs formed in milling the rack teeth.

STEP 18 Disassemble the setup if necessary. Place the attachments in appropriate storage compartments. Return the cutter and arbor, other accessories, and tools to appropriate storage areas.

HELICAL GEARING

Helical grooves, teeth in helical gears, flutes, and worm threads are often milled on a horizontal milling machine. Two conditions must be met in order to mill a helix:

- The machine setup must permit a rotating workpiece to move past a revolving cutter. (The lead is produced through change gears positioned between the worm shaft on the dividing head and the table feed screw).
- The workpiece and table must be set at the angle of the helix.

How to Mill Helical Gear Teeth

STEP 1 Swivel the table to the helix angle and for the direction of the helix.

STEP 2 Calculate or use handbook tables for the change gears that will produce the required lead. Use the formula:

$$\frac{\text{lead of helix to be cut}}{\text{lead of table feed screw}} = \frac{\text{product of driven gears}}{\text{product of driver gears}}$$

STEP 3 Mount the required change gears between the table feed screw and the worm gear shaft on the dividing head. Use an idler to change direction, if needed.

STEP 4 Mount the workpiece between centers or hold it securely in a chuck or other work-holding attachment on the dividing head.

Note: The workpiece should be positioned so that the large-diameter end of the mandrel is centered on the dividing head center.

STEP 5 Center the cutter approximately over the groove, flute, or tooth layout.

STEP 6 Set the spindle speed, cutting feed, and table trip dogs according to job requirements. Adjust the cutting fluid nozzle.

STEP 7 Raise the table to the work depth of the tooth. Lock the knee.

STEP 8 Hand-feed until the complete setup is checked for secureness and correct cutting action. Then, engage the power feed.

STEP 9 Return the cutter to starting position.

STEP 10 Index for the next tooth. Reposition the cutter at the work depth. Machine the tooth.

STEP 11 Stop the spindle. Clean the workpiece. Remove burrs from the machined surface.

STEP 12 Measure the normal circular pitch with a gear tooth vernier caliper. Make a working depth adjustment if needed.

STEP 13 Repeat step 10 until each tooth is milled.

FACTORS TO CONSIDER IN BEVEL GEAR CUTTING

Generating-type bevel gear cutting machines are widely used to produce correctly formed bevel gear teeth. Each tooth has the same cross-sectional shape throughout its length, except for the uniform size reduction from the large to the small end of each tooth.

There are many cases where straight bevel teeth must be produced by milling. Applications of milled gears, however, are limited in comparison to gears generated by other methods. For example, milled gears, having teeth that are not as precisely machined as generated gears are not suited to high-speed mechanisms or where angular motion is to be transmitted within a high degree of accuracy.

The bevel gear cutter has the same profile as the shape of the teeth at the small end. Bevel gear milling cutters are the same as the cutters used for milling spur gears. However, they are thinner to conform to the tooth size at the small end.

Figure 14–3 Setup for Machining Bevel Gear Teeth with a Gear Blank Mounted on a Dividing Head Spindle

One common setup technique for machining bevel gear teeth is to mount the bevel gear blank on an arbor. A straight-shank arbor is usually held in the dividing head spindle (Figure 14–3).

How to Mill Teeth on Miter Bevel Gears

STEP 1 Center the cutter (Figure 14–3) over the center line of the gear blank. Set the knee graduated collar at zero. Clear the cutter and gear blank.

STEP 2 Raise the table to the whole tooth depth at the large end of the teeth. Lock the knee.

STEP 3 Mill the first tooth. Return the cutter to starting position.

Caution: Lock the dividing head spindle before each cut is taken.

STEP 4 Index and cut the next tooth. Stop the spindle.

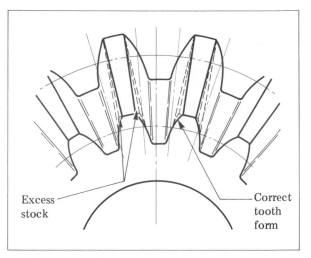

Figure 14–4 Excess Stock to Be Trimmed from Both Sides of the Bevel Gear Tooth Form

Note: Many milling machine operators rough machine all teeth. The sides are then trimmed to size.

STEP 5 Measure the tooth thickness at the large and small ends to determine the additional amount of material that may need to be trimmed from both sides of the tooth form (Figure 14–4).

STEP 6 Move the gear blank *off center* the distance calculated for the table offset. (Figure 14–5).

STEP 7 Index and cut each successive tooth until all teeth have been milled on the one side. Stop the spindle.

STEP 8 Unlock the cross slide. Return the table and indexing plate to the original setting.

STEP 9 Move the table the calculated offset distance in the opposite direction.

STEP 10 Turn the index crank in the opposite direction the same amount as was required for trimming during the first cut.

STEP 11 Take a trial cut just far enough to permit measuring the large end of the tooth. Make whatever adjustments are needed.

STEP 12 Proceed to finish cut the first tooth. Recheck the measurement of the small end. Adjust the setup if required.

STEP 13 Index and trim the second tooth side until all teeth are milled.

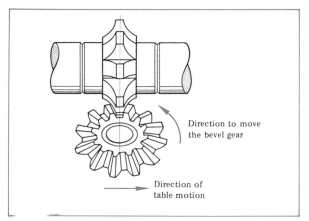

Figure 14–5 Dividing Head and Cutter Setup for Table Offset

WORM THREAD MILLING

Worm threads may be milled on a plain or universal milling machine. A dividing head is used for indexing and to serve as the mechanism for turning the workpiece past a revolving thread milling cutter. The required helix movement is produced by the gear train mounted between the table feed screw and the worm shaft on the dividing head.

When the helix angle to be milled exceeds the swivel angle of the machine table, a universal *spiral milling attachment* is used (Figure 14–6).

How to Mill a Worm (Thread)

STEP 1 Calculate the lead, pitch, whole depth, and thread angle. Determine the required change gears and the indexing. Select the thread milling cutter.

STEP 2 Position and secure the dividing head and footstock.

Caution: Disengage the index plate lock.

STEP 3 Assemble the gear train.

STEP 4 Secure the spiral milling attachment (Figure 14–6).

STEP 5 Mount the selected thread milling cutter on an arbor.

STEP 6 Swing the attachment to the required worm helix angle and the correct direction of the worm lead.

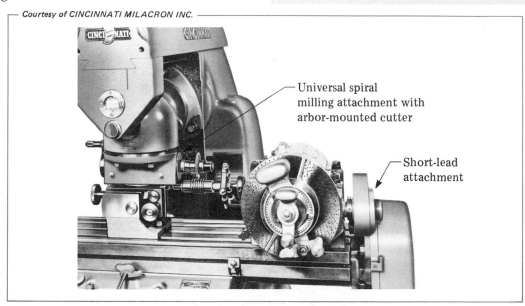

Universal spiral milling attachment with arbor-mounted cutter

Short-lead attachment

Figure 14–6 Application of Universal Spiral Milling Attachment for Cutting a Worm Thread

STEP 7 Align the centers of the cutter and work-piece.

STEP 8 Start the cutter. Bring it into contact with the workpiece.

STEP 9 Clear the cutter and worm blank. Raise the table to thread depth. Lock the knee.

STEP 10 Mill the worm thread. Hand-feed until the setup is checked. Then, engage the automatic feed.

STEP 11 Stop the machine. Clean the workpiece. Measure the worm thread.

STEP 12 Disassemble the setup. Remove the burrs produced during the milling process.

WORM GEAR MILLING

Worm gears are mass-produced on hobbing machines. Replacements of worm gears are often machined on a miller. The teeth are first indexed and cut by *gashing*. A formed cutter and a dividing head are used for gashing. The teeth of the worm gear (wheel) are then finish milled to size and shape by *hobbing*. A hob is used with cutting teeth conforming to the shape of the worm teeth.

How to Mill a Worm Gear

STEP 1 Set the gear blank assembly between the dividing head and footstock centers.

STEP 2 Swivel the table to the helix angle and thread direction of the worm.

STEP 3 Align the center line of the cutter with the vertical axis and throat radius of the mounted gear blank.

STEP 4 Start the spindle and cutting fluid.

STEP 5 Feed the cutter to within 0.015'' (0.4mm) of the whole tooth depth.

STEP 6 Lower the table. Index for the next tooth. Gash the tooth.

STEP 7 Continue to index and gash until all teeth are cut. Stop the spindle.

STEP 8 Replace the gear cutter with the required hob. Remove the dog from the mandrel. Adjust the centers so that the mandrel turns easily (Figure 14–7).

STEP 9 Center the hob with the throat radius and outside diameter of the gashed gear.

STEP 10 Start the hob revolving. Carefully engage the hob teeth in the tooth spaces. Feed to the required depth.

Courtesy of CINCINNATI MILACRON INC.

Hob centered with the gear radius and centerline of the gear

Mandrel free to revolve

Figure 14--7 Milling Machine, Setup for Hobbing the Teeth on a Worm Gear

Safe Practices for Setting Up and Milling Gear Teeth

- Use a dividing head dog with the proper fit in the driving fork. Secure the mandrel assembly by tightening the setscrews in the driving fork.
- Align the center line of a formed spur gear cutter with the center line of the gear blank. The teeth sides will then be milled axially for correct mesh.
- Mill bevel gear teeth so that the cutting action and forces are toward the dividing head.

- Clear the formed gear tooth cutter or stop the spindle rotation when the cutter is returned to the starting position for the next tooth.
- Stop the spindle, clean the workpiece, and remove any machining burrs before a precise gear tooth measurement is taken.
- Standard machine safety precautions must be observed. Trip dogs are required to control table movement. Backlash must be compensated for in helical gear cutting and when reversing the direction for settings. Setup and adequate machining clearance are essential.

TERMS USED IN MACHINE SETUPS AND PRACTICES FOR GEAR MILLING

Gear cutter set (involute form) — A series of formed cutters for milling tooth sizes from 1 to 48 diametral pitch. A set of eight full-size (#1 to #8) and seven half-size (#1 1/2 to #7 1/2) cutters. Commercially available cutters for machining from 12 to 135 teeth in a gear.

Driving fork — A slotted plate for mounting on a dividing head. A device for holding a dog securely and driving a workpiece setup.

Straight pitch line — An imaginary line on a rack corresponding to the pitch circle on a mating gear. A line located one addendum (English standard) or module (metric) below the top face of the tooth.

Linear pitch — The distance on a rack equal to the circular pitch of a gear.

Rack milling attachment — A milling machine attachment mounted for driving by means of the spindle on a horizontal milling machine. A mechanism for changing the position of and driving a gear cutter. An attachment for right-angle and other angle positioning of a head to machine a straight rack or to take cuts at other helix angles.

Rack indexing attachment — A mechanical device for quickly and accurately indexing a workpiece. A mechanism for advancing a workpiece and positioning it at a specified distance corresponding to the linear pitch of the gear teeth.

Table movement increments (rack gear cutting) — Standard distances a milling machine table may be moved during each indexing. A uniform distance a milling machine table may be moved using a rack indexing attachment. Uniform table movement controlled by the ratio of the change gears.

Table offsetting (bevel gear cutting) — The amount a milling machine table is moved (offset) transversely to permit machining bevel gear teeth accurately.

Trimming bevel gear teeth   The process of removing excess material from the sides of milled teeth. Finishing gear teeth to more closely conform to the required tooth profile.

Gashing worm gear teeth   Knee-feeding a worm gear blank into a revolving gear cutter. Up-feeding to cut a gear tooth to the whole tooth depth.

Hobbing (worm gear teeth)   Use of a formed multiple-width tool cutter to track in gashed teeth. Milling worm gear teeth to shape and size by using a hobbing cutter. Machining as a hob automatically engages each tooth on a free-to-rotate gashed gear.

SUMMARY

- In the English module system, standard involute gear cutter sizes range from 1 to 48 diametral pitch. A similar range of sizes is available in the SI Metric module system.
  - The tooth form on gear cutters changes slightly from the #1 to #8 cutters. The shape variations permit gear and pinion combinations with the same diametral pitch (module), but with different numbers of teeth, to mesh accurately.
- The gear cutter size to select depends on the diametral pitch (module) and the number of gear teeth.
  - Roughing cuts are taken when large amounts of material are to be removed. All teeth are roughed out before the finish cut is set at the working tooth depth.
- Increased spindle speed and/or decreased feed rate permit milling teeth to a higher quality of surface finish.
  - Table trip dogs are positioned to reduce excess table travel and machining time and to safeguard against cutting into any parts in the setup.
- Gear tooth measurements on a gear rack are taken along the straight pitch line. The linear pitch corresponds to the circular pitch.
  - A rack milling attachment provides greater flexibility for positioning the cutter to mill grooves or straight gear teeth or gear teeth at a helix angle.
- A rack indexing attachment simplifies and speeds up the positioning of the table for successive teeth. At the same time, the use of the attachment prevents indexing errors. The table movement is controlled by the ratio of the change gears that are positioned between the table feed screw and an indexing plate.
  - Short lengths of rack teeth may be milled conventionally with either transverse or longitudinal table movement. Longer lengths require the use of a rack milling attachment.
- Precision-milled gears usually require a roughing cut and a finish cut at the working tooth depth.

- Helical gear teeth are milled by setting a gear cutter at the helix angle. The workpiece is rotated under the gear cutter so that it advances the required lead distance. The lead and its direction are controlled by the change gear combination between the table feed screw and the worm shaft on the dividing head.

- Bevel gear teeth are trimmed by removing excess material from the sides of each gear tooth. After the gear blank is offset, the gear teeth are rotated a distance that permits cleaning up the excess stock.

  - The operator must continuously check against any backlash in the longitudinal or transverse movements of the table or from indexing. Checking is especially important when the direction of movement is reversed.

- Short helix angle worm threads may be milled by swiveling the table to the required angle. Steep helix angles are milled by using a special milling attachment. The thread lead is produced by the change gearing rotation of the dividing head and the workpiece.

  - Worm gears are milled in a jobbing shop by gashing, followed by hobbing.

- A gear-cutting hob of the same shape and size as the worm engages in the rough cut teeth. Cutting continues as the teeth on the free-to-rotate gear blank are hobbed to working tooth depth.

UNIT 14 REVIEW AND SELF-TEST

1. Give one reason why a set of full- and/or half-size gear cutters is required for each diametral pitch or metric module.
2. Set up a work processing plan to mill a spur gear.
3. State two milling problems that may be overcome by using a rack milling attachment instead of a conventional horizontal spindle setup.
4. Cite the principal advantage for using a rack indexing attachment instead of indexing teeth with a dividing head.
5. Identify two milling machine setups for milling helical gears that differ from the milling of spur gears.
6. Assume that all teeth on a miter bevel gear are rough cut to depth. List the work processes for finish trimming the gear teeth.
7. Give one advantage to the milling of a coarse-pitch, double-lead worm thread in comparison to cutting the same thread on a lathe.
8. Explain the difference between gashing and hobbing the teeth on a worm gear.
9. State three special safety precautions to follow when setting up and milling gears.

Helical and Cam Milling

Many machined parts are designed to include a helical form. A *helical form* is a groove that advances longitudinally at a uniform rate as the part into which it is cut rotates about its axis. The groove (path) is generated on a cylindrical or cone-shaped surface by a cutting tool. This tool is fed lengthwise at a uniform rate. At the same time, the cylinder or cone on which the helical groove is being formed also rotates at a uniform rate.

Helical forms serve many purposes. The grooves, or flutes, on cutting tools such as machine reamers, drills, and taps provide surfaces along which chips flow out of a workpiece. The relieved areas of cutting tools also permit the design of efficient cutting edges. In noncutting applications such as drum cams, the helical groove actuates other machine parts to move at specified rates. Helical forms are widely used in gear design. Here the smooth, sliding action of mating

gears that are cut at a helix angle quietly transmits motion and power.

This unit deals with the principles, terminology, and processes related to the milling of helical forms. The cutting of helical forms on universal, horizontal, and vertical milling machines and dividing heads is representative of jobbing shop work. Machine and dividing head setups, cutter positioning, computation of change gears, and techniques of machining are examined.

The unit also deals with common cam and follower motions, designs, terms, machining setups and procedures, calculations for angle settings of vertical-spindle and dividing head axes, and the end milling of a uniform rise cam. A *cam*, classified for design purposes as a machine element, is a device applied to a moving machine part to change rotary motion to straight-line (reciprocating) motion. The cam transmits the motion through a follower.

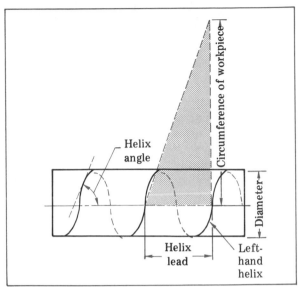

Figure 15–1 Calculating a Helix Angle to Establish Correct Table Setting

LEAD, ANGLE, AND HAND OF A HELIX

The *lead of a helix* is the distance the helix advances longitudinally on a part as it makes one complete revolution about its axis. The lead of a helix generated by a single-point cutting tool is illustrated in Figure 15–1.

Each helix has an angle. The *helix angle* is the angle formed by the groove (helix) and the axis of the workpiece. The cutting of a helix is influenced by the diameter of the workpiece. Figure 15–2 shows how, if the helix angle remains constant but the work diameter changes, the lead (and the helix angle for a specified lead) changes. The lead of a helix varies with the angle of the helix and the diameter of the part.

Unless the table is swiveled, the helix contour will be milled incorrectly although the lead will be correct, as illustrated in Figure 15–3. The helix contour varies from the incorrect setting and width (Figure 15–3A) to the table setting at the required helix angle (Figure 15–3B). This setting permits milling the correct groove width and form.

RIGHT- AND LEFT-HAND HELIX

The direction of a helix is called the *hand of the helix*. A *right-hand helix* (when it is viewed from the end) wraps around a workpiece in a clockwise direction. A *left-hand helix* (when seen with the axis of the workpiece running in a horizontal direction) advances in a counterclockwise direction and slopes down to the left (Figure 15–1).

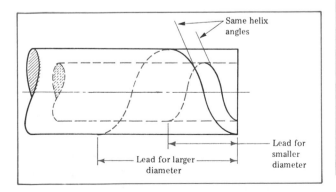

Figure 15–2 Effect of Work Diameter on Helix Lead

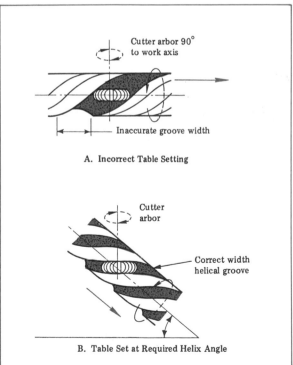

Figure 15–3 Effect of Angle Table Setting on Helix Form

DIRECTION FOR SWIVELING THE TABLE

A left-hand helix requires swiveling the milling machine table in a clockwise direction when viewed from the front of the machine. Similarly, a right-hand helix requires swiveling the table counterclockwise to the required helix angle.

Calculating the Helix Angle. The helix angle to which the table is set is calculated in relation to the helix lead and workpiece circumference (Figure 15–1).

$$\text{Tangent of Helix Angle} = \frac{\pi \times \text{Diameter}}{\text{Lead}}$$

Example: A 8.16″ lead is to be cut on a 1.500″ diameter machine part. The angle for setting the table is to be established by using the formula.

$$\text{Tangent of Helix Angle} = \frac{3.1416 \times 1.500}{8.16}$$

$$= 0.57736$$

The tangent value of 0.57736 represents a 30° angle. Thus, in this example, the milling machine table is set at a 30° angle.

GEARING FOR A REQUIRED LEAD

A helix is produced by advancing a workpiece past a revolving cutter according to the helix angle and lead. The universal dividing head is generally used to turn the workpiece. The workpiece is fed into the cutter by advancing the table, which has been positioned at the helix angle.

The ratio between the rotation of the dividing head spindle and the longitudinal movement of the table requires the use of *change gears*. The change gears provide the required gearing between the worm shaft of the dividing head and the table feed (lead) screw.

Gearing for the dividing head worm shaft and the table feed screw is illustrated in Figure 15–4. An idler gear is inserted in the gearing setup for a right-hand helix. The idler is not a driver or driven gear and is not included in the calculations for change gears. By using an addi-

tional idler gear, the direction of rotation of the workpiece is reversed for a left-hand helix.

CALCULATING CHANGE GEARS

The standard dividing head ratio is 40:1. The standard milling machine table feed screw has four threads per inch. The gear on the dividing head worm shaft is driven by the gear on the table feed screw. One revolution of the feed screw moves the dividing head spindle 1/40 of a revolution. It takes 40 revolutions of the feed screw to turn the workpiece (mounted in the dividing head) one complete revolution. In the process, the 40 revolutions of the feed screw advance the table 10″ (40 revolutions × 0.250″ lead). With equal driver and driven gears—assume, for example, two 48-tooth gears—a lead of 10″ is produced.

The ratio of the change gears required to produce a specific lead may be calculated by using a simple formula:

$$\text{Gear Ratio} = \frac{\begin{array}{c}\text{Lead of Required Helix}\\ \text{(Driven Gears)}\end{array}}{\begin{array}{c}\text{Lead (Feed) of Machine}\\ \text{(Driver Gears)}\end{array}}$$

A series of change gears is used in a gear train to accommodate different ratios of gears on the worm shaft and the feed screw. A set of change gears is provided as standard equipment with universal dividing heads and universal milling machines. The gears and the number of teeth on each gear are as follows: 24, 26, 28, 32, 40, 44, 48, 56, 64, 72, 86, and 100. The combination of gears that are available to produce a required ratio may be determined by using the following formula:

$$\frac{\text{Lead of Helix}}{\text{Lead of Machine}} = \frac{\text{Product of Driven Gears}}{\text{Product of Driver Gears}}$$

Producing a Helix by Simple Gearing. To produce a helix by simple gearing, the gear ratio is found by dividing the required lead by the machine lead.

Courtesy of CINCINNATI MILACRON INC.

Dividing head worm shaft gear (driven gear)

Second gear (driver gear) on stud

First gear on stud (driven gear)

Idler gear (one for right-hand helix, two for left-hand)

Feed screw gear (driver gear)

Figure 15–4 Change Gearing for Dividing Head Worm Shaft and Table Feed Screw

Example: A helix lead of 12″ is to be milled on a workpiece. Standard change gears are available. Calculate the gear ratio for the driven and driver gears. Use the formula,

$$\text{Gear Ratio} = \frac{\text{Lead of Required Helix (Driven Gears)}}{\text{Lead (Feed) of Machine (Driver Gears)}}$$

The lead of a standard milling machine is 10″ for each revolution of the workpiece (lead of helix). By inserting the lead values in the formula,

$$\text{Gear Ratio} = \frac{12}{10}$$

There are no 10- or 12-tooth gears to provide this ratio. However, if each value is multiplied by the same number, appropriate gears may be found. The ratio remains the same. By using 4 as a multiplier,

$$\frac{12 \times 4}{10 \times 4} = \frac{48 \text{ (Driven Gear)}}{40 \text{ (Driver Gear)}}$$

Thus, a 40-tooth gear on the feed screw driving a 48-tooth gear on the worm shaft will produce a 12″ helix lead. The gearing setup is called *simple gearing*.

A word of caution is in order for setting up change gears. The two driver gears and the two driven gears may be interchanged without changing the gear ratio. However, *the gear ratio is changed if one driver gear is interchanged with one driven gear.*

Producing a Helix by Compound Gearing. Many helixes require more complicated setups that involve *compound gearing* in which a number of gears are formed into a *train*.

Instead of computing change gears for different leads, the craftsperson refers to handbook tables. Change gear combinations are given for leads ranging from 0.670″ (1.7mm) to 60.00″ (1,524mm). Table 15–1 provides examples of change gears for different helix leads. For instance, note the gears recommend for a 36″ lead. A 40-tooth driver gear is secured at the end of the feed screw. This gear drives a 64-tooth gear in the second position on the stud. The 32-tooth gear in the first position on the stud serves as a driver for the 72-tooth driven gear on the worm shaft.

GEARING FOR SHORT-LEAD HELIXES

Short-lead helixes, having leads smaller than 0.670″, require a different gearing setup. Short leads are produced by disengaging the dividing head worm and worm wheel. Gearing is direct from the table feed screw to the dividing head spindle. The gears in a direct setup produce a lead that is 1/40 of the lead given in a handbook table. That is, the dividing head spindle turns the workpiece at an increased number of revolutions in relation to the table lead.

In the case of a standard table feed screw having four thread per inch,

$$\text{Gear Ratio} = \frac{\text{Lead to be Milled}}{0.250} = \frac{\text{Driven Gears}}{\text{Driver Gears}}$$

MILLING A HELIX

Helical grooves may be machined on a miller by using a double-angle, concave, fluting, or other formed cutter or end mill.

The setup and milling of a plain helical-tooth milling cutter are used to demonstrate how design features and part specifications (Figure 15–5) are applied to the processes involved in milling the helix.

Table 15–1 Examples of Change Gears for Different Helix Leads

Helix Lead (in ″)	Driven Worm Shaft	Driver 1st Gear on Stud	Driven 2nd Gear on Stud	Driver Gear on Feed Screw
		Position		
35.83	86	32	64	48
36.00	72	32	64	40
36.36	100	44	64	40
36.46	100	48	56	32
36.67	48	24	44	24
36.86	86	28	48	40

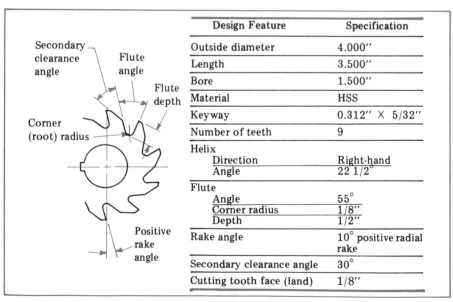

Design Feature	Specification
Outside diameter	4.000″
Length	3.500″
Bore	1.500″
Material	HSS
Keyway	0.312″ × 5/32″
Number of teeth	9
Helix	
Direction	Right-hand
Angle	22 1/2°
Flute	
Angle	55°
Corner radius	1/8″
Depth	1/2″
Rake angle	10° positive radial rake
Secondary clearance angle	30°
Cutting tooth face (land)	1/8″

Figure 15–5 Milling Cutter Design Features and Specifications

How to Mill a Helix on a Universal Milling Machine

STEP 1 Lay out the helix angle of 22 1/2° on the circumference and the 10° positive radial rake angle (Figure 15–6) on the workpiece.

STEP 2 Select an index plate and set the sector arms to permit indexing the nine required teeth.

STEP 3 Calculate the helix lead (31.10") by formula.

STEP 4 Refer to a handbook table of change gears for helical milling different leads. Select the closest lead to 31.10". In this example, the lead is 31.11".

STEP 5 Note the number of teeth and position, mount, and secure each gear in the gear train.

gear on feed (lead) screw: 48 teeth (driver)
second gear on stud: 56 teeth (driven)
first gear on stud: 24 teeth (driver)
gear on worm shaft: 64 teeth (driven)

STEP 6 Swivel the table in a counterclockwise direction to the helix angle of 22 1/2°.

STEP 7 Mount the flute milling cutter (55° angle with 1/8" radius).

STEP 8 Adjust the angle position of the workpiece in relation to the edge of the fluting cutter.

Note: The cutter blank is rotated until the 10° radial layout line of the flute is aligned (same angle) with the angle face of the fluting cutter (Figure 15–6).

STEP 9 Move the table transversely until the center line of the cutter intersects the center of the layout (Figure 15–6).

STEP 10 Return the table (counterclockwise) to the 22 1/2° helix angle. Secure this angle.

STEP 11 Move the cutter to starting position to mill the first tooth. Raise the table to the 0.500" depth.

STEP 12 Take a trial cut for a short distance. Stop the machine. Check the helix, tooth angle, and depth of helix. Then, continue to mill the first flute.

STEP 13 Index, reset the cutter, and machine each successive flute.

How to Machine the Secondary Clearance Angle

The secondary clearance angle (30°) in the example in Figure 15–5 extends from the point of intersection with one angular side of the flute to the land. The clearance angle is machined to provide a 1/8" wide land.

The workpiece must be indexed to rotate it so that the 30° angle is in a correct relation to the side of the flute:

$$\text{Indexing} = 90° - 30° + \frac{\text{Flute Angle of } 55°}{2}$$
$$= 32\ 1/2° \text{ or } 32°30'$$

The 32°30' indexing requires 3 full turns and an additional 11/18 of a revolution.

STEP 1 Replace the fluting cutter with a plain helical milling cutter.

STEP 2 Index the 3 and 11/18 turns.

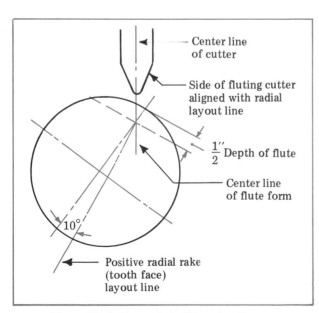

Figure 15–6 Cutter Axis Aligned at Center of Tooth Layout

STEP 3 Adjust the cutter so that it clears the cutting face of the flute form.

STEP 4 Raise the table to cut the 30° angle face to depth.

> **Caution:** Stop the cutter rotation to check the width of the land.

STEP 5 Continue to machine. Index to mill the secondary clearance angle on each tooth.

MILLING A HELIX BY END MILLING

Cams and other parts—for example, actuators for transmitting motion—are often designed with helical grooves that have vertical sides. End mills are commonly used to mill such grooves on either plain or universal milling machines. Since the width of the groove produced by the cutter does not vary with the helix angle, it is not necessary to swivel the table.

The helical groove shape is determined by the form of the end mill. Standard square-face, angular-face, or ball-end end mills may be used. The helix may be milled with the end mill mounted in the spindle of a horizontal or a vertical milling machine.

How to Machine a Helix by End Milling

STEP 1 Determine the change gear setup required to produce the helix lead.

STEP 2 Set up the change gears between the table feed screw, studs, and the worm shaft of the dividing head.

STEP 3 Select the dividing head plate. Set the sector arms for any fractional part of a turn.

STEP 4 Determine the appropriate cutting feed. Set the table feed accordingly.

> **Caution:** The longitudinal movement, spindle, and cutter setup must be checked for clearance at the end of the cut. For safety, the table trip dogs are positioned to control the overall movement of the setup.

STEP 5 Position the table. Align the centerline (axis) of the end mill at the axis of the helical groove.

STEP 6 Start the spindle and flow of cutting fluid. Adjust the table setting so that the cutting edge just grazes the work surface. Set the graduated collar at zero.

STEP 7 Set the end mill to cut to depth. Take a trial cut for a short distance and check.

STEP 8 Engage the power longitudinal feed after the cut is started by hand-feeding if the cutting conditions warrant.

STEP 9 Clear the end mill at the end of each cut. Return the setup to the starting position.

STEP 10 Index for the next helical groove.

STEP 11 Position the end mill for the depth of cut. End mill the second helical groove.

STEP 12 Continue to index, position the cutter, and mill the remaining grooves.

FUNCTIONS OF CAMS IN CAM MILLING

As a machine element, a cam generates a desired motion. The motion is transmitted to a follower by direct contact. The combination of cam and follower produces a given motion, velocity, or acceleration during a specific portion of a cycle and in a particular direction.

Cam motion depends on timing and type of movement in a part of a cycle or a whole cycle. *Motion* refers to the rate of *movement* or *speed*. The speed of the cam follower is related to the speed of rotation of the cam. *Displacement* refers to *distance*. Displacement is the distance a cam follower moves in relation to the rotation of the cam.

The three most common cam motions deal with:

- Uniform (constant velocity) motion,
- Parabolic motion,
- Harmonic motion.

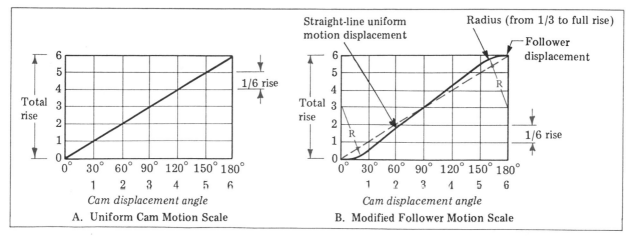

Figure 15-7 Layout of Uniform and Modified Cam and Follower Motion

DESCRIPTIONS OF CAM MOTIONS

UNIFORM (CONSTANT VELOCITY) MOTION

A mechanism, cutting tool, or device that must rise and fall at a constant rate of speed requires a cam that produces a *uniform motion*. The rate of movement of the follower is the same from the beginning to the end of the stroke.

For example, if a cutter is to advance 24mm in one-half revolution of a cam and return to zero during the remaining 180°, the follower would feed the cutter at a uniform rate of one-sixth of 24mm (4mm) for each 30° the cam rotates. The motion is uniform. The straight-line graph of Figure 15-7A illustrates uniform cam motion. Figure 15-7B shows the modified motion of the follower.

PARABOLIC MOTION

Parabolic motion is identified as uniformly accelerated or decelerated motion in 180°. A curve similar to the curve for accelerated motion follows for the retarded (decelerated) motion in the remaining 180° part of one cycle or revolution.

In Figure 15-8, points are plotted every 30°. The six divisions increase and decrease by a ratio of 1:3:5:5:3:1. Assume a follower is to rise 72mm in 180° and points are plotted every

30° in six proportional divisions of 1:3:5:5:3:1 (18 divisions). In the first 30°, the follower rises 1/18 of 72mm, or 4mm. In the second 30°, the follower rises 3/18 (1/6) of 72mm, or 12mm. The follower rise in the third 30° is 5/18 of 72mm. or 20mm. Similarly, the fourth, fifth, and sixth 30° rises are 20mm, 12mm, and 4mm, respectively.

Uniformly accelerated and decelerated motion starts slowly and then accelerates or decelerates at a uniform rate. The follower is permitted to come to a slow stop before reversal. A cam that produces parabolic motion has smooth operating characteristics and is adaptable for high-speed applications.

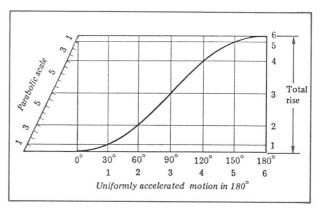

Figure 15-8 Layout for Parabolic Motion of a Cam

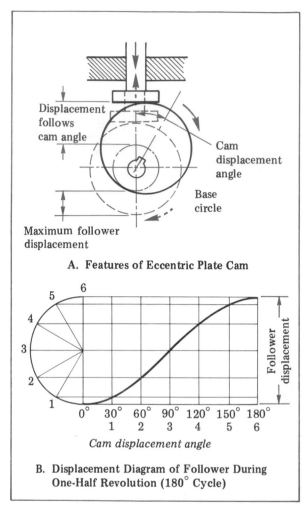

A. Features of Eccentric Plate Cam

B. Displacement Diagram of Follower During One-Half Revolution (180° Cycle)

Figure 15-9 Features and Layout of Harmonic Motion for an Eccentric Plate Cam

HARMONIC MOTION

Harmonic motion relates to a cam and follower that produce a true eccentric motion. The cam moves a follower in a continuous motion at a constant velocity (Figure 15-9A).

The cam displacement angle and follower displacement in Figure 15-9B are at 30° intervals. Harmonic motion cams are used in high-speed mechanisms when uniformity of motion is not the essential design requirement.

COMMON TYPES OF CAMS AND FOLLOWERS

POSITIVE AND NONPOSITIVE TYPES OF CAMS

Plate or bar cams transform linear motion into the reciprocating motion of the follower. Templates that are used to control motion on a tracer miller; profiler; or contour, surface-forming lathe are another form of cam. Cams that impart motion are referred to as *positive* or *nonpositive*.

On a positive design cam, the follower is positively engaged at all times. The cylindrical (drum) cam (Figure 15-10) and grooved recessed plate cam are examples of positive-type cams. Engagement between the cam and follower is accomplished by a pin or roller riding in the groove.

The plate cam and knife-edge follower and the toe and wiper cam and follower (Figure 15-11) are common examples of nonpositive types. The cam produces the direction of motion and acceleration. The follower is pushed against the cam by gravity, a spring, or some other outside force.

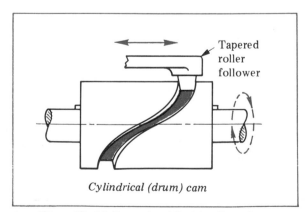

Cylindrical (drum) cam

Figure 15-10 Example of Positive-Type Cam

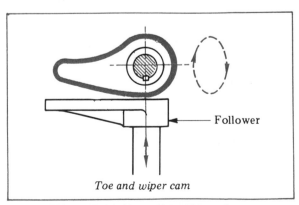

Toe and wiper cam

Figure 15-11 Nonpositive Type of Cam

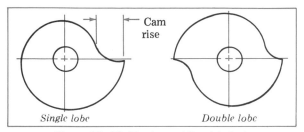

Figure 15-12 Single- and Double-Lobe Uniform Rise Cams

BASIC FOLLOWER FORMS

There are four basic cam follower designs. The *roller (flat) type* has a free, rolling action that reduces frictional drag to a minimum. The *tapered roller type* is tapered to form a cone shape. The angle sides ride in a corresponding groove form. The *flat or plunge type* is widely used to transmit heavy forces in actuating a mechanism through a particular motion, and time sequence. The *knife-edge and pointed type* is used for intricate and precise movements that require a limited contact surface.

GENERAL CAM TERMS

While the field of cam and follower designs and applications is extensive, only six terms used daily in the shop are described here:

- *Lobe* refers to a projecting area of a cam; the lobe imparts a reciprocal motion to the follower; and single- and double-lobe uniform rise cams (Figure 15-12) are common examples.

- *Rise* is the distance one lobe on a revolving cam raises or lowers the follower.

- *Uniform rise* is the distance a follower moves as generated at a uniform rate around the cam.

- *Lead* is the total distance a uniform rise cam (with only one lobe in 360°) moves a follower in one revolution.

- *Cam profile* is the actual contour of the working surface of a cam.

- *Base circle* is the smallest circle within the cam profile.

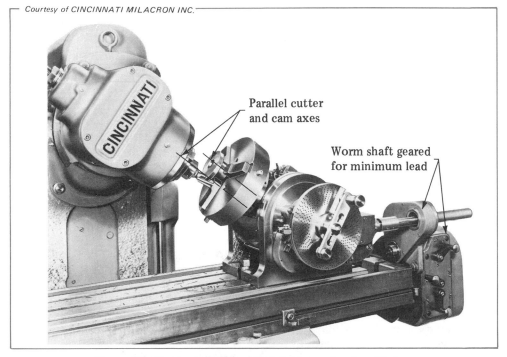

Courtesy of CINCINNATI MILACRON INC.

Parallel cutter and cam axes

Worm shaft geared for minimum lead

Figure 15-13 Vertical Milling Machine Setup for Cam Milling

BASIC CAM MACHINING PROCESSES

MACHINING UNIFORM RISE CAMS

Drum-shaped cams with a uniform rise may be machined as described in the first part of this unit. The groove may be formed on a horizontal miller using a preformed double-angle milling cutter or a square or special-formed end mill.

Uniform rise cams may also be produced on a vertical milling machine. One common machining method is to use a vertical head, the uniform rotation of the dividing head spindle, and the uniform feed of the machine table. In the vertical position, only cams with the same lead as the change gear setup lead may be cut.

Other uniform rise cams may be machined by positioning the axis of the vertical head or attachment (and the end mill) parallel to the axis of the dividing head spindle (workpiece). With the vertical milling machine setup shown in Figure 15–13, any required cam lead may be cut provided the lead is less than the forward feed of the table in one revolution of the work. That is, the required cam lead is less than the lead for which the milling machine is geared.

In milling a uniform rise helix with an end mill at a right angle to the workpiece (using equal gears on the table feed screw and the worm shaft of the dividing head), the table advances 10″ during one revolution of the workpiece. The 10″ results from dividing 40 (turns of the dividing head crank to revolve the spindle one revolution) by 0.250″ (lead of the table feed screw).

Assume an imaginary setup in which the end mill and work axes are parallel and in a horizontal (0°) plane (Figure 15–14). It is obvious that if any cut were taken, the cutter would remain at a fixed distance around the axis of the workpiece. As the workpiece is rotated and fed at a particular lead, the end mill theoretically produces a circular form. No lead would be generated.

The 90° and the theoretical 0° settings represent the range of leads that may be milled. The range is from a maximum of 90°, which the milling machine change gears permit, to zero, as produced at the 0° setting. Thus, it is possible to

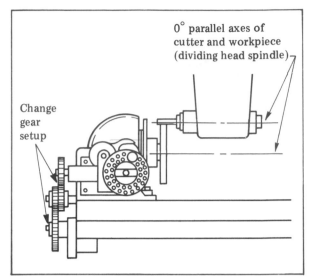

Figure 15–14 End Mill and Work Axes at 0°

machine a specific lead within the range by positioning the cutter and workpiece axes at a given angle between 0° and 90°. The angle setup for cam milling may be computed from a cam drawing and other specifications. The following formula is used to find the lead of a specified uniform rise cam:

$$\text{Lead} = \frac{\text{Lobe Rise in Inches for the Full Cam Circumference}}{\text{Space Occupied by the Lobe in Degrees}}$$

$$= \frac{\text{Lobe Rise (″)} \times 360°}{\text{Lobe Space (° of Circumference)}}$$

Design drawings often represent the circumference as divided into 100 equal parts. Thus,

$$\text{Lead} = \frac{\text{Lobe Rise (″)} \times 100}{\text{Lobe Space (Percent of Circumference)}}$$

CALCULATING CUTTER AND WORK ANGLES FOR MILLING SHORT-LEAD CAMS

The shortest possible lead to mill with standard change gears is 0.670″ (17mm). Any shorter leads require the vertical spindle and the divider head (cutter and work) to be positioned at the

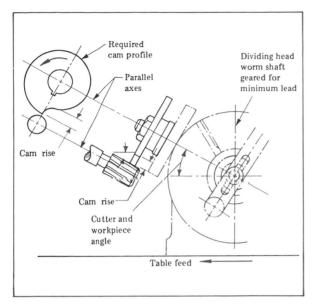

Figure 15–15 Factors Affecting the Angular Setup for Cam Milling

same specific angle. The relation between cutter and dividing head axes, angle of inclination, and cam and table lead are shown in Figure 15–15. Factors affecting the angular setup for cam milling include (1) table feed, (2) dividing head worm shaft geared for minimum lead, (3) cam rise, and (4) parallel cutter and work axes.

Example: If a lead of 0.400″ is to be cut and the shortest possible lead is 0.670″, the fractional value is the sine of the angle to which the cutter and workpiece are set. Thus,

$$\text{Sine of Angle} = \frac{\text{Required Lead}}{\text{Shortest Machine Lead}}$$
$$= \frac{0.400}{0.670}$$
$$= 0.5970$$

From trigonometric tables, a sine value of 0.5970 represents an angle of 36°39′. The vertical head and the dividing head spindles are positioned at this angle.

By setting up the change gears (24, 86, 24, and 100) and positioning the cutter and dividing head at 36°39′, as the table advances 0.670″ in one revolution and the workpiece turns, the cutter mills a cam with a uniform rise of 0.400″.

How to Mill a Uniform Rise Cam

Example: Mill a double-lobe uniform rise cam. Each lobe occupies 180°. Each lobe has a rise of 0.250″.

STEP 1 Calculate the cam lead. Use the formula,

$$\text{Lead} = \frac{\text{Lobe Rise} \times 360°}{\text{Lobe Space (° of Circumference)}}$$
$$= \frac{0.250 \times 360}{180} = 0.500″$$

STEP 2 Refer to a handbook table of change gears for cam milling. Select the gearing for the smallest lead.

Note: A table lead of 0.670″ is produced by positioning an 86-tooth gear on the feed screw (driver), a 24-tooth gear in the first position on the stud (driven), a 100-tooth gear in the second position (driver), and a 24-tooth gear on the worm shaft (driven).

Note: The locking device for the index plate must be disengaged.

STEP 3 Color the cam face with dye. Lay out a center line to indicate the starting point of each lobe.

STEP 4 Set up the workpiece in the dividing head and an end mill in the vertical head spindle.

Note: The end mill must be long enough to provide adequate clearance at the start and finish of each cut.

STEP 5 Calculate the required angle for the dividing head and the vertical head spindle. Use the formula,

$$\text{Sine of Angle} = \frac{\text{Required Lead}}{\text{Shortest Machine Lead}}$$
$$= \frac{0.500}{0.670}$$
$$= 0.7463 \ (48°16′)$$

STEP 6   Set the dividing head at 48°16′. Set the vertical head spindle at the same angle.

STEP 7   Align the center line of the cutter with the scribed center line of the workpiece. Adjust the table until the cutter grazes the underside of the cam blank.

Note: A more rigid setup is produced when the cut is taken from this position. Also, chips flow away faster and layout lines remain visible.

STEP 8   Set the vertical feed collar at zero. Start the flow of cutting fluid. Feed the cutter through 180° by rotating the cam blank with the index crank.

STEP 9   Lower the table at the end of the cut so that the cutter and workpiece clear. Disengage the gear train or the dividing head worm.

STEP 10   Move the table back to the starting position.

STEP 11   Position the work so that the center line on the circumference is aligned with the axis of the cutter.

STEP 12   Reengage the gear train or the dividing head worm.

STEP 13   Mill the second lobe.

STEP 14   Shut down the machine. Check the dimensional accuracy of each lobe.

MACHINING IRREGULAR RISE PLATE CAMS

A number of plate cam applications require an uneven, irregular motion. For these applications, the cams are generally layed out. The cam form is produced by *incremental cuts*. The cam blank is rotated a number of degrees called an *angular increment*. A series of cuts is taken so that each splits the layout line. Each cut is continuously adjusted as necessary to produce the desired form in the angular increment. The ridge irregularities between the milling cuts are then filed and polished.

Safe Practices in Setting Up and Milling Helixes and Cams

• Check the position of each driver or driven gear in a change gear setup for milling. Changing the position of driver gears or driven gears does not change the lead produced. However, interchanging only one driver and one driven gear does change the ratio.

• Check the length of the end mill to be sure it is long enough to produce the helical curve or cam form for short-lead machining.

• Smooth out the small indentations produced in cam milling and polish the surface for smooth operation and longer wear.

• Take successive cuts to rough out deeply formed grooves, particularly if the form is being end milled. Use the largest possible size of end mill.

• Reduce the force against the sharp-angle cutting edges of a formed cutter at the start of a helical milling process. Avoid forcing or bringing the cutter teeth into sharp contact with the workpiece.

• Use cutting fluids to flow away chips and to aid in producing a quality surface finish.

• Hand-feed small end mills and formed cutters until all cutting conditions have been checked and the processes may be carried on by power feed.

• Stop the spindle and cutter before any measurement is taken.

• Check the clearance between the work-holding, cutter, and machining setups and the helix angle setting of the universal table before a cut is started.

• Set trip dogs to ensure that the table will not be moved accidentally beyond the place where the cutter clears the workpiece.

• Disengage the index plate locking device in short-lead milling when the index crank is used to rotate the workpiece.

TERMS USED IN HELICAL AND CAM DESIGN, SETUPS, AND PROCESSES

Cam	A machine element through which a particular pattern of motion is transmitted to a follower.
Follower	A roll, edge, or point on a device that transmits the motion of a cam to the movement of another.
Helix angle	The angle formed by the angular path of the helix and the centerline of the part containing the helix.
Change gear tables (lead milling)	Handbook information of change gear combinations for different leads from 0.670″ (17mm) to 60″ (1,524mm).
Short-lead helixes	Helixes having leads smaller than 0.670″. Smaller helix leads than it is possible to generate with standard milling machine change gears.
Secondary clearance machining setup (milling)	Positioning the workpiece by indexing to permit a plain milling cutter to mill clearance between a flute and a land.
Positive and nonpositive cam types	Cam designs in which the follower is positively engaged at all times or in which an outside force is needed to keep the follower in contact with the cam.
Lobe	A design feature on a cam over which a follower rides to transmit motion.
Incremental cuts (milling)	Machining a cam form by taking a series of cuts. Cuts that are stepped off to conform to the required cam shape at different angular positions of the cam.
Milling short-lead cams	The process of setting up the work and cutter with axes parallel. Machining a cam having a lead smaller than the minimum dimension lead that a standard change gear set may accommodate.

SUMMARY

- A helix has a lead and a helix angle. The distance a groove advances in one complete revolution represents the helix lead. The angle formed by the helix and the axis of the part is the helix angle.

  - Compound change gearing for helix milling involves a gear train of four gears: feed screw (driver), first (inside) position on stud (driven), second (outside) position on stud (driver), and worm shaft (driven).

- Helixes with a lead of 0.670″ (17mm) and larger may be milled with arbor-type cutters by swiveling the table at the helix angle. Helixes may be end milled without swiveling the table.

  - A helix is generated by gearing the dividing head. The workpiece is turned at a particular rate as the table (work) feeds past a cutter the distance of the lead in one revolution.

- Short-lead helixes are produced by direct gearing between the table feed screw and the dividing head spindle.
    - Secondary clearance angles are milled by rotating a workpiece a number of degrees equal to 90° – the secondary clearance angle + 1/2 the included angle of the flute.
- Helical grooves may be milled with arbor-type formed cutters or end mills on a horizontal miller or with end mills on a vertical milling machine.
    - Motion deals with rate of movement. Displacement refers to distance.
- Uniform motion is constant velocity motion in which motion and displacement are constant.
    - Parabolic motion is uniformly accelerated or decelerated during each interval of a cycle.
- Harmonic motion produces a continuous motion at a constant velocity.
    - Cams are of the positive type if they are positively engaged; they are nonpositive if a force is required to keep the cam and follower in contact.
- Followers are designed as straight or tapered rollers, flat or plunge type, or knife edge and pointed types for following precise contours.
    - Uniform rise short-lead cams are milled by setting the workpiece and cutter parallel and at the same angle setting. Change gearing is set for the smallest possible lead.

UNIT 15 REVIEW AND SELF-TEST

1. State three basic functions served by helical forms that are milled into a cylindrical surface.
2. a. Describe the lead of a helix.
   b. Tell how the hand of a helix affects the table setting of a milling machine when an arbor (hole) type of milling cutter is used.
   c. Calculate the helix angle setting for a 50.8mm diameter part that has a lead of 254mm.
3. a. Determine the gear ratio for milling a machine part with an 18″ lead helical groove.
   b. Select a pair of gears for a direct-drive (simple) gearing setup to cut the 18″ lead.
4. a. Refer to a handbook table of change gears for milling different helix leads.
   b. Determine the number of teeth and position of each gear in the change gears that will produce a lead of 190.5mm.
5. Set up a vertical milling machine or prepare a work production plan to mill a three-lobe (120° apart) uniform rise cam. The rise on each lobe is 0.188″.
6. State two personal safety precautions to observe when milling helical flutes or cams.

PART 5 Vertical Band Machines

Band Machine Technology and Basic Setups

Vertical band machines are used for intricate and precise sawing and filing processes and for fast, economical removal of material. This section deals with:

- Principles of band machining and types of machines;
- Functions of major band machine components;
- Factors that affect cutting speeds and feeds;
- Machine and saw band setups for preparing the saw band, setting up the band machine, and regulating the speeds and feeds.

UNIT 16

Band Machine Characteristics, Components, and Preparation

OBJECTIVES

After satisfactorily completing this unit, you will be able to:

- Point out advantages of band machining over other machining processes.
- Describe structural and operational features of vertical band machines and major machine components and types of band cuts.
- Identify and operate the saw band welding components.
- Understand the relationship of jaw gap, jaw pressure, and annealing to the butt welding of saw bands.
- Review common saw band welding problems, causes, and corrective action.
- Perform each of the following processes.
  - Preparing the Saw Band.
  - Setting Up the Saw Band Machine for Operating.
  - Regulating Speeds and Feeds.
- Use technical tables for band sawing, filing, grinding, and polishing.
- Follow *Safe Practices* dealing with personal, machine tool, and workpiece protection and correctly use related *Terms*.

ADVANTAGES OF THE BAND MACHINE

Band machining has a number of advantages over other methods of removing material to produce finished parts. These advantages are:

- Limited material is cut away because a comparatively narrow saw band is used (the amount of material normally wasted as chips are produced);

- Less time is required to remove excess material;

- Angle cutting permits removal of a blank with enough material left to finish machine it as a mating part. The punch and die block of a blanking or forming die set are a typical example;

- Intricate shapes may be cut on a single machine with no limit to the angle or direction and length of a cut;

- The wear on the cutting teeth is distributed uniformly over the great number of teeth on the file band;

- The uniform chip load per tooth and narrow tooth kerf require a relatively light cutting force (less horsepower is required);

- Cutting time is reduced;

- Band saw teeth dissipate the cutting heat faster. Each tooth is in contact with the workpiece for only a short time and can cool before the next cutting pass.

MACHINING SPECIFICATIONS

Band machine specifications include dimensions for *throat* and *thickness* and *transmission speeds*. The dimensions indicate the work space bounded by the column, head, and table. The throat indicates the maximum width or radius of work that may be accommodated. The throat is the distance from the column to the saw band.

The thickness is the maximum work thickness that may be handled.

Some manufacturers use model numbers. For example, in the model number 2014-3, the first two digits indicate a throat of 20" (508mm). The next two digits give the maximum work thickness of 14" (3.56mm). The last digit refers to a three-speed transmission. Other important information besides machine size is provided on the manufacturer's name plate. For example, the length of saw band or file band to use and other facts such as lubrication are given.

TYPES OF BAND MACHINES AND CUTS

Band machines are used primarily for sawing functions such as cutting off, shaping, and slotting. Figure 16-1 illustrates how, by combining internal and external straight and corner radius cuts in two planes, it is possible to produce a part that otherwise would be complicated to machine. In this illustration, the term *three-dimensional cutting* is used. This term means that sawing cuts are made in two or more planes to alter the shape of a part.

Slicing refers to a cutoff operation that is usually performed with a knife-edge blade. *Slabbing* is another cutoff operation. It refers to cutting a thin section or several thin pieces from heavy workpieces.

Band machine shaping refers to cuts that alter the shape of a part by removing excess portions.

SIZES AND TYPES OF BAND MACHINES

General-purpose band machines (Figure 16-2) usually have nonpower-feed work tables. The table can be tilted for angle cutting. Angle cuts up to 10° may be taken when the table is tilted 10° counterclockwise. On large models, the cutting angle is limited to 5°. The table is designed for settings up to 45° clockwise. The workpiece may be fed by hand or power. A *hydraulic tracing accessory* is also available. The cutting path is guided by a stylus as it follows a pattern or template.

Heavy-duty band machines have power-feed work tables. The table is hydraulically fed. The path of the cut is produced according to a steering mechanism setup. The operator guides and controls the movement by turning a handwheel.

Figure 16–1 Three-Dimensional Band Sawing to Produce a Complicated Machined Part

Heavy-duty production band machines are developed to accommodate special needs.

High-tool velocity band machines (also called high-speed band machines) are adapted to the cutting and trimming of nonmetallic products such as nonferrous metals, plastic laminates, paper, wood, and other fibrous materials. Band speeds range as high as 15,000 feet per minute (fpm) to cut these comparatively soft materials. High-speed band machines are also adapted to continuous friction sawing applications. These machines are designed for vibration-free operation at high tool speeds.

Manufacturers' Recommendations. Table 16–1 is a partial table of band machine tool and cutting tool manufacturer's recommendations for conventional sawing. Data is provided for selected band machining high-speed steel and water-hardening carbon tool steel parts. The working data relates to information the machine operator must know for each basic machining process. This information is provided for parts ranging in thickness from 0″ to 12″ (0mm to 304mm).

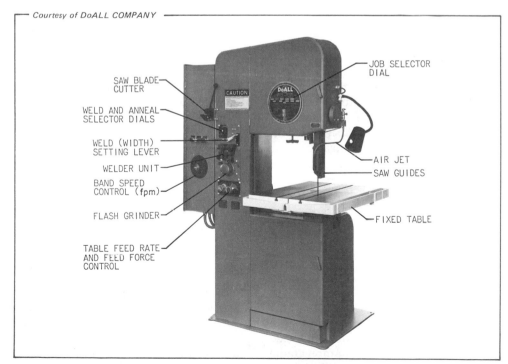

Figure 16–2 Basic Design Features of the Fixed-Table Band Machine

Table 16-1 Partial Table of Manufacturers' Recommendations for Band Machining
High-Speed Steel and Carbon Tool Steel Parts

Material	Work Thickness		Conventional Sawing								Coolant/ Lubricant#
			High-Carbon Saw Bands				High-Speed Steel Saw Bands				
			Tooth Form	Pitch	Band Speed		Tooth Form	Pitch	Band Speed		
	(in)	(mm)			(fpm)	(m/min)			(fpm)	(m/min)	
High-speed steel											
	0-1/4	0-6.4	P	18	140	43					240
	1/4-1/2	6.4-12.7	P	14	110	34	P	10	160	49	240
Tungsten base types T₁, T₂	1/2-1	12.7-25.4	P	10	90	27	P	8	135	41	240
	1-3	25.4-76.2	P	8	70	21	P	6	110	34	240
	3-6	76.2-152.4	C	3	55	17	P	4	85	26	240
	6-12	152.4-304.8	C	3	50	15	C	3	60	18	240
	. . .				. . .				. . .		
Carbon tool steel											
Water hardening	0-1/4	0-6.4	P	18	175	53					240
	1/4-1/2	6.4-12.7	P	14	150	46	P	10	270	82	240 HD-600
W-1 special	1/2-1	12.7-25.4	P	10	125	38	P	8	225	69	240 HD-600
	. . .				. . .				. . .		

Note: mm values are rounded to one decimal place; m/min values, to the nearest whole number
Tooth Form   P = Precision   B = Buttress   C = Claw tooth

FUNCTIONS OF MAJOR MACHINE COMPONENTS

BASIC DRIVE SYSTEMS

Fixed-speed machines have a belted or a direct drive. Intermediate speeds may be set in steps of 1,000 feet per minute (fpm) between low and high range. Variable-speed machines may be set for intermediate band speeds within either a low-speed or a high-speed range.

The controls for setting the band speed are shown in Figure 16-3. The band speed is set by first positioning the *gearshift control lever* to the range of speeds appropriate for the job. The *variable-speed control handwheel* is then turned to increase or decrease the speed within the range. The band fpm (m/min) to which the controls are set is read directly on the *tachometer*.

WORK TABLES

The work table is mounted on and secured to the base by a *trunnion*. A protractor scale and a pointer are attached to the trunnion cradle for angle settings of the table (Figure 16-4).

Power-table band machines are designed with forward and reverse table workstops. The work table rests on hardened and ground steel *slide rods*.

Table feed controls (Figure 16-5) are located in the column. The controls include a *vernier control knob* (feed-pressure) and a *positioning lever*. The control knob regulates feeds from 0 to 12 fpm (0 to 3.6 m/min) on lighter machines and 0 to 8 fpm (0 to 2.4 m/min) on heavy-duty models. Table movement is controlled by turning a directional handle to the power-feed or stop position. The forward position is used for rapid forward traverse; reverse is for rapid reverse traverse away from the work.

CARRIER WHEELS AND BAND ASSEMBLY

The saw band, as the cutting tool, forms an endless band that rides on the *upper* and *lower carrier wheels*. The band is driven by the lower carrier wheel. The upper wheel is free wheeling.

The wheels are flanged and tapered and have a replaceable flanged back. The tapered face permits *tracking*. When correctly adjusted, the

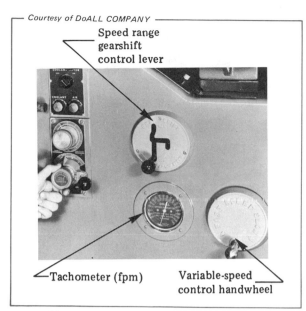

Figure 16–3 Range, Speed Control (fpm), and Speed Indicator

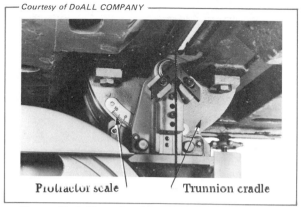

Protractor scale   Trunnion cradle

Figure 16–4 Trunnion Cradle and Protractor Scale Used for Angle Settings of the Table

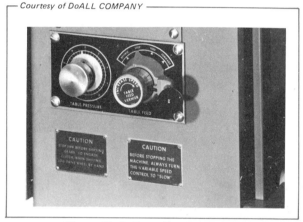

Figure 16–5 Table Feed Controls

saw band rides up the taper, onto the rim, and against the flange at the back of the wheels.

Filing, polishing, and special bands work only on rubber-tired wheels. These wheels are standard equipment on fixed-table and small sizes of band machines. They are available for power-table machines. The rubber tire has a convex shape and is crowned toward the center. As the carrier wheels revolve, the band moves toward and correctly rides on the top of the crown.

Heavy-duty production machines require steel-rimmed, flat-steel, flanged wheels. These wheels permit the saw band to be tightened for more positive drive and nonslip gripping.

Tracking and Saw Band Adjustments. The upper carrier band wheel tilts in or out to position and track a saw band. The tilt angle is adjusted by turning the tilt adjusting screw until the belt tracks. The upper carrier band wheel may also be moved up or down to change the band tension. A removable hand crank is turned clockwise to move the carrier band wheel upward to tighten the tension of the band.

Accurate band machining partly depends on band tension. The amount of and changes in band tension are given on the *band tension*

indicator (Figure 16–6). The indicator face shows three sets of values.

SAW GUIDE POST AND SAW BAND GUIDES

The *saw guide post* serves two prime functions. It supports the upper saw band guide. The post is adjustable up or down to permit machining workpieces of different sizes. Some posts have a calibrated scale to indicate the distance between the table and the section of the saw guide. The post is positioned close to the width of the workpiece in order to provide maximum support for the saw band.

As its name implies, the *saw band guide* functions to provide a bearing for a saw band. During sawing, the saw band is subjected to the force applied by feeding the work into the band. A second

Courtesy of DoALL COMPANY

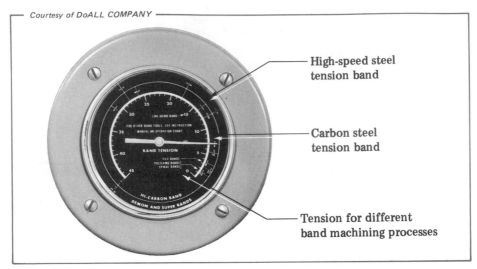

Figure 16–6 Band Tension Indicator

force is produced by the resistance of the workpiece to the cutting action. This force tends to bend (laterally deflect) the saw band sideways. The saw band guides help to overcome these two forces.

The guide must be able to support the back edge of the saw band without causing damage. Therefore, the guide is designed with a *backup bearing.* This bearing takes up the back feed thrust and permits the band to move at high speed. The lateral deflection is controlled by *saw band guide inserts.* They are brought in close to the saw band, but with enough space left to permit the saw band to travel freely during a cut. The two common types of saw band guides are *roller* and *insert.*

Courtesy of DoALL COMPANY

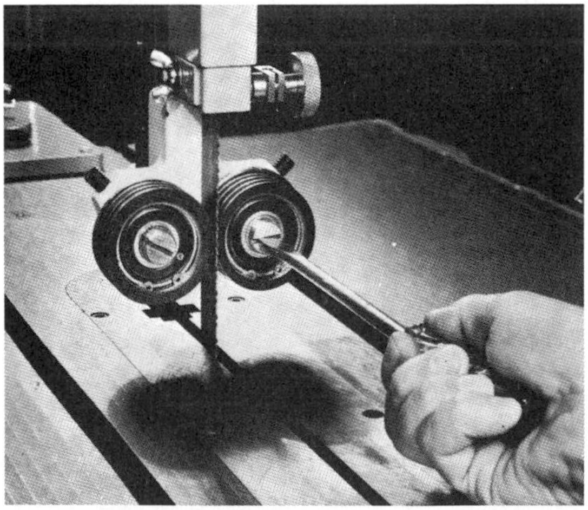

Figure 16–7 Adjusting Roller-Type Saw Band Guide to Accommodate Saw Band Thickness

Roller-Type (High-Speed) Saw Band Guides. Roller-type saw band guides (Figure 16–7) are preferred for continuous high-speed sawing. Rollers are available for use with different widths of saw bands from 1/4″ to 1″ (6.35mm to 25.4mm) and for speed ranges from 6,000 to 10,000 fpm (1,820 to 3,048 m/min).

Insert-Type Saw Band Guides. The *light-duty* saw guide insert is a precision sawing guide for bands from 1/16″ to 1/2″ (1.5mm to 12.7mm) wide. It operates at comparatively low band speeds not exceeding 2,000 fpm (610 m/min).

The *heavy-duty* carbide-faced saw guide insert (Figure 16–8) is used with bands from 5/8″ to 1″ (15.8mm to 25.4mm) wide for heavy-duty continuous sawing.

The *high-speed* saw guide insert is used for band sizes from 1/16″ to 1/2″ (1.5mm to 12.7mm) wide and for band speeds in the 2,000 to 6,000 fpm (610 to 1,830 m/min) range. Guides are available for all different band sizes. The guides on the high-speed type are constructed with a large antifriction bearing and

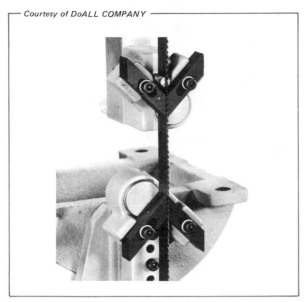

Figure 16-8 Heavy-Duty Model of Insert-Type Saw Band Guides

a hardened, wear-resistant steel, thrust roller cap.

COOLANT SYSTEMS

Heat generated in band machining may be dissipated by either reducing the band speed and cutting feed or using a cutting fluid. In some applications where cutting oils are not required, the chips may be removed by a jet air stream.

Where a mist is preferred, the coolant and air are mixed in the same manifold. The controlled mist is forced by air pressure onto the saw band teeth and workpiece. The nature and amount of coolant flow is regulated by turning the knob on the spray manifold.

SAW BAND WELDING

Four separate components are needed to weld a saw blade. They include a *cutoff shear*, *butt welder*, *annealing unit*, and *grinder*.

CUTOFF SHEARING

The manufacturer's name plate on the machine provides information about the required length of band. This length is layed out and accurately marked. The saw blade is placed (teeth facing outward) in the throat of a cutoff shear. A continuous, firm force is applied to the handle to make a sharp, clean cut.

With the blade cut to length, it is necessary to grind off one or more teeth on each side of the cut. This grinding is required on blades of 4 to 10 pitch because about 1/4" (6.35mm) of

Table 16-2 Position of Shear Cut and Required Teeth to Be Ground for Proper Spacing after Butt Welding (Examples: 4 and 10 pitch saw blades)

Saw Blade Pitch	Equivalent Teeth to Be Ground to Gullet Depth*	Position of Cut Relative to the Gullet of the Teeth		
4	1	1/2 tooth → ←→ ← 1/2 tooth	Area of blade consumed during butt welding	1 tooth ground off
10	2 1/2	1 1/4 teeth → ← ← 1 1/4 teeth		2 1/2 teeth ground off

*Tooth form is ground equally 1/8" (3mm approximately) on both sheared blade ends.

Table 16–3 Control Setting Chart for Jaw Gap, Jaw Pressure, and Annealing Settings

Carbon Alloy Steel Bands						
Width		Gage		Jaw Pressure	Jaw Gap	Annealing Setting
(mm)*	(inch)	(mm)*	(inch)			
1.6	1/16	.6	.025	2	1	1
6.4	1/4	.6	.025	3	4	1
9.5	3/8	.6	.025	3	4	1
12.7	1/2	.6	.025	3	4	1
15.9	5/8	.8	.032	3	4	2
19.1	3/4	.8	.032	3	5	2
25.4	1	.9	.035	4	5	3
25.4	1	1.0	.050	5	6	3

High-Speed Steel Saw Bands						
6.4	1/4	.6	.025	2 turns less than maximum	6	1
12.7	1/2	.6	.025	maximum	6	1

Friction Saw Bands						
12.7	1/2	.8	.032	3	5	2
19.1	3/4	.9	.035	4	5	2
25.4	1	.9	.035	4	5	2

Spiral Band/Rod Welding				
Diameter		Jaw Pressure	Jaw Gap	Annealing Setting
(mm)*	(inch)			
1.6	1/16	2	1	1
2.4–3.2	3/32–1/8	3	3	1
4.8	3/16	4	4	2
6.4	1/4	6	6	3
7.9	5/16	6	6	3

*mm values rounded to one decimal place

Figure 16–9 Positioning the Jaw Pressure Selector at a Required Control Chart Number

material is consumed in making the butt weld. Table 16–2 shows the position of the cut relative to the gullet and the number of teeth to grind off for blades of 4 and 10 pitch.

BUTT WELDING

The edges of the upper and lower welding jaw inserts are beveled. The narrow beveled edge is used for blades up to 1/2" (12.7mm) wide and fine-pitch blades 3/4" (19mm) and wider. The wide, beveled edge accommodates coarse-pitch blades of 3/4" (19mm) or wider. The upper and lower jaw inserts, with matching bevels, are positioned to hold the saw blade segments with the teeth facing inward toward the column.

During the butt welding process, particles of excess metal are blown in the area of the welding jaws and inserts. Therefore, the jaws, clamping assembly, and inserts must be cleaned after every weld.

Using a Control Setting Chart. A *control setting chart* (Table 16–3) is provided by band machine manufacturers. The chart gives the jaw gap and jaw pressure for the width, gage, and material on the saw band. The setting for annealing the welded area is also given.

A number on the *jaw pressure selector* corresponds to a number on the chart for the blade to be welded. Figure 16–9 shows the jaw pressure selector being turned to a particular setting.

The control knob on the opposite side of the butt welder is called the *jaw gap control*. The control knob is turned for the proper jaw gap according to the width, gage, and material in the saw blade.

Making the Weld. If an internal cut is to be made, the blade is moved through the workpiece. The blade is aligned so that the ends are clamped as they touch and are centered between the two jaws.

Caution: Use safety goggles or a shield. Step to one side to avoid the welding flash.

The weld switch is pressed in to its stop. The weld is made automatically. The movable jaw of the butt welder advances the jaw and blade.

At the time welding takes place, there is a consistent flash. It is sharp and bright in color. A dull color or *sputtering* indicates an error in setting the welder controls.

The *reset lever* is raised when the weld is completed. This lever releases the spring tension. The saw band clamp handles are turned down to release the clamping pressure. The saw band is then removed and inspected at the weld.

Inspecting and Treating the Weld. The *flash* is the buildup material that is compressed during heating and flows on both sides of the band at the weld position. The color of the upset material should be blue-gray and of equal intensity across the flash. The spacing of the teeth must be uniform. The weld should be located in the center of the gullet. After grinding, the straightness of the band is checked with a straight edge. If the blade sections are misaligned, the band must be broken at the weld. The edges are ground square and the ends are rewelded. Common saw band welding problems, causes, and corrective action are given in Table 16–4.

ANNEALING THE WELD

During saw band welding, the metal around the joined area hardens and becomes brittle. The band must be annealed in the area. Annealing restores the joint to its original heat-treated hardness.

The butt-welded area is clamped in the butt welder jaws. The teeth are positioned to the rear. The welded section is in the gap midway between the two jaws. The reset lever is moved to the anneal position and then back toward the weld position. This movement positions the movable jaw about 1/16'' (1.5mm) toward the stationary jaw and allows for expansion of the band upon heating.

The *anneal selector switch* is set to the annealing heat number found in the control setting chart. The anneal switch button (Figure 16–10) is jogged intermittently until the saw band reaches a dull cherry red for both carbon and high-speed steel bands.

Note: Overheating, which causes the band to harden and become brittle, must be avoided.

The switch button is then released. As the welded area starts to cool, the switch button is again jogged and released. The cooling process slows down and the annealed area is thus produced.

Table 16–4 Common Saw Band Welding Problems, Probable Causes, and Corrective Action

Problem	Probable Cause	Corrective Action
Overlapped or crooked welds	—Misalignment	—Check and adjust jaw alignment
	—Dirty or worn jaw inserts	—Clean or replace inserts
	—Wrong jaw pressure setting	—Recheck control setting chart
Brittle welds; hard spots in weld; spots of lighter color	—Incorrect jaw gap setting	—Increase jaw gap setting by 1/2 notch
	—Low voltage	—Check incoming voltage
Incomplete or partially joined weld	—Low voltage	—Check incoming voltage
	—Burned or pitted weld switch contacts	—Replace contacts
	—Wrong electrical cutoff timing	—Check timing control
	—Wrong jaw pressure setting	—Reset according to control setting chart

Figure 16–10 Jogging the Anneal Switch Button for Gradual Temperature Buildup and Slow Cooling Rate

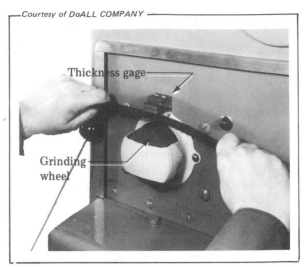

Figure 16–11 Band Grinder Unit

GRINDING THE WELD FLASH

A grinder unit is included on the band machine. The unit has a grinder wheel and a thickness gage (Figure 16–11). After annealing, the excess flash is removed. The welded area is ground to the blade thickness. The band is held with the teeth facing outward. Care must be taken to prevent grinding into the teeth. The band is moved continuously to prevent overheating or burning the ground area. If the saw band is burned, it will need to be rehardened.

Grinding starts at the tooth gullet. The band is inclined at a slight angle with the teeth higher than the back edge. The band is brought into contact with the bottom of the grinding wheel on its front edge. The band is moved forward with a light, uniform force. The positioning and grinding are repeated until the flash is removed.

The band is then flipped over and turned around. The flash is ground off the other side. The blade weld is checked for thickness as the ground area approaches the band thickness. The blade is checked with the thickness gage.

How to Prepare the Saw Band

Preparing the Blade

STEP 1 Determine the saw blade requirements. Check the machinability of the material, machine capability, and job specifications.

STEP 2 Select the saw blade that closely meets the manufacturer's data.

STEP 3 Check the required blade length on the machine name plate.

STEP 4 Pull out slightly more than the required length of saw blade from its container.

Caution: Use gloves to protect the hands from the sharp blade teeth.

STEP 5 Place the saw blade in the throat of the cutoff shear.

Note: The teeth face outward and the blade is held against the squaring bar.

STEP 6 Position the blade at the required length. Apply a firm, continuous force on the shear handle. Cut the blade to length.

STEP 7 Grind off the number of teeth on each end so that the teeth will be uniformly spaced after welding.

Making the Weld

STEP 8 Read the control setting chart for the jaw gap, jaw pressure, and annealing settings.

STEP 9 Set the jaw gap.

STEP 10 Turn the jaw pressure selector knob to the number given on the control chart.

STEP 11 Set the correct beveled edges of the jaw inserts in the inner position.

> Note: The welding jaws must be clean. Any protective coating on the saw blade must be removed. An oil film prevents good electrical contacts.

STEP 12 Position the jaw inserts for the kind of blade, blade width, and pitch.

STEP 13 Place one end in the right jaw insert. Point the teeth to the rear and against the back aligning surface of the jaw.

> Note: The end of the blade must be centered between the two jaws. Then the blade may be clamped in the jaw.

STEP 14 Place the second end in the left jaw insert. Butt both ends. Clamp.

> Note: The blade ends must be checked to be sure they are held securely, touch, and are located midway between the two jaws. The ends should not be offset or overlap.

STEP 15 Press the weld switch button in to a full mechanical stop. Then, immediately release the switch button.

> Caution: Stand away from the machine to avoid the flash and sparks.

STEP 16 Remove the saw band from the vise jaws. Inspect the weld for a uniform flash across the band, a blue-gray color, uniform teeth spacing at the weld, and straightness.

> Note: Any spattered metal particles must be wiped or scraped from the jaws and inserts after the weld.

Annealing the Weld

STEP 17 Move the reset lever to the anneal position. Then, move it back about 1/16" (1.5mm) toward the weld position.

STEP 18 Position the band so that it is centered between the jaws. The saw teeth point inward. Clamp the band.

STEP 19 Set the anneal selector switch at the required annealing heat. Push and jog the anneal switch button in at short intervals until the band reaches the correct color (dull cherry red) and temperature.

STEP 20 Press the anneal switch button occasionally to slow down the cooling process. This produces the right temper. Remove the saw blade.

Grinding and Inspecting

STEP 21 Grind the excess flash off the top of the band. Use the bottom side of the machine grinding wheel.

STEP 22 Repeat grinding the flash on the second side. Check the blade thickness at the weld. Regrind and recheck if needed.

How to Set Up the Machine for Operating

Installing Saw Guides and Inserts

STEP 1 Remove the upper and lower saw guides (Figure 16-12). Clean and inspect the backup bearing.

> Note: Any chips around the bearing must be removed.

STEP 2 Select a set of inserts to match the band size. The insert size appears on the side.

> Note: Inserts that are too wide will damage the saw teeth set; too narrow, will not provide adequate side support.

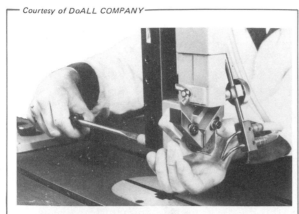

Upper guide

Lower guide (below saw table)

Figure 16–12 Removing Upper and Lower Saw Guides

STEP 3   Mount and set the left-hand insert (Figure 16–13). Turn the holding screw just far enough to prevent sliding.

STEP 4   Place a saw gage corresponding in thickness to the saw band to be used in the right-hand slot. Unloosen the holding screw. Adjust the insert until it touches the edge of the saw gage.

STEP 5   Replace the upper and lower saw guides on the machine. Add the right-hand insert. Adjust to within 0.001″ to 0.002″ (0.02mm to 0.05mm) of the band. Secure.

Note: Roller-type guides are mounted with a flanged backup roller and a flangeless side roller in each block. The rollers are brought up to a correcting tracking band by turning the eccentric bearing shaft. Adjustment is made until the rollers are free to turn without touching the band.

Tracking and Applying Correct Tension

STEP 6   Stop all machine motion. Open the upper wheel door, post saw guard, and the lower wheel door; remove the filler plate.

Note: The table connecting bar on power-table models must be removed.

Caution: Check the position of the gear-shift lever before opening the doors.

STEP 7   Lower the upper wheel with the band tension control.

STEP 8   Grasp the saw band with both hands. The teeth face out toward the operator. The teeth on the right-hand side face downward.

Caution: Wear protective gloves to handle saw blades and bands.

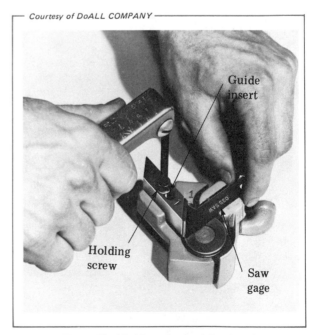

Figure 16–13 Mounting and Setting the Left-Hand Insert

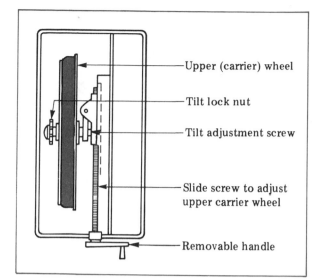

Figure 16–14 Features for Adjusting the Upper Carrier Wheel

STEP 9 Place the saw band under the lower and over the upper carrier wheels.

Note: The band is centered to ride on the center of the wheel crown.

STEP 10 Position the band between the upper and lower guide inserts. The back of the band must touch the backup bearing.

STEP 11 Turn the tension hand crank to take up the slack in the band.

STEP 12 Check the band for correct tracking. Adjust the upper carrier band wheel by turning the tilt handwheel screw clockwise to move the band toward the backup bearings (Figure 16–14); counterclockwise, to move the band away from the bearings.

STEP 13 Close the post saw guard and the upper and lower wheel guard doors. Replace the filler plate on the connecting bar (depending on the machine model).

STEP 14 Read the required tension on the band tension indicator or chart. Turn the band tension crank until this tension is reached. Lock the upper wheel at this position.

Caution: Check the tension indicator regularly, especially with new saw bands that may stretch.

STEP 15 Lower the saw guide post as close as practical to the workpiece.

Selecting and Using Cutting Fluids

STEP 16 Select the cutting fluid, if required, from the manufacturer's tables.

STEP 17 Check and make necessary changes in the cutting fluid and level in the coolant tank.

STEP 18 Position the air and/or coolant nozzle(s) to direct the flow to the saw band.

Note: Grease and solid-type lubricants are used on machines that do not have coolant systems.

STEP 19 Turn the air and coolant mist valve to clear chips away from layout lines. Leave the coolant motor switch off if a cutting fluid is not recommended.

STEP 20 Turn on the coolant motor and the coolant valves. Adjust the quantity of flow.

FACTORS AFFECTING FEEDING FORCE

Feeding the work with too great a force produces three basic problems:

- The band may be deflected (unless the cut is through the workpiece, a curved cut results);
- The backup bearings may be damaged;
- The saw band may be bent back.

Saw pitch affects feeding force. The total force required to penetrate the work depends on the number of teeth cutting at the same time. Too fine a pitch requires a greater feeding force and creates the possibility of inducing a welding action of chips in the gullets. Too coarse a pitch produces an added load on each tooth. This condition may cause the teeth to break or dull faster than is necessary.

How to Regulate Speeds and Feeds

Setting Band Speed

STEP 1 Check the band tension and saw band tracking. Move the gearshift lever to the range recommended on the job selector.

STEP 2 Press the start button for the drive motor.

STEP 3 Turn the variable-speed control handwheel until the band reaches the required speed. Read the speed on the indicator.

Using Table Feed Controls

Pressure Control Method

Note: This method is used for general light-duty sawing of irregularly shaped parts.

STEP 4 Move the table feed control handle to the stop position.

STEP 5 Turn the table pressure knob counterclockwise to zero.

STEP 6 Rotate the table feed vernier knob clockwise to the end point.

STEP 7 Move the table feed control handle to the required feed position.

STEP 8 Turn the table pressure knob clockwise. Adjust the knob to produce a normal chip formation.

Vernier Feed Control

STEP 9 Move the table feed control handle to the stop position.

STEP 10 Turn the table pressure knob to the maximum setting.

STEP 11 Turn the table feed vernier knob to the zero setting.

STEP 12 Move the table feed control handle to the feed position.

STEP 13 Rotate and adjust the table feed vernier knob until the required feed rate is reached and chip is formed.

Safe Practices in Setting Up the Vertical Band Machine

- Close the guard on the saw guide post and upper and lower wheel door guards before applying tension or starting the machine band.
- Position the upper and lower saw guides close to the work to provide maximum support against cambering or deflection during machining.
- Adjust the cutting feed to prevent work hardening, vibration, and bending of the saw band.
- Maintain a feed rate that produces a chip with a tight curl.
- Set the band speed within the manufacturer's recommendations for the job at hand.
- Set the forward and reverse table workstops on power-table machines. These workstops prevent table movement beyond fixed limits.
- Point the teeth of a saw band down toward the machine table in the direction for cutting. The saw teeth dull quickly if reversed.
- Track the saw band to ride centrally on center-crowned wheels or against the rim of the upper carrier wheel.
- Remove particles of excess metal deposited on the jaw inserts during butt welding.
- Inspect each butt weld at the flash joint for cracks and uneven welding of the ends.
- Stay within the recommended cutting feed rate. With experience, it may be possible to safely increase the feed.
- Keep the band tension constant as recommended. Read the tension indicator during machining. Adjust as needed.
- Check each flash area after welding and grinding. The butt weld must be solid across the width of the blade. The blade color must remain unchanged. A crack or incomplete weld or softening of the blade is dangerous and requires corrective action.
- Use protective gloves for handling saw stock or bands. Safety goggles or eye shields are required during machining.

TERMS USED FOR BAND MACHINE FEATURES AND PROCESSES

Band tension indicator Gage that indicates the tension on the cutting tool band during setup and machining.

Saw guide post Vertically adjustable post to which saw band guides are attached.

Cutoff shear Device mounted on the column frame for cutting saw stock square and to length.

Control setting chart (butt welder) Manufacturer's chart for jaw gap and jaw pressure to use in butt welding different widths, thicknesses (gage), and saw band materials. A chart guide that gives the settings for a jaw pressure selector and jaw gap control knob.

Flash Uniformly ridged excess metal formed on both sides of a saw band by butt welding the ends of saw stock.

Annealing (butt weld) Heating the flash area of the saw band to a dull cherry red color and cooling it slowly. The process of reducing blade hardness.

Tracking Positioning a saw band on the upper and lower carrier wheels so that it rides centrally on the upper crowned wheel. Correct positioning of a saw band. (The back edge just grazes the thrust bearings on the upper and lower saw guides.)

Tensioning Applying force to a correctly tracking saw band. Adjustment of the upper carrier wheel to permit it to transmit motion and prevent damage to the saw band for cutting.

Cambering Arcing or bending of the cutting edge, or back edge, of a saw band. Positive camber is backward arcing of the cutting edges of a saw band. Negative camber is forward arcing of the cutting edges.

Chip welding Fusing of a chip to a tooth face. Extreme heating between the teeth and the material in the part.

SUMMARY

- The vertical band machine provides for the fast, economical removal of material. Straight-line or contour sawing, filing, grinding, polishing, friction sawing, and electroband machining are prime processes.
  - Band machines include features for mounting, guiding, and driving different cutting tool bands. A welding, annealing, and grinding attachment is provided to form a saw band.
- The two basic forms of carrier wheels are (1) the center-crowned, rubber-tired wheel and (2) the flat-steel wheel with a flanged inner end.
  - The saw guide post provides a track for the band and an adjustable support for the saw band guides.
- Saw band guides provide a bearing (against which the back of a saw band rides) and flat inserts or rollers to guide and to support the sides of the cutting tool.
  - Coolant and air supply systems permit the use of air to blow chips away at the cutting area, to produce a mist, or to use the air jet together with a cutting fluid.

■ Jaw inserts on the welder are beveled to accommodate the tooth set. The jaw gap and jaw pressure for butt welding depend on the width, gage, and blade material.

  ■ The flash is annealed and ground to blade thickness. Grinding begins in back of the tooth and at the gullet.

■ Band machine setup processes include: selecting and installing the correct size and type of saw guides and inserts; inserting the saw band, tracking, and applying the correct tension; replacing the band guide and upper and lower wheel guards; positioning the guide and inserts and adjusting them; and selecting the appropriate cutting fluid, adjusting the air and coolant pressure and flow, and positioning the nozzles.

UNIT 16 REVIEW AND SELF-TEST

1. Inspect a vertical band machine. Identify three safety features that are incorporated in the design of the machine.

2. a. State the purpose served by grinding off teeth at both ends of a blade where the blade is to be butt welded.

   b. Tell how many teeth are to be ground on a (1) 6-pitch blade, (2) 8-pitch blade, and (3) 10-pitch blade in preparation for butt welding.

3. a. State one function of a roller-type and another function of an insert-type saw band guide.

   b. Identify two construction features that distinguish the roller-type from the insert-type saw band guide.

4. Tell what purpose is served by annealing the weld joint area of a saw band.

5. The four parts listed below are to be band sawed and filed. The material and thickness of each part is also listed.

	Material in Part	Thickness of Part
Part A	High-speed steel T–2	20mm
Part B	High-speed steel M–2	1 1/2″
Part C	Carbon tool steel W–1	8mm
Part D	Carbon tool steel W–2	1 1/2″

   a. Prepare a chart with five vertical columns in addition to the three above. Three of the new columns relate to conventional sawing with HSS saw bands. Mark these three columns: Tooth Form, Pitch, and Band Speed. The two remaining new columns are for recording the Band Tool and Band Speed for filing the parts.

   b. Refer to a manufacturer's table of recommendations for band machining high-speed and carbon tool steel parts to complete the chart.

6. Describe briefly how a mist spray is produced and regulated.

7. State four precautions to take for grinding the flash on a butt-welded saw band.

SECTION TWO

Band Machining: Technology and Processes

This section applies the technology, basic principles, and setup procedures of vertical band machining to straight and contour sawing processes, filing, polishing, and grinding. Spiral band cutting, the use of diamond-edge saw bands, friction sawing, line grinding, and electroband machining are also described. Tables are included on common band machining problems; setups for cutting internal and external sections in one operation; and band velocity, pitch, and cutting rates according to work thickness.

Band Machine Sawing, Filing, Polishing, and Grinding

OBJECTIVES

After satisfactorily completing this unit, you will be able to:

- Assess common band machine sawing problems, possible causes, and corrective steps to take.
- Set up for slotting, slitting, and radius and other contour cutting.
- Understand friction sawing principles, saw band characteristics, and high-speed sawing.
- Discuss spiral-edge saw band cutting and cutting with diamond-edge saw bands.
- Select file bands, speeds and feeds, grit sizes for abrasive filing and polishing bands, and band guide machine parts.
- Perform each of the following processes.
  - Making External and Internal Cuts on a Band Machine.
  - Sawing Out an Internal Section.
  - Setting Up and Band Filing.
  - Grinding and Polishing with an Abrasive Band.
  - Line Grinding.
- Follow each recommended *Safe Practice* in setting up and operating band machines, accessories, and workpieces and correctly use related *Terms*.
- Use technical tables relating to sawing problems, layouts and setups for machining internal and external sections, cutting band characteristics, and other operational data.

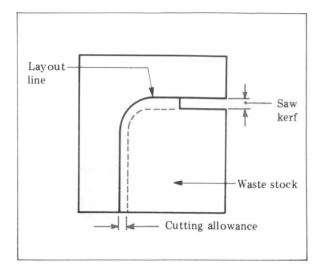

Figure 17–1 Cutting Allowance for Saw Kerf

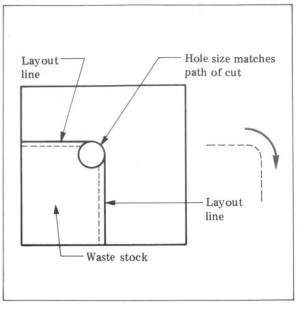

Figure 17–2 Hole Drilled to Match Corner Radius and Tangent to Horizontal and Vertical Layout Lines

EXTERNAL SAWING

Before sawing to a layout line, the machine operator must read the part print to determine the degree of accuracy of the sawed surface. If the specified quality can be met by sawing alone, the cut is taken to the layout line. Allowance is made for the saw kerf from the waste side (Figure 17–1). If the surface is to be further machined, an allowance of about 1/64″ (1.5mm) is made for finish filing; 0.010″ to 0.015″ (0.25mm to 0.4mm), for grinding. Allowance for the width of the saw kerf and an additional width for machine finishing are taken on the waste side.

For cutting corners that have a radius, a hole is drilled to match the radius. If the hole size is large enough, the workpiece may be fed to cut to the first layout line. The workpiece is turned at the corner formed by the radius of the drilled hole (Figure 17–2).

A square corner may be cut to shape by either drilling a hole or making straight cuts. The excess stock at the corner radius is removed by *notching*. The same square corner may also be produced by cutting along one layout line almost to the second line. The workpiece is carefully removed and fed so that a cut is taken along the second layout line.

Figure 17–3 illustrates the sawing technique for cutting a square section without first drilling a corner hole. The waste material is removed. Then, the excess metal left by the curvature of the first cut is removed by notching.

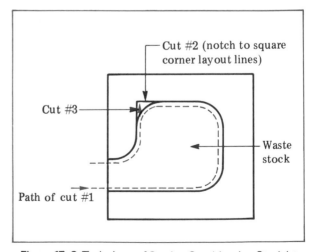

Figure 17–3 Technique of Sawing Combination Straight and Curved Surfaces, Including a Square Corner

How to Make External Cuts

STEP 1 Determine the correct type of saw band teeth, pitch, blade material, and saw width.

STEP 2 Mount, track, and apply the necessary tension to the saw band.

STEP 3 Lower the upper saw guide until it clears the workpiece by approximately 1/4" (6mm). Secure the guide at this position.

STEP 4 Check the job selector for the saw band speed and feed. Set the speed at the required fpm (m/min).

STEP 5 Determine the correct cutting fluid, if required. Position the air jet and the coolant nozzle.

Note: The air stream blows the chips away from the layout lines. The cutting fluid keeps the blade and workpiece cool and adds to cutting efficiency.

STEP 6 Move the workpiece up to the moving saw band.

STEP 7 Start the cut on the waste material side.

STEP 8 Feed the workpiece into the saw band.

Caution: Use a pusher block on small parts. Slow down the feed rate toward the end of the cut.

STEP 9 Stop the machine. Use a brush to remove chips. Burr the workpiece.

INTERNAL SAWING

Internal sawing refers to the removal of an internal section of a workpiece. The cut begins and ends on the inside. The steps in contour sawing inside a workpiece are shown in Figure 17–4. One or more starting holes are drilled tangent to the layout lines. The drill diameter must be large enough to permit the correct width blade to be used.

The saw blade is cut and threaded through a starting hole. The blade ends are butt welded and the flash area is ground to saw band thickness. The saw band, which is free to move through the workpiece, is tracked and then tensioned.

The saw cut is started parallel with the layout line. In cases of internal contour sawing (Figure 17–4), the stock is first notched out. Notching permits feeding the workpiece so that the cut follows the curved layout line.

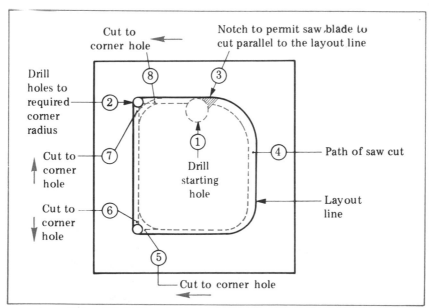

Figure 17–4 Steps in Contour Sawing Inside a Workpiece

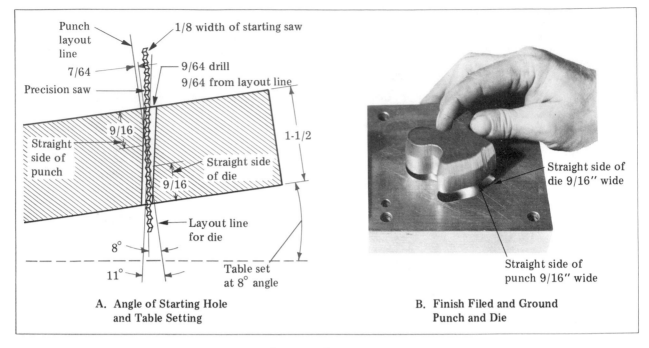

Figure 17–5 Internal Band Sawing to Produce a Punch and Die Block from the Same Piece of Stock

The corners are then notched to remove the excess material. A stock allowance is made for machine finishing the inside surface. When the machine is stopped, the tension is released and the saw band is removed. The saw band is cut at the point of the weld to completely remove the weld.

CONTOUR SAWING MATING PARTS (INTERNAL AND EXTERNAL SAWED SECTIONS)

The internal cutting process makes it possible to use a contour-sawed inner section. By cutting at an angle, enough material is left to finish mating parts to the same size. For example, internal band machining is used extensively for sawing a punch and die block from the same workpiece in one operation (Figure 17–5). The process requires that a starting hole be drilled at a specified angle and that the table be set at a smaller angle (Figure 17–5A). Note that the starting hole is eliminated in the straight section of the punch and die. After finishing to size, the straight sides

of the punch and die are both 9/16″ (14.3mm) (Figure 17–5B).

A starting hole is first drilled at a specified angle. Table 17–1 provides the specifications for laying out, drilling, machine setting, and band sawing internal and external sections from one workpiece. Full information about the drill angle and saw cut angle is given for die thicknesses from 1/2″ (12.7mm) to 6″ (152.4mm). The starting hole begins inside the die layout line. Theoretically, it intersects the layout line at the center of the die block. The hole emerges from the die block on the opposite side of the layout line.

For a 1 1/2″ (38.4mm) thick die block, Table 17–1 gives the distance from the die layout line to the center of the starting hole as 9/64″ (3.6mm). A 9/64″ (3.6mm) starting drill is used. The workpiece is set at an angle of 11°. After the hole is drilled, a 1/8″ wide starting saw blade is inserted through the hole, welded, and ground to band thickness. The band machine table is set to an angle of 8°. A sawing path is followed inside the layout line of the punch. When the inner section is completely sawed at the 8° angle,

Table 17–1 Specifications for Laying Out, Drilling, Machine Setting, and Band Sawing Internal and External Sections from One Workpiece (Partial Table).

Die Thickness		Distance from Die Layout Line to Starting Hole		Diameter of Drill		Angle of Starting Hole (deg.)	Distance from Die Layout Line to Center of Saw Kerf		Angle for Saw Cut (deg.)	Width of Starting Saw		Amount of Straight Sides of Punch and Die		Minimum Outside Radius		Minimum Inside Radius	
(in.)	(mm)	(in.)	(mm)	(in.)	(mm)		(in.)	(mm)		(in.)	(mm)	(in.)	(mm)	(in.)	(mm)	(in.)	(mm)
1/2	12.7	3/32	2.4	1/8	3.2	21	5/64	2.0	18	3/32	2.4	3/16	4.8	9/32	7.1	1/8	3.2
3/4	19.1	1/8	3.2	1/8	3.2	18	3/32	2.4	15	3/32	2.4	9/32	7.1	5/16	7.9	1/8	3.2
1	25.4	1/8	3.2	9/64	3.6	14	3/32	2.4	11	1/8	3.2	3/8	9.5	7/16	11.1	1/4	6.4
1-1/4	31.8	1/8	3.2	9/64	3.6	12	3/32	2.4	9	1/8	3.2	15/32	11.9	7/16	11.1	1/4	6.4
1-1/2	38.4	9/64	3.6	9/64	3.6	11	7/64	2.8	8	1/8	3.2	9/16	14.3	15/32	11.9	1/4	6.4
2	50.8	3/16	4.8	13/64	5.2	10	1/8	3.2	7	3/16	4.8	13/16	20.6	11/16	17.5	7/16	11.1
3	76.2	1/4	6.4	17/64	6.7	9	5/32	4.0	6	1/4	6.4	1-1/8	28.6	7/8	22.2	9/16	14.3
4	101.6	9/32	7.1	17/64	6.7	8	7/32	5.6	6	1/4	6.4	1-5/8	41.3	1-1/32	26.2	9/16	14.3
5	127.0	5/16	7.9	17/64	6.7	7	1/4	6.4	6	1/4	6.4	2-1/4	57.2	1-1/16	27	9/16	14.3
6	152.4	5/16	7.9	17/64	6.7	6	1/4	6.4	5	1/4	6.4	2-1/2	63.5	1-1/8	28.6	9/16	14.3

Note: All mm dimensions are rounded off to one decimal place.

there is excess material inside the die and on the outside of the punch to machine the sides straight for 9/16″ (14.3mm). The starting hole is almost eliminated from the straight sides of both the punch and die.

How to Saw Out An Internal Section

STEP 1 Secure a band machine table of angle settings for sawing internal and external sections. Determine the layout distance from the layout line to the center of the starting hole.

STEP 2 Drill the starting hole with the diameter drill and to the angle given in the table.

STEP 3 Set up the table of the band machine at the required angle for the saw cut.

STEP 4 Thread the precision saw blade of specified width (as given in the table) through the workpiece.

STEP 5 Weld and grind the flash to blade thickness. Anneal and inspect.

STEP 6 Mount, track, and tension the saw band.

STEP 7 Replace the table filler plate.

STEP 8 Set the band speed according to the job selector.

STEP 9 Direct the air and coolant nozzles and start the flow.

Caution: Use protective goggles or a safety shield.

STEP 10 Feed the workpiece carefully so that the blade cuts at the specified distance from the layout line.

Note: A cutting line is usually layed out to assist in cutting at the correct distance from the layout line.

Note: Unless enough material is left to machine straight sides on the two parts that are to be accurately fitted together, one or both parts may be spoiled.

STEP 11 Band file and band grind the two sections to fit within the specified tolerance and surface finish.

STEP 12 Stop the machine at the end of the cut. Remove the blade.

STEP 13 Cut the blade at the weld. Use the cutoff shear.

COMMON BAND MACHINE SAWING PROBLEMS

The band machine operator must be aware of problems caused by incorrect pitch, velocity of the saw band, and feed rate. Saw band tracking and tension, the rigidity with which the workpiece is held, and the type and features of the saw band all affect quality and machining accuracy.

Factors which may cause premature dulling of the teeth, vibration in the cut, teeth chipping or loading, welding of chips, and other damage to the saw band or the workpiece are given in Table 17-2. Probable causes and corrective action the machine operator may take are included.

SLOTTING

Slotting is an adaptation of the band sawing process. It requires the use of a saw band having a kerf equal to the required width of the slot. Slots may be sawed to a tolerance of +0.002" (+0.05mm). The advantage in slotting on a band machine over a milling machine is the shorter distance the table travels to complete the cut.

SLITTING

Slitting is another typical band sawing process. It is used widely to separate parts of castings or forgings. The separated parts are then further machined and held together with fasteners. Other parts such as bearings and bushings are first machined as a single piece and then slit. Fixtures

Table 17-2 Selected Common Band Machine Sawing Problems, Probable Cause, and Corrective Action

Problem	Probable Cause	Corrective Action
Saw band vibrating in cut	—Velocity of saw band is too slow or too fast	—Adjust fpm (m/min) control according to job requirements
	—Feed rate is too light or too heavy	—Adjust rate of speed
Saw band breaking prematurely	—Velocity of saw band is too slow	—Adjust fpm (m/min) control according to job requirements
	—Feed rate is too heavy	—Adjust rate of speed
	—Saw band tension is too great	—Turn tension adjusting handwheel to correct tension
Teeth dulling prematurely	—Velocity of band is too fast	—Adjust fpm (m/min) control according to job requirements
	—Feed rate is too light	—Adjust rate of speed
	—Saw band pitch is too coarse	—Use finer-pitch saw blade
Saw band teeth chipping or breaking	—Velocity of band saw is too slow	—Adjust fpm (m/min) control according to job requirements
	—Feed rate is too heavy	—Adjust rate of speed
	—Saw band pitch is too coarse	—Use finer-pitch saw blade
Loading of gullets of saw band teeth	—Velocity of saw band is too fast	—Adjust fpm (m/min) control according to job requirements
	—Saw band pitch is too fine	—Use coarser-pitch saw blade
Welding of chips to saw band teeth	—Feed rate is too heavy	—Adjust rate of speed
	—Wrong type of cutting fluid and/or improper flow	—Change cutting fluid to meet job requirements
Wandering of saw band	—Feed rate is too heavy	—Adjust rate of speed
	—Saw band improperly tracking	—Adjust carrier wheel tracking nut

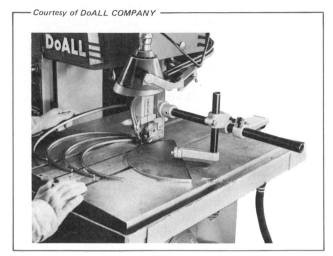

Figure 17-6 Disk-Cutting Attachment for Band Machining Radii

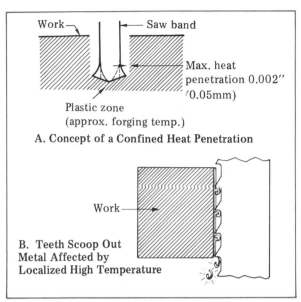

A. Concept of a Confined Heat Penetration

B. Teeth Scoop Out Metal Affected by Localized High Temperature

Figure 17-7 Heating and Cutting Action in Friction Sawing

are used in such applications to seat and locate the workpiece and hold it securely.

RADIUS AND OTHER CONTOUR CUTTING

Contours may be sawed by using any one of three common methods: (1) hand-feeding, (2) sawing with a disk-cutting attachment, and (3) cutting with a contour sawing accessory.

The *disk-cutting attachment* (Figure 17-6) simplifies single-radius contour sawing. This attachment has an adjustable center point. The center is positioned at the radius distance from the saw band. The workpiece center provides a bearing surface for the attachment center. The workpiece is fed at the radius distance past the saw band.

The *contour sawing accessory* combines the forward hydraulic table motion and the rotary motion produced by the turning of the control handwheel.

FRICTION SAWING

PRINCIPLES OF FRICTION SAWING

Friction sawing requires the instant generation of heat immediately ahead of the saw teeth. The heat produces a temperature that causes a breakdown of the crystal structure of the metal. The cutting edge of the saw band then removes the material affected by the friction heat. The

saw teeth, traveling at extreme velocities between 6,000 to 15,000 fpm (1,828 to 4,572 m/min), generate the heat. During friction sawing, sparks and extremely hot metal chips are produced.

Figure 17-7 illustrates the heating and cutting action that occur in friction sawing. Figure 17-7A shows the confined heating area. The maximum heat penetration into the sides of the cut is 0.002" (0.05mm). The heat is further confined to the small area ahead of the kerf. The metal immediately ahead of the saw teeth is referred to as the *plastic zone*. In this zone, the approximate forging temperature of the workpiece is reached. The action of the saw blade in forming the chips is shown in Figure 17-7B.

APPLICATIONS OF FRICTION SAWING

Friction sawing is used to cut hard-to-machine steel alloys without annealing them. Other alloys that work-harden during conventional sawing may be cut efficiently by friction sawing. Friction sawing also replaces slower grinding and chemical machining processes on ferrous metals having a high hardness rating.

Friction sawing often replaces cutting methods that produce distortion in the workpiece. Friction sawing may be used on thin-walled

Blade turned 45°

Figure 17–8 High-Speed, Roller-Bearing Angle Saw Band Guides

tubes, thin parts, and thin-sectioned workpieces. This sawing process may be applied to straight, radius, and intricate contour shapes.

LIMITATIONS OF FRICTION SAWING

Friction sawing does produce a 1/32″ to 1/16″ (0.8mm to 1.6mm) thick, sharp burr on the underside of the work.

Some metals, such as aluminum, copper, and brass, most thermoplastic materials, steels containing tungsten, and most cast irons will not friction saw.

Work thickness is also a limiting factor. Friction sawing is practical for workpieces 1″ (25.4mm) and smaller in thickness. Within this range, it is easy to control the high unit forces between the saw band and the work. Larger sizes may be accommodated if the workpiece may be rocked.

FRICTION SAWING BAND MACHINES

DESIGN OF SAFETY FEATURES

The high band speeds required for friction sawing require the saw band to be completely guarded to the point of the cut and at all times. Heavy-constructed telescoping saw band guards are designed and provided to cover the space between the upper carrier wheel and the work table.

Hydraulic brakes are included on both the upper and lower carrier wheels. The saw band guides, which provide adequate backing and side support for the band tool, are mounted on a heavy, precision-ground post. A slide block provides the bearing on which the post is mounted. A set of high-speed, roller-bearing saw guides is shown in Figure 17–8.

FRICTION SAW BAND CHARACTERISTICS

BLADE SELECTION

The features of the friction saw blade (Figure 17–9) enable it to overcome the extreme flexing action, stresses at high speeds, and heavy feed forces.

The standard widths for friction saw blades are 1/2″, 3/4″, and 1″ (12.7mm, 19mm, and

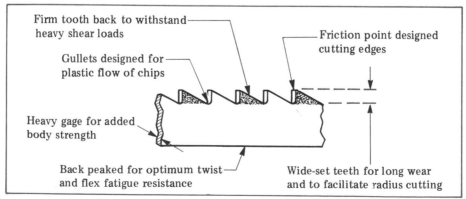

Firm tooth back to withstand heavy shear loads

Gullets designed for plastic flow of chips

Heavy gage for added body strength

Back peaked for optimum twist and flex fatigue resistance

Friction point designed cutting edges

Wide-set teeth for long wear and to facilitate radius cutting

Figure 17–9 Principal Features of a Friction Saw Blade

Table 17–3 Examples of Band Velocity, Pitch, and Cutting Rate for Friction Sawing According to Work Thickness for Selected Metals

Material (AISI–SAE Designation)	Work Thickness in Inches (mm)										
	1/16-1/4 (1.6-6.4)	1/4-1/2 (6.4-12.7)	1/2-1 (12.7-25.4)	1/16-1/4 (1.6-6.4)	1/4-1/2 (6.4-12.7)	1/2-1 (12.7-25.4)	1/16 (1.6)	1/4 (6.4)	1/2 (12.7)	3/4 (19.1)	1 (25.4)
	Saw Band Velocity			Saw Band Pitch			Approximate Linear Cutting Rates in/min (m/min)*				
Carbon steel (plain) 1010-1095	6,000	9,000	12,500	14	10	10	1,400 (35.6)	60 (1.5)	30 (0.8)	8 (0.2)	6 (0.2)
Free-machining steel 1112-1340	6,000	9,000	12,500	14	10	10	1,400 (35.6)	60 (1.5)	25 (0.6)	7 (0.2)	3 (0.1)

*Rounded to nearest tenth of a meter

25.4mm). A standard carbon alloy, precision-type blade is used for applications requiring a narrower saw band width than 1/2″ (12.7mm).

PITCH

Most friction sawing is done with the 10-pitch blade. A 14-pitch blade is used for thin work. The teeth are *raker set*. The pitch is affected by the work thickness and the amount of frictional heat to be generated in relation to the number of teeth that are removing the chips.

Table 17–3 provides examples of the kind of information the operator needs for friction sawing.

FRICTION SAWING SETUP PROCEDURES

Procedures for setting up the friction sawing band machine are similar to the procedures used in conventional sawing. There are two changes, however. First, no coolant is used. Second, roller guides are required unless only an occasional single part is to be sawed.

Extra safety precautions must be observed. The plastic guard must be positioned around the saw band and guides. The machine operator should wear a helmet with transparent visor. A pusher block should be used to feed small pieces. The friction-sawed chip is stubby, wrinkled, and has numerous cracks. The chips differ from the uniformly curled chips produced by conventional sawing.

HIGH-SPEED SAWING

High-Speed sawing refers to the speed of the saw band, usually within the range of 2,000 to 6,000 fpm (610 m/min to 1,830 m/min).

APPLICATIONS OF HIGH-SPEED SAWING

High-speed sawing is practical for such soft materials as plastics, wood, paper, and other fibrous products. It is also widely used on non-ferrous metals, such as aluminum, brass, bronze, and magnesium. Common forms that are cut include bar stock and plate and sheet stock, as well as extruded, cast, and tube forms.

Insert-type guides are used for short runs of high-speed sawing. These guides permit the greatest accuracy. However, production high-speed sawing requires the use of roller-type guides. Since high-speed sawing is used largely in production, special holding, positioning, and feeding fixtures are used.

SPIRAL-EDGE BAND SAWING

Spiral-edge band sawing requires the use of a round, continuous spiral-tooth band of small diameter. The spiral-edge saw band is unique. A continuous tooth is formed by a spiral cutting edge. The saw band is round as opposed to conventional flat-back blades. The spiral tooth design provides 360° of cutting. Intricate-pattern contours may be formed in thin-gage metals and

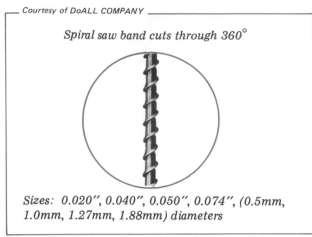

Spiral saw band cuts through 360°

Sizes: 0.020", 0.040", 0.050", 0.074", (0.5mm, 1.0mm, 1.27mm, 1.88mm) diameters

Figure 17–10 Four Basic Sizes of Spiral Saw Bands

other soft materials. Spiral-edge saw bands are practical for cutting small-radius contours that cannot be sawed by a conventional saw band because of the saw band width.

SIZES AND DESIGN FEATURES OF SPIRAL SAW BANDS

There are four basic sizes for spiral saw bands. The outside diameters are 0.020", 0.040", 0.050", and 0.074" (0.5mm, 1.0mm, 1.27mm, and 1.88mm). A close-up of the tooth form is illustrated in Figure 17–10. Spring-tempered spiral-edge saw bands are used for plastics, wood, and soft materials. Hard spiral-edge saw bands are adapted to cutting metals. The sharp, spiral cutting edges permit the cutting of true, precise contours to the minimum radius of the spiral band.

Spiral-edge bands are used only with center-crowned, rubber-tired carrier wheels. A special saw guide design is required.

DIAMOND-EDGE BAND SAWING

Diamond-edge saw bands, like other diamond-impregnated cutting tools, are designed to cut superhard materials. Straight, angular, and contour cutting is done on abrasive plastics, ceramic products, hard carbons, quartz, glass, granite, and other difficult-to-cut materials.

PREPARATION FOR DIAMOND-EDGE SAWING

Diamond-edge saw bands operate at maximum velocities between 2,000 and 3,000 fpm

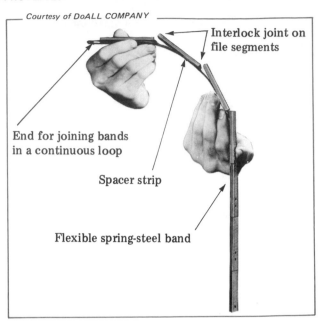

Interlock joint on file segments

End for joining bands in a continuous loop

Spacer strip

Flexible spring-steel band

Figure 17–11 Design Features of a File Band

(600 to 900 m/min). The saw bands are available in 1/4", 1/2", 3/4", and 1" (6.4mm, 12.7mm, 19.1mm, and 25.4mm) widths. Tapered, rubber-tired, flanged carrier wheels and roller-type saw band guides are required.

The diamond band is welded in a manner similar to a conventional saw blade.

BAND FILING

The vertical band machine is used extensively for band filing. *Band filing* refers to the filing of internal or external surfaces to a uniformly smooth, accurate surface. Band filing may be done on almost any kind of metal.

The cutting tool consists of a series of file segments. These segments are riveted on one end to a flexible spring-steel band (Figure 17–11). As the flexible band travels around the carrier wheels, the loose ends of the file segments lift away. When the flexible band returns to the vertical plane, the loose end of one file segment fits into the next segment. A special interlocking joint between file segments permits the formation of a continuous, flat filing band.

KINDS OF FILE BANDS

Cut, as related to file bands, refers to the short angle or bastard pattern; grade (such as coarse,

Table 17-4 Sample File Band Segment Cuts and Applications

| Sample File Segment | Cut of File Segments | | | Width | Application |
	Tooth Form	Grade	Number of Teeth		
Flat	Short angle	Coarse	10	3/8", 1/2" (9.5mm, 12.7mm)	Aluminum, brass, copper, cast iron, zinc
Short angle, coarse cut, 10 teeth — 1/2" (12.7mm)	Bastard	Medium-Coarse	14	1/2" (12.7mm)	General use on steel
		Coarse	12	3/8" (9.5mm)	General use on cast iron and nonferrous metals
Bastard, medium-coarse cut, 14 teeth — 1/2" (12.7mm)		Medium-Coarse	16		General use on tool steel
	Medium	20		3/8" (9.5mm)	Medium finish on tool steel
				1/4" (6.3mm)	General use on tool steel

medium-coarse, or medium); and the number of teeth. The job selector is used to identify the type of file. For example, a short-angle file is used for aluminum, brass, cast iron, copper, and zinc products.

Table 17-4 identifies information which is available for flat file band segments. The two other common segments are the oval and half-round. The widths of these files are 1/4", 3/8", and 1/2" (6.3mm, 9.5mm, and 12.7mm).

The file cuts illustrated in Table 17-4 are for 10, 12, 14, 16, and 20 teeth. The short-angle coarse cuts with 10 teeth are for band filing soft materials; the medium-coarse cuts, for general use on mild steels; and the medium cuts for filing tool steels.

BAND FILING SPEEDS AND FEEDING FORCE

Band speeds between 50 to 100 fpm (15.2 to 30.5 m/min) are used for all-purpose band filing applications. However, recommendations on the job selector and in manufacturers' tables on band filing should be followed.

A continuous, steady, light force (pressure) is applied. The workpiece is moved with an even movement across the width to be filed. When the pattern of file lines on the workpiece shows perpendicular lines, the cutting action is correct. Diagonal and criss-cross patterns indicate the

workpiece is being moved too fast from side to side.

Heavy and excessive feed pressure may load the tooth gullets, break or stall the file band, or groove the lower carrier wheels. Light pressure produces a better surface finish in the same time.

Band filing operations are performed from the right side of the table. The file segment design requires them to flex from a spring-steel band. The cutting teeth face the right side. All cutting is done on the face of the file.

How to Set Up and Band File

Replacing the Saw Band and Guides

STEP 1 Remove the center plate for sawing.

STEP 2 Release the tension in the saw band. Remove and store the band in its proper storage area.

STEP 3 Remove the saw band guides.

STEP 4 Mount the adapter and file guide support on the keeper (lower) block (Figure 17-12).

Note: The slot in the backup support must accommodate the width of the file band.

STEP 5   Lower the upper guide post to within working distance of the workpiece.

Note: If the work thickness is more than 2" (50.8mm) for a 1/4" (6.2mm) file band or more than 4" (101.6mm) for a 3/8" or 1/2" (9.5mm or 12.7mm) file band, a special set of larger guides must be inserted.

STEP 6   Install the upper file guide. Secure it to the saw band guide post.

Installing the File Band

STEP 7   Select the correct file band to meet the job requirements. Grasp the file band in both hands. One hand holds the painted end.

STEP 8   Hold the two ends at an angle (Figure 17–13).

STEP 9   Set the rivet head into the slotted hole. Slide the rivet head toward the small end of the elongated slot.

Note: For internal filing, the band section is threaded up through the workpiece. The two file band ends are then joined at the machine.

STEP 10   Straighten the file band to allow the spring-steel end to snap over the dowel.

Note: The ends of the file segments must be checked to be sure they interlock properly.

STEP 11   Place the file band on the carrier wheels so that the band is aligned with the file guides.

STEP 12   Install the center plate used for band filing.

STEP 13   Track the file band in the center of the carrier wheels.

Note: Center-crowned, rubber-tired wheels are recommended for band filing.

STEP 14   Close the carrier wheel and the guide post guards.

STEP 15   Apply the band tension specified by the blade manufacturer.

Note: A light band tension produces more accurate filing results.

Caution: Too much tension may cause the file segment rivets to shear when a heavy filing force is applied.

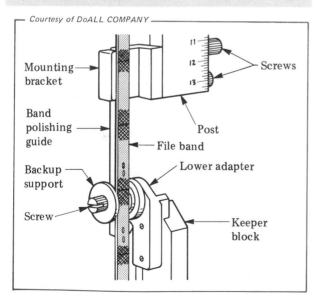

Courtesy of DoALL COMPANY

Mounting bracket

Screws

Band polishing guide

Post

File band

Backup support

Lower adapter

Screw

Keeper block

Figure 17–12 File Band Guides and Supports Mounted on Keeper Block

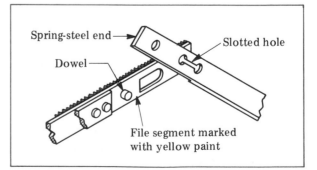

Spring-steel end

Slotted hole

Dowel

File segment marked with yellow paint

Figure 17–13 Position of File Band Segments for Joining the End Sections

STEP 16 Shift the carrier wheel (file band) speed to the low gear range. Start the drive motor. Check the band tracking.

Note: A correctly tracked file band runs freely in the file guide channels.

Band Filing

STEP 17 Set the file band speed according to the job requirements.

STEP 18 Position the air jet, coolant nozzle, or spray mist combination.

Note: The workpiece and file band must be kept cool. The teeth and grooves in the file segments must be clean.

STEP 19 Start the machine. Stand at the right side. Apply a steady, light force and move the workpiece across the area being filed.

Note: If the teeth load, stop the machine. Use a file card to remove the file particles from each segment.

Note: Loaded teeth cause irregular filing and produce a scratched surface.

STEP 20 Stop the machine at the end of the job. Shift the transmission to neutral. Inspect and clean the file band with a file card.

Note: Every file segment must be cleaned.

STEP 21 Clean the machine table. Remove and return the file band to storage.

BAND GRINDING AND POLISHING

Band grinding and/or *polishing* produces an excellent grained or polished surface in contrast to the coarser surface produced by band sawing or filing. Abrasive polishing bands are used. The bands are 1″ (25.4mm) wide.

APPLICATION OF GRIT SIZES

The 50 grit size of aluminum oxide abrasive bands is adapted to heavy stock removal operations. The operating band speed should be checked on the job selector. The 80 grit size is for general surface finishing or coarse polishing operations. Band speeds for these operations run to 1,000 fpm (305 m/min). The 150 grit size is for light stock removal and high polish applications. Band speeds range between 800 and 1,500 fpm (244 and 457 m/min) for these abrasive bands.

ABRASIVE BAND GUIDE COMPONENTS

Abrasive bands require a rigid backup support and an adapter. The abrasive band adapter replaces the grooved face adapter used for the file band support. Figure 17–14 shows how the band polishing guide (which is secured in the upper mounting bracket) rides against the polishing guide support. The guide usually has a graphite-impregnated facing, which serves as a dry lubricant. A graphite powder is usually rubbed in the polishing band fabric to lubricate it and to increase wear life.

SILICON CARBIDE ABRASIVE GRINDING BAND

Silicon carbide abrasive bands are used for grinding tough, hard metals such as carbide-tipped cutting tools. Grit sizes of 50, 80, and 150 are common. The band width is 1″ (25.4mm). A coolant is generally used.

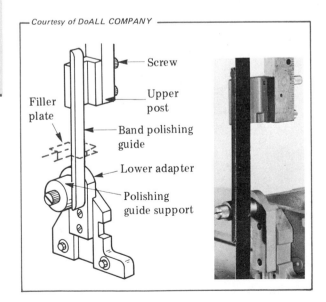

Courtesy of DoALL COMPANY

Screw
Upper post
Filler plate
Band polishing guide
Lower adapter
Polishing guide support

Figure 17–14 Guide Components for Band Polishing

Since a steady, heavier force is required, a *tool finisher attachment* is available. The tool is gripped and held securely in the attachment. The movement of the cutting tool is controlled against the abrasive band. The combination of cutting with a silicon carbide abrasive band, holding the part securely, and applying a constant force across the area being polished produces a superior finish.

How to Grind and/or Polish with an Abrasive Band

STEP 1   Select the abrasive grain and grit size appropriate for the grinding or polishing process and quality of surface finish.

STEP 2   Set up the rigid backup band polishing guide and the lower adapter support.

STEP 3   Replace the center plate with the polishing band center plate.

STEP 4   Lower the saw band guide post to within 4″ (100mm) of the work table.

STEP 5   Mount and track the abrasive band.

STEP 6   Check a band tension chart for the correct abrasive band tension. Adjust the tension control handwheel.

STEP 7   Set the band speed according to the job selector or manufacturers' recommendations.

STEP 8   Bring the part to be ground or polished to the abrasive band. Apply a steady force as the workpiece is moved slowly across the surface area.

Note: The tool or workpiece is held and fed with a tool finisher attachment when a silicon carbide band is used and a superior surface finish is required.

LINE GRINDING

Line grinding requires a line-grind band. This band tool has a continuous abrasive cutting edge bonded to a steel band. The abrasive coating may be silicon carbide or aluminum oxide. Carbide-coated bands are recommended for grinding quartz, granite, ceramic, glass, and other hard nonferrous materials. Bands coated with aluminum oxide are best for grinding metals, heat-treated alloy steels, heat-resistant steels, wear-resistant nonferrous alloys, and other difficult-to-machine materials.

QUALITIES OF LINE-GRIND BANDS

Line-grind bands produce a clean, smooth finish surface. The grinding capacity (width) is limited only by the working depth of the machine.

Tapered, rubber-tired carrier wheels and roller-type saw guides are used with line-grind bands. The bands are welded the same as diamond bands. The coated edge is protected by the beveled jaw inserts that face the operator. Before welding, 1/8″ (3.2mm) of abrasive is removed from each end of the band. When this amount of the metal band is consumed in welding, the abrasive edge forms a continuous band.

HEAT GENERATION AND BAND SPEEDS

Considerable heat is generated by the cutting action of the abrasive grains. Therefore, a flood coolant application is needed. Special qualities are required in the coolant. It may be necessary to flush the system to change to the coolant recommended for line grinding a particular metal.

Band speeds for line grinding between 3,000 to 5,000 fpm (900 to 1,500 m/min) are best. Bond abrasives are sharpened by dressing the band with a diamond dressing stick. The stick is moved lightly across the abrasive cutting edge.

How to Line Grind

STEP 1 Select the type of abrasive and grit size according to the job requirements.

STEP 2 Cut the abrasive band to length. Grind, butt weld, remove flashing, and anneal.

Note: The band is threaded through the section when an internal area is to be ground.

STEP 3 Install the band around the carrier wheels. See that it tracks correctly around the tapered, rubber-tired carrier wheels and the roller guides.

STEP 4 Set the band speed according to the material to be ground and the nature of the operation.

STEP 5 Position the coolant nozzle and control the flow to deliver a heavy stream.

STEP 6 Use a light hand-feeding force. Grind to the required layout line or dimension.

Note: A steady force is applied as the workpiece is moved across the width of the section being ground.

STEP 7 Dress the abrasive band when the grains dull and for final finish cuts.

ELECTROBAND MACHINING

Electroband machining operates on the scientific principle of disintegrating the material in front of a cutting edge without having a knife-type band ever touch the material. A low-voltage, high-amperage current is fed into a knife-type band. The electrical input causes the knife edge to discharge a sustained arc into the material being cut. The arc disintegrates the material at the same instant a flood of coolant quenches the arc. The coolant prevents burning and damaging the material.

ELECTROBAND TOOL CHARACTERISTICS, OPERATION, AND FEED RATES

The band tool used in electroband machining must have properties to withstand arcing with-out cracking and to maintain long flex life. Electrobands are welded in the same manner as knife-edge blades.

The electroband machine tables are hydraulically powered. The band speed is fixed at 6,000 fpm (1,828 m/min). Splash guards contain the coolant without obstructing the operation.

Surfaces requiring a fine finish may be electroband machined at a rate ranging from 5 to 50 square inches (32.2 to 322 sq. cm) per minute. This rate can be increased (when the quality of the work edge is not important) to 150 to 200 square inches (968 to 1,290 sq. cm) per minute.

KNIFE-EDGE BANDS AND PROCESSES

Thus far, blades with cutting teeth have been described and applied primarily to metal-working jobs and processes. *Knife-edge blades*, as the name implies, have a knife edge. The edge may be straight (knife), wavy, or scalloped. When formed into a band, the knife-edge blade is especially adapted for cutting soft, fibrous materials. Other types of cutting edges would tear or fray such materials.

The knife-edge saw blade is produced with a single- or double-bevel cutting edge. The wavy-edge and scallop-edge blades have a double-bevel cutting edge.

The three types of saw blades come in different widths and gages. The sizes of the knife-edge blades range from 1/4″ to 1 1/2″ (6.4mm to 38.2mm) wide and from 0.018″ to 0.032″ (0.46mm to 0.8mm) gage thickness. The sizes of the scallop- and wavy-edge saw blades are from 0.015″ to 0.032″ (0.4mm to 0.8mm) thick and from 1/4″ to 2″ (6.4mm to 50.8mm) wide.

Knife-edge bands cut efficiently at relatively high band speeds. The bands must withstand maximum *flexation* (bending) over the carrier wheels. In addition, the blades must be ground to razor sharpness and be able to maintain it over long usage. Most knife-edge cutting is done dry and at high speeds. The knife-edge bands slice through to separate material without producing a kerf. As a result, no chips are produced.

Wavy- and scallop-edge forms are used on materials that offer increased resistance to cutting.

Safe Practices in Setting Up and Performing Basic Band Machining Operations

- Use the machine roller guide shield on high-speed saw cuts. In addition, wear protective safety glasses when working at or near a band machine.
- Replace the table filler plate with the plate designed for the operation (sawing, filing, grinding, and so on) to be performed.
- Wear gloves for handling saw bands during mounting and dismounting. Then, remove the gloves unless the condition of the workpieces or operation justifies their use.
- Check each cutting band before mounting for possible weld defects, the condition of the teeth or cutting edge, and signs of band fatigue.
- Recheck band tracking and the application of the recommended band tension on all band machine cutting tools. Close all doors and secure all guards before starting the machine.
- Adjust the automatic hydraulic carrier wheel brakes for the band tool being used.
- Use only center-crowned, rubber-tired carrier wheels for driving spiral-edge saw bands.
- Apply less force when hand-feeding as the blade starts to cut through. Use a block or pusher on short workpieces.
- Check the feed rate at which the machine table or feed accessory is set against the job selector.
- Remove chips from the grooves of each segment of a file band. Otherwise, scratches and a poor-quality filed surface will be produced.
- Apply a continuous, light force when band filing or grinding. Light force is preferred to excessive force, which may load the file band teeth or abrasive band or cause the file band units to shear.
- Use a flood coolant to remove the excessive heat produced with line-grind bands.
- Set the splash guards to collect the flood of coolant from electroband machining.

TERMS USED IN BASIC BAND MACHINE PROCESSES

Friction sawing	A sawing process where heat is instantly generated ahead of the saw teeth. Scooping out metal at forging point temperature.
Plastic zone	The metal in a limited area immediately ahead of the saw teeth. Metal heated by friction to a temperature where the crystals break down and the plastic metal is removed by the saw teeth.
Thermal conductivity	The ability of a material to conduct heat. (The low thermal conductivity of steels permits limited heat penetration to 0.002″ (0.05mm) on the kerf sides of a cut).
High-speed sawing	A method of band sawing at cutting speeds within a range of 2,000 to 6,000 fpm (610 to 1,830 m/min).
Spiral-edge saw band	A circular saw band formed with a continuous spiral cutting edge. A round, spiral-cutting saw band for sawing a precise contour with a minimum radius.
Abrasive band	A continuous fabric band coated with grains of aluminum oxide or silicon carbide. A band used for polishing a previously file finished surface.
Band grinding/polishing	Use of an abrasive band for purposes of grinding and polishing a surface to a required degree of accuracy.

Rigid backup support and adapter	The support in back of the abrasive band. An abrasive band guide that permits grinding a flat surface. An adapter below the work table against which the backup support rests.
Diamond-edge saw band	A uniformly distributed concentration of industrial diamonds fused to a steel band.
Line grinding	A grinding process using a steel band that has one edge coated with either silicon carbide or aluminum oxide abrasive grains.
Electroband machining	The discharge from a saw band producing an arc that disintegrates the material to be cut at the same instant the arc is cooled by a flow of coolant.

SUMMARY

- An allowance of 1/64″ (1.5mm) is made on a sawed surface for finish filing; 0.010″ to 0.015″ (0.25mm to 0.4mm), for band grinding.
  - The band machine operator must determine the following saw band requirements from a part print and the job selector: the type of saw band teeth, pitch, kind of blade material to use, and the blade width and thickness.
- Knowing the kind of material, thickness, and surface finish requirements, the operator needs additional information relating to the machine setup as follows: the cutting speed (fpm or m/min) of the saw band, the band tension, the recommended feed rate, the type of liquid and/or air coolant, and quantity (flow) of coolant (when used).
  - The technique of contour sawing mating parts from the same workpiece requires a starting hole to be drilled at a specified angle.
- Slotting requires the use of a saw band having a kerf the same size as the required width of the slot.
  - Friction sawing is done by using a special saw band traveling at velocities of 6,000 to 15,000 fpm (1,828 to 4,572 m/min). The heat generated by friction produces a breakdown of the crystal structure immediately ahead of the saw band teeth.
- High-speed sawing is used primarily on soft and fibrous materials. Saw band speeds range from 2,000 to 6,000 fpm (610 to 1,830 m/min). The process produces a heavy chip formation and requires a rapid chip flow.
  - Spiral-edge band machining requires the use of a round, continuous spiral-tooth band of small diameter (0.020″, 0.040″, 0.050″, and 0.074″ (0.5mm, 1.0mm, 1.27mm, and 1.88mm). Spiral-edge bands are used with center-crowned, rubber-tired roller guides and with specially grooved jaws on the butt welder.
- The three common shapes of file segments are square, half-round, and oval. The file widths are 1/4″, 3/8″, and 1/2″ (6.3mm, 9.5mm, and 12.7mm). Different angles, tooth forms, and pitches are available.

- Diamond-edge saw bands have a concentration of industrial diamonds distributed uniformly and fused on a band. Straight, angular, and contour cuts are possible on superhard and abrasive-action ferrous and nonferrous materials.

- Band grinding and polishing produce finely grained surfaces as a second operation. Abrasive bands, coated with either aluminum oxide or silicon carbide grains, are used. Three basic grain sizes (50, 80, and 150) produce coarse, medium, and fine surface finishes.

UNIT 17 REVIEW AND SELF-TEST

1. Referring to a job selector on a band machine, list the kind of information an operator needs to know about conventional sawing a 25.4mm thick block of free-machining steel.

2. List three differences between high-speed sawing and friction sawing.

3. a. Identify two distinguishing design features of a spiral-edge saw band.
   b. Cite two types of jobs for which spiral-edge band sawing is adapted.

4. Compare electroband machining with conventional band sawing in terms of (a) principles of machining and (b) a typical application.

5. a. Study the following working drawing of a Form Guide. This part is to be band machined.
   b. List the work processes required (1) to saw the square corner, relief, and the contour; (2) to file; and (3) to grind the 0.750 ±0.001 area. Assume the workpiece has the one edge ground to $\frac{32}{}$, the plug end is square, and the part is layed out.

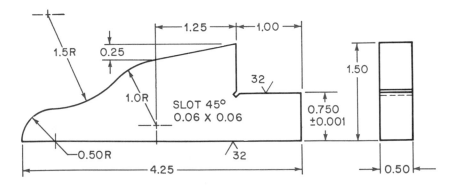

6. a. State one similarity of a line-grind band and an abrasive band.
   b. Give two major differences between the two bands.

7. Give three personal safety precautions prior to operating a band machine.

PART 6 Shaping, Slotting, Planing Machines and Broaching

SECTION ONE

Machine Tool Technology and Processes

Shapers, planers, slotters, and broaching machines are used for machining internal and external plane and contour surfaces. Shapers, slotters, and planers generate a required linear surface or form by a reciprocating straight-line cutting motion. Material may be removed to produce a horizontal, vertical, or angular plane surface or a varying-contour form. This section provides an overview of shaping, slotting, and planing machine design features, fundamental tool geometry, cutting speed and feed computations, and basic setups, and other broaching machining processes.

UNIT 18

Shapers, Slotters, Planers, and Broaching Machines

OBJECTIVES

After satisfactorily completing this unit, you will be able to:

- Understand work, tool positioning and tool holding, and movable components of vertical and horizontal surface forming machine tools and machining processes.
- Describe setups and cuts: horizontal, vertical, angular, contour, groove, spline, and internal, as related to shaping, slotting, planing, and broaching.
- Explain major features of planer setups, and planing operations using a rail head and a side head.
- Describe tool geometry and applications of adjustable, serrated planer tool bits and tool holders.
- Apply formulas and tables of cutting speeds, feed rates, and depth of cut for different cutter materials.
- Trace step-by-step procedures for performing the following processes.
  - Setting Up and Machining a Part Square and Parallel.
  - Machining a Vertical or Angular Surface.
  - Machining Internal Surfaces and Features.
  - Broaching a Keyway.
- Follow *Safe Practices* in shaper, slotter, planer work, and broaching operations.
- Correctly use related *Terms*.

259

BASIC TYPES OF SHAPERS

There are two basic types of shapers: *vertical* and *horizontal*. As the names indicate, the movement of the cutting tool is in either a vertical plane or a horizontal plane.

THE VERTICAL SHAPER (SLOTTER)

The vertical shaper is sometimes called a *vertical slotter*. It is designed primarily to shape internal and external flat and contour surfaces, slots, keyways, splines, special gears, and other irregular surfaces. The design features of a vertical shaper are shown in Figure 18-1.

A rotary table is mounted on the carriage. The table is also movable longitudinally. The combination of three directions of movement permits the table and workpiece to be positioned and fed for straight-line and rotary cuts.

Motion for cutting is produced by the reciprocating movement of the *vertical ram*. The ram and mechanism for producing motion are housed in the column. Cutting tools are secured in a toolholder on the forward section of the ram. The ram position is adjustable. The cutting tool is set in relation to the starting and ending of a cut. The distance the ram travels is called the *length of stroke*. The ram on some machines may be adjusted at an angle to take angular shaping cuts on vertical surfaces. Power feed is provided for longitudinal and transverse feeding. Generally, the rotary table is turned manually. Cutting speeds are controlled through drives in different speed ranges.

The vertical shaper is of heavier design and construction than most machine tools used in toolrooms and jobbing shops. Workpieces are usually strapped to the work table. Many of the same types of straps, clamps, and T-bolts used

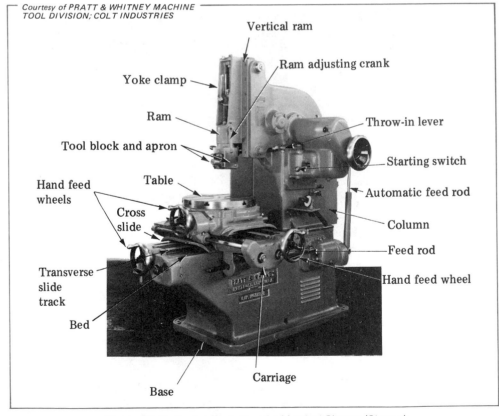

Courtesy of PRATT & WHITNEY MACHINE TOOL DIVISION; COLT INDUSTRIES

Vertical ram
Ram adjusting crank
Yoke clamp
Ram
Throw-in lever
Tool block and apron
Starting switch
Hand feed wheels
Table
Automatic feed rod
Cross slide
Column
Transverse slide track
Feed rod
Hand feed wheel
Bed
Carriage
Base

Figure 18-1 Design Features of a Vertical Shaper (Slotter)

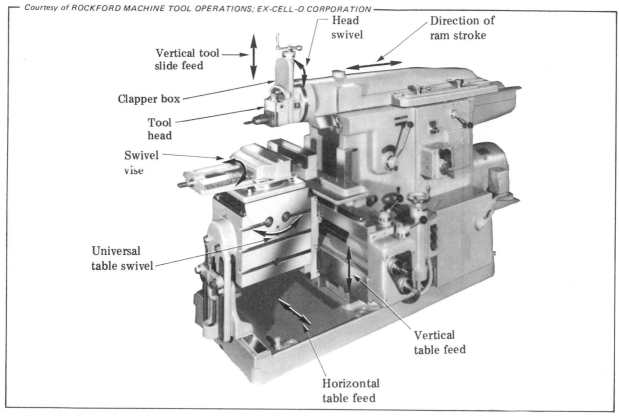

Figure 18–2 Positioning, Feeding, and Cutting Movements on a Universal Horizontal Shaper

for milling machine setups are used with the vertical shaper/slotter. Parallels, step blocks, angle plates, and adjustable jacks are required for positioning and supporting overhanging and other sections. Special end-cutting tools and other tools similar to shaper cutting tools are used for vertical cuts.

THE HORIZONTAL SHAPER

In contrast to the reciprocating cutting movement of the vertical shaper, cutting action takes place on the *horizontal shaper* by feeding the workpiece for the next cut on the noncutting return stroke. The cutting tool removes metal on the forward longitudinal stroke of the *horizontal ram*. A surface is shaped by taking a series of successive cuts. For each cycle of the ram, cutting is done on the forward stroke; feeding, on the return stroke. Shaping of small workpieces is usually done on *bench shapers*.

Positioning, Feeding, and Cutting Movements. The different positioning, feeding, and cutting movements on a universal horizontal shaper are shown in Figure 18–2. Note on this model that the *universal table* may be rotated through a 90° arc.

The *tool head* may be swiveled to feed the cutting tool at a required angle. The angle settings are read directly from the angle graduations on the end of the ram. The clapper box is adjustable to permit the cutting tool to clear the workpiece on the return stroke.

The table is movable vertically and transversely. The cutting tool may be moved up or down by turning the handwheel on the vertical tool-head slide.

BASIC TYPES OF SHAPER CUTS

SHAPING VERTICAL SURFACES

The tool head and clapper box are positioned vertically for machining flat and regular curved

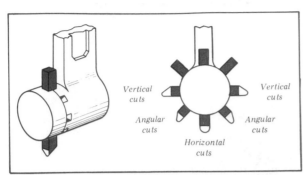

Figure 18–3 Adjustable-Type Universal Cutting-Tool Holder

surfaces. However, the clapper box and tool head must be swiveled for angular and vertical cuts. Otherwise, the toolholder and/or the cutting tool may not clear the workpiece. Swiveling the head prevents the cutting tool from dragging along and scoring the machined surface during the return stroke.

SHAPING SIMPLE AND COMPOUND ANGLES

Angular surfaces are shaped either by positioning the workpiece or the work-holding device or by setting the tool head at the required angle.

Some parts require the machining of surfaces at a compound angle. The vise is set at the first angle; the standard table, at the second angle; and the universal table, at the third angle. The compound angle is then produced by horizontal shaping.

In addition to standard straight and offset shaper toolholders, an adjustable-type universal cutting-tool holder is available (Figure 18–3). The design permits setting a tool bit for horizontal, vertical, angular, and contour cuts.

SHAPING DOVETAILS

When dovetails are cut on shapers, the tool head is swiveled at the angle of the dovetail. The cutting tool is ground at a smaller included angle than the angle of the dovetail. The tool is ground for cutting on the front face and either the left- or right-hand edges, depending on whether a right- or left-hand dovetail is to be cut. The front cutting edge is formed to the required corner size and shape. Usually, the corners of an internal dovetail are relieved.

SHAPING CONTOUR SURFACES

Contour surfaces may be shaped by using a shaper tracing attachment or by combining the vertical feed on the tool slide and the longitudinal feed of the table and workpiece. In general practice, power transverse feed and manual vertical feed are used for roughing cuts. Finish cuts are taken by synchronizing the table and cutter movements. Finish cutting tools and finer feeds are used.

SHAPING INTERNAL FORMS

One common practice in machining a regular form inside a workpiece is to use a broach. The process is fast. The size and shape of the form may be held to close tolerances. When a single part is to be produced, the shaper is a practical machine tool to use for the occasional shaping of regular or other internal forms. One type of general-purpose toolholder for internal shaping cuts includes an extension bar and cutting tool which may be turned for vertical and angular shaping cuts. The bar is adjustable, depending on the length of the cut. A slower ram speed is used than for external shaping. The internal toolholder is not as rigid as a solid tool or toolholder.

CARBIDE CUTTING TOOLS

Carbide cutting tools permit machining at faster cutting speeds and about the same feed rates as with most high-speed steel tool bits. However, the greater cutting efficiency depends on three conditions:

- The cutting speed of the shaper must exceed 100 fpm (30 m/min);
- A constant speed and feed rate must be maintained;
- The tool head should be fitted with a tool lifter to permit the tool to clear the work on the return stroke.

Cutting speeds of carbide cutting tools are usually two or three times faster than for high-speed steels. In using carbides, it is better practice to take heavy, deep cuts and to use a lighter feed. This practice distributes the chip force over a longer cutting area.

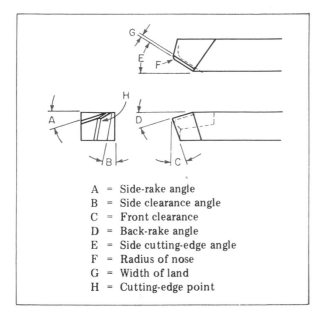

A = Side-rake angle
B = Side clearance angle
C = Front clearance
D = Back-rake angle
E = Side cutting-edge angle
F = Radius of nose
G = Width of land
H = Cutting-edge point

Figure 18–4 Tool Geometry of a Carbide Cutting Tool

TOOL GEOMETRY FOR CARBIDE TOOLS

Two designs of carbide shaper cutting tools are in general use. Brazed-tip tools are applied to light-duty cutting and fine cuts. The carbide tip is brazed onto a solid shank. Tools with replaceable insert tips are used for deep cuts, large-area cuts, and interrupted cuts.

The tool geometry relating to the shape and angles of a carbide shaper cutting tool is shown in Figure 18–4. Tool manufacturers furnish tools having specified angles and tool forms. These tools produce efficiently, machine to particular surface finishes, and have maximum tool life.

Like the high-speed steel tool bits, the front clearance on carbide cutters is 4°. A 0° to 20° negative back-rake angle is used for roughing cuts on steel and cast iron. The back-rake angle permits the cut to start above the cutting edge. This action protects the cutting edge (as the weakest part) from the shock of starting or intermittent cuts. The back-rake angle is increased to a positive angle on finish cutting tools. The positive angle prevents tool chattter when light finish cuts are taken. A side cutting-edge angle of 30° to 40° with a large nose radius should be used on roughing tools.

MACHINING INTERNAL FORMS: BROACHING

Broaching is a cost-effective machining method for producing geometrically and dimensionally accurate standard and other special-shaped internal holes and keyways. Broaching requires the use of an accurately ground, preformed, multiple-tooth, cutting tool.

Starting at a minimum hole diameter size, each tooth along the body of a broach is ground slightly larger than the preceding tooth until the required hole size is reached.

As the broach is pulled or pushed through the workpiece, the cutting tool (broach) takes a series of incremental cuts. By the time the broach nears the end of the cutting action, the final finish dimension cut is taken.

CHARACTERISTICS AND FEATURES OF BROACHES

The versatility of broaching is shown by the examples displayed in Figure 18–5. Beyond the wide range of sizes and forms, custom-made broaches are also widely used. Broach design depends on the following factors.

• Type of material to be broached.
• Minor hole size.

Figure 18–5 Examples of Internal Hole Forms Machined by Broaching

• The finish dimensions and shape (geometry) of the hole or surface to be broached.

• Allowable tolerances.

• Type and capacity of the broaching machine.

Broaches are made of a high quality high speed steel. These cutting tools are available with TiN (titanium nitride) coating for improved machinability qualities and to extend the cutting edge wear life of the broach.

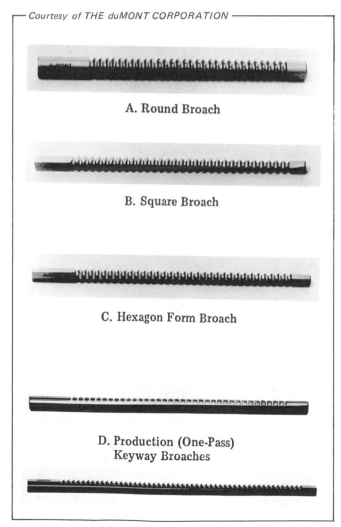

A. Round Broach

B. Square Broach

C. Hexagon Form Broach

D. Production (One-Pass)
Keyway Broaches

Figure 18-6 Common Forms of
Hole Forming Broaches

COMMON FORMS OF HOLE FORMING BROACHES

Four common shapes of hole forming broaches are illustrated in Figure 18-6 at A, B, C and D. Other special forms of broaches are also available. A hand arbor press is functional for push broaching operations in custom jobbing shops. General production broaching machines are commercially produced in sizes ranging from 5 to 20 tons of force and ram speeds of 0 to 23 feet per minute (fpm), or equivalent 0-6 m/min). Stroke ranges are from 0 to 26″ (0-660mm); throat depths, 8″-9″ (200-228mm). There are adjustable controls for pressure, broaching speed, and stroke length.

Figure 18-7 illustrates a heavy-duty 50 ton production pull-down model hydraulic broaching machine. This particular model has a 90″ (2300mm) stroke, dual semi-automatic positioning and clamping fixture stations, and two special broaches for *helical broaching* operations.

PRODUCTION KEYWAY BROACHING SETS

Keyway broaching sets are adaptable for short-run production. A *broach set* consists of three or more broaches and 6, 8, 18, or more different size bushings. Each bushing is machined lengthwise to receive the back of the keyway broach. The groove serves to guide the broach and to control the keyway depth (Figure 18-8).

Keyway combination sets are available for keyway sizes starting at 1/16″ to a 3/4″ (2mm through 8mm) large, heavy-duty size. Bushing diameter sizes range from 1/4″ to 3″ (6mm to 65mm). The combination sets permit multiple use of broaches and bushings. For example, a set of four broaches and 18 bushings (keyway sizes of 1/8″, 3/16″, 1/4″ and 3/8″ and with bushing diameter sizes from 1/2″ to 1 9/16″) may be used to machine 36 different sizes of keyways.

OTHER SURFACE MACHINING AND BROACHES

Just a few common internal hole forming broaches have been discussed. *Surface broaches* are available for producing flat, concave, or convex external surfaces; irregular forms, serrations, cam shapes, dovetails, and other special forms. There are still other internal thread-cutting

Figure 18–7 Production Broaching Machine Setup for Helical Broaching Operations

broaches. *End-cutting* broaches are used on screw machines. These broaches produce simple hexagon or other internal multiple surfaces while the work spindle is operating. *External broaching dies* produce external forms on screw machine workpieces.

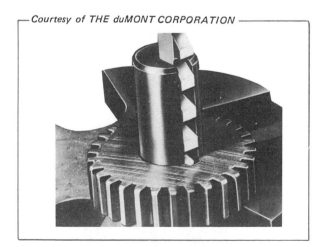

Figure 18–8 Broaching a Keyway

How to Machine Broach a Keyway

STEP 1 Select the appropriate ram speed from a machinability table for the material to be cut. Set the ram speed.

STEP 2 Insert the appropriate diameter bushing in the correct work bore diameter.

STEP 3 Insert the correct size broach for the desired width of keyway into the bushing slot.

Caution: At least two broach cutter teeth must be continuously engaged.

STEP 4 Place the assembly on an arbor or hydraulic press. Check the alignment.

STEP 5 Select the correct lubricant. Lubricate the back of the broach and brush or flow the lubricant on the cutter.

Note: Brasses and cast iron are broached dry.

STEP 6 Push the broach through the workpiece.

STEP 7 Clean the broach and workpiece. Remove the fine machining burrs produced by broaching.

MODERN PLANER DESIGN

Some planers are designed with their upright columns joined together at the top to increase machine rigidity. Such planers are called *double-housing planers*. Other planers are designed with a single column on the right side. These *open-side planers* permit parts to be machined that are wider than the width of the platen. However, the size is still specified by the width, height (depth), and length measurements on planers with two permanent columns. Some open-side planers are designed with a support column accessory to provide additional machining rigidity. Figure 18–9

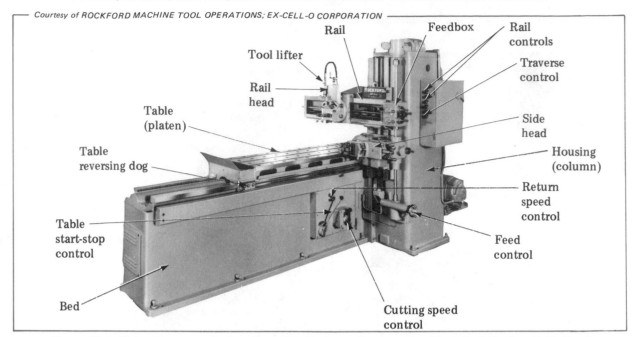

Figure 18–9 Major Components of an Open-Side Hydraulic Planer with a Single Rail Head and Side Head

shows the major components of an open-side hydraulic planer with a single rail head and side head.

One machine tool manufacturer identifies open-side planers that are under 42″ × 42″ (1m × 1m) as *shaper-planers.* They are available with single or double rail heads and with or without a side head mounting. Typical horizontal and vertical roughing cuts may be taken by both rail and side heads simultaneously.

Milling and grinding heads are available as accessories. These heads provide the added capability to perform milling and grinding processes, respectively. Template and contour following, indexing, and footstock accessories are also available. Cutting tools are sometimes set up on multiple heads for gang planing. Some cutters are actuated by a stylus to permit two- and three-axis duplicator processes to be carried on.

One of the greatest advantages of a planer is its versatility and lower costs for certain machining operations. Single-edge cutting tools that are primarily used for planer operations may be easily modified. These tools often replace more highly expensive multiple-point cutting tools when one part or a small number of parts are to be machined.

Universal planers permit cutting during both the forward stroke and the return stroke of the platen. On the forward stroke, one cutting tool edge is firmly held against the tool block during the cut. The second tool edge is then automatically indexed at one end of the stroke. With the second cutting edge in position, a cut is taken during the return stroke. Machining time is reduced as the lost return stroke time is eliminated.

MACHINE AND WORKPIECE SETUPS IN PREPARATION FOR PLANER WORK

PLACING THE WORKPIECE

Single or multiple workpieces are placed on the table so that planer stops may be inserted in holes that are just in front of the workpieces to be machined (Figure 18–10). Poppets and planer pins are placed behind the workpieces. The poppets and pins force the workpieces to the front stops. The parts are then secured by using T-bolts and appropriate shapes and sizes of straps.

ADJUSTING THE RAIL

The rail is set at a height that permits the greatest support for the cutting tool with minimum overhang of the tool head. There must,

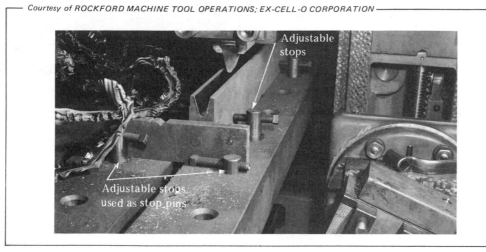

Figure 18-10 Applications of Stops and Stop Pins against the Sides and Front End of a Workpiece

however, be adequate, safe clearance between the workpiece and the cutter. When a rail height needs to be changed, the rail locking clamps and bolts need to be tightened. Open-side shaper-planers have back rail clamps that are loosened for rail adjustment and tightened to rigidly support the cross rail apron to the front of the column. Other planing machine designs require the tightening of bolts to lock the cross slide apron at a fixed height. Consideration is given to taking up backlash on the elevating screw. The rail is lowered to slightly lower than the required height and then raised to position and locked.

ADJUSTING THE TOOL HEAD

The tool head is moved downward almost to the surface to be planed. Once the table stops and cutting speeds are set, the tool slide is moved to locate and to feed the cutting tool to the depth of the first cut. The same procedure is followed to position the cutting tool on a second tool head.

Somewhat similar steps are taken to mount, secure, position, and locate the cutting tool on a side head. The difference is that the cutting action will be vertical and with down-feed. This vertical cut is usually combined with a right (hand) horizontal cut.

SETTING THE CUTTING SPEEDS AND TABLE STOPS

With the table stops positioned with adequate clearance for the starting and ending of

each cut, the cutting and return speeds are next set. Some hydraulic planers have adjusting valves on two sides of a hydraulic pump. The valve on one side controls the rate of the cutting stroke. The independent valve on the other side regulates the return stroke. Other hydraulic planers have a single cutting-speed control lever.

ADJUSTING THE FEED RATE

Feed rate on a hydraulic planer is controlled by a valve. The feed control valve on the side of the machine column is adjusted to the required feed rate.

PLANING FLAT SURFACES

Roughing cuts are taken to within 0.005" (0.12mm) of the finished dimension. As the roughing cuts are completed, the tool slides are raised. The cutting tools are replaced by finish cutting tools that have a broad contact surface.

Finish cuts at a depth of about 0.005" (0.12mm) are taken. The feed rate is increased to about two-thirds of the cutting face width. Usually, the same table speed is used for roughing and finish cuts. If a cutting oil is to be used, the entire surface is coated before there is any table movement. The combination of a flat tool form, fine depth of cut, and wide feed rate results in a smooth, finely finished plane surface.

SIMULTANEOUS PLANING WITH A SIDE HEAD

When a side head is used, steps must be taken to position the workpiece against side stops. The surface to be machined must be parallel to the edge (side face) of the planer table. The front stops are used to prevent any forward movement under heavy cutting forces. The side stops position the workpiece in relation to the cutting action of the side head cutting tool. Once the workpiece is positioned, the side head cutting tool feeds and speeds are similar to the feeds and speeds used with the rail heads.

PLANER CUTTING TOOLS AND HOLDERS

Tool geometry for planer cutting tools is similar to shaper tools. Clearance angles are reduced to a minimum of 4° to 5°. This angle provides maximum support behind the cutting edge. Carbide tools are ground from 0° to negative top rake angles away from the working area of the cutting tool edge.

Planer toolholders are generally drop forged of nickel chromium steel. The holders are heat treated for maximum toughness. Some holder

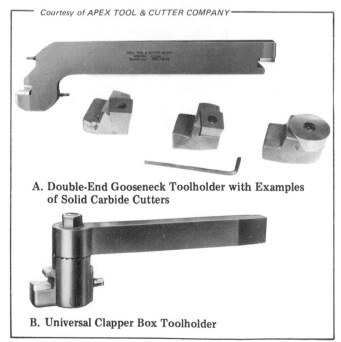

A. Double-End Gooseneck Toolholder with Examples of Solid Carbide Cutters

B. Universal Clapper Box Toolholder

Figure 18-11 Basic Designs of Planer Toolholders

Figure 18-12 Simultaneous Planing Using Rail Head and Side Head on an Open-Side Shaper-Planer

designs have a separate *taper seat*. The seat provides a reference surface for holding, adjusting, and replacing a tool bit as the cutting tool. The seat is adjusted to pull the tool bit into position until it is securely locked on three sides: back, top, and bottom. The seat is *serrated* with a series of evenly spaced V-grooves.

Straight planer toolholders assure rigidity and permit the heaviest cuts to be taken. *Gooseneck* planer toolholders (Figure 18-11A) provide an "underhung" tool height. This design of toolholder permits the cutting tool to lift under severe cutting forces and to minimize chatter.

Spring-type planer toolholders are used for finishing operations. A chatter-free finish is produced because the holder design absorbs vibration. *Universal clapper box* planer toolholders (Figure 18-11B) are adjustable to any angle.

Other planer toolholders, classified as special-purpose holders, are designed for specific processes. *Adjustable* toolholders are used for planing T-slots. *Multiple* toolholders permit gang planing or simultaneously planing machine bed V-ways. *Double* toolholders provide for cutting at two depths at the same time. *Reversing* toolholders are used for planing on the forward and return strokes on planers designed for such cutting action.

ADJUSTABLE, REPLACEABLE, SERRATED PLANER TOOL BIT GEOMETRY

The tool bits for planer toolholders with serrated seats have a corresponding series of grooves. The grooves permit a tool bit to be adjusted sideways in steps of 0.060" (1.5mm) and to compensate for wear. Also, the tool bit can be changed without disturbing the original setup.

Tool bits are available in high-speed steel, cobalt high-speed steel, cast alloys, and all grades of carbides. Standard steel serrated tool bits are hardened, tempered, and ground. The end of each tool bit is preformed to a particular shape. Tool bits are supplied in sets in sizes such as #4, #6, #7, #8, and so on. The higher the number, the larger the features are proportionally. The design of each numbered size provides for interchangeability with all styles of holders of the same size. Each tool bit may be inserted and held in a standard straight, gooseneck, or spring-type holder. The cutting tools in a set include right- and left-hand forms, a straight parting tool; full-width, flat-nose tool; diamond-point tool; and straight round-nose tool.

The manufacturer's specifications for three selected size #8, replaceable, serrated-base planer tool bits are shown in Figure 18–13. Clearance and rake angles, form, and sizes are indicated to show the tool geometry of a right-hand, hog-nose tool bit; an offset, flat-nose tool bit; and a right-hand, dovetail tool bit. The cutting tool materials are high-speed steel and cobalt steel.

Full-radius and other special-form tool bits are available in high-speed steels or cobalt steels or as carbide-tipped tool bits. The term *plug* is used with carbide inserts (cutting tool bits). The plugs are generally square or round and are designed with either positive or negative clearance.

CUTTING SPEEDS FOR PLANER AND SHAPER WORK

The cutting speed in fpm (or m/min) is equal to the product of the stroke(s) per minute and the length (ℓ), of the table stroke in feet (or meters).

Cutting Speed (fpm/m/min) =(s/min) $\times(\ell$/ft or /m)

Machine operator judgments must be made about cutting speeds based on the following factors:

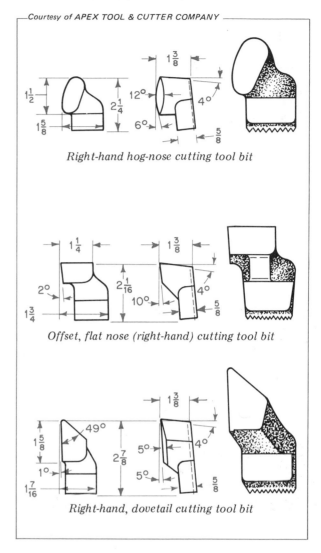

Courtesy of APEX TOOL & CUTTER COMPANY

Right-hand hog-nose cutting tool bit

Offset, flat nose (right-hand) cutting tool bit

Right-hand, dovetail cutting tool bit

Figure 18–13 Manufacturer's Specifications for a Selected Replaceable, Serrated Planer Tool Bits

- Rigidity and power of the planer;
- Size and nature of the workpiece to withstand great cutting forces;
- External and internal surface condition and the structure of the casting, forging, weldment, or other heavy-mass part;
- Rigidity of the setup;
- Shape and size of the cutter and holder;
- Tool contact.

How to Set Up the Workpiece and Planer and Machine Parallel Surfaces

Preparing to Machine the First Surface

STEP 1 Determine the safest method of handling and placing the workpiece.

STEP 2 Clean and remove any burrs or grind away any surface imperfections from the surface opposite the one to be machined.

Note: If required, a protective metal or hard plastic material is placed over the platen area in which a rough-surfaced part is to be located.

STEP 3 Attach clamps to a heavy workpiece when there is no other way of lifting it. Use a cable or heavy rope sling. Handle and place the workpiece on the platen by using a crane.

Caution: Attach the cable so that the eyes swivel freely to avoid twisting the clamps.

STEP 4 Place one or more stops in front of the position at which the workpiece is to be located.

STEP 5 Position the workpiece on the platen as close as possible to the stops.

STEP 6 Disassemble the sling and clamps when used. Locate poppets and planer pin screws in back of the part. Tighten the pins to force the part against the front stops.

Note: If vertical cuts are to be taken, the part must be aligned parallel to the side of the platen before the poppet screws are tightened.

STEP 7 Select suitable straps, T-nuts and bolts, and packing and step blocks. Place them between workpieces and at positions where the greatest amount of force may be applied.

Setting the Tool Head

STEP 8 Set the tool head vertically for general horizontal cuts. If angle cutting is required, set the tool head at the specified angle.

STEP 9 Select the appropriate cutting tool and holder, if necessary. Place and secure the cutting tool to the clapper box.

STEP 10 Move the cutter away from the workpiece. Release the rail locking clamps.

STEP 11 Lower the rail so that the cutting tool clears the top surface of the workpiece. Then, raise the rail slightly to take up backlash on the adjustment screw.

STEP 12 Tighten the back rail clamps on the cross rail locking bolts to secure the cross rail apron to the front of the column.

Setting the Overall Travel Stops and Platen Speeds

STEP 13 Set the cutting and return stroke speeds at a minimum. Start the planer.

STEP 14 Set the stop at the beginning of the stroke with adequate clearance to permit feeding the tool before each cut begins.

STEP 15 Set the return stop so that the cutter clears the end of the cut.

STEP 16 Start the platen in motion.

STEP 17 Adjust one hydraulic cylinder control valve for the cutting rate. Turn the other control valve for the rate of return.

Tooling Setups and Feed Rate

STEP 18 Continue with the platen in motion to set the rate of feed. On hydraulic planers, adjust the feed rate valve until the dial registers the required feed.

STEP 19 Stop the planer. Feed the rail head manually across the workpiece to clear the starting position. Turn the tool slide handwheel to lower the cutting tool to a roughing cut depth.

STEP 20 Lock the tool slide in place. Start the planer. Take the first cut.

STEP 21 Continue the cut across the second part if a number of workpieces are mounted side by side.

STEP 22 Stop the machine after the roughing cuts. Replace the roughing cutter with a broad (slightly curved cutting edge) finish cutting tool. Reset the cutter and holder.

STEP 23 Change the feed rate to from one-half to two-thirds the width of the cutter. Leave the platen speed at the same rate as for the roughing cuts.

STEP 24 Spray or coat the cutter surface with a cutting oil, if required. Then, start the planer. Take the finish cut.

STEP 25 Stop the planer. Clean the workpiece and machine area. File any burrs produced by machining.

Planing a Parallel Surface

STEP 26 Remove all clamps, poppets and planer pins, and all other work-holding accessories.

STEP 27 Rig the workpiece for removal to a wooden platform. Turn the workpieces over. Remove the platen covering. Return the workpiece to the platen.

STEP 28 Use stops, straps, and work supports and secure the parts to the platen.

STEP 29 Proceed to take roughing cuts and a finish cut. Follow steps 19 through 25.

STEP 30 Remove all work-holding tools and accessories. Return each to its proper storage compartment. Set the table travel and feed rates to minimum as a safety precaution for the next operator.

Safe Practices in Tooling Up and Preparing for Shaping, Slotting, Planing, or Broaching

- Stop the machine. Under no circumstances are workpiece adjustments or checking to be made while the machine table is moving.
- Stand clear of a heavy or hard-to-handle part that is being moved to or set on the platen. Keep hands away from the underside of the workpiece. Be ready to instantly move safely away from the setup in the event of danger.
- Use stops against the front end of a workpiece to take up the cutting thrust.
- Protect the platen, table, or other work-holding device against scratching, burring, or denting.
- Position the safety dogs on a planer to keep the table from moving past the table length.
- Move the shaper or slotter ram or planer table manually (or at an extremely slow speed) through at least one forward and reverse or up and down stroke. Check the clearance of the cutting tool at the start and end of the cutting cycle. Make a special check for the clearance between the tool head and/or ram head, the workpiece, and/or parts of the machine when making angular machining setups.
- Check for excessive cutting speeds, particularly on long strokes, where tremendous momentum and forces act to shift shaper, slotter or planer setups or move the cutting tool, resulting in possible damage to the workpiece and/or machine.
- Remove all excess tools, fastenings, and other parts from a table, vise, or platen prior to manually moving through the first cycle or the startup of an operation.
- Keep areas around each machine tool clear of chips and free of oil. Use a liquid absorbent compound on the floor around each machine to keep the area dry and free from slippage.
- Use a protective screen for shaper, slotter, and planer operations.
- Wear safety glasses or a protective shield during setup and machining processes.

───── TERMS USED FOR SHAPER, PLANER, SLOTTER, AND BROACHING MACHINE ─────
DESIGN FEATURES, TOOLING, AND BASIC PROCESSES

Vertical shaper/slotter	A machine tool for shaping parts by taking a series of cuts with a cutting tool solidly mounted to a reciprocating vertical ram. A machine tool designed to machine internal and external flat, angular, circular, and contoured surfaces by controlling the longitudinal, transverse, and rotary movement of the workpiece that is mounted on a movable table.
Universal horizontal shaper	A shaper model having a secondary table that may be rotated to position a workpiece at a compound angle to the horizontal plane.
Contour machining	Producing a dimensionally accurate formed surface by shaping, planing, slotting or broaching. Controlling the down-feed and/or up-feed of the cutting tool in relation to cross feeding a workpiece (planing or shaping).
Double-housing planer	A planer with the two vertical columns connected at the top.
Open-side shaper-planer	A planer with one vertical column and a horizontal cross rail. A planer adapted to machining wider workpieces than can be accommodated by a double-housing planer.
Platen	A planer table fitted with T-slots to receive T-nuts and bolts and with holes in which stops and poppets are placed.
Rail and side heads	Machine components each consisting of an apron, tool head, tool slide, and clapper box. A vertical-mounted cutter-holding head and travel unit and a side-mounted head.
Speed control relief valve	A valve that controls the cutting and return speeds of the platen.
Feed control valve	A hand adjustment valve for controlling the cutting tool feed rate on a shaper or planer.
Universal clapper box toolholder (planer)	A toolholder that provides for the adjustment of a replaceable cutting tool from 0° to 90° to the plane of the platen. An adjustable toolholder used on planers for taking horizontal, vertical, angular, and contour form cuts.
Broach	A flat, long, multiple-tooth, incrementally-increased form cutter that is pulled or drawn through a workpiece to produce an accurate internal hole form or to machine a flat or other broach shape on external surfaces.
Serrated tool bit	A replaceable, specially formed cutting tool with a serrated bottom face. A tool bit with serrations to match the serrations of the seat.

SUMMARY

- The vertical shaper/slotter is adapted to shaping flat, angular, round and contour shapes by cutting in vertical planes.

    - The four basic adjustments of a shaper include vertical movements of the saddle and table, transverse movements of the table, horizontal movement of the ram, and vertical and angular movements of the tool slide.

- Broaches and the broaching process and machines are used to accurately machine regular hole forms (round, square, hexagon, etc.); multisurface internal forms (like splines, serrations, and tooth shapes); and flat and contoured external surfaces.

    - Vertical surfaces are shaped on workpieces that generally are held in a vise. The cutting tool is fed vertically for roughing and finish cuts. The toolholder and apron are set to permit a cut to be taken and the cutting tool to clear the machined surface on the return stroke.

- Contour shaping requires the positioning of the cutting tool in relation to the transverse movement of the table and workpiece. The down-feed or up-feed of the cutting tool is coordinated with the transverse movement of the workpiece.

    - Compound angle shaping requires a combination of angle settings of a swivel table and/or the universal table rocker block.

- Angle cuts are produced by swiveling the tool head on a shaper to the required angle, offsetting the apron at a small arc, and down-feeding.

    - Planers are designed with single or double rail heads, a side head, and milling and grinding heads for milling and grinding in addition to the regular planer processes.

- Hydraulic systems provide smooth driving power for table (platen) movement, cross feeds, vertical feeds, and side head feeds. Simultaneous cuts may be taken from one or more rail heads and a side head.

    - Stop pins and poppets are placed in holes in the platen. Workpieces are held against the pins with toe dogs to prevent movement during the cutting stroke.

- Cutting speeds can be computed by using standard turning or milling handbook tables and applying depth of cut and feed rate factors.

    - Flat surfaces are finished planed with a broad, slightly crowned cutting edge. A coarse feed that is about two-thirds the width of the cutting edge, is used. The cutting speed is the same as for a roughing cut.

- Adjustable, replaceable, serrated planer tool bits are furnished in numbered sets. Tool bits of the same number are interchangeable in any type of holder within the set.

    - Planer tool bits are preformed and ground and are available in standard grades of high-speed steel and cobalt high-speed steel and as carbide-tipped cutters.

UNIT 18 REVIEW AND SELF-TEST

1. List three operations that are best performed on each of the following machine tools: (a) shaper, (b) planer, (c) slotter, and (d) broaching machine.
2. a. Name three major parts of a tool (swivel) head on a shaper.
   b. List the steps to follow in swiveling a tool head to cut a 30° angle surface.
3. Explain why most carbide tools used for roughing shaper and planer cuts have a negative back-rake angle.
4. a. Prepare a work production plan for shaping the Cast Iron Angle Form Block. A standard shaper is used. The workpiece is to be positioned and held in a swivel vise.
   b. Determine the angle for setting the vise.
   c. Calculate (1) the cutting speed in fpm and (2) the ram strokes per minute.

   The casting size allows for one approximately 3/16″ deep roughing cut and a 0.010″ finishing cut on each surface. The feed for the roughing cut is 0.060″. The feed factor of 1.00 may be used for computing the cutting speed of the final cut. High-speed roughing and finish cut cutting tools are used.

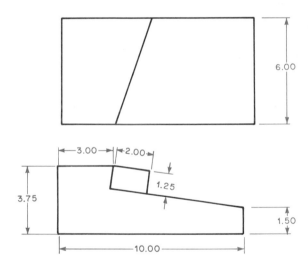

5. a. Name the four principal construction features of a shaper-planer.
   b. Describe the function of each feature.
6. a. Identify two different types of planer toolholders that may be used to prevent chatter.
   b. Provide a brief explanation of how chatter is overcome with selected tool-holders.
7. Give two advantages for using replaceable, serrated planer tool bits instead of older, solid-forged cutting tools.
8. Cite one machine or tool safety precaution to take for each of the following conditions: (a) setting the ram or platen stroke, (b) setting a solid-shank tool or toolholder, (c) swiveling a tool slide at a steep angle on a shaper, and (d) machining at an angle on a shaper.
9. State three personal safety precautions the operator follows after completing a shaper, planer, slotter, or broaching machine job.

Grinding Machines, Abrasive Wheels, and Cutting Fluids

This section deals with the types and construction features of grinders used in abrasive machining and precision grinding processes. The characteristics of abrasives and feeding techniques to accommodate a wide variety of work parts are described. Principles of cutting action relating to surface, centerless, cylindrical, form, plunge, and tool and cutter grinding are examined. Microfinishing, honing, lapping, and superfinishing processes are treated.

Also covered is the technology of grinding wheels, ANSI representation and dimensioning, and wheel speeds and feeds. General grinding problems, causes, and corrective action are summarized. Further, a description of cutting fluids and the selection, dressing, truing, and maintenance of grinding wheels is provided.

UNIT 19

Grinding Machines and Grinding Wheels

OBJECTIVES

A. GRINDING MACHINES: FUNCTIONS AND TYPES

After satisfactorily completing this unit, you will be able to:

• Visualize and appreciate "sustained, incremental, historical contributions" in continuous abrasive machine tool and abrasive product developments leading up to modern abrasive machine tool technology and processes.

• Describe the uniqueness and advantages of grinding processes, cutting actions, and surface finish qualities.

• Visualize surface grinding, internal and external cylindrical grinding, plunge and form grinding, centerless grinding, and tool and cutter grinding processes.

• Describe functions of vibratory finishing and free-abrasive grinding machines.

• Understand microfinishing, polishing, and buffing techniques.

B. GRINDING WHEEL CHARACTERISTICS, STANDARDS, AND SELECTION
- Assess conditions of use that affect the selection of a grinding wheel; depth of cut, rate of feed, sfpm of wheel, coolants, workpiece material, machine condition, worker expertise, and required surface texture.
- Recognize symbols used for abrasive type, grain size, grade, structure, and bond.
- Identify grinding wheels for horizontal- and vertical-spindle surface grinding, cylindrical, tool and cutter, and offhand grinding.
- Recognize common grinding problems and take corrective steps.
- Use formulas for computing cutting (wheel) speed, table travel rates, and work (surface) speed.
- Follow the American National Standards Institute system of representing and dimensioning basic types of grinding wheels.
- Observe *Safe Practices* for grinding machines and abrasive grinding processes and correctly use related *Terms.*

EARLY TYPES OF ABRASIVES, GRINDING MACHINES, AND ACCESSORIES

Grinding, today, meets the industrial tests of cutting efficiency, productivity, and quality because of developments in the United States during the period between 1860 and 1900. Dr. Edward Acheson discovered silicon carbide in 1896. In 1897, Charles Jacobs fused bauxite to produce aluminum oxide. *Silicon carbide* and *aluminum oxide* provide artificial abrasives with more reliable abrasive grains than natural abrasives. Also, in 1896 O.S. Walker invented the magnetic chuck for surface grinding machines.

A grinding machine that produced accurately machined cylindrical unit pieces with excellent surface finish was first made in the United States in 1860. Charles Norton's *cylindrical grinder* was capable of grinding cylinders and tapered cylindrical workpieces. In the 1880s, Norton suggested the manufacture of a 1′ diameter × 1″ wide grinding wheel. With the manufacturing of artificial abrasives (silicon carbide and aluminum oxide) and with better bonds and bonding materials, Norton developed a production cylindrical grinder in the 1890s. However, on June 9, 1862, the Brown & Sharpe Manufacturing Company had already patented a *universal grinding machine.* The universal features extended the applications of external and internal cylindrical grinding and tool and cutter grinding to one machine. An 1876, overhead flat-belt driven universal grinder is pictured in Figure 19-1.

Early in the 1900s, grinding machines became more ruggedly built and more versatile. At the same time, industrial production was moving toward extremely fine surface finishes held to close dimensional tolerances. Rapid improvements of grinding machines, abrasives, grains and wheels, and machine accessories made it possible to meet precision high-production needs of automotive and other industries. The outside diameter of parts were production ground on a *centerless grinder* developed around 1915 by L.R. Heim

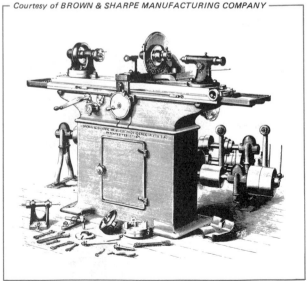

Courtesy of BROWN & SHARPE MANUFACTURING COMPANY

Figure 19-1 Flat-Belt (Overhead Countershaft) Driven Universal Grinder of 1876

The *vertical-spindle*, *rotary-table grinder* was developed in 1900. The grinding was done by the flat face of a grinding wheel. The principle of grinding on a vertical-spindle, rotary grinder is shown in Figure 19–2. Large sizes of grinding wheels are designed with segments of abrasive blocks. They are strapped to a wheel (chuck) that is attached to the vertical machine spindle. This type of grinder is a widely used tool for the fast grinding (rapid removal) of large volumes of metal and other materials. The rotary table is a work-holding device.

Developments of the horizontal spindle surface grinder also occurred in the last quarter of the 1800s. This machine had many similar design features to the milling machine. The Brown & Sharpe surface grinder of 1887 (Figure 19–3) had a knee elevating screw and a handwheel for adjusting the table height. The table crank permitted comparatively rapid longitudinal movement of the workpiece past the grinding wheel. The handwheel at the top of the column controlled the traverse feed of the wheel. This grinder evolved into the modern Type I surface grinder.

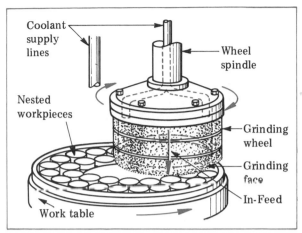

Figure 19–2 Principle of Grinding on a Vertical-Spindle, Rotary Grinder

edges may come in contact with the workpiece each minute.

ABRASIVE MACHINING

Abrasive machining refers to heavy metal-removing operations that previously were performed with conventional cutting tools and

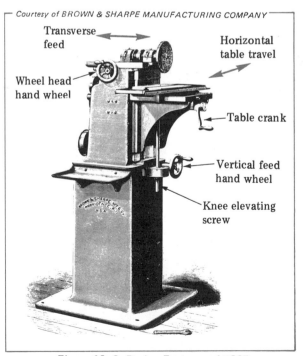

Figure 19–3 Design Features of 1887 Brown & Sharpe Surface Grinder

---

A. GRINDING MACHINES: FUNCTIONS AND TYPES

CHARACTERISTICS OF ABRASIVES AND GRINDING

Abrasive cutting tools have thousands of cutting edges (grains). Each sharp grain that comes in contact with the workpiece cuts away a chip. Chips magnified 350X are shown in Figure 19–4. The many cutting edges provide continuous cutting points. Abrasive cutting tools are capable of generating a smoother surface than any other machine tool. Also, the surface generated is dimensionally precise; therefore, grinding processes are widely used in finishing workpieces.

Grinding wheels operate at speeds of 6,000 to 18,000 sfpm (1,830 to 5,490 m/min). These speeds are roughly equivalent to 1.1 to 3.4 miles (1.8 to 5.4km) per minute. Usually, the distance between abrasive grains is less than 1/8" (3mm). On a 1" (25mm) wide grinding wheel, over 4,500,000 (at 6,000 sfpm) tiny, sharp cutting

Figure 19-4 Chips Produced during Grinding
(Magnified 350X)

machines. Abrasive machining may be done using abrasive grinding wheels or coated abrasive belts.

ADVANTAGES OF ABRASIVE MACHINING

Six advantages are generally cited for abrasive machining:

- Machine tools are designed with *automatic sizing* of the abrasive wheel to compensate for cutting tool wear.
- Parts handling, loading, and unloading costs are reduced.
- The costs of fixtures may be reduced or eliminated.
- Abrasive wheels cut through the outer scale of castings and forgings, hard spots, and burned and rough edges due to welding processes.
- Thin-walled areas of workpieces may be abrasive machined with less springing or breakage.
- Casting and forged part design may be simplified. Normal allowances of extra stock in order to cut under the scale may be reduced.

BASIC PRECISION GRINDING

Precision grinding processes are generally grouped into six categories: surface, cylindrical (external and internal), form, plunge, centerless, and tool and cutter grinding.

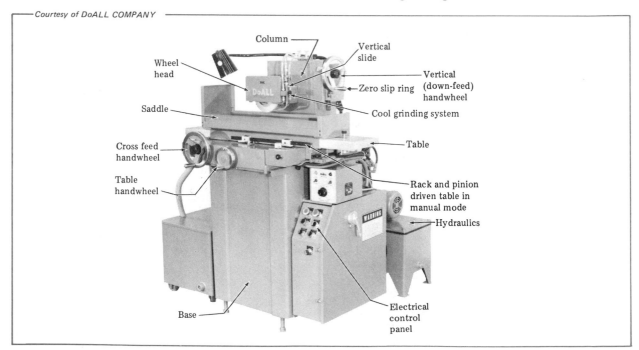

Figure 19-5 Basic Construction Features of a Saddle-Type, General Purpose 6″ X 12″
(Type I) Horizontal Surface Grinder

SURFACE GRINDING MACHINES

Essentially, *surface grinding* deals with the generating of an accurate, finely finished, flat (plane) surface.

The four basic types of *surface grinders* are:

- *Type I* (horizontal spindle, reciprocating table);
- *Type II* (horizontal spindle, rotary table);
- *Type III* (vertical spindle, reciprocating table);
- *Type IV* (vertical spindle rotary table).

TYPE I (HORIZONTAL-SPINDLE, RECIPROCATING-TABLE) SURFACE GRINDER

The type I surface grinder (Figure 19–5) is the most commonly used surface grinding machine. Flat, shoulder, angular, and formed surfaces are ground on the machine. Workpieces are generally held by a magnetic chuck on the table. The table is mounted on a saddle. Cross feed for each stroke is provided by the transverse movement of the table. The workpiece is moved longitudinally, under the abrasive wheel on the reciprocating table. The grinding wheel downfeed is controlled by lowering the wheel head. The amount of feed is read directly on the graduated micrometer collar.

Depending on the material used and nature of the operation, grinding may be done dry or wet. All general-purpose machines are designed either with or for the attachment of a coolant supply system. Dry grinding requires an exhaust attachment to carry away dust, abrasives, and other particles.

Most type I surface grinders (Figure 19–5) are 6″ × 12″ (150mm × 300mm). This designation indicates that workpieces up to 6″ (150mm) wide × 12″ (300mm) long may be surface ground. Standard heavy-duty type I machines are produced to grind surfaces 16′ and longer.

By dressing the wheel, the face and edges may be cut away in reverse form to grind to a required shape. Radius- and other contour-forming devices and diamond-impregnated formed blocks are used. The formed wheel then reproduces the desired contour.

TYPE II (HORIZONTAL-SPINDLE, ROTARY-TABLE) SURFACE GRINDER

The type II surface grinder produces a circular scratch-pattern finish. This finish is desirable for metal-to-metal seals of mating parts. Workpieces are held on the magnetic chuck of a rotary table. The parts are nested so that they rotate in contact with one another. The grinding wheel in-feed is measured by direct reading on the wheel feed handwheel micrometer collar. The traverse movement of the wheel head feeds the grinding wheel across the workpieces. The principle of grinding on a type II surface grinder is shown in Figure 19–6. The table may be tilted to permit grinding a part thinner either in the middle or at the rim.

TYPE III (VERTICAL-SPINDLE, RECIPROCATING-TABLE) SURFACE GRINDER

The type III surface grinder, designed with a *vertical spindle* and a *reciprocating table*, grinds on the face of the wheel. The work is moved back and forth under the wheel. The positioning of workpieces on a magnetic table and the cutting action of a type III surface grinder are pictured in Figure 19–7.

Type III grinders are capable of taking comparatively heavy cuts. The wheel head is designed so that it may be tilted a few degrees from a vertical position. This design permits faster cutting and metal removal as the rim of the grinding wheel contacts the workpiece. In a vertical position, the surface pattern produced is a uniform

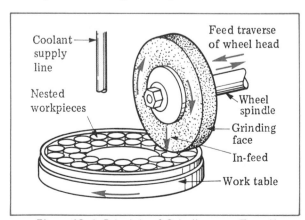

Figure 19–6 Principle of Grinding on a Type II Surface Grinder

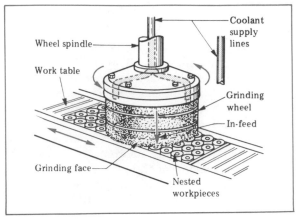

Figure 19–7 Positioning of Workpieces and Cutting Action of a Type III Surface Grinder

series of intersecting arcs. A semicircular pattern is generated when the wheel head is tilted.

ADAPTATIONS OF TYPE III SURFACE GRINDERS

Figure 19–8 shows examples of surface grinding processes using auxiliary spindles positioned vertically, at an angle, or with the wheel dressed. Straight, angular, and dovetailed surfaces may be ground. Auxiliary spindles are used largely with heavy-duty or special multiple-operation surface grinders.

TYPE IV (VERTICAL-SPINDLE, ROTARY-TABLE) SURFACE GRINDER

The type IV surface grinder, designed with a *vertical spindle* and a *rotary table*, also grinds on the face of the wheel to generate a flat (plane) surface.

GENERAL-PURPOSE CYLINDRICAL GRINDING MACHINES

Cylindrical grinding basically involves the grinding of straight and tapered cylindrical surfaces concentric or eccentric in relation to a work axis. Such grinding is generally performed on either *plain* or *universal cylindrical grinding machines.*

Form grinding refers to the shape. The grinding of fillets, rounds, threads, and other curved shapes are examples. Straight form grinding processes are performed on surface grinders.

Cylindrical grinders are also used for producing straight, tapered, and formed surfaces by moving the grinding wheel into the workpiece without table traverse. The process is referred to as *plunge grinding*.

UNIVERSAL CYLINDRICAL GRINDING MACHINES

The universal cylindrical grinder has the capability to perform all the operations of the plain cylindrical grinder. In addition, both external and internal cylindrical grinding operations, face grinding, and steep taper and other fluted cutting tool grindings may be performed.

The design features of a toolroom cylindrical grinder are shown in Figure 19–9. A strong, heavy base adds to the stability of the machine. The table may be swiveled for taper grinding. The headstock and wheel spindle head may also be swiveled for grinding steep angles and

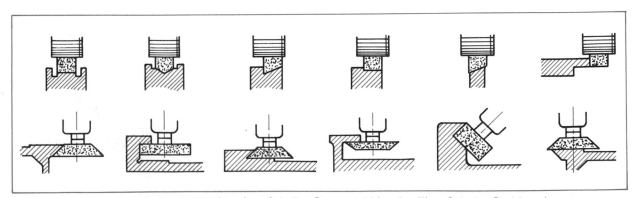

Figure 19–8 Examples of Surface Grinding Processes Using Auxiliary Spindles Positioned Vertically, at an Angle, or with the Wheel Dressed

Figure 19–9 Design Features of a Production and Toolroom Universal Cylindrical Grinder

faces. The footstock is provided with a center for holding work between centers. On some machines the internal grinding unit is mounted directly above the wheel spindle head. The versatility of the machine is extended by a number of work-holding devices and machine accessories.

Plain and universal toolroom and commercial shop cylindrical grinders are capable of grinding diameters to within 25 millionths of an inch (0.0006mm) of roundness and to less than 50 millionths of an inch (0.001mm) accuracy. Surface finishes to 2 microinches and finer are produced in the cylindrical grinding of regular hardened steel as well as tungsten carbide parts.

INTERNAL CYLINDRICAL GRINDING MACHINE

The setup for and principle of internal cylindrical grinding are shown in Figure 19–10.

Concentric workpieces may be held in collets and chucks. Irregularly shaped parts, such as an automotive connecting rod, may be positioned and held in a special fixture. Internal grinding may be done on small-diameter holes with mounted grinding wheels. The diameter should be at least three-quarters of the finished hole diameter. Frequent dressing is required unless a *cubic boron nitride* abrasive is used. This abrasive is a recently developed product of the Specialty Materials Department, General Electric Company. The abrasive bears the trademark of BOROZON™ CBN. While cubic boron nitride abrasives are almost as hard as diamonds, they are an exceptionally good abrasive on hardened steels. Cubic boron nitride wheels require infrequent dressings.

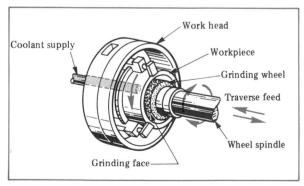

Figure 19–10 Machine Features and Principle of Internal Cylindrical Grinding

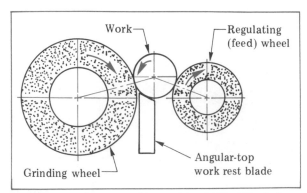

Figure 19–11 Principle of Controlling RPM of Workpiece on a Centerless Grinder (Work Set above Center)

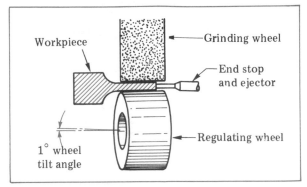

Figure 19–12 Principle of In-Feed Centerless Grinding

CENTERLESS GRINDING METHODS AND MACHINES

Centerless grinding machines are primarily used for production machining of cylindrical, tapered, and multiple-diameter workpieces. The work is supported on a *work rest blade.* The blade is equipped with guides suited to the design of the workpiece.

The principle of centerless grinding requires the grinding wheel to rotate and force the workpiece onto the work rest blade and against the regulating wheel (Figure 19–11). This wheel controls the speed of the workpiece. When the wheel is turned at a slight angle, it also regulates the longitudinal feed movement of the workpiece. The rate of feed is changed by changing the angle of the regulating wheel and its speed.

The center heights of the grinding wheel and the regulating wheel are fixed. The work diameter is, thus, controlled by the distance between the two wheels and the height of the work rest blade. There are three common methods of grinding on centerless grinders:

- In-feed centerless grinding,
- End-feed centerless grinding,
- Through-feed centerless grinding.

IN-FEED CENTERLESS GRINDING

In-feed centerless grinding is used to finish several diameters simultaneously or to grind tapered and other irregular profiles. The principle of in-feed centerless grinding is illustrated in Figure 19–12.

END-FEED CENTERLESS GRINDING

End-feed centerless grinding is used to grind tapered parts. The principle of end-feed centerless grinding is shown in Figure 19–13.

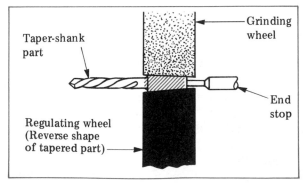

Figure 19–13 Principle of End-Feed Centerless Grinding

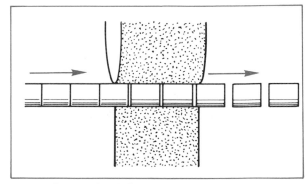

Figure 19–14 Principle of Through-Feed Centerless Grinding on a Cylindrical Grinder

THROUGH-FEED CENTERLESS GRINDING

Through-feed centerless grinding involves feeding the work between the grinding and regulating wheels, as illustrated in Figure 19–14.

ADVANTAGES OF CENTERLESS GRINDING

The four prime advantages of using centerless grinding processes and machines over other methods are as follows:

• Long workpieces that would otherwise be distorted by other grinding methods may be ground accurately and economically by centerless grinding. Axial thrust is eliminated;
• Generally, less material is required for truing up a workpiece. The work *floats* in a centerless grinder, in contrast to the eccentricity of many workpieces that are mounted between centers;
• Greater productivity results as wheel wear and grind time are reduced (with less stock removal);
• Workpieces of unlimited lengths may be ground on a centerless grinder.

TOOL AND CUTTER GRINDERS

Tool and cutter grinders are especially designed for grinding all types of milling cutters, reamers, taps, drills, and other special forms of single- and multiple-point cutting tools.

UNIVERSAL AND TOOL AND CUTTER GRINDER

The *universal and tool and cutter grinder* (or *universal tool and cutter grinder*) is a general-purpose grinding machine.

Many attachments are available for the machine. Magnetic chucks, collet chucks, universal scroll chucks, index centers, and straight and ball-ended mill sharpening attachments are produced for tool and cutter grinding. While many cutter sharpening operations are performed dry, wet grinding attachments are available.

The general features of grinding machines are modified to meet special job requirements. For example, one particular type of tool and cutter grinder is capable of generating helical surfaces and of sharpening ball-ended mills.

GEAR AND DISK GRINDING MACHINES

Gear grinding machines are usually of two types. The first type, a *form grinder*, has the grinding wheel dressed to a slope opposite to the slope of the gear tooth form. The second type of gear grinding machine generates the tooth form. The form is produced by the combination movement of the workpiece and the cutting action of the abrasive wheel. Gear grinding machines are designed for the range of general straight-tooth spur gears to hypoid gears.

OTHER ABRASIVE GRINDING MACHINES

Many surfaces are finished by using abrasive belts. Workpieces, as pictured in Figure 19–15, are moved under the abrasive belts by a conveyor drive unit. The abrasive finishing process is efficient on soft metal, plastic, and other parts requiring a smooth surface.

VIBRATORY FINISHING (DEBURRING) MACHINE

In addition to the bonding of abrasives into sheet, roll, grinding wheel, and other forms, the grains themselves are used in many production finishing and deburring processes. For example, machined parts may be deburred in a *vibratory finishing machine*. The parts and abrasive grains are vibrated. During this process, the tumbling abrasive reaches and deburrs inaccessible areas of workpieces.

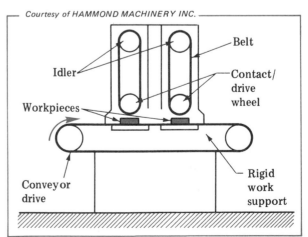

Figure 19–15 Schematic of Simple, Double-Belt, Coated-Abrasive Finishing Process

FREE-ABRASIVE GRINDING MACHINE

Flat surfaces may be generated by using free abrasive grains in a free-abrasive grinding machine. The grains are made into a slow moving *slurry* and fed onto a hardened steel plate. The plate is water cooled. The grains circulate along the plate. As the grains roll around the workpiece, they cut into the material.

MICROFINISHING PROCESSES AND MACHINES

The term *microfinishing* is used here to identify four selected high-quality surface finishing processes. Each process requires the removal of limited amounts of material. The ultimate finely finished and highly accurate surface is specified in microinches.

Honing, lapping, and superfinishing processes and machines are discussed. Polishing is included although the process is not intended to control the size or shape of a finished part.

HONING PROCESSES AND MACHINES

Honing is a grinding process in which two or three different low-pressure cutting motions of a honing stone produce a microfinished surface. The stock removal rate of honing, up to 0.030″ (0.8mm), is the highest of the microfinishing process rates. Honing can correct out-of-roundness and taper and axial distortion. Usually, from 0.004″ (0.1mm) to 0.010″ (0.25mm) is left for honing. The general range of honing finishes is from 8 to 10 microinches. However, a fine surface finish to 1 microinch is obtainable. Honing may be employed on soft or hardened materials.

HONING STONE HOLDERS AND ABRASIVES

Honing requires the use of bonded abrasive *honing stones.* Some honing stones are held loosely in holders. They may be regularly spaced or interlocked. The interlocked design provides a continuous honing surface. An example of a typical honing tool (holder and stones) is shown in Figure 19–16.

The abrasive stones are expanded against a bored hole. This action is produced mechanically or by hydraulically operated mechanisms. In production, automatic controls to within 0.0001″ (0.0025mm) to 0.0003″ (0.0075mm) of a specified size are used with hydraulic types of honing tools.

Aluminum oxide, silicon carbide, and diamond grain abrasives are used. The usual bonds are resinoid, shellac, vitrified, or metallic. The cutting action and wear life of the stones are increased and controlled by using a "fill" of sulfur, resin, or wax.

A coolant is required in honing. The general coolants include kerosene, turpentine, lard oil, and manufacturers' trade brand names. Water-soluble cutting oils are not recommended.

Completely round holes or interrupted holes such as keyways may be honed, provided the stones overlap the cutaway area. Honing may be used on ferrous and nonferrous parts, carbides, bronzes and brass, aluminum, ceramics, and some plastics.

LAPPING PROCESSES AND MACHINES

Lapping is a process of *abrading* (removing) material by using a soft-metal, abrasively charged lap against the surface to be finished. Loose-grain *abrasive flows* are used.

Courtesy of MICROMATIC HONE CORPORATION

Figure 19–16 Typical Honing Tool (Holder and Stones)

CHARACTERISTICS OF LAPS AND LAPPING

Laps are made of a material that is softer than the parts to be lapped. Soft cast iron, brass, copper, lead, and soft steels are commonly used materials. They easily receive and retain the abrasive grains. Soft, close-grained gray cast iron laps are common.

An open-grained lap will cut faster than a close-grained lap. The harder the lap, the slower the cutting action. Also, the harder the lap, the duller the finish and the faster the wear. The faces of laps are serrated (grooved) to permit the working surface to remain clean and free of abrasive and other particles.

Lapping is used as a finishing process, not as a metal-removing process alone. The amount of material removed in rough lapping is only 0.003″ (0.08mm). In finish lapping, the material removed can be as little as 0.0001″ (0.002mm). The common practice in commercial lapping processes is to grind to within 0.001″ (0.02mm) of the basic size.

Lapping is used to control flatness, parallelism, roundness, surface finishes to within 0.4 a.a. microinch, and dimensional tolerances to ±0.000002″.

ABRASIVES AND LAPPING WORKHOLDERS

The three most commonly used abrasives for lapping include aluminum oxide, silicon carbide, and diamond flours. Emery, rouge, levigated alumina, and chromium oxide are also used. Aluminum oxide is employed for soft metals. Silicon carbide is a practical abrasive for hardened parts. Diamond flours are used for small precision work and extremely hard workpieces.

The abrasives are used with a *vehicle*, or lubricant. The lapping workholder is known as a *spider* or *retainer*. The workholder is usually a flat plate designed with holes to accommodate workpieces of a particular shape. In addition to nesting workpieces, the retainer confines them.

The upper wheel (lap) on the lapping machine and the bottom wheel rotate in opposite directions. The wheel speeds are approximately one-tenth the surface feet per minute of regular grinding wheels. The combination of movements produces an ever-changing cutting pattern and an even wear on the laps. On some machines, the retainers provide a planetary motion to the workpiece.

SUPERFINISHING PROCESSES AND MACHINES

Superfinish™ is the trade name of the process that was invented and developed by the Chrysler Corporation. *Superfinishing* is a refined honing process in which bonded abrasive stones at low cutting speeds and constantly changing speeds produce a scrubbing action. Forces applied on the abrasive are low. From three to ten different motions are involved. As a result, a random, continuously changing surface pattern is produced. Superfinishing to 2 and 3 microinches is common practice.

SUPERFINISHING MACHINES

The three major features of a superfinishing machine include a headstock, tailstock, and a vertical spindle. The workpiece is rotated around a horizontal or vertical axis, as required. The abrasive stone is mounted above the work. It is oscillated and moved by short strokes across the workpiece. The abrasive stone may be shaped with a diamond-type boring bar. The vertical spindle superfinishing machine is usually used for flat surfaces on round parts.

POLISHING AND BUFFING AS SURFACE FINISHING PROCESSES

POLISHING

Polishing wheels are used in production work. These wheels are made of leather, canvas, felt, and wool. The abrasive grains are glued to the wheel face. Workpieces are held against the revolving polishing wheels and rotated until the desired polished surface is obtained.

BUFFING

Buffing produces a high luster and removes a minimum amount of material. Buffing follows polishing. The luster is produced by feeding and moving the surface areas of a workpiece to a buffing wheel revolving at high speed.

Buffing wheels consist of a number of layers of soft fabric. They may be linen, cotton, wool, or felt. The wheels are charged with extremely fine (flour size) grains of rouge or other abrasives. The abrasive and a wax are formed into a stick or cake. In this form, the abrasive is applied by force against the revolving buffing wheel. Buffing is not a widely used jobbing shop process.

B. GRINDING WHEEL CHARACTERISTICS, STANDARDS, AND SELECTION

CONDITIONS AFFECTING A GRINDING WHEEL

Conditions of use effect the nominal grade of a grinding wheel. Conditions of use cause a grinding wheel to act harder or softer than its nominal grade. If the cutting forces on the individual abrasive grains increase, the wheel acts softer, and the grains tear away more rapidly. If the cutting force is decreased, the wheel retains the grains longer and thus acts harder than the nominal grade. The cutting force is partly controlled by the grain depth of cut.

The rate of feed also causes changes in the action of the wheel. As the feed increases, a greater force is applied by the grains as they cut. Again, this force produces a softer-acting wheel.

The diameter of the grinding wheel also affects the cutting action. If constant spindle RPM is maintained, as the wheel wears the sfpm decreases. The cutting force on each grain then increases. The grinding wheel acts softer.

Conditions of use, as they affect cutting forces and cutting action, are illustrated in Figure 19–17. The depth of cut, chips, and feed are greatly enlarged. An abrasive grain, represented at A, moves from A to B during its cutting action. The distance is called the *arc of contact*. The *grain depth of cut* is from C to the arc of contact.

The feed of the workpiece during the arc of contact is from C to B. The shaded area represents the chips taken by the force of the grain against the workpiece. The *wheel depth of cut* is from C to D.

OTHER CONDITIONS AFFECTING THE GRADE

Four other conditions must be considered by the worker as follows:

- Area of contact (area represented by the arc of contact and the width of the cut);
- Kind and application of a coolant;
- Hardness, softness, and composition of the workpiece;
- Required surface finish.

GRADE AND ABRASIVE TYPE

No one factor may be isolated when determining the wheel characteristics for a particular job. Grade must be related to the kind of abrasive. Grain size, bond, and structure must also be considered.

Iron, steel, steel alloys, and other ferrous metals usually require wheels with aluminum oxide. Copper, brass, aluminum, and other soft nonferrous metals, and ceramic parts are ground with silicon carbide wheels. Cemented carbides and nonmetallic materials such as granite, glass, quartz, and ceramics are ground with diamond wheels.

Cubic boron nitride wheels are efficient for grinding hard steel and other high-speed steels. Cubic boron nitride wheels are usually more expensive than aluminum oxide wheels. Therefore, they are used only when there is a demonstrated advantage, as for example, the grinding of very hard steel alloys. Many cubic boron nitride wheels are made of a CBN layer around a core of another material.

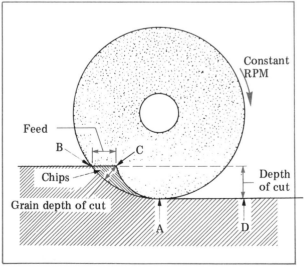

Figure 19–17 Conditions of Use Affecting Cutting Forces and Cutting Action

WHEEL TYPES, USES, AND STANDARDS

HARDNESS OF MATERIAL

Coarse and hard-grade abrasive grains are used on ductile materials. The grain size for general-purpose grinding is usually in the 36 to 60 range. Finer grain sizes in the 80 to 120 range and softer grains are used for hard materials. The hardness range of soft wheels for ordinary grinding is from F to I. The medium-hard wheel range includes J, K, L, and often M and N. Grade P, Q, and R are hard wheels.

Grinding dry or wet influences the grain hardness. Generally, when a coolant is applied, a one-grade-harder grinding wheel may be used. However, if a wheel is too hard, the workpiece may be *burned* due to overheating. Burning produces surface discoloration and may cause internal stresses, distortion, or checking.

REQUIRED SURFACE FINISH

Grinding wheel selection is based on the amount of stock to be removed and the required degree of surface finish. For heavy stock removal, a very coarse, resinoid-bonded wheel is recommended. Where fine production finishes are specified, and a coolant is used, a fine-grain, rubber- or shellac-bonded wheel is suggested. It is practical, with proper dressing, to produce an excellent surface finish by using a coarser grain.

AREA OF WHEEL CONTACT

Grinding wheels are used in both *side grinding* (grinding with the *side* of the wheel) and *peripheral grinding* (grinding with the *face* of the wheel). The area of wheel contact, however, varies. Cup wheels and segmented wheels are side grinding wheels. Coarser and softer grain sizes—for example, 301 and 401—are used.

In peripheral grinding, the face of the wheel makes (compared to side grinding) a small line contact with the workpiece. Cylindrical, tool and cutter, and surface grinders provide examples of peripheral grinding. The area of contact, however, varies in each instance. A small area of contact is made in cylindrical (peripheral external) grinding. The area is generated by the arc of contact on the periphery of the workpiece and

the wheel. Grain sizes from 54 to 80 and hardness grades of K, L, and M are used in general cylindrical grinding processes.

There is a larger area of contact in surface grinding than in external cylindrical grinding. The area is produced by the arc of the wheel in relation to a flat surface. A coarser grain size, such as 46, and a softer grade, such as I or J, may be used.

On internal cylindrical grinding, the area of contact is formed by the inside diameter of the workpiece and the grinding face of the wheel. The area of contact is larger than in external grinding. A softer grinding wheel is usually used.

WHEEL SPEED

In all instances, wheel speed is a factor. Every grinding wheel is marked with the safe limits of sfpm or RPM. The safety range of vitrified wheels is around 6,500 sfpm. Resinoid-, shellac-, and rubber-bonded wheels may be operated at speeds up to 16,000 sfpm.

WHEEL MARKING SYSTEM

The grinding wheel marking system incorporates a series of letters and numbers. The standard marking system for abrasive (wheel) types A, B, and C is summarized in Table 19–1. The kind of information a manufacturer generally gives appears on the blotter of a grinding wheel. This information relates to operator safety checkpoints, the safe operable RPM, wheel composition (according to ANSI codes), wheel dimensions, and a repeated safety precaution.

PERIPHERAL GRINDING WHEELS

Types 1, 2, 5, and 7 are straight wheels used primarily for peripheral grinding. A number of standard straight *wheel faces* are obtainable from grinding wheel manufacturers. Twelve standard shapes of wheel faces are pictured in Figure 19–18. The standard code of letters which is used in relation to dimensional features on all types of grinding wheels appears in Table 19–2.

SURFACE GRINDING WHEELS

The design features of three types of peripheral grinding wheels used for surface grinding are

Table 19-1 Standard ANSI Marking System for Abrasive Types A, B, and C

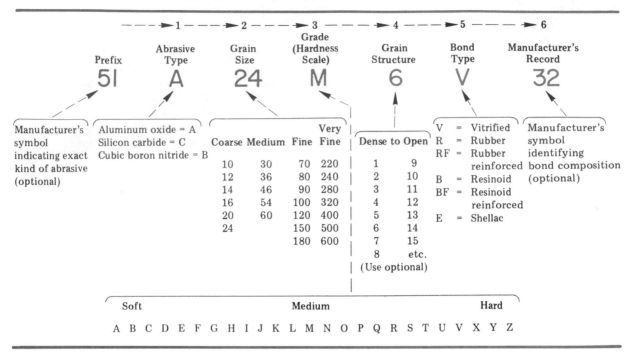

			Grade			
Prefix	Abrasive Type	Grain Size	(Hardness Scale)	Grain Structure	Bond Type	Manufacturer's Record
51	A	24	M	6	V	32

| | 1 | 2 | 3 | 4 | 5 | 6 |

Manufacturer's symbol indicating exact kind of abrasive (optional)

Aluminum oxide = A
Silicon carbide = C
Cubic boron nitride = B

Coarse	Medium	Fine	Very Fine
10	30	70	220
12	36	80	240
14	46	90	280
16	54	100	320
20	60	120	400
24		150	500
		180	600

Dense to Open
1	9
2	10
3	11
4	12
5	13
6	14
7	15
8	etc.
(Use optional)

V   = Vitrified
R   = Rubber
RF = Rubber reinforced
B   = Resinoid
BF = Resinoid reinforced
E   = Shellac

Manufacturer's symbol identifying bond composition (optional)

Soft   Medium   Hard

A B C D E F G H I J K L M N O P Q R S T U V X Y Z

Table 19-2 Standard Code of Letters for Dimensions on All Types of Grinding Wheels

Letter Code	Dimension
A	Radial width of flat at periphery
B	Depth of threaded bushing in blind hole
D	Outside or overall diameter
E	Thickness of hole
F	Depth of recess on one side
G	Depth of recess on second side
H	Diameter of hole (bore)
J	Diameter of outside flat
K	Diameter of inside flat
N	Depth of relief on one side
O	Depth of relief on second side
P	Diameter of recess
R	Radius
S	Length of cylindrical section (plug type)
T	Overall thickness
U	Width of edge
V	Face angle
W	Wall (rim) thickness of grinding face

shown in Figure 19-19. The wheel types include types 1, 5, and 7.

CYLINDRICAL GRINDING WHEELS

Many grinding wheels have one or two sides *relieved*—that is, the side of the wheel is cut away. The side tapers from the diameter of the inside flat (K) to the radial flat at the periphery (A). The design features of four types of peripheral

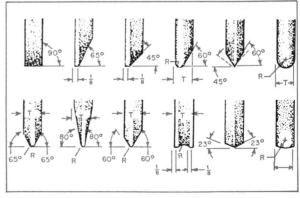

Figure 19-18 Twelve Standard Shapes of Wheel Faces

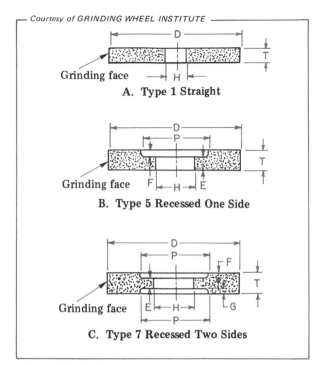

A. Type 1 Straight

B. Type 5 Recessed One Side

C. Type 7 Recessed Two Sides

Figure 19–19 Types and Design Features of Peripheral Grinding Wheels Used for Surface Grinding

grinding wheels used for cylindrical grinding are shown in Figure 19–20. The wheel types include types 20 through 26. Note that some wheels are relieved on two sides. Other wheels are relieved and recessed.

SIDE GRINDING WHEELS

Side grinding, to repeat, means grinding with the side of the wheel in contrast to grinding with the face as in peripheral grinding. Three of the most widely used side grinding wheels used primarily on vertical-spindle surface grinders are identified as types 2, 6, and 11. The design features of these types of side grinding wheels are shown in Figure 19–21.

COMBINATION PERIPHERAL AND SIDE GRINDING WHEELS

The type 12 dish wheel, illustrated in Figure 19–22, is designed for straight peripheral (face) grinding in addition to side (wall) grinding. This wheel is similar to a flaring-cup wheel. However, there are two essential differences. The

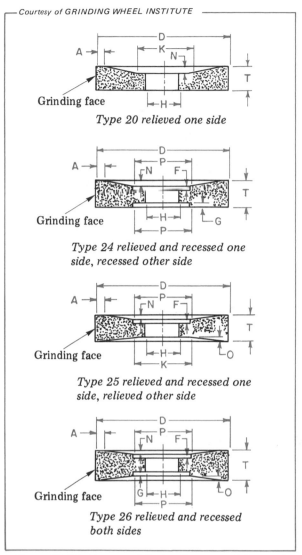

Type 20 relieved one side

Type 24 relieved and recessed one side, recessed other side

Type 25 relieved and recessed one side, relieved other side

Type 26 relieved and recessed both sides

Figure 19–20 Types and Design Features of Selected Peripheral Grinding Wheels Used for Cylindrical Grinding

dish wheel is shallower. The periphery is dressed as a secondary cutting face.

MOUNTED WHEELS (CONES AND PLUGS)

Types 16 through 19 are referred to as *cones* and *plugs*. They are manufactured to be threaded on a bushing or they are made with a standard-diameter shank. The cones and plugs may be formed to any shape. Figure 19–23A shows the

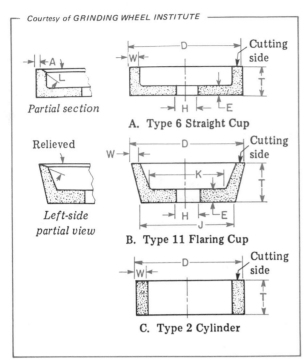

Courtesy of GRINDING WHEEL INSTITUTE

Partial section

A. Type 6 Straight Cup

Relieved

Left-side partial view

B. Type 11 Flaring Cup

C. Type 2 Cylinder

Figure 19–21 Types and Design Features of Side Grinding Wheels (Vertical-Spindle Surface Grinding)

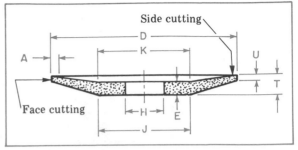

Figure 19–22 Face (Peripheral) and Side Cutting Type 12 Dish Wheel

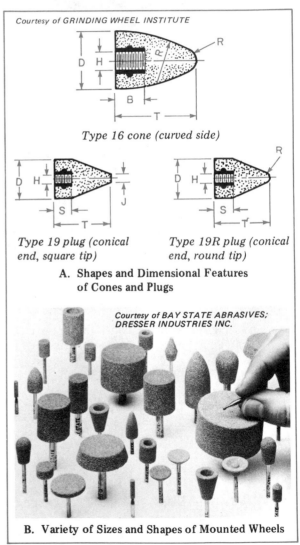

Courtesy of GRINDING WHEEL INSTITUTE

Type 16 cone (curved side)

Type 19 plug (conical end, square tip)

Type 19R plug (conical end, round tip)

A. Shapes and Dimensional Features of Cones and Plugs

Courtesy of BAY STATE ABRASIVES; DRESSER INDUSTRIES INC.

B. Variety of Sizes and Shapes of Mounted Wheels

Figure 19–23 Cones, Plugs, and Mounted Abrasive Wheels

shapes and dimensional features of cone and plug types 16 and 19. An assortment of mounted wheels used for burring and general grinding are illustrated in Figure 19–23B. Cutting is done on any face. Some cones and plugs are used for internal grinding. Most often, they are used in portable grinders for deburring, breaking the edges, or rough grinding.

COMPUTING WORK SPEED AND TABLE TRAVEL

WORK (SURFACE) SPEED

Work (surface) speed relates to the feet (or meters) per minute that a workpiece moves past the area where cutting takes place.

The RPM of the headstock for cylindrical grinding processes is equal to the work speed (fpm) divided by the circumference of the work in feet:

$$\text{RPM} = \frac{\text{Work Speed (fpm)}}{\text{Work Circumference (in feet)}}$$

If the work speed is given in meters per minute,

$$RPM = \frac{\text{Work Speed (m/min)}}{\text{Work Circumference (in meters)}}$$

RATE OF TABLE TRAVEL

Table travel rates of grinding machines are generally given in inches per minute (ipm). The rate of table travel varies from two-thirds of the width of the grinding wheel down to one-eighth. The smaller rate of travel is used to produce a very smooth surface finish. Longitudinal table travel rate changes are made by adjusting the table speed-selector knob.

The rate of table travel in inches is determined (for cylindrical grinding processes) by multiplying the desired work speed in RPM by the distance (in inches) the workpiece should travel each revolution:

$$\text{table travel} = \frac{\text{work speed (RPM)} \times \text{table}}{\text{travel per revolution}}$$

Manufacturer's tables are used for selecting the grain sizes of grinding wheels. Recommended coarse-to-medium grain sizes for commonly machined metals are given in Table 19–3.

FACTORS TO CONSIDER IN SELECTING GRINDING WHEELS

Two major production factors govern the selection of a grinding wheel for manufacturing purposes. These factors relate to quality of finish and the cost involved. The surface finish should be no better than is necessary for the proper functioning of the part. The cost of producing the surface should be as low as possible.

However, there are a number of other production factors to consider in the selection of a grinding wheel. These factors are as follows:

- Material to be ground;
- Amount of stock to be removed;
- Area of contact between the wheel and the work;

Table 19–3 Grain Sizes for Grinding Selected Metals

Metal	Recommended Coarse-to-Medium Grain Sizes
Soft steel	20 to 46
Hardened steel	16 (roughing cuts) 36 to 46 (finishing cuts)
High-speed steel	36 to 46
Cast iron	16 to 36
Aluminum	20 to 46
Brass	20 to 24
Copper	36 to 54

- Type and condition of the grinding machine;
- Wheel speed, work speed, and required finish;
- Nature and application of a coolant during the grinding process.

The ability and experience of the operator are as important as these production factors. The worker must interpret production factors precisely to determine the size and the abrasive of which the wheel is made and the grain, grade, structure, and bond.

Manufacturers' recommendations provide specifications based on successful use in production work. However, any change in wheel specifications must be based on the user's knowledge of the factors governing wheel performance. Most difficulties in grinding are due to a lack of knowledge of the principles of grinding wheels and cutting action.

In selecting a wheel, a final choice should not be made until the grinding wheel has been tested under actual working conditions. The product (ground workpiece) that conforms to the specified quality of surface finish and dimensional accuracy must be considered first. Then, maintenance, wheel cost, and labor costs are considered.

RECOMMENDED ABRASIVE APPLICATIONS

An aluminum oxide abrasive is used for materials of high tensile strength. Metals usually ground with this type of abrasive are as follows: alloy steels, stellite, monel metal, annealed steels, case-hardened steels, hardened steels, high-carbon steels, high-speed steels, and manganese steels; forgings and steel castings; malleable iron, wrought iron, and hard bronze.

Regular aluminum oxide wheels are used for general-purpose grinding operations on surface, cylindrical, and tool and cutter grinding machines.

Friable aluminum oxide wheels are used for grinding operations on extra-hard, heat-sensitive tool steels. The heat generated at the area of contact is reduced because the grains of these wheels break down readily.

Semi-friable aluminum oxide wheels are used for finish grinding hard steels. These wheels are adaptable for grinding operations where there is a large area of contact.

Heavy-duty aluminum oxide wheels are applied in rough foundry snagging processes or when heavy cutting forces are involved.

Silicon carbide abrasive wheels are usually used to grind cast iron, gray iron, and chilled iron; tungsten, tantalum, and carbide alloys; glass and ceramic and other nonmetallic materials; and nonferrous metals such as brass, soft bronze, copper, and aluminum.

Black silicon carbide wheels are used for general grinding purposes and the grinding of nonferrous and nonmetallic materials such as marble, stone, and rubber.

Green silicon carbide wheels are widely used for grinding single- and multiple-point cemented carbide cutting tools.

How to Select a Grinding Wheel

Determining the Shape

STEP 1 Use a type 1 or straight wheel for grinding flat horizontal surfaces or slots or for cutting-off processes.

STEP 2 Use a straight wheel with a formed face for radius work, profile grinding, or angular surface grinding.

STEP 3 Use a straight wheel with a relieved side for taking light cuts on a vertical surface. This same wheel may be used for grinding adjoining vertical and horizontal surfaces.

STEP 4 Use either a straight-cup or flaring-cup wheel for grinding a vertical surface.

STEP 5 Use a type 12 dish wheel for grinding square shoulders. This wheel may also be used to grind adjoining vertical and horizontal surfaces.

Determining the Size

STEP 1 Check the wheel size with the specifications on the blotter. The dimensions are given in the following order: outside diameter, thickness, and bore (hole) diameter.

STEP 2 Measure the diameter of the wheel. A worn wheel will not run at the correct circumferential speed unless the spindle RPM is increased.

STEP 3 Measure the maximum and minimum distance between flanges. The thickness of the wheel must be within these limits.

STEP 4 Check the bore diameter of the wheel with the size of the wheel collet or the wheel spindle.

Selecting the Type of Abrasive

STEP 1 Examine the blueprint or sketch of the workpiece. Determine the material to be ground and the general specifications of the ground surface(s).

STEP 2 Check with a wheel manufacturer's chart or specifications. Determine whether a regular aluminum oxide abrasive or a modified form of this abrasive is recommended.

STEP 3 Select a silicon carbide abrasive wheel for grinding materials of low tensile strength. The sharp cutting edges of silicon carbide grains cut brittle materials rapidly without breaking down prematurely.

STEP 4 Select a cubic boron nitride wheel if there are a number of very hard tool steel alloy or high-speed parts to be ground. CBN has the advantage over aluminum oxide and silicon carbide because of its hardness, sharp and cool cutting action, and long wheel wear life.

Selecting the Grain (Grit)

STEP 1 Use a coarse-grained wheel for coarse finishes and rough grinding and for soft ductile materials.

STEP 2 Use the coarse-to-medium grain (grit) sizes for grinding the common metals indicated in each instance in Table 19–3. This table gives the coarse-to-medium grain sizes for grinding a few selected metals.

STEP 3 Select a fine-grained wheel on tungsten and tantalum carbides. Use this grain size for both finish grinding and the rapid removal of stock.

Note: Rapid removal of stock is accomplished by deep penetration of the grains into the material being ground. Hard, brittle, close-grained material prevents deep penetration of the coarser abrasive grains. Consequently, there is no advantage gained by using these grains for rapid removal of stock.

STEP 4 Use a fine-grained wheel for fine finishes, for form and angular grinding, or whenever it is important to retain the shape and corners of the wheel.

Note: A fine-grained wheel may be used with an oil-base lubricant.

Selecting the Grade (Hardness)

The general rule for grade selection is to use a wheel in the soft range for hard materials; in the medium range, for soft materials.

STEP 1 Use a wheel from E to G grade for surface grinding cemented carbides; from H to N grade for high-speed steel, hardened steel, and cast iron; and from J to N grade for soft steel.

STEP 2 Select a harder-grade wheel for cutting-off operations. For example, use a wheel from M to S grade for cutting off aluminum parts; from P to R grade for cast iron; from O to Q for hardened steel; and from P to W grade for soft steel.

STEP 3 Determine whether the amount of stock to be removed is a factor in selecting the grade of the wheel.

Note: Deeper-than-average cuts used for roughing out increase the area of contact. As a result, the wheel acts hard. Decreasing the wheel speed or decreasing the work speed tends to compensate for this condition. Frequently, a one-grade-softer wheel is used for roughing operations.

STEP 4 Determine the sfpm of the wheel in relation to the table speed. Remember that the wheel acts softer as it wears.

STEP 5 Select a wheel of slightly harder grade when it is used on a machine of light construction or when worn parts cause vibration.

STEP 6 Use a wheel of softer grade on heavy, rigid machines that are in excellent condition.

STEP 7 Select a slightly harder-grade wheel when a coolant is used.

Determining the Structure

STEP 1 Use wide grain spacings for soft, ductile materials. Closer grain spacings are used for hard, brittle materials. Cemented carbides are the exception to this practice. Cemented carbides require a wide grain spacing.

STEP 2 Check the structure numbers recommended by the wheel manufacturer.

STEP 3 Select a wheel with an open structure (such as number 12) for the rapid removal of stock. The open structure provides adequate chip clearance and cool cutting action. A denser structure is used for heavy, plunge cuts.

STEP 4 Determine the condition of the machine. Use a denser wheel on light machines.

STEP 5 Use a close (dense) grinding wheel for fine surface finishes.

Determining the Bond or Bonding Process

STEP 1 Select a wheel made by the vitrified process for general precision metal grinding.

Note: This bond is not suitable for wheels over 30″ in diameter, for thin wheels, or for wheels where lateral stresses may cause wheel breakage.

STEP 2 Use wheels bonded with resinoid, rubber, or shellac for grinding narrow grooves and for work involving lateral stresses.

Note: Wheels bonded with shellac are not suitable for heavy grinding operations where considerable heat is generated.

STEP 3 Select wheels with resinoid, rubber, or shellac bonds for cutting-off operations.

STEP 4 Use wheels bonded with resinoid, rubber, or shellac for high finishes.

STEP 5 Use vitrified wheels for speeds up to 6,500 sfpm. Use resinoid, rubber, and shellac bonds for speeds above 6,500 sfpm.

STEP 6 Use silicate-bonded wheels for milder or cooler cutting action, such as tool and cutter grinding. Silicate wheels are practical for wet grinding operations when water or an emulsion is used as a coolant.

STEP 7 Select a vitrified wheel if an oil-base cutting fluid is used. Resinoid wheels are substituted in some cases.

Table 19–4 Selection Factors and Wheel Characteristics Affected

Factors Influencing Grinding Wheel Selection	Wheel Characteristics Affected				
	Abrasive Type	Grain Size	Grade	Structure	Bond Type
Type of grinding machine and operation (surface, cylindrical, tool and cutter; form, shoulder, and so on)	X	X	X	X	X
Characteristics of material to be ground (tensile strength and hardness)	X	X	X	X	
Machine features and condition (heavy-duty, light; worn bearings, loose fitting parts)			X		
Operator work practices (tendency toward light or heavy cuts)			X		
Area of contact		X	X	X	
Stock removal (severity of heavy or light cuts)		X	X	X	X
Wheel speed (sfpm)			X	X	X
Work speed (ipm)			X		
Feed rate		X	X	X	
Quality of surface finish required		X	X	X	X
Wet or dry grinding process			X	X	X

SUMMARY OF FACTORS INFLUENCING THE SELECTION OF GRINDING WHEELS

Table 19–4 indicates a number of major factors influencing the selection of grinding wheels and the grinding wheel characteristics that are affected.

RECOGNIZING GRINDING PROBLEMS AND TAKING CORRECTIVE ACTION

Three groups of conditions must be recognized as contributing to grinding problems:

• *Machine systems.* The surface texture quality and dimensional accuracy of a workpiece are partially controlled by the correct functioning and precision of the pneumatic, electrical, and mechanical systems;
• *Machine processes.* A number of problems are grouped around the condition of the grinding wheel, coolant system, removal of chips, and basic processes;
• *Operator responsibilities.*

Many of the common problems encountered in grinding, their probable causes, and corrective action are summarized in Table 19–5.

Table 19–5 Common Grinding Wheel and Workpiece Problems, Probable Causes, and Corrective Action

Probable Cause and/or Corrective Action	Surface Texture					Dimensional Accuracy			Wheel		Spindle
	Chatter Marks	Scratches	Spiral Marks	Burning/Checking	Burnishing	Not Ground Flat	Out-of-Parallel or round	Not Sizing Uniformly	Loading	Glazing	Spindle Running Too Hot
Grinding Wheel											
Grain size											
Too fine				X	X				X	X	
Too coarse		X									
Structure											
Too dense				X					X	X	
Too hard	X			X	X	X			X	X	
Grade too soft	X	X									
Dressing too fine				X		X			X	X	
Out-of-balance	X										
Dull, glazed, loaded	X			X							
Not trued properly	X	X									
Speed too high				X							
Workpiece											
Work speed too slow				X							
Out-of-balance	X										
Not adequately supported	X										
Centers worn or require lubrication	X						X	X			
Insufficient number or improper adjustment of back rests	X						X	X			
Work speed or table travel too high	X	X									

Safe Practices for Machine Grinding

- Store grinding wheels in a storage rack. Use protective corrugated shims between wheels. Place each wheel in its proper compartment.
- Handle wheels carefully. Avoid hitting a wheel, particularly on its edges.
- Check the outside diameter of the wheel (if worn) and the recommended maximum sfpm on the wheel blotter. Adjust the spindle to keep within the manufacturer's range.
- Use a guard that covers at least half the diameter of the wheel. Adjust the spark guard and work rest on a high-speed grinder.
- Stand to one side of the grinding wheel, particularly during start-up.
- Bring the wheel speed up to operating speed. Run at this speed for about a minute before taking a cut.
- Use only the recommended face(s) of a grinding wheel as designed and specified for each process.
- Check the grinding wheel (or spider), workpiece, and machine rotating parts before start-

ing the machine. All should clear without interference.
- Check all work-holding accessories, particularly magnetic chucks. They must be held securely in place to receive a workpiece. In turn, test the workpiece to be certain the chuck is "on."
- Set the wheel head as near as possible to the position at which it will operate.
- Dress and true a wheel for dry grinding dry.
- Use a dust collector system when dressing, truing, or grinding dry.
- Use plenty of coolant on the wheel when truing, dressing, or wet grinding.
- Test the coolant characteristics. Be sure the coolant is clear of dirt and chip particles and clean.
- Shut off the coolant at the end of the operation before stopping the spindle to prevent the coolant from collecting at the bottom of the wheel. Such a collection produces an out-of-balance condition.
- Use safety goggles or a protective shield. Position safety glass shields (when they are provided) on the grinding machine.

TERMS USED WITH GRINDING WHEELS AND ABRASIVE MACHINING

Abrasive machining	The faster removal of material to a specified finish and dimensional sizes, shapes, and tolerances than by using conventional cutting tools.
Types I, II, III, and IV surface grinders	Four basic types of grinders used to generate plane, angular, and contoured surfaces on flat workpieces. Classifications of horizontal or vertical grinding spindles with reciprocating or rotary tables.
Centerless grinder principle	A variation in the speed of two turning bodies (grinding wheel and regulating wheel) that produces and controls the direction and speed (RPM and sfpm) a workpiece rotates.
Universal cylindrical grinding machine	A versatile cylindrical grinder for helical, internal, thread forming, and other grinding processes, in addition to processes done on a plain cylindrical grinder.
Plunge grinding	Direct feeding of a formed grinding wheel into a revolving workpiece.

Tool and cutter grinder	A grinding machine with a headstock, footstock, wheel head, adjustable table, and accessories for grinding tools and cutters.
Microfinishing	The removal of limited amounts of material to produce a finely finished and dimensionally precise surface; honing, lapping, and superfinishing processes and machines.
Honing	A grinding process using two or three different motions of a honing stone in relation to a workpiece under low pressure to cut away material and produce a dimensionally accurate and high-quality surface.
Lapping	Abrading (removing) material by moving an abrasively charged lap in a random pattern in relation to the surface of a workpiece.
Superfinishing	A refined honing process used to remove minute quantities of material to produce a smooth, fine crystalline finish.
Peripheral grinding	Grinding with the grains at the periphery of the grinding wheel.
Side (wall) grinding	Grinding on the side or wall of the grinding wheel.

SUMMARY

- During the period between 1860 and 1900, the cylindrical and universal grinder, horizontal-spindle surface grinder, and magnetic chuck were developed within the United States.

  - The discovery and production of artificial aluminum oxide and silicon carbide abrasives accelerated the applications of grinding processes.

- Some of the advantages of grinding include: elimination of many work-holding fixtures and reduced handling costs, ability to cut through the outer scale of castings and hardened parts and maintain efficient machining, and simplification of parts design because thin-walled sections are machined with less spring than other machining processes.

  - Precision parallel and circular surface finish patterns are produced by surface grinding. Flat, angular, and straight formed surfaces are commonly ground.

- Plain and universal cylindrical grinders are used for straight, taper, shoulder, and form grinding. The universal cylindrical grinder is also adapted to internal grinding operations.

  - The universal and tool and cutter grinder combines the features of the cylindrical grinder, surface grinder, and the tool and cutter grinder.

- Straight, end, helical, tapered, form, and other multiple-tooth cutters may be ground on a tool and cutter grinder.

  - The development and use of cubic boron nitride abrasives, particularly on mounted grinding wheels, produce high wear resistance and cutting qualities, which are both important in internal grinding.

■ Centerless grinders are a production machine for grinding cylindrical, tapered, and multiple-diameter workpieces.

■ Microfinishing relates to finish grinding processes whereby limited quantities of material are removed. Dimensionally precise and high-quality surface finishes are produced and are measured in microinches.

■ The honing process can be used to correct out-of-roundness and taper and axial distortion in a bored hole. Rotating, bonded abrasive stones and two or three reciprocating motions produce the cutting actions. Microfinishes in microinches are possible.

■ Lapping requires the use of a retainer to move the workpiece between the abrasive-impregnated surface of a soft-metal lap.

■ Superfinishing requires the use of a formed abrasive stone, low cutting forces, low cutting speeds, and an oscillating movement with short strokes.

■ Aluminum oxide grains are usually used for grinding ferrous metals. Silicon carbide wheels are practical for grinding soft, nonferrous metals and ceramics parts.

■ Very hard alloy steels require the tougher cubic boron nitride grains. These grains are often included as a layer around a wheel core.

■ The grain size for general-purpose grinding ranges from 36 to 60.

■ Hard materials are ground with softer grains within the 80 to 120 grain sizes.

■ The soft wheel hardness range for ordinary grinding is from F to I. The hard wheel range extends from J through N.

■ A grinding wheel acts softer than its nominal grade when the cutting forces on each grain are increased. Depth of cut, rate of feed, diameter of the wheel, and sfpm are factors that produce a softer or harder cutting action.

■ Grinding wheel symbols relate to (1) abrasive type, (2) grain size, (3) grade or hardness, (4) grain structure, and (5) bond type.

■ Vitrified wheels are generally used for precision grinding; resinoid wheels, for rough grinding.

■ Within the 28 types of ANSI-coded grinding wheels are included peripheral and side grinding wheels. These wheels may be straight or recessed on one or two sides. Some wheels are of a cup type; others, of a dish type.

■ The following formulas are used to compute wheel speed, work speed, and rate of table travel:

$$\bullet \text{RPM} = \frac{\text{sfpm of wheel}}{\text{circumference of wheel (in feet)}}$$

• sfpm = RPM × circumference of wheel (in feet)

• table travel = work speed (RPM) × table travel per revolution

UNIT 19 REVIEW AND SELF-TEST

A. GRINDING MACHINES: FUNCTIONS AND TYPES

1. State two differences in the cutting action of abrasive grains as compared to the cutting action of multiple-tooth milling cutters.

2. a. Differentiate between in-feed and end-feed centerless grinding.
   b. Give one application of in-feeding and another of end-feed centerless grinding.

3. Describe briefly what is meant by (a) a soft, abrasively charged lap and (b) the lapping process.

4. Refer to the Appendix Table on the microinch range of surface roughness for selected manufacturing processes.
   a. Indicate the general manufacturing surface finish roughness height for the following processes: (1) finish grinding, (2) honing, (3) lapping, and (4) superfinishing.
   b. Compare the surface textures produced by finish turning and by commercial grinding in terms of (1) measurement and (2) quality of the surface texture.

5. State two machine and/or cutting tool precautions the operator must take when starting up the spindle of a grinding machine.

B. GRINDING WHEEL CHARACTERISTICS, STANDARDS, AND SELECTION

1. State three conditions of use that affect the selection of a grinding wheel. (As a result, the wheel for a particular job may differ from a manufacturer's standard recommendations.)

2. Tell what effect the area of contact has in selecting a grinding wheel for cylindrical grinding as compared to surface grinding.

3. a. List the kind (category) of information that is covered by each of three major symbols in ANSI standards for grinding wheel selection.
   b. Describe the wheel construction features for each symbol.

4. a. Compute the wheel speed (in meters per minute) of a 450mm diameter grinding wheel that is traveling at 1,500 RPM. Use $\pi = 3.14$.
   b. Determine the spindle speed for a 12″ diameter grinding wheel that is traveling at 6,200 fpm. Use $\pi = 3.14$.

5. a. Compute the RPM at which a 6″ diameter workpiece is to be turned on a cylindrical grinder to produce a work speed of 84 fpm. Use $\pi = 3.14$.
   b. Determine the table travel rate for a 1 1/2″ wide wheel that is operating at 400 RPM. The table advances one-sixth of the wheel face width each revolution.

6. State three machine and/or personal safety precautions to follow while operating a grinding machine.

Grinding Wheel Preparation and Grinding Fluids

Dimensionally accurate parts may be produced with high-quality surface finish only when efficient cutting conditions are maintained. The grinding wheel must be mounted, trued, dressed, and balanced. This unit examines the principles and practices related to this preparation. Several wet coolant systems and procedures for maintaining the quality of the grinding fluid are described. Many key items in the ANSI "Safety Code for the Use, Care, and Protection of Abrasive Wheels" are summarized.

OBJECTIVES

After satisfactorily completing this unit, you will be able to:

- Describe form dressing techniques using crush rolls, diamond roll form dressers, and diamond-plated block dressers.
- Dress concentric and parallel faces with single-point and cluster diamond dressers.
- Correct any negative effect of traverse feed on the cutting action of a grinding wheel.
- Apply information about properties, features, wheel marking systems, conditioning, and preparation of diamond and cubic boron nitride (CBN) wheels.
- Carry on wheel balancing with both parallel and overlapping disk balancing ways.
- Analyze truing, dressing, and balancing problems (using tables) for probable causes, and take corrective action.
- Determine applications of water-soluble chemical/oil and straight oil grinding fluids.
- Understand flood, through-the wheel, and mist coolant systems; dry grinding exhaust systems, and cyclonic and magnetic separator systems.
- Perform or simulate each of the following processes.
  - Truing and Dressing Aluminum Oxide and Silicon Carbide Grinding Wheels using Diamond Dressers (Surface Grinder and Cylindrical Grinder Wheels).
  - Preparing Borazon (CBN) and Diamond Abrasive Wheels.
  - Checking and Maintaining the Quality of a Grinding Fluid.
- Use tables to convert sfpm and RPM values to metric meters per second (m/s).
- Follow the ANSI Safety Code and other recommended *Safe Practices* and correctly use related *Terms.*

ABRASIVE WHEEL PREPARATION

ALUMINUM OXIDE AND SILICON CARBIDE ABRASIVE WHEELS

During a grinding process, the forces between the workpiece and the grains in an abrasive wheel produce a cutting action. Material is cut away and most of the dulled abrasive grains are torn out of the wheel. However, after considerable use, a wheel may become dull. Such wheels (grains) are then sharpened by *dressing*. Dressing is particularly necessary with silicon carbide and aluminum oxide wheels. Cubic boron nitride and diamond wheels do not require as frequent or as severe a dressing. The grains on these abrasive wheels are harder. They do not wear away as fast as aluminum oxide and silicon carbide.

In the design of aluminum oxide and silicon carbide wheels, the few thousandths of an inch tolerance between the bore (hole) size and the spindle diameter is enough to produce an out-of-true wheel. In the case of diamond or cubic boron nitride wheels, the hole in the wheel core is fitted to the spindle to a closer tolerance, which is another reason why these wheels do not require severe dressing or truing. An out-of-true, or out-of-balance, condition produces surface irregularities.

Precision dressing is usually done with either a *single diamond* or a *diamond cluster*. A cluster-type dresser contains a number of small-size diamonds that are imbedded in a metal matrix for dressing a flat face. A single diamond is usually traversed across the face of a grinding wheel. The cluster-type dresser is often wider than the grinding wheel. In this case, the wheel is dressed without traversing the dresser.

WHEEL DRESSING TECHNIQUES AND ACCESSORIES

FORM DRESSING

More and more form work is being precision ground to shape and size. Slots, grooves, and contours may be ground by *form dressing* the grinding wheel.

Wheels may be form dressed by a crush roll or with diamond rolls or dressing blocks.

CRUSH FORMING

A *crush roll* is a duplicate form of the required shape to be ground. The crush roll is brought against a slowly revolving grinding wheel (100 to 300 sfpm) with considerable force. The roll *crushes the form* into the grinding face of the wheel. Figure 20–1 illustrates the setup for forming a surface grinder wheel with a crush roll.

DIAMOND ROLL FORM AND PLATED-BLOCK DRESSER

The action of a *diamond roll* form dresser is not as severe as a crush roll. As a result of less force, the abrasive grains are cut instead of crushed. The form generated on the wheel by using a diamond roll form dresser produces a finer-quality surface finish.

A *diamond-plated dressing block* is often used for form dressing a wheel. The dressing block is preformed to the desired shape. It is then positioned on the table of the surface grinder. The revolving grinding wheel is brought down to depth while being traversed forward and backward over the block.

CONCENTRIC AND PARALLEL FACE DRESSING (TRUING)

SINGLE-POINT AND DIAMOND CLUSTER DRESSERS

The *single-point diamond dresser* is widely used to produce a true wheel face. The face is usually concentric and parallel with the spindle centerline. The *single-point diamond dresser* may be mounted in a simple holder (Figure 20–2A). Some surface grinders have the diamond mounted in an *overhead dresser* (Figure 20–2B). The diamond is fed to depth by means of a micrometer dial control. A dresser traverse handle is moved to the right and left to produce a true and parallel face.

Courtesy of DoALL COMPANY

Duplicate form on grinder wheel

Crush roll

Crush Forming a Surface Grinder Wheel

Figure 20–1 Application of a Crush Roll

Courtesy of DoALL COMPANY

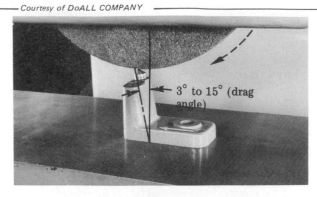

A. Diamond Dresser Mounted in a Simple Holder

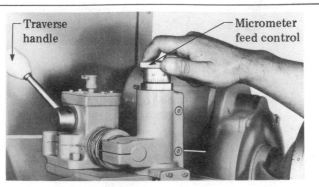

B. Overhead Dresser with Micrometer Feed Control and Traverse Handle

Figure 20-2 Applications of a Single-Point Diamond Dresser

GENERAL PROCEDURES FOR USING DIAMOND DRESSERS

How to True and Dress a Surface Grinder Wheel

STEP 1 Select a sharp single-point diamond.

STEP 2 Examine the diamond to see that it is not flat, cracked, or loose in its setting.

STEP 3 Mount the diamond in a block or wheel truing fixture. The diamond mounting should bring the center line of the diamond and wheel at a 10° to 15° intersecting angle.

STEP 4 Mount the grinding wheel (or combination unit) on the spindle.

STEP 5 Position the fixture or diamond holder on a magnetic chuck.

STEP 6 Energize the magnetic chuck. Test to see that the holder is properly seated and secure.

STEP 7 Replace and secure the wheel guard.

Caution: Be sure to use safety goggles or a protective shield.

STEP 8 Start the grinder. Run it free for a minute or so. Position the grinding wheel so that the center line is 1/8" (3mm) to 1/4" (6mm) to the right of the diamond.

STEP 9 Bring the grinding wheel down slowly until contact is made at the high point.

Note: In wet grinding, use a coolant. In dry grinding, be sure the diamond is air cooled after each two or three passes.

STEP 10 Feed the diamond across the wheel face by using the cross feed handle.

STEP 11 Continue to feed the diamond 0.0005" (0.01mm) for each pass across the wheel face. The final feed for precision grinding may be reduced to 0.00015" (0.004mm).

Note: In the last pass or two across the wheel face, there is no wheel feed—that is, the wheel is allowed to *spark out* as it is traversed over the diamond.

STEP 12 Dress the wheel corners with a hand dresser.

STEP 13 Stop the coolant flow a minute or more before the machine.

How to True and Dress a Cylindrical Grinder Wheel

STEP 1 Secure the diamond holder on the table or footstock.

STEP 2 Set the center of the holder on the horizontal center line of the wheel. The vertical angle of the holder should be between

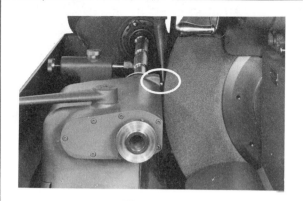

Figure 20-3 Diamond Dresser Setting in Footstock for Dressing a Cylindrical Grinder Wheel

3° to 15°. The position of a diamond holder in a footstock is shown in Figure 20-3.

STEP 3 Replace all guards. Start the machine.

Caution: Use an eye protection device. Stand to one side. Allow the bearings and machine to reach operating temperature.

STEP 4 Adjust the wheel until the highest point is in contact with the diamond.

Note: The coolant supply must be flowing before there is any contact. Otherwise, it is possible to crack and damage the diamond.

STEP 5 Feed the diamond 0.0005″ (0.01mm) each pass.

STEP 6 Continue to feed and traverse the wheel each pass.

EFFECT OF TRAVERSE RATE ON WHEEL CUTTING ACTION

The same wheel may be dressed for rough grinding and finish grinding. The faster the dresser traverses the face of a revolving grinding wheel, the sharper and more open the grains that are produced. Sharp grains are better suited for fast cutting action and heavy cuts. The slower the traverse, the finer the grains. With a slow traverse, the diamond dresser cuts the abrasive and dulls the grains slightly. Such grains cut less and produce a finer surface finish.

In general practice, traverse rates from 12 to 25 ipm produce sharp cutting grains. A medium cutting grain results from using traverse rates from 6 to 12 ipm. A fine cutting grain is produced at a traverse rate of 2 to 6 ipm.

PROPERTIES AND FEATURES OF CBN AND DIAMOND WHEELS

CUBIC BORON NITRIDE (BORAZON CBN) WHEELS

The hardness range of CBN is between silicon carbide and the diamond. CBN has a Knoop hardness reading of 4,700. This reading compares to 7,000 for the diamond, which is the only material harder than CBN. The extremely hard CBN crystals have sharp corners, remain sharper longer, and normally do not load. Therefore, in many machine grinding operations where CBN wheels are used, the following note appears: **USE AS IS. DO NOT DRESS THE WHEEL.** The cooler cutting action of CBN produces less thermal damage to the workpiece.

Like diamond grinding wheels, CBN wheels consist of a core. This core is surrounded by a bond in which abrasive crystals are impregnated. The core is precision machined to fit the spindle within a closer tolerance than is allowed for conventional wheels. The close fit helps the wheel to stay in balance.

DIAMOND GRINDING WHEELS

Diamond grinding wheels are generally produced in resinoid, metal, and vitrified bonds. The core is molded into one of the wheel type forms. The core material may be steel; an aluminum-filled resin composition (aluminoid); or bronze, copper, or plastic. Three common bonds are generally used in the manufacture of diamond grinding wheels:

• *Resinoid bond*, which produces fast and cool cutting action for grinding carbide cutters and for other precision grinding processes,

- *Metal bond*, which is a harder bond that resists the wearing effect produced, for example, by offhand grinding single-point parts and tools;
- *Vitrified bond*, which combines grinding characteristics of both the resinoid bond and the metal bond to produce an intermediate fast and cool cutting action and a high resistance to wear.

DIAMOND WHEEL MARKINGS

The depth of the diamond section varies from 1/16″ (1.6mm) to 1/4″ (6.4mm). The greater depths are used for heavy stock removal, large areas of wheel contact, and heavy production demands. Diamond and cubic boron nitride wheel markings differ from the markings used on conventional wheels. Diamond and CBN grinding wheel markings include a *concentration number* which represents the ratio between the diamond grains and the bonding material. A high concentration (100) is desirable for heavy stock removal. A low concentration (25) is adequate for light cuts and where low heat generation is a requirement.

Marking standards for diamond and cubic boron nitride (CBN) grinding wheels are provided in Table 20-1.

CONDITIONING DIAMOND AND BORAZON (CBN) GRINDING WHEELS

Great care must be taken in conditioning diamond grinding wheels. If the wheel is not concentric, the edges tend to chip, greater wheel wear occurs, and surface finish and dimensional accuracy are impaired. Therefore, diamond wheels—and borazon(CBN) wheels—must run true. As a gen-

Courtesy of DoALL COMPANY

Figure 20-4 Brake-Controlled Dresser for Truing Diamond Wheels

eral rule, the tolerance for outside diameter wheel runout in cylindrical grinding must be within 0.0005″ (0.01mm). The tolerance for extremely precise work is 0.00025″ (0.006mm). The grinding face of a cup wheel should be held to 0.0001″ (0.002mm) minimum. Testing the trueness requires the use of a *tenths dial indicator*.

Diamond and cubic boron nitride wheels are never trued or dressed with a diamond dresser. A coolant should always be used when truing a diamond wheel. A small amount of water as a spray is applied to CBN. Three common devices for truing diamond wheels are used. A standard dressing stick or brake-controlled dresser (Figure 20-4) is used on surface grinder applications. On cylindrical and tool and cutter grinding, a toolpost grinder is occasionally used for truing the wheel.

Table 20-1 Diamond and Cubic Boron Nitride Grinding Wheel Markings

	ASDC	100	N	75	B	69	1/8

Abrasive	Grit Size	Grade	Concentration	Bond Type	Bond Modification	Depth of Diamond Section
D = Natural Diamond SD = Manufactured ASD = Armored (nickel) ASDC = Armored (copper) CB = Borazon	24 120 500 36 150 500S 46 180 600S 60 220 800S 80 240 1200S 100 320 1500S 100S 400 2000S 400S	Res. Metal Vit. H R L R J R J Q* N L T L T* P N N W* Q P (*For Borazon only)	Low = 25 50 75 High = 100 (Not shown for Borazon)	B = Resinold M = Metal MC = Metal carbide V = Vitrified BA = Borazon, wet BB = Borazon, dry	Numeral to designate special bond modification. Example Resinoid—76 and 69. (Symbol optional)	1/16 1/8 1/4

How to Prepare Borazon (CBN) and Diamond Abrasive Wheels

Truing and Conditioning Diamond and Borazon (CBN) Wheels

STEP 1 Mount the diamond or borazon wheel, as for a conventional wheel.

STEP 2 Secure the wheel flange nut.

STEP 3 Position a 0.0001" (0.002mm) dial indicator to read the trueness of either the circumference or the face, as required.

STEP 4 Move the indicator toward the wheel. Adjust until the pointer is ready to record any variation. Set the reading at zero.

STEP 5 Turn the grinding wheel slowly by hand through a complete revolution. Note the location and amount of runout, if any.

STEP 6 Tap the high spot on the circumference of the wheel carefully and lightly. Use a soft block against the wheel.

Note: Steel or brass shims are used for correcting runout on face grinding wheels. The shims are placed between the shoulder flange and the back of the wheel core.

Note: If the indicator is not removed, then tapping is done away from the indicator to prevent damage.

STEP 7 Recheck the wheel until no runout movement is registered.

STEP 8 Tighten the wheel flange nut securely. Recheck with the indicator.

Note: A diamond wheel must be trued if the runout exceeds the allowable limits and the condition cannot be corrected according to these eight steps.

Using the Dressing Stick Method to True Diamond and CBN Wheels

STEP 1 Mount a standard dressing stick in a holding device.

STEP 2 Start the spindle and run for a minute.

STEP 3 Position the coolant nozzle. Direct a flow of coolant between the wheel and dressing stick.

STEP 4 Bring the wheel so that the high spot just touches the surface of the dressing stick.

STEP 5 Take cuts of 0.0001" (0.002mm) each pass.

STEP 6 Continue until the wheel face is trued.

Note: Resinoid-bonded diamond wheels are sometimes trued by grinding a piece of low-carbon steel.

Using a Brake-Controlled Device to True Diamond and CBN Wheels

STEP 1 Select the appropriate grade and grain size to accommodate the diamond or CBN wheel to be trued and dressed.

Note: A metal- or vitrified-bonded diamond wheel is trued with a 100 grain, M grade, silicon carbide vitrified wheel. A resinoid-bonded diamond wheel is trued with a 100 grain, M grade, aluminum oxide vitrified wheel. A resinoid-bonded borazon (CBN) wheel is trued with an aluminum oxide or silicon carbide stick or wheel. A fine grit (400 or finer) and G hardness bond are recommended.

STEP 2 Position the brake-type truing device. The holder must have about a 30° angle with the vertical center line of the wheel.

STEP 3 Run the spindle for a minute. Move the diamond or CBN wheel into contact with the brake-controlled dresser wheel.

STEP 4 Retard the dressing wheel speed by applying the spindle wheel brake. This action results in scrubbing and truing of the diamond or CBN wheel face.

Note: This truing is usually followed by light stick dressing. This dressing improves the cutting action of the wheel.

Figure 20–5 Parallel Balancing Ways

Mount with Two Balancing Weights

Figure 20–6 Positioning of Weights to Balance a Grinding Wheel

FUNCTIONS AND TECHNIQUES OF WHEEL BALANCING

In *wheel balancing*, the weight of a grinding wheel is distributed evenly so that no centrifugal forces are set up at high speed. Grinding wheels are balanced to eliminate vibration and the resulting bad effects. From a safety point of view, the forces in an out-of-balance wheel may cause it to fracture. Damage to the operator, machine, and workpiece are possible.

PARALLEL AND OVERLAPPING DISK BALANCING WAYS

Parallel and overlapping disk balancing ways are two common devices used in balancing grinding wheels. The *parallel balancing ways* (Figure 20–5) consist of two parallel knife-edge surfaces. These ways are part of a solid frame. The device is leveled so that the two parallel ways are on a horizontal plane.

A grinding wheel is balanced by placing the wheel and its adapter on a *balancing arbor*. The assembly is placed carefully on the ways. The wheel is allowed to turn slowly until it comes to rest. The heavy spot is on the underside of the arbor. This spot is marked with chalk. Two other lines are then drawn at a right angle to the heavy spot.

Balancing weights in the flanges of the adapter (Figure 20–6) are positioned for a *trial balance*. The wheel is retested. Further adjustments of the balancing weights are made until the wheel remains at rest in any position on the balancing ways. Wheel mounts are available with two, three, or more balancing weights.

The *overlapping disk balancing ways* consist of four perfectly balanced and free-turning overlapping disks. The grinding wheel assembly and balancing arbor are placed across the overlapping edges of the disks. The assembly is allowed to turn until the heavy spot comes to rest at the lowest point. The balancing weights are then adjusted the same as when parallel balancing ways are used.

GRINDING PROBLEMS RELATED TO TRUING, DRESSING, AND BALANCING

Any one or combination of grinding wheel, machine, and operator variables affects grinding. Eight common truing, dressing, and balancing problems with probable causes and corrective actions appear in Table 20–2. If the problem is not corrected, other variables must be checked. These variables deal with the characteristics of the grinding wheel, the workpiece, cutting fluid, machine condition, and work processes.

Table 20-2 Truing, Dressing, and Balancing Problems with Probable Causes and Corrective Action

Problem	Probable Cause	Corrective Action
Chatter marks	—Out-of-balance	—Rebalance after truing operation —Balance carefully on own mounting
	—Out-of-round	—True before and after balancing
	—Acting too hard	—Use faster dressing feed and traverse
Scratching of the work	—Foreign matter in wheel —Coarse grading	—Dress the wheel
Checking or burning of the work	—Wheel too hard	—Use a faster dress rate to open the wheel face
Diamond lines in work	—Dressing too fast	—Slow dress the wheel face
Inaccuracies in work	—Improper dressing	—Check alignment of dressing process
Rate of cut too slow	—Dressing too slow	—Increase the dressing rate to open the wheel face
Wheel acting too soft (not holding size)	—Improper dressing	—Slow down the traverse rate —Use lighter dressing feed
Wheel loading	—Infrequent dressing	—Dress the wheel more often

GRINDING FLUIDS

KINDS OF GRINDING FLUIDS

The terms *grinding fluids*, *grinding coolants*, and *cutting coolants* are used interchangeably in the shop.

Three general kinds of grinding fluids are used. *Water-soluble chemical grinding fluids* are the most common. They are particularly useful in medium to heavy stock removal processes. These fluids are transparent (a valuable aid to the operator) and have excellent cooling properties.

Rust inhibitors and detergents are added to the solutions to provide good adhesion and cleaning qualities. The lubricating properties are improved by adding water-soluble polymers. Fluid life is increased by adding bactericides and disinfectants in the solutions. They control bacteria growth and retard rancid conditions in the fluids.

Water-soluble oil grinding fluids combine the cooling qualities of water and the lubricating advantages of oil. These oil/water fluids form a milky solution. An emulsifying agent is added to the water and oil (either natural or synthetic).

Caution: The additives must be controlled. Too strong a disinfectant solution may produce skin irritation or possible infection.

Water-soluble oil grinding fluids are used for light to moderate stock removal processes. Parts may be held to reasonably accurate dimensional and surface finish tolerances.

Straight oil grinding fluids are the most costly of the three fluids. These fluids provide excellent lubrication. However, they are not as effective as the water-soluble fluids in dissipating heat. Straight oil grinding fluids are used primarily on very hard materials requiring high dimensional accuracy and surface finish and long wheel wear life. Oil grinding fluids are especially practical in thread and other heavy form grinding processes.

COOLANT SYSTEMS

WET GRINDING METHODS

There are three principal methods of supplying grinding fluids at the point of cutting action between the workpiece and the grinding wheel. These methods of wet grinding include the *flood*, *through-the-wheel*, and *mist coolant systems*.

Flood Coolant System. The grinding fluid is forced from a reservoir through a nozzle. The fluid floods the work area and is then recirculated through the system. Both the volume and the force of the coolant are controlled to overcome air currents around the wheel. Otherwise, these currents tend to force the fluid away from

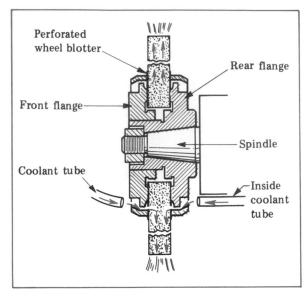

Figure 20–7 Features of Through-the-Wheel Coolant System

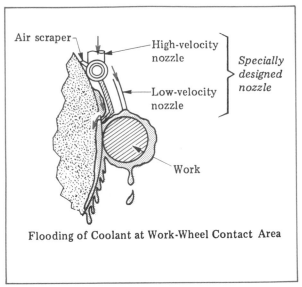

Figure 20–8 Effect of Coolant Nozzle Design on Coolant Application at Point of Contact

the required point of contact. The fluid system provides for straining out foreign particles and producing a recycled, clean coolant.

Through-the-Wheel Coolant System. The features of the through-the-wheel coolant system are illustrated in Figure 20–7. The grinding fluid is pumped through two coolant tubes. These tubes are on both sides of the grinding wheel. The wheels have flanges. As the fluid is discharged into a recess on each flange, the centrifugal force of the revolving wheel carries the fluid through a series of holes in the flanges, washers, and blotters to the center of the wheel. The flow continues with force to the periphery of the wheel. The porous feature of vitrified wheels is applied in flowing the coolant through the wheel.

This system keeps the wheel face flushed free of chips and abrasive particles so that there is a cool cutting action.

Mist Coolant System. This system works on an atomizer principle. Air at high velocity flows through a T-connection. One branch includes the coolant supply. By passing the air through the connection, a small volume of liquid is atomized into a large mass with air. The cool-

ant stream emerges from the nozzle as a mist. Cooling action is produced by the air and the evaporation of vapor. The material and abrasive particles are blown from the grinding area. One advantage of this system is that the operation is unobstructed.

IMPORTANCE OF COOLANT NOZZLE DESIGN

With a standard nozzle the air flow produced by a highly revolving grinding wheel causes an air pocket or bubble to form. This bubble causes the fluid to be blown away from the area where it is needed. With a specially designed nozzle consisting of a high-velocity nozzle, a low-velocity nozzle, and an air scraper (Figure 20–8), flooding of the work-wheel interface and the immediate adjacent area occurs. This condition is needed for effective cooling, lubricating, and chip removal.

EXHAUST SYSTEMS FOR DRY GRINDING

Dry grinding refers to operations that are carried on without a liquid fluid. An exhaust system is needed to remove grinding dust, abrasive material, and chips from the work, grinding wheel, and surrounding area. This system is important for the protection of all machines in the work area. Removal of the dust from the air also constitutes an industrial safety requirement to protect the operator.

The exhaust system usually consists of a nozzle attached to the wheel guard. A flexible tubing leads from another part of the guard to the exhaust attachment. The foreign particles that are removed (drawn) from the grinding area and around the wheel and guard are sucked through a spiral separator.

DESIRABLE PROPERTIES OF GRINDING FLUIDS

A summary of the desirable properties of a good grinding fluid follows. The fluid:

- Has good wetting properties and lubricates the work-wheel interface and surrounding areas;

- Absorbs and dissipates heat at a rapid rate to maintain a comparatively cool temperature under working conditions;

- Permits the settling out of chips, abrasive grains, and other particles fast enough to prevent their recirculation;

- Provides no health hazard to the machine operator or possible damage to the machine parts or grinding wheel;

- Provides a protective, noncorrosive coating for the surface of the ground part and other machined surfaces;

- Breaks down foam rapidly for easy flow-off and recycling through the coolant system;

- Emulsifies readily and without the application of heat;

- Is noninflammable so that there is no possibility of any burning action;

- Has high resistance to becoming rancid or producing objectionable cutting or storing odors.

These properties may be extended to meet special requirements for dimensional and surface accuracy.

How to Check and Maintain the Quality of the Grinding Fluid

STEP 1 Check the odor of the grinding fluid. If it smells sour or rancid, it is necessary to drain the supply from the system.

STEP 2 Determine whether the coolant is clean, dirty, or old. A simple test is to rub a drop of the solution between the fingers. The presence of foreign particles may be felt easily.

STEP 3 Check the amount of chips and other foreign particles in the settling compartment. If needed, clean the filters, nozzle, tank, and trays and other parts of the system.

STEP 4 Cycle the coolant through the system. Check for the proper filtering and settling of abrasive grains, chips, and other foreign matter. Recheck the coolant to see that it is clean.

STEP 5 Check the coolant level. Losses result during machine operation and from evaporation. If the level is not maintained, the fluid circulates faster and becomes warmer. There is also the possibility of recirculating swarf.

STEP 6 Check the strength of the solution with a cutting fluid gage, particularly when water-based solutions are used. Water and/or concentrates should be added to bring the solution up to strength and coolant level.

Note: Too rich a solution may affect cutting efficiency and surface quality. Too lean a solution may produce higher scrap losses.

STEP 7 Keep a continuous check to be sure machine lubricating oils are not being fed into the coolant system, especially when water-based coolants are used. Small quantities of oil severely affect coolant efficiency and the surface texture.

STEP 8 Adjust the coolant nozzle to reach the point and plane of contact between the workpiece and wheel under actual conditions.

Safe Practices in Grinding Wheel Preparation

The following safe practices are summarized from the ANSI "Safety Code for the Use, Care, and Protection of Abrasive Wheels."

- Fit the wheel to the adapter mounting flange or spindle. A close sliding fit is needed for close wheel balance. All contacting surfaces of the wheel and mounting must be burr free and perfectly clean.
- Place a compressible washer wheel blotter of proper thickness and diameter between each flange and each side of the wheel.
- Tighten any retaining screws on an adapter so that they hold the wheel securely.
- Secure the wheel guard. It must cover more than half the wheel.

- Ring test each grinding wheel. Check the recommended sfpm. Be sure the spindle RPM does not exceed the speed limitations.
- Stand to one side and out of line with the wheel when starting a grinding wheel.
- Wear only approved safety glasses or other face protective devices.
- Grind only on the face(s) that is specifically designed for a particular grinding operation.
- Stop the coolant flow a short time before the grinding wheel is stopped to prevent a heavy side of the wheel from developing and throwing the wheel out of balance.
- True and dress a wheel with a diamond or other dresser pointed away from the centerline of the wheel (at a $10°-15°$ angle in the direction of rotation of the wheel).
- Start to true and dress a wheel by carefully bringing the wheel face so that its high point makes contact. Successive cuts are then taken to produce a concentric wheel face.

TERMS USED FOR GRINDING WHEEL PREPARATION AND GRINDING FLUIDS

Form dressing	Process of forming a wheel in a reverse contour to the form to be ground. Preparation of a grinding wheel to grind grooves, slots of varying shapes, and other profiles.
Crush roll	A duplicate roll that crush forms a desired shape in a grinding wheel.
Diamond-plated dressing block	A block preformed to conform to a required contour. A formed block with diamond grains. A block used to produce a reverse form in a grinding wheel for purposes of form grinding.
Diamond cluster dressers	Multiple-point diamond dressers used primarily on production grinding machines.
Drag (negative) angle	The angle formed by the centerlines of a diamond holder and the grinding wheel. The angle at which a dresser is held in relation to the centerline of the wheel.
Wheel sleeve (adapter) mounting	The practice of mounting a grinding wheel permanently (during its wear life) on a wheel sleeve (adapter). A technique that saves setup and truing time and extends wheel productivity.
Dressing traverse rate	The speed with which a dressing tool is moved across the face of a grinding wheel. A factor in controlling the sharpness of abrasive grains and the quality of surface finish.

Diamond and CBN wheel mark	A system of standard wheel symbols and values. Manufacturer's specifications of the characteristics and structure of diamond and CBN wheels.
Concentration	A ratio between the diamond grains and bonding material. Four common ratios (25, 50, 75, and 100) for designating the relationship of grains to bond in a diamond wheel.
Conditioning diamond and CBN wheels	Steps dealing with mounting, testing, and correcting any runout of diamond or cubic boron nitride wheels.
Brake-controlled dressing device	A grinding wheel truing device that removes abrasive grains by a scrubbing action and that retards the spindle speed of a grinding wheel that is in contact with such a device.
Work-wheel interference	Generally refers to the immediate area of cutting action between a grinding wheel and a workpiece.
Wheel balancing	Distributing the weight (mass) of a grinding wheel so that no centrifugal forces set up at high speed. Neutralizing a heavy spot in a grinding wheel. An internal condition of a grinding wheel where it revolves in balance. (A grinding wheel that is balanced cuts effectively and produces a dimensionally accurate surface.)

SUMMARY

- Dressing is the process of sharpening a wheel. A glazed or loaded surface is cut away. Sharp, new abrasive grains are exposed.

  - A grinding wheel should be trued on its adapter (mounting) before being balanced.

- A trued and dressed wheel is essential to grinding accuracy, good surface finish, and high production.

  - A ring test should be performed before a new or used wheel is mounted.

- Grinding wheels are generally form dressed by crush forming. Preshaped rolls, a diamond roll, or plated block dressers are used.

  - Single-point and diamond cluster dressers are widely used for concentric and parallel face dressing.

- Diamond dressing tools are set at a 3° to 15° negative angle and either at the center (cylindrical grinding) or beyond the center (surface grinding) for truing.

  - Wheels are dressed under operating conditions. Coolants are applied for wet grinding. In dressing with a diamond for a dry grinding operation, the diamond must be air cooled to prevent fracturing and breaking down.

■ Silicon carbide and aluminum oxide wheels are generally trued with a diamond or an abrasive holding device. Diamonds are not used to true other diamond or CBN wheels.

    ■ Diamond and CBN wheels are trued with a dressing stick or a brake-controlled device and conventional abrasives and by grinding against a low-carbon steel.

■ Diamond and CBN wheels have metallic cores. The diamonds are bonded around the core by a resinoid, metal, or vitrified bond.

    ■ Diamond wheel markings include the type of diamond abrasive, grain (grit) size, grade, concentration, bond type (and sometimes bond modification numeral), and depth of diamond section.

■ Parallel ways and overlapping disk balancing ways are two common balancing devices and methods.

    ■ Three general kinds of grinding fluids are used: Water-soluble chemical, water-soluble oil, and straight oil. Rust inhibitors, detergents, bactericides, and other additives are mixed in different solutions to improve the properties of grinding fluids and for hygienic reasons.

■ Wet grinding fluid systems include the flood, wheel penetration (through-the-wheel), and mist coolant systems.

    ■ Three basic systems attached to the machine for recirculating clean grinding fluid include centrifugal, cyclonic, and magnetic separators for removing swarf.

UNIT 20 REVIEW AND SELF-TEST

1. Explain briefly why cubic boron nitride and diamond wheels require less frequent dressing than aluminum oxide and silicon carbide abrasive wheels.

2. State the difference between crush forming and preformed diamond dressing block contour forming on a grinding wheel face.

3. List the steps for truing and dressing an aluminum oxide abrasive wheel for grinding flat surfaces with a horizontal surface grinder.

4. Describe briefly how the grinding machine operator may change the cutting action of an abrasive wheel during dressing.

5. a. State a probable cause of each of the three following grinding problems: (1) scratches in the surface finish, (2) dimensional variations of the finished surface (the dimension is not held uniformly), and (3) chatter marks.
   b. Give the correct action to take for each problem.

6. a. Name three additives to cutting fluids used in grinding.
   b. Make a general statement about the functions performed by the cutting fluid additives.

7. State three checks the surface grinder operator makes to prevent damage to a workpiece due to grinding wheel faults.

Surface Grinders and Accessories: Technology and Processes

This section deals with the major construction and control features of standard and automated precision surface grinders. The technology and shop practices are related to work-holding accessories, machine setups, and flat grinding processes. Accessories and procedures for grinding angular and vertical surfaces and shoulders, truing and dressing devices, and methods of form (profile) grinding and cutting-off, are covered in detail.

UNIT 21

Design Features and Setups for Flat Grinding

OBJECTIVES

After satisfactorily completing this unit, you will be able to:

- Identify and describe the functions of basic components of horizontal-spindle surface grinders.
- Understand operating controls and handwheel increment movements, including hydraulic table controls, down-feed mechanisms, and the over-the-wheel dresser.
- Describe operating features of a fully automatic precision surface grinder with automatic digital position readout.
- Use wheel and magnetic and nonmagnetic workholding accessories.
- Diagnose problems related to discontinuous, interrupted surfaces, stresses within a part due to production methods, and distortion.
- Establish machining stock allowances for rough and finish grinding.
- Use down-feed and cross-feed handwheels for dimensional grinding to size; precision vises for grinding edges square and parallel.
- Perform or simulate each of the following processes.
  - Setting Up a Surface Grinder and Taking a First Cut.
  - Grinding Edges Square and Parallel.
  - Grinding Ends Square with the Axis (Round Work).
  - Grinding Thin Workpieces.
- Follow *Safe Practices* relating to personal safety as well as the protection of the surface grinder, accessories, setups, and workpieces.
- Correctly use each new *Term.*

313

DESIGN FEATURES OF THE HORIZONTAL-SPINDLE SURFACE GRINDER

The horizontal-spindle surface grinder, regardless of whether it is hand operated or partially or totally automatic, is designed with five basic construction features: *base, column, saddle* (or other cross feed movement mechanism), *wheel head,* and *table.* These construction features are shown on the horizontal-spindle, reciprocating table, surface grinder in Figure 21–1. A one-shot lubrication system automatically delivers oil to all ways, screws, gears, and bearings.

MACHINE BASE

The base provides a rigid foundation for the machine. Some bases are made of a nickel-alloy cast iron. Others use a rigid preformed steel base. An expanding concrete is cast into these structural steel members. Tests show that this newer construction has about four times the vibration-absorbing capacity of cast iron. This construction combines the strength and versatility of steel with the mass and stability of concrete. The vibration absorption, strength, and stability characteristics add to the smoothness of the grinding process.

SADDLE

The saddle is a heavily ribbed, H-shaped casting. The saddle is fitted to the bed ways. Cross feed movement is included in the design features. The top of the saddle has another set of precision ways. These ways are machined at a 90° angle and provide for the longitudinal movement (traverse) of the table.

The cross feed movement of the saddle toward or away from the column may be controlled manually or automatically. A cross feed handwheel is used to move the saddle manually. Saddle movement on machines equipped for automatic operation is produced by the power cross feed control. Larger and production surface grinders permit the quick positioning of the table and workpiece by a rapid traverse attachment to the cross feed.

COLUMN

The column serves two main purposes. It supports the spindle housing and wheel head. It permits vertical movement for positioning and controlling the depth of cut and performing all grinding operations.

TABLE AND CROSS FEED FEATURES

The main components and operating handwheels on a horizontal-spindle, reciprocating-table surface grinder are identified in Figure 21–1. Surface grinders employ both inch and SI metric units of measure.

The usual table travel of a 612 model is 14″ (356mm). Table feeds with hydraulic controls for this size of machine are infinitely variable from 0 to 78 sfpm (0 to 21.3 m/min). The range of automatic cross feed increments is from 1/64″ to 1/4″ (0.4mm to 6mm). Cross feed reference graduations are usually in 0.001″ (0.025mm). However, a fine-feed cross feed attachment is available. The attachment is graduated in 0.0001″ (0.002mm) increments.

WHEEL HEAD

The down-feed handwheel on toolroom grinders is graduated in 0.0001″ or 0.002mm (Figure 21–2). A handwheel may be fitted with a *zeroing slip ring* or mounted pointer, which permits the operator to set the handwheel to a zero reference point. This setting simplifies the dimensional information the operator must remember. The zeroing ring also reduces setting errors.

FEATURES OF THE FULLY AUTOMATIC PRECISION SURFACE GRINDER

Plunge and straight surface grinding may be done automatically on production runs. Dimensional accuracies may be held to within ±0.0001″ (±0.002mm) with finishes of 8 to 10 microinch a.a. The five basic features included in an automatic precision surface grinder are as follows:

- A complete digital control system,
- Automatic down-feed (plunge grinding),
- Automatic surface grinding,
- Automatic wheel dressing,
- Automatic digital position readout.

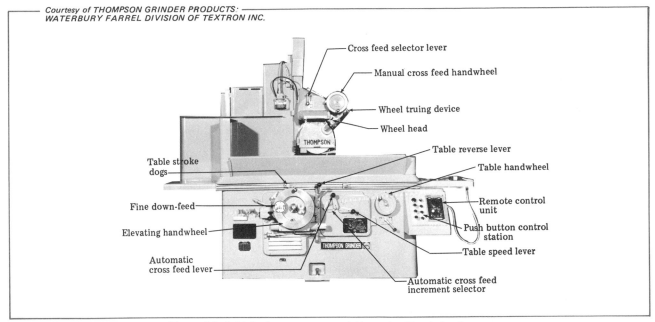

Courtesy of THOMPSON GRINDER PRODUCTS:
WATERBURY FARREL DIVISION OF TEXTRON INC.

Cross feed selector lever
Manual cross feed handwheel
Wheel truing device
Wheel head
Table reverse lever
Table handwheel
Table stroke dogs
Fine down-feed
Remote control unit
Elevating handwheel
Push button control station
Table speed lever
Automatic cross feed lever
Automatic cross feed increment selector

Figure 21–1 Main Components and Operating Handwheels on a Horizontal-Spindle, Reciprocating-Table Surface Grinder

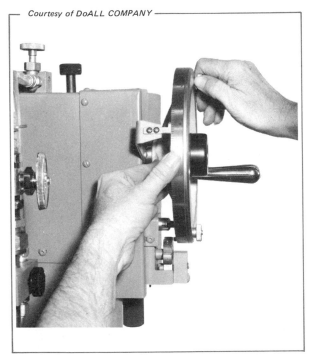

Courtesy of DoALL COMPANY

Figure 21–2 Down-Feed Handwheel and Zeroing Slip Ring (0.0001″ Graduations)

DIGITAL CONTROL SYSTEM

The digital control system panel on a fully automatic precision surface grinder (Figure 21–3) provides the following controls:

- Vertical feed, a four-position switch for manual, automatic, down, and up movements;
- Single step, a control that provides up or down wheel head movement at 0.000050″ (0.0001mm) increments;
- Full feed down, a control that permits the operator to conserve on setups and on productive machine time;
- Down-feed table, the down-feed selector control for plunge grinding;
- Down limit set, a control used for referencing the down-feed counter to the zero plane;
- Auto dresser, an off-on selector switch;
- Single step offset, a control that moves the wheel head in single 0.000050″ (0.0001mm) increments to correct any accumulated error;
- Reset up, a control that raises the wheel head to the up limit;
- Cycle start and cycle stop controls.

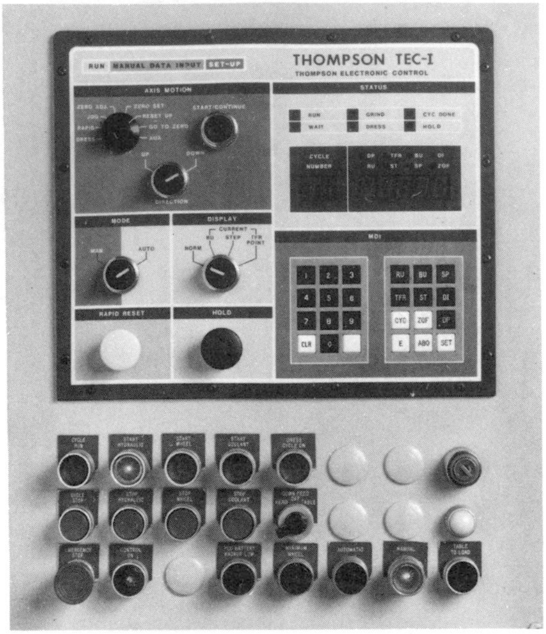

Figure 21-3 Digital Control System Panel on Fully Automatic Precision Surface Grinder

Other switches on the control panel are used with the coolant system, lubricant control, and for an emergency stop.

AUTOMATIC DOWN-FEED

Down-feed may be controlled for each table reversal or every other table pass. The feed increments may range from 0.0001'' (0.025mm) to 0.0099'' (0.25mm).

The grinding wheel is first referenced to the zero plane. The up limit thumb switches on the console panel are set to the desired dimension. The coarse-feed increment (0.0001'' to 0.0099'' or 0.02mm to 0.25mm) is set. Then the fine-feed increment (0.0001'' to 0.0009'' or 0.02mm to 0.025mm) and the point at which the fine feed is to begin are set.

Additional thumb switches are set for the number of spark-out passes, automatic wheel dressing and compensation (if required), and when the down-feed is to occur. The vertical feed selector is set to auto. Finally, the cycle start button is pushed.

AUTOMATIC SURFACE GRINDING

Selector switches are used for machine cycling. The down-feed increments and maximum depth are established as for plunge grinding. Both the cross feed reversal and wheel down-feeds are controlled automatically. They may be set for each or every other reversal.

AUTOMATIC WHEEL DRESSING

The operator selects when the grinding wheel is to be automatically dressed. The digital control is set to the number of feeds to wheel dress. The dressing may take place at any given number of grinding passes from 1 to 999.

The diamond dressing tool in the over-the-wheel dresser is fed automatically down to a preselected depth. Following this dressing pass, the wheel height is set to lower automatically. This lowering is to compensate for the reduction of wheel diameter resulting from dressing. Normally, increments of from 0.0002'' (0.005mm) to 0.001'' (0.02mm) are used.

DIGITAL POSITION READOUT

The position of the freshly dressed wheel above the table top or any other reference surface is shown on the digital position readout. The accuracy of a readout is within 0.0001'' (0.002mm).

WHEEL AND WORK-HOLDING ACCESSORIES

General grinding wheel accessories include wheel adapters, a wheel balancing arbor, and parallel ways or an overlapping disk balancing stand.

Magnetic chucks are the most versatile work-holding devices for surface grinding ferrous metals. Magnetic chucks may be of the *permanent magnet* or *electromagnetic* type. The common chuck shape for horizontal-spindle surface grinders is rectangular. However, a rotary magnetic chuck is adapted to operations that require a circular, scratch pattern. This chuck has an independent power and magnetizing source. Applications of magnetic chucking principles are to be found in magnetic V-blocks, vises, and parallel blocks, and sine chucks.

Magna-vise clamps, the toolmaker's vise, and the three-way vise are commonly used in combination with magnetic accessories. The three-way vise may be set quickly at any compound angle.

ELECTROMAGNETIC CHUCKS

Electromagnetic chucks stand higher on the table than permanent magnet chucks. The top face is laminated and usually has holding power across its entire area. Ferrous metal workpieces may be held directly on the face or by using other accessories such as magnetic V-blocks and parallels and vises.

Electromagnetic forces are controlled by an on-off position switch. Direct current is used to produce the electromagnetic effect. Residual magnetism remaining in the workpiece is removed on some machines by using a *selective chuck control.*

MAGNA-VISE (PERMA-)CLAMPS

Nonmagnetic workpieces that do not have a large bearing surface may be held for grinding with magna-vise clamps. Each clamp is made of a flexible, comb-like, thin, flat bar. The bar is attached to a solid bar by a spring-steel hinge.

The clamps, also called *perma-clamps,* are used in pairs (Figure 21–4). The solid edge of one clamp is positioned against the backing plate of the magnetic chuck. The work is brought against the toothed edge of the clamp. The toothed edge of the second clamp is fitted snugly against the side of the workpiece. As the magnetic chuck is energized, the toothed edges

of the perma-clamps pull down. The force exerted by the perma-clamps holds the workpiece against the chuck face.

LAMINATED PLATES, PARALLELS, AND BLOCKS

Magnetic chuck parallels are referred to as *laminated blocks.* Usually, narrow strips of low-carbon steel, alternated with thin separators (spacing strips), are welded into a positive non-shift unit. The chuck parallels are then precisely ground for parallelism.

Round, square, and irregular-shaped workpieces are sometimes held on precision laminated V-blocks, also called *chuck V-blocks.* An adapter (auxiliary top plate) is used to hold small and thin workpieces securely.

Backing plates on the sides and ends of magnetic chucks provide ideal locating surfaces for many grinding setups.

VACUUM CHUCKS

Vacuum chucks hold work against a plane surface by exhausting the air from between the part and the chuck face. This technique of holding applies to magnetic and nonmagnetic materials. The vacuum chuck has been found to be a practical device for holding paper-thin (0.002″ to 0.003″) workpieces.

PROBLEMS ENCOUNTERED IN PRODUCING FLAT SURFACES

DISCONTINUOUS SURFACES

Flat surfaces are *discontinuous* if they include holes, ridges, grooves, or slots. When many workpieces having a space between each one are ground at the same setting, the flat surface is interrupted. A discontinuous surface grinding setup is shown in Figure 21–5. As the grinding wheel approaches an interrupted section, the reduced area of wheel contact permits the wheel to cut slightly deeper. This dipping of the wheel is due to the fact that a grinding wheel normally exerts force during the cutting action. Dimensional and surface flatness inaccuracy may be produced.

The problem of discontinuous surfaces is overcome by (1) keeping the wheel sharp and cool cutting at all times and (2) taking light finish cuts.

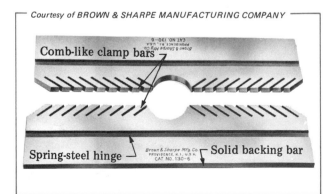

Figure 21–4 Magna-Vise (Perma-) Clamps

Courtesy of DoALL COMPANY

Figure 21–5 Discontinuous Surface Grinding Setup Requiring Operator Judgment and Control

WORK STRESSES AND DISTORTION

Castings; forgings; and heat-treated, cold-formed, extruded, or other manufactured parts have internal stresses and strains produced by chilling or mechanically working the material. Such stresses and strains are relieved, or unlocked, when the top surface or skin is machined away. A relieved surface changes shape. The grinder operator needs to recognize that the workpiece is *distorted* (warped or bent) away from its nominal shape. Compensation must be made to overcome this distortion (warp or bend). This may be done at the machine or by subjecting the part to heat treatment. Severely distorted parts require straightening by force on arbor and other presses.

WARPED WORKPIECES

Thin shims are often used to facilitate holding and grinding a warped piece. The top surface is carefully ground to flatness. The part is then turned over and held without the use of shims. The second surface, when ground, should be parallel to the flat top surface.

One of the easiest ways of checking the flatness of a surface is to place a precision straight edge on the ground surface.

Often, the workpiece is placed on a surface plate. If a feeler gage can be inserted at any point between the workpiece and the surface plate, the ground surface is not perfectly flat.

DOWN-FEEDS AND CROSS FEEDS FOR ROUGH AND FINISH GRINDING CUTS

Coarse, rough cuts are taken at the beginning of the grinding operation. The depth of successive cuts is reduced. Sufficient stock is left for finish grinding. The surface is generally rough ground by using cross feed increments of from 0.030″ to 0.050″ (0.8mm to 1.2mm). The rough-cut down-feed per pass on general jobbing shop work is 0.003″ to 0.008″ (0.1mm to 0.2mm) for average rough grinding. A heavier down-feed may be used on cast iron and soft steel. The down-feed may be increased still further if the work area being ground is small.

Fine down-feeds are used for the finishing passes. An excessive number of fine down-feed passes tends to load the wheel face. The cutting action is then inefficient and overheating may result. It is important to redress the wheel for light finish cuts of 0.0005″ (0.01mm) and finer. Where the down-feed is very light, fast table speeds and large cross feed increments are used. The workpiece must be passed under the grinding wheel fast enough to eliminate the possibility of overheating.

How to Set Up a Surface Grinder and Take a First Cut

STEP 1 Select the best available grinding wheel to suit the job and machine requirements.

STEP 2 Clean the wheel spindle and wheel bore. Remove any burrs, if necessary.

STEP 3 Ring test the wheel. Mount on the wheel spindle.

STEP 4 Replace the wheel guard. True and dress the wheel.

Note: The diamond or other wheel dresser must be positioned ahead of the centerline of the wheel at a 10° to 15° drag angle. A coolant is used if the part is to be ground wet.

STEP 5 Wipe the chuck face. Remove any burrs. Rub the palm of the hand across the chuck to remove all traces of dust.

STEP 6 Clean, burr, and place the workpiece as near to the center of the chuck as practical.

Note: In shoulder grinding and other flat grinding, the part may be aligned against the side or end plates on the chuck. In other cases, the part is accurately aligned with the longitudinal travel of the table by indicating.

STEP 7 Turn the chuck control lever or switch "ON."

Caution: Check the workpiece to be sure it is held securely. Make certain all feed control levers are disengaged and are in the neutral position.

STEP 8 Set the automatic trip dogs to accommodate the longitudinal travel of the workpiece and table.

STEP 9 Start the motors for the spindle, coolant system, and hydraulic fluid pump for the longitudinal feed. If dry grinding, turn the exhaust motor on.

Caution: Stand out of line with the plane of rotation of the grinding wheel.

STEP 10 Down-feed the grinding wheel carefully until it just touches (sparks) the high spot. Then, move the wheel toward one edge of the workpiece. Set the depth of cut for the first rough grinding pass.

STEP 11 Turn the table traverse feed control knob to the selected feed. In a similar manner, set the horizontal table travel speed to correspond with the sfpm recommended for the material being ground and the required surface finish.

STEP 12 Engage the power cross feed and longitudinal table feed.

Note: During the first pass, the operator must check the clearance between each end of the work and the centerline of the wheel. This checking is usually done manually, without the power feed.

STEP 13 Stop the table after a few strokes. Check the quality of the surface finish and the cutting action.

STEP 14 Take the first cut (pass). Feed the down-feed handwheel from 0.008″ (0.2mm) to 0.003″ (0.1mm) for successive roughing cuts. Leave from 0.002″ to 0.005″ (0.05mm to 0.12mm) for finish grinding.

STEP 15 Shut down the coolant flow. Then, stop the wheel at the end of the cutting stroke.

GRINDING EDGES SQUARE AND PARALLEL

In addition to having flat surfaces ground, most rectangular workpieces require that the edges be ground parallel and square. Such workpieces are usually rough machined slightly oversize. Stock is left to correct any inaccuracy in squareness, parallelism, and machining surface irregularity and to permit the sides to be ground square and to the required dimensional accuracy for surface finish and size. The general shop practice is to allow 0.010″ (0.2mm) on each side for rough and finish grinding.

A common method of grinding a workpiece square involves the use of a precision vise. The vise base surface is first checked for parallelism in relation to the spindle. The dimensional accuracy requirements (length, width, and height) for grinding the edges of a workpiece square and parallel are illustrated in Figure 21–6. The workpiece is placed on a parallel on the vise base. The first edge ① is ground. The adjacent edge ② is ground next. The squareness of its position in the vise is checked vertically (at 90°) with a dial indicator.

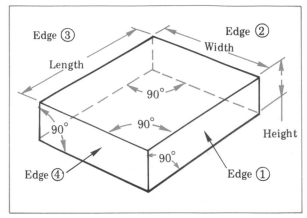

Figure 21–6 Dimensional Accuracy Requirements for Grinding Edges Square and Parallel

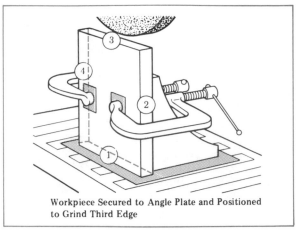

Workpiece Secured to Angle Plate and Positioned to Grind Third Edge

Figure 21–7 Position and Holding of Workpiece to Grind Third and Fourth Edges

After these two edges are ground square, each edge is used as a reference plane. Edge ① is placed on a parallel. The workpiece is secured and edge ③ is ground to the required dimension. Then edge ② is positioned on a parallel and clamped securely. Edge ④ is then ground parallel, square, and to the specified dimension. The squareness of edges ① and ② may be checked with a solid-steel square. A more precise measurement may be taken by using a cylindrical square.

A second method of grinding square edges is to use an angle plate (Figure 21–7). The large, flat surfaces are first ground parallel. The faces then act as plane reference surfaces. The use of a V-block and angle plate for holding a round part is illustrated in Figure 21–8.

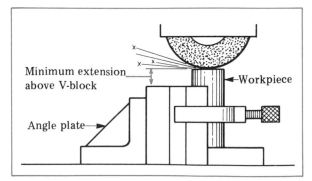

Figure 21–8 Vertical Setup for Round Part Using V-Block and Angle Plate on a Magnetic Chuck

How to Grind Edges Square and Parallel

Using an Angle Plate

STEP 1 Check the size of the workpiece. There must be sufficient stock left to clean up the sides so that they may be squared and brought to size and required finish.

STEP 2 Clean the workpiece, magnetic chuck, and angle plate. Remove any burrs.

STEP 3 Place a piece of paper under the angle plate and between it and the magnetic chuck.

STEP 4 Center one of the ground faces against the angle plate. Position the workpiece by resting one edge on a parallel or directly on the magnetic chuck.

STEP 5 Place a protecting shim between the outside finished face of the workpiece and either a C-clamp or a pair of parallel clamps. Secure the workpiece.

STEP 6 Magnetize the setup. Check to be sure the angle plate and workpiece are secure.

STEP 7 Start the spindle, hydraulic fluid system, and the coolant system. Position the coolant nozzle.

STEP 8   Set the trip dogs. Start the table. Bring the wheel down until it contacts the work.

STEP 9   Set the down-feed handwheel at a zero reading. Feed the wheel for the first roughing cut.

STEP 10   Stop the table at the end of the first few strokes. Check the dimension, cutting action, and the quality of surface finish.

STEP 11   Continue to take roughing cuts. Allow from 0.002″ to 0.005″ (0.05mm to 0.12mm) for the finish cut.

STEP 12   Dress the wheel before taking any finishing cuts (if necessary).

STEP 13   Increase the table speed and traverse feed. Set the down-feed for each finish cut. Machine to the required dimension.

STEP 14   Stop the machine. Wipe all surfaces dry and clean. De-energize the magnetic chuck.

STEP 15   Remove the clamps from edge ②. Relocate them on edge ①. Remove the clamps on edge ④. Turn the angle plate 90° so that the adjacent edge rests firmly on the magnetic chuck.

STEP 16   Energize the setup. Rough and finish grind the second edge.

STEP 17   Stop the machine. Wipe the surfaces dry and clean. Disassemble the setup.

STEP 18   Turn the workpiece 90° to position the third edge ③ (Figure 21–7).

STEP 19   Stop the machine when an area of the workpiece is ground and a measurement is possible. Measure the part. Set the graduated dial at zero.

STEP 20   Recheck the workpiece for dimensional accuracy. The surface finish may be checked against a comparator specimen.

STEP 21   Shut down the machine. Break the edges of the workpiece, if required. Clean the machine.

How to Grind Ends Square With the Axis (Round Work)

STEP 1   Select a V-block that will accommodate the diameter of the workpiece.

STEP 2   Clamp the workpiece in the V-block so the ground end extends 1/16″ to 1/8″ (2mm to 3mm) above the block.

STEP 3   Move the chuck energizing control knob to the "ON" position. Check on how securely the assembly is held.

STEP 4   Adjust the table reversing (trip) dogs. Position the coolant nozzle. Start the machine, hydraulic fluid system, and coolant system.

STEP 5   Lower the grinding wheel until it grazes the workpiece. Position and move the wheel down and start the cut.

STEP 6   Measure the length of the workpiece. Set the down-feed handwheel collar at zero.

STEP 7   Take roughing and finishing cuts to clean the end to a particular size.

STEP 8   Shut down the operation. Remove the workpiece and file or stone any fine wire edge around the ground end.

STEP 9   Reverse the workpiece in the V-block. Position the second end above the top of the V-block. Use an accessory block against the V-block, if required.

STEP 10   Energize the magnetic chuck. Check the setup to be sure it is secure.

STEP 11   Start the machine and coolant flow.

STEP 12   Position the wheel for depth. Set for the first roughing cut. Turn the handwheel setting to zero. Take the cut. Stop the machine.

STEP 13   Measure the workpiece. Take successive cuts to grind the piece to size.

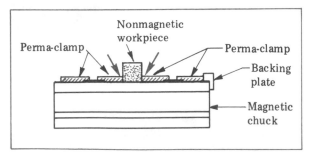

Figure 21–9 Forces Exerted by Perma-Clamps to Hold Nonmagnetic Workpieces Securely

HOLDING THIN WORKPIECES

USE OF MAGNA-VISE CLAMPS

Serrated magna-vise clamps provide a convenient work-holding device for small, thin parts (that are slightly thicker than the height of the clamps). A thin part should be shimmed, if necessary. With this setup, the workpiece is mounted parallel to the side faces of the magnetic chuck. Nonferrous metals are also held securely in a magnetic chuck with magna-vise clamps (Figure 21–9).

ADAPTER PLATES: FINE POLE CHUCKS

An adapter plate is specially designed to hold small, thin workpieces. The adapter plate is a precision ground rectangular plate. It is set directly on a magnetic chuck as illustrated in Figure 21–10. The laminations run lengthwise and are finely spaced to allow more, but weaker, lines

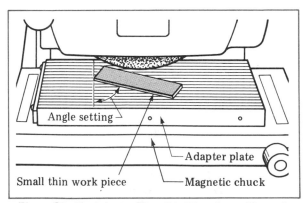

Figure 21–10 Adapter Plate Set Directly on Magnetic Chuck for Holding Small, Thin Workpieces

of force to act on the workpiece. The adapter plate is energized by turning the magnetic chuck on. Small, thin workpieces are positioned at an angle of 15° to 30° to minimize the heat generated during each pass.

How to Grind Thin Workpieces

STEP 1 Select a cool-cutting, open-grained wheel to produce the required surface finish.

STEP 2 Mount, true, and dress the grinding wheel. Relieve part of the wheel, if required.

STEP 3 Prepare the machine for grinding.

STEP 4 Test the workpiece for straightness. Determine whether there is adequate material left for grinding (if a distorted part is shimmed).

STEP 5 Straighten the work, if necessary. Mount, support, and secure the work on a magnetic chuck or an adapter plate.

STEP 6 Set the necessary machine controls, speeds, feeds, and coolant flow.

STEP 7 Feed the wheel slowly down to the workpiece. When the wheel sparks at the high point, feed it from 0.0005″ (0.01mm) to 0.001″ (0.02mm). Use a finer cross feed.

STEP 8 Continue with fine cuts until the surface is cleaned up.

STEP 9 Demagnetize the chuck and workpiece. Test for flatness.

Note: It may be necessary to grind the second side almost to size and go back over the first side if the work is still distorted.

STEP 10 Stop the coolant flow, all controls, and the motors. Stone or use a fine file to remove any grinding burrs.

Safe Practices in Grinding Flat Surfaces and Thin Workpieces

- Add a square block or angle plate in front of a V-block to give added support against the cutting forces of a grinding wheel. Nest small workpieces against end and side blocks.
- Check the clearance between the workpiece, mounting device, wheel, and other machine parts before starting the machine.
- Bring the grinding wheel into contact at the high spot of a surface to be ground by slowly feeding the grinding wheel.
- Recheck how securely the diamond holder is held on the magnetic chuck for dressing and truing operations. Maintain a sharp diamond dresser by rotating the point from 10° to 20° after each dressing. Use a coolant during dressing with a diamond.

- Place thin workpieces that may be warped as a result of heat generated during grinding, diagonally across the magnetic chuck or adapter plate. It may be necessary to relieve the face of the wheel to reduce the area of contact.
- Increase the traverse rate per pass and the table speed when using a dressed wheel. This combination is essential to producing a high-quality surface finish.
- Replace the machine guard and wear protective goggles or a shield. Observe general machine safety precautions. Stand out of line of the wheel travel.
- Check the position of splash guards. The coolant must be contained on the machine and flow back to the reservoir. Any spilled grinding fluid or other oils are to be wiped from the machine area.

TERMS USED WITH DESIGN FEATURES, SETUPS, AND FLAT SURFACE GRINDING PROCESSES

Basic machine components The base, column, saddle, wheel head, and table, which comprise the machine tool. In addition, the control handwheels and mechanisms that produce circular, vertical, horizontal, and traverse movements.

Wheel head The wheel spindle and drive mechanism, which control the grinding wheel. Mechanisms that are housed as a single unit. The spindle and drive component that is adjustable to the work height and for down-feeding.

Fine-feed cross feed attachment A further refinement of the micrometer dial (collar) on the cross feed. Cross feed movements are dial graduated in vernier 0.0001″ or 0.002mm increments rather than standard increments.

Digital position readout Usually, a numerical value that tells the operator the relationship between a cutting tool and a reference plane.

Discontinuous surface A plane surface that is interrupted by a slot, groove, ridge, hole, or other cutaway section.

Distortion (warped part)	The twist or bend of a part away from its nominal shape and size.
Increment (down-feed or cross feed)	The amount (distance) the grinding wheel and/or the workpiece is fed uniformly at the completion of each pass over the work face.
Dimensional grinding	Controlling the dimension to which a part is to be ground by feeding to depth (size) with the handwheel graduations. Grinding to within 0.0001″ (0.002mm) tolerance directly from calibrated handwheel settings.
Sparking out	Uniformly traversing the ground surface of a workpiece after the last cut without further down-feed.

SUMMARY

- The five basic construction features of a surface grinder are the base, column, saddle, wheel head, and table.
  - Surface grinder movements are controlled by three sets of handwheels and systems that relate to down-feed, cross feed (transverse), and longitudinal (horizontal) travel.
- Handwheels may be equipped for vernier ("tenth": 0.0001″ or 0.002mm) settings. A zeroing slip ring permits the operator to set a handwheel to a zero reference point.
  - Hydraulic table systems provide infinitely variable table feeds such as 0 to 70 sfpm (0 to 21.3 m/min).
- Hydraulic cross feed systems permit feeds within a 1/64″ to 1/4″ (0.4mm to 6.0mm) range.
  - A fully automatic precision surface grinder includes five supportive systems: complete digital control, automatic down-feed increments from coarse to fine feed, automatic cross feed reversal and wheel feed controls, automatic wheel dressing, and automatic digital position readout.
- Common work-holding accessories for surface grinder work may be magnetic or nonmagnetic. Magnetic chucks, adapters, parallels, V-blocks, magna-vise clamps, and sine blocks are widely used. Toolmaker's and conventional universal vises, straps and clamps, and vacuum chucks are also common.
  - The dimensional accuracy of a ground flat surface is influenced by such factors as surface continuity, distortion, release of internal stresses produced by certain manufacturing methods, and general grinding practices.
- Grinding discontinuous surfaces requires machine and work rigidity and a sharp cutting wheel.
  - The down-feed is decreased to 0.0005″ (0.01mm) and finer for final finish cuts. For extremely light down-feeds, the table speed and cross feed on the final cut are increased.

- The trip dogs are set so that the workpiece clears the wheel from 1/2″ (12mm) to 1″ (25mm) at each end.

    - The spindle RPM is checked before the operation. It is brought as near as possible to the RPM required to develop the recommended grinding wheel sfpm.

- In dimensional grinding, the down-feed handwheel graduated dial (zeroing slip ring) is set at zero at the start of the cut. The wheel is fed over a series of cuts the distance required to machine to a specified dimension.

    - When edges are ground square, the first two edges are ground at 90°. These edges may be positioned directly on the magnetic chuck. The third and fourth edges are then ground parallel and square, respectively.

- A V-block is supported against an angle plate or square block when a round workpiece is held vertically and the end is to be ground square.

    - Distortion, buckling, and warpage are conditions that must be corrected or compensated for in grinding thin workpieces.

- Diagonal positioning and relieving part of the grinding wheel face reduce the heat generated during the grinding of small, thin workpieces.

UNIT 21 REVIEW AND SELF-TEST

1. a. List three design features of a fully automatic surface grinder, exclusive of the digital control system.
    b. Describe briefly the function of each of the design features listed.

2. a. Tell what common practices are followed with table speeds and traverse feeds when fine finish cuts of 0.0005″ (0.01mm) are taken.
    b. Identify two problems encountered by the machine operator (1) when an excessive number of fine cuts (passes) are taken and (2) what corrective steps may be followed.

3. Identify the steps for setting up and grinding the two ends square and to accurate length for a short-length die block. An angle plate and magnetic chuck setup are to be used.

4. a. Tell what effect relieving part of a wide-face abrasive wheel has on the generation of heat during grinding.
    b. Explain how the opposite faces of thin parts that are bent or warped may be ground parallel.

5. State two machining techniques that may be used to prevent the distortion that may be produced when thin workpieces are ground.

6. List six safety precautions the surface grinder operator must check for the setup, prior to taking the first cut.

Angular, Vertical, Shoulder, and Profile Grinding and Cutting-Off Processes

Common accessories that are used to hold workpieces at an angle and to generate a flat (angular) surface are grouped as follows:

- Vises (plain, adjustable, universal, and swivel);
- Angle plates (adjustable and sine angle plates);
- Magnetic work-holding accessories (V-blocks, toolmaker's knee, swivel chucks, and sine table);
- Grinding fixtures for production work.

The first part of this unit describes these accessories with applications to angular surface grinding. Step-by-step procedures are given for machine setups and the actual grinding of straight, filleted, and angular shoulders and vertical surfaces.

Form grinding (as described in Part B) deals with dresser accessories and dressing tools, machine setups, and form dressing processes. Grinding wheels may be formed by hand. Simple abrasive dressing sticks of aluminum oxide, silicon carbide, and boron carbide are used for simple hand forming and dressing. These hand-guided dressing tools are practical for roughly shaping a wheel to a limited degree of accruacy. Parts that are to be ground within close dimensional tolerances for shape and size require that the wheel be precision dressed.

OBJECTIVES

A. GRINDING ANGULAR AND VERTICAL SURFACES AND SHOULDERS

After satisfactorily completing this unit, you will be able to:

- Select appropriate plain, adjustable, universal, and swivel vises; solid right angle, adjustable, and sine angle plates; and magnetic work-holding accessories, for single and compound angle grinding.
- Describe square, filleted, and angular shoulders and techniques of forming and relieving wheels for shoulder grinding.
- Set up and perform the following processes.
  - Grinding Angular Surfaces Using Magnetic and Nonmagnetic Work-Holding Devices.
  - Grinding an Angle with a Right Angle Plate, Sine Bar, and Gage Chuck Setup.
  - Setting Up the Magnetic Sine Table to Grind a Compound Angle.
  - Form Dressing a Grinding Wheel for Angular Grinding.
  - Grinding Shoulders.

B. PROFILE (FORM) GRINDING AND CUTTING-OFF PROCESSES

- Select appropriate commercial wheel forms for specific form grinding.
- Understand functions and setups of radius dressers to generate concave, convex, and continuous contour wheel forms.
- Use the sine plate, angle plate, and parallel method of setting up to dress a wheel at an angle.
- Recognize when cone, full-ball radius, and other forms of diamond point dressing tools are used.
- Describe how work and reference rolls are formed and their use with idler and power-driven units for crush forming and form control.

- Relate design features of the microform grinder.
- Provide product information about fully and zone-reinforced cutoff grinding wheels; wheel specifications, and properties of rubber, resinoid, shellac and metal bonds for aluminum oxide, silicon carbide, and diamond cutoff wheels.
- Apply handbook tables on gage block constants, cutting speeds, and feeds.
- Perform or simulate the following processes.
  - Rough Forming a Wheel.
  - Forming a Concave or Convex Radius by Using a Radius Dresser.
  - Form Dressing a Wheel with a Radius and Angle Dresser.
  - Form Grinding with a Crush Dressed Wheel.
- Follow *Safe Practices* related to personal safety, work-holding accessories, workpieces, surface grinder, and grinding processes.
- Correctly use each new *Term*.

A. GRINDING ANGULAR AND VERTICAL SURFACES AND SHOULDERS

MAGNETIC WORK-HOLDING ACCESSORIES

MAGNETIC TOOLMAKER'S KNEE

The magnetic toolmaker's knee (Figure 22–1) is especially designed to hold odd-shaped parts for grinding. The model illustrated is a permanent magnetic tool. The magnetic surface is 4″ × 6 1/2″ (100mm × 165mm) and is parallel within 0.002″ (0.05mm) overall. The tolerance for squareness is 0.00005″ per inch.

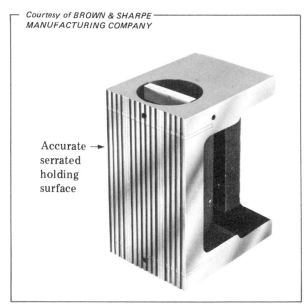

Courtesy of BROWN & SHARPE MANUFACTURING COMPANY

Accurate → serrated holding surface

Figure 22–1 Permanent Magnetic Toolmaker's Knee for Inspection, Layout, and Grinding Setups

SWIVEL-TYPE MAGNETIC CHUCK

This horizontal magnetic chuck has a swivel feature. This feature makes it possible to position the chuck at an angle from the usual horizontal plane. The top face on which the workpiece is positioned and held by magnetic force is set at the required angle. The accuracy of the angle may be checked with a vernier bevel protractor. Finer angle setups are checked against an angle sine block combination.

MAGNETIC SINE TABLES

Plain and compound electromagnetic sine chucks are shown in Figure 22–2. A plain electromagnetic sine chuck (Figure 22–2A) has two hinged sections. The first section includes a base and angle plate. Angles from 0° to 90° may be set. Gage blocks are used to establish the vertical height for a required angle.

The second section is hinged to the accessory table. Again, gage blocks are used to set the second table (a magnetic table) at a specified angle from 0° to 90°. Workpieces are positioned directly on the magnetic chuck face.

A compound electromagnetic sine chuck (Figure 22–2B) includes a third hinged section. This design feature permits setting the electromagnetic chuck face at a compound angle. Each section is set precisely at one of the required angles by using gage blocks.

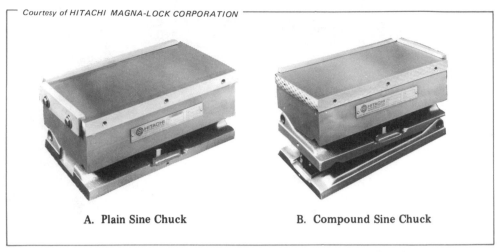

A. Plain Sine Chuck   B. Compound Sine Chuck

Figure 22–2 Plain and Compound Electromagnetic Sine Chucks

How to Grind Angular Surfaces

Using a Plain Vise

STEP 1 Select an appropriate size and type of vise. Clean and burr all locating surfaces.

STEP 2 Place the vise on the magnetic chuck. Use a parallel to position the vise jaws at a right angle to the wheel spindle. Position and secure the vise (Figure 22–3).

Note: The method of locating the work at a specific angle depends on the nature of the job and the required accuracy. A Surface gage or a vernier height gage may be used to locate layout lines.

STEP 3 Proceed to select the best wheel available. Mount, true, and dress the wheel.

STEP 4 Set the speeds, feeds, length of table movement, and so on.

STEP 5 Start the machine and coolant system. Bring the wheel head down until the grinding wheel just grazes the area to be ground.

STEP 6 Rough and finish grind the angular surface. Check for dimensional accuracy and surface finish.

STEP 7 Proceed to remove any grinding burrs. Disassemble the work-holding setup. Clean and replace each part in its proper storage place.

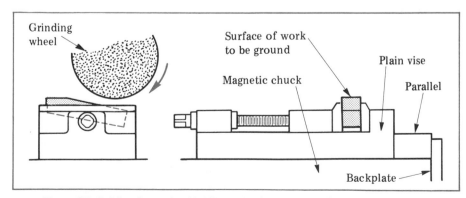

Figure 22–3 Vise Setup for Holding a Workpiece to be Ground at an Angle

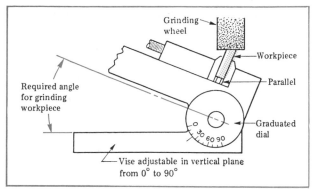

Figure 22-4 Adjustable Vise for Angular Grinding Applications

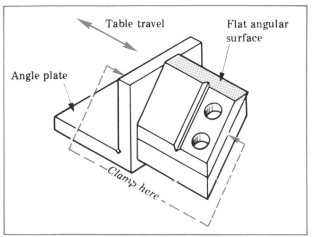

Figure 22-5 Angle Plate Setup for Grinding an Angle

Using an Adjustable Vise (Single Angle)

STEP 1 Remove the keys from the vise base. Position and hold the vise on the magnetic chuck.

STEP 2 Set the workpiece on a parallel in the vise. Secure the workpiece between the vise jaws.

STEP 3 Loosen the hinged-base clamping nut. Swing the vise to the desired angle. Use the graduated quadrant to obtain a direct reading (Figure 22-4).

STEP 4 Secure the clamping nut. Recheck the angle setting.

Note: The angle setting is often checked by using a dial indicator in combination with angle gage blocks.

STEP 5 Proceed with the standard machine setup. Take roughing and finish grinding cuts.

STEP 6 Measure the angle and surface finish. Shut down the operation after the part is ground to size, deburred, and measured.

Using a Right Angle Plate

STEP 1 Select an angle plate of a suitable size. Set the angle plate on the chuck. Align the vertical face.

Note: The surface to be ground should be set parallel with the table travel.

STEP 2 Place the surface of the work against the one face of the right angle. Position the work at the required angle. Use a vernier bevel protractor, gage blocks, or other measuring device. The angle plate setup for grinding an angle is shown in Figure 22-5.

STEP 3 Clamp securely. Check to be sure the wheel clears any clamps or other accessory.

STEP 4 Set up the machine. Take roughing and finishing cuts. Check the angular measurements. Grind to size and surface finish.

Using a Magnetic V-Block

STEP 1 Select the magnetic V-block that accommodates the part to be ground. Clean the base and the work contacting surfaces.

STEP 2 Locate the magnetic V-block centrally on the chuck. Position it parallel with the side of the chuck. The setting and use of a magnetic V-block to position and hold a workpiece are illustrated in Figure 22-6.

STEP 3 Nest the workpiece in the magnetic V-block. Turn the magnetic chuck control lever to the "ON" position.

Caution: The side of the wheel must clear the vertical surface. During the last pass or two, disengage the automatic traverse feed to provide better machine operator control in feeding the wheel as it contacts the side of the workpiece at the shoulder.

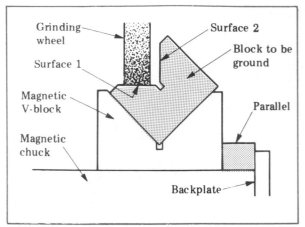

Figure 22–6 Setting and Use of Magnetic V-Block to Position and Hold Workpiece

STEP 2 Mount the workpiece on the angle sine bar. Clamp in position.

STEP 3 Lubricate and prepare the machine. Take roughing and finish cuts.

STEP 4 Check the dimensional accuracy of the machined surface.

STEP 5 Deburr the workpiece, if necessary. Wipe the machine, accessories, and workpiece clean and dry. Place each item in its proper storage container.

How to Grind an Angle With a Right Angle Plate, Sine Bar, and Gage Block Setup

STEP 1 Place the angle plate on a surface plate. Set up the angle sine bar combination for the required angle. Figure 22–7 shows the sine bar and angle plate setup commonly used for laying out, holding, and grinding a precision angular surface. As shown, the workpiece is set at a 15° angle.

How to Set Up the Magnetic Sine Table to Grind a Compound Angle

STEP 1 Select and bring together the gage block combination for the vertical height of the first angle setup.

Note: The sine formula or a sine table may be used to establish the gage block height.

Caution: Use wear blocks as the end blocks.

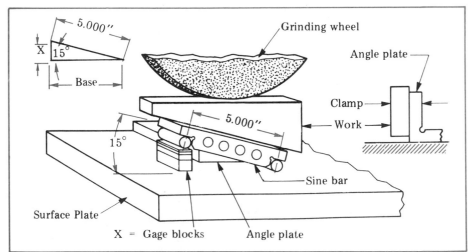

Figure 22–7 Sine Bar and Angle Plate Setup for Grinding a Precision Angular Surface

STEP 2
Place the gage blocks between the sine table base and the hardened end roll. Bring the sine table down so that it rests on the gage blocks. Tighten the hinge swivel clamping nut.

STEP 3
Align the sine table with the side plate of the magnetic chuck. The sine table face will then be parallel to the horizontal movement of the table. Strap the base to the table.

STEP 4
Mount the burred and cleaned workpiece on the sine chuck face.

STEP 5
Energize the magnetic chuck and the sine chuck. Check to be sure that the workpiece is held securely to the sine chuck. Similarly, the sine chuck must be secure on the magnetic chuck.

STEP 6
Lubricate the grinder. Select the appropriate grinding wheel. Determine feed rates, table speed, and coolant composition and flow rate.

STEP 7
Set up the surface grinder. Take roughing and finishing cuts.

STEP 8
Check for dimensional accuracy and surface finish.

STEP 9
Disassemble the setup after the machine is shut down. Return each tool and instrument to its proper storage place.

FIXTURES FOR PRODUCTION GRINDING WORK

Grinding fixtures serve the same functions as fixtures used in milling machine, drilling machine, and other machine setups. They are designed as work-positioning and work-holding devices. Fixtures are simple to operate and are fast, accurate, and dependable. They may be operated manually, pneumatically, or electronically or by combining hand and automatic controls. Fixtures are essential in production work, particularly where a large number of parts require precision grinding to the same size, shape, and surface finish.

Some fixtures for ferrous parts contain special magnetic holding devices. The workpiece may be located accurately, held securely by magnetic force, and released quickly and easily. Fixtures are widely used to nest irregularly shaped workpieces where, because of the shape, it is difficult to position and hold the part in a standard work-holding accessory.

UNDERCUTTING FOR SHOULDER GRINDING

Wherever practical, the juncture of the two flat planes of a square shoulder should be undercut by prior machining processes to eliminate a sharp corner (Figure 22–8). Undercutting is important for the following reasons:

- To preserve the shape of the wheel at the edge.

- To permit the wheel to produce a true surface over the entire area without leaving a slight ridge in the corner.

- To help to eliminate wheel stresses at the corner contact area.

- To increase the life of the abrasive wheel and thus decrease the time consumed in truing and dressing operations.

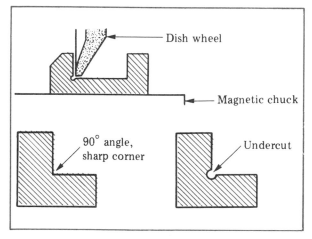

Figure 22–8 Undercut Features on a Square Shoulder

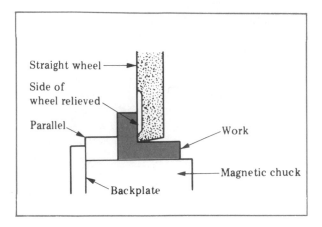

Figure 22-9 Setup of Workpiece for
Square Shoulder Grinding

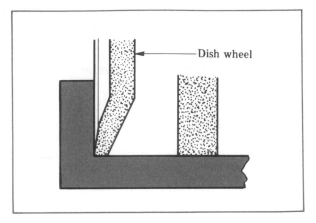

Figure 22-10 Use of Dish Wheel to Grind
Vertical Surface

TRUING AND DRESSING DEVICES FOR GROUND SHOULDERS

The straight-face and the dish-type wheels illustrated in Figures 22-9 and 22-10 are commonly used for grinding square and vertical face shoulders. The redressing or relieving of the side of the wheel may be done by using an abrasive stick offhand. A diamond dresser or other truing and dressing tool, mounted in a suitable holding device, is used for more accurate truing and dressing.

Figure 22-11 illustrates the grinding of square, angular, and filleted shoulders by using straight wheels dressed to shape.

Similarly, one or both sides of a straight-face wheel may be dressed to the required angle for an angular shoulder. The correctness of the angle is usually checked with a bevel protractor or angle gage. When a wide horizontal surface and angle are being ground simultaneously, the face (periphery) is sometimes relieved as shown in Figure 22-11.

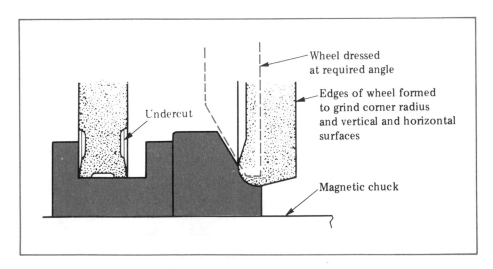

Figure 22-11 Grinding Square, Angular, and Filleted Shoulders by
Using Straight Wheels Dressed to Shape

How to Form Dress a Grinding Wheel for Angular Grinding

STEP 1 Select the wheel dressing tool. Mount it in a fixture or other holding device on the magnetic chuck.

Note: The dressing tool or stick is set in a fixture at the designated angle. The angular side of the wheel is formed by feeding the wheel truing diamond or other abrasive cutter across the side face.

STEP 2 Increase the dressing tool contact a few thousandths of an inch at a time. Continue to dress the side of the wheel until the correct angle or radius is formed and the cutting faces of the wheel are trued.

STEP 3 Stop the wheel spindle. Check the radius with a radius gage or template. Check the angular side face with a bevel protractor or other gage.

How to Grind Shoulders

Rough Grinding Shoulders

STEP 1 Start the table reciprocating. The coolant is turned on, if the work is to be wet ground.

STEP 2 Feed the wheel down to start rough grinding the horizontal surface. Take from 0.001″ to 0.002″ (0.02mm to 0.05mm) per table stroke as a cut.

Note: This operation requires the manipulation of both the vertical and crossfeed handwheels to set the depth of cut and to traverse the area to be ground, respectively.

STEP 3 Rough grind the vertical surface. Feed the work face toward the grinding wheel. Continue to rough grind the vertical surface to within 0.001″ to 0.0015″ (0.02mm to 0.04mm) of finish size.

Note: A dish-type wheel is preferred over a straight-face wheel when a square shoulder with a narrow adjacent horizontal face is ground.

Finish Grinding Shoulders

STEP 1 Traverse the table to the extreme end. True and dress both the face and side of the wheel to true the wheel, produce the necessary relief, and restore sharp abrasive cutting edges.

STEP 2 Repeat the steps used for rough grinding to position the wheel for finish grinding.

STEP 3 Feed the grinding wheel down about 0.0005″ (0.01mm) after the first sparks and grind the horizontal surface.

STEP 4 Feed the vertical face toward the wheel at 0.0002″ to 0.0003″ (0.005mm to 0.008mm) per stroke.

STEP 5 Continue to machine to size and shape according to the required degree of dimensional and surface finish accuracy.

STEP 6 Allow the wheel to spark out in the final finish grinding step.

B. FORM (PROFILE) GRINDING AND CUTTING-OFF PROCESSES

GENERAL TYPES OF WHEEL FORMING ACCESSORIES

The three general types of wheel forming accessories are as follows:

- Radius truing and dressing devices which form a concave or convex radius;

- Angle truing and dressing devices which form angles, shoulders, and bevels;

- Combination radius and angle truing devices which form both radii and angles as used singly or in combination.

RADIUS DRESSER

While differing in design and construction, all radius truing and dressing devices of the *cradle type* follow the same principles. Each device supports a diamond dressing tool that is

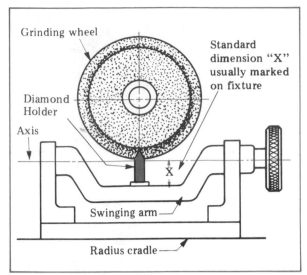

Figure 22–12 Relationship of the Diamond Position and the Axis of a Radius Dresser

used to form a required radius on the grinding wheel. The radius is produced by rotating a curved arm that extends from an upright member on the dresser body. The curved arm may be rotated on its axis to swing the diamond in an arc.

The concave or convex radius produced is determined by the relationship of the diamond nib (point) to the axis of the arm. A concave form is generated in the wheel face when the diamond is set above the axis and is swung in an arc while in this position (Figure 22–12). Conversely, a convex form is produced when the diamond is set below the axis.

How to Form a Convex or Concave Wheel by Using a Radius Dresser

STEP 1   Select an appropriate grinding wheel for the job. Mount the wheel. Prepare the grinder for operation.

STEP 2   Select a suitable cone-point diamond dressing tool to fit the radius dresser.

STEP 3   Align the radius dresser squarely on the magnetic chuck.

STEP 4   Set the stops on the dresser to control the arc over which the radius is to be swung. (Figure 22–13 illustrates a wheel that is to be formed with a 0.250″ radius over 90° of the wheel face and edge. The stops in this case are set 90° apart.)

STEP 5   Position the diamond point at the required radius.

STEP 6   Locate the positioned diamond under the center of the wheel. Lock the grinder table in this position.

STEP 7   Start the machine and flow of coolant. Rotate the dresser arm so that the diamond is in a horizontal position.

STEP 8   Move the diamond carefully toward the wheel until it just touches the side. Then, lock the saddle.

STEP 9   Raise the wheel head until the diamond clears the wheel. Start rotating the radius dresser to make contact with the face.

STEP 10   Feed the wheel from 0.002″ to 0.003″ (0.05mm to 0.08mm) for each swing (pass) of the diamond. Continue to form the radius.

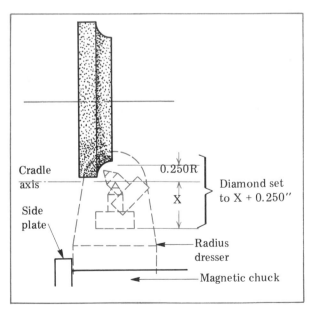

Figure 22–13 Radius Dresser Setup to Form a Corner Concave Radius

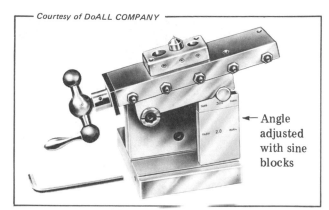

← Angle
adjusted
with sine
blocks

Figure 22-14 Precision Sine Dresser

ANGLE DRESSER

Another method of producing an angular surface is to form the wheel at a required angle with an *angle dresser*. The *precision sine dresser* shown in Figure 22–14 is adjustable to 45° either side of a horizontal (0°) position. This device is set by using gage blocks to accurately obtain the required angle.

COMBINATION RADIUS AND ANGLE DRESSER

A *radius and angle dresser* combines the processes performed by the separate dressers. A radius and angle dresser is designed to form convex and concave outlines and angles on grinding wheels. Moreover, this device makes it possible to generate combinations of radial and angular shapes that are continuous (tangential to each other). On this type of dresser, the diamond dressing tool must be located at the horizontal center line of the wheel. The wheel is formed with the diamond and holder in a right-side position instead of under the wheel.

The principal parts of a precision radius and angle dresser are identified in Figure 22–15. The base carries a sliding platen that is moved horizontally by a micrometer collar. This movement permits the dressing of angles. Concave and convex radii are formed by using a swivel base movement. The stop pins are located to permit dressing each radius and angle according to specified dimensions.

The radius and angle dresser is adaptable to the precision grinding of punches, sectional dies, cutting tools, special forming tools, and other work involving hardened steels.

How to Form Dress a Wheel with a Radius and Angle Dresser

Setting the Dresser to Form a Convex or Concave Radius

STEP 1   Measure the distance from the selected diamond dressing tool point to the micrometer plate on the dressing device (Figure 22–15).

STEP 2   Add the required radius to the micrometer reading.

Note: A convex form is generated by subtracting the required radius.

STEP 3   Turn the micrometer base lead screw until the measurement over the micrometer pins equals the combined radius plus micrometer reading (Figure 22–15).

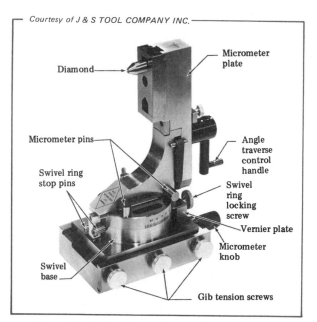

Diamond →

Micrometer plate

Micrometer pins

Swivel ring stop pins

Angle traverse control handle

Swivel ring locking screw

Vernier plate

Micrometer knob

Swivel base

Gib tension screws

Figure 22-15 Radius and Angle Dresser Features

Setting the Dresser to Form
One or More Angles

STEP 1 Loosen the swivel ring locking screw.

STEP 2 Bring the two swivel ring stop pins together. Move them until the desired angle for the lower pins is read on the vernier plate.

STEP 3 Move the dresser in the opposite direction to the second required angle. This movement sets the position of the upper swivel ring stop pin.

STEP 4 Lock the pins in the desired position by tightening the swivel ring locking screw.

Note: The positions of the stop pins are the points at which the angles are tangent to the radius.

Dressing the Wheel
(to a Radius and Two Angles)

STEP 1 Align the dresser on the magnetic chuck. Secure it.

STEP 2 Position the grinding wheel spindle so that its axis coincides with the horizontal axis of the dresser diamond.

STEP 3 Adjust the machine table so that the diamond almost touches the periphery of the wheel. Lock the table in position. Position the diamond in relation to the width of the wheel.

STEP 4 Lock the saddle. Start the machine and coolant flow if the wheel is to be dressed wet.

Caution: Use the exhaust system if the wheel is to be dressed dry.

STEP 5 Move the diamond until it grazes the wheel. The micrometer base lead screw knob is turned to produce this movement.

STEP 6 Feed the diamond for 0.002" to 0.003" (0.05mm to 0.08mm) and swivel the dresser against one stop.

STEP 7 Dress one angle by turning the angle dressing handle in the direction the diamond is to move. Then, return the handle to the center position.

STEP 8 Dress the radius by swiveling the dresser upright to the other stop.

STEP 9 Produce the other angle by turning the angle dressing handle.

STEP 10 Take successive cuts, repeating steps 6 through 9, until the full form is produced on the wheel.

Note: The feed and speed (movement of the diamond) are reduced for the last two or three passes, if a finely ground surface finish is required.

FORMING A WHEEL WITH CRUSH ROLLS

In addition to the single high-speed steel or carbide crush roll, two crush rolls are preferred for quantity production. One roll serves as a work roll. The second roll, a *reference roll* is used to form dress the work roll when it becomes worn.

CONSIDERING GRINDING WHEEL CHARACTERISTICS

All grinding wheels cannot be crush formed. It is impractical to apply this process to shellac-, rubber-, or resinoid-bonded wheels. The grains tend to break out in clusters. Thus, crush forming is possible only with vitrified-bonded aluminum oxide and silicon carbide grinding wheels. Organic-bonded wheels are too elastic to permit building up enough force between the roll and the wheel to crush form them.

The radius of the form is important in selecting the size of grit. The grit size must be smaller than the smallest radius. The general grit size of grinding wheels falls within the 120 to 400 medium-grade range.

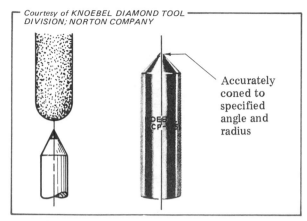

Accurately coned to specified angle and radius

Figure 22-16 Cone Point for Precision Dressing Intricate Radii and Combination Forms

Crush dressing produces a sharp cutting wheel. The grains are sharp pointed for fast removal of material. However, the quality of the surface finish is not as high as the finish produced by diamond dressing a wheel.

DIAMOND DRESSING TOOLS FOR FORMING GRINDING WHEEL SHAPES

The size and shape of both the diamond and holder depend on the contour to be generated in the wheel, the wheel diameter and characteristics, and the nature of the application (light, medium, heavy, or extra-heavy). A *cone-point* diamond cutting tool, as pictured in Figure 22-16, is used for the precision dressing of intricate radii and complex combination forms. Cone-point dressing tools are ordered by the type of cone point, the included angle on the diamond, the radius of the diamond, and the application. This information is required by the manufacturer.

The *full-ball* radius dressing tool (diamond dresser) is designed to plunge dress a full 180° concave radius in a grinding wheel. This dressing tool is available in standard radii between 0.010″ to 0.050″ (0.25mm to 1.25mm). Smaller and larger radii dressers are available to meet special requirements. Frequent turning of the tool is necessary to provide longer tool life. When a diamond becomes worn with use, it may be relapped.

There are three common forms of *radius diamond dressing tools*. The first form is

available in six standard sizes to form concave radii from 0.010″ to 0.250″ (0.25mm to 6.25mm). The second form generates convex radii on small-diameter grinding wheels. There are three general sizes of this dressing tool. The sizes accommodate radii from 0.020″ to 0.500″ (0.5mm to 12.5mm). The third form is applied when dressing half-circle concave radii. Radii of 0.032″ (0.8mm), 0.062″ (0.16mm), and 0.125″ (3mm) are standard sizes. Designs are available for use with radius, angle and combination dressers, and different pantographs.

IRREGULAR FORM DRESSING WITH A PANTOGRAPH DRESSER

The forming of a grinding wheel to produce complex forms or contours in a workpiece requires the use of a *pantograph dresser* (Figure 22-17). This accessory is mounted on a surface grinder. It consists of a tracer that follows

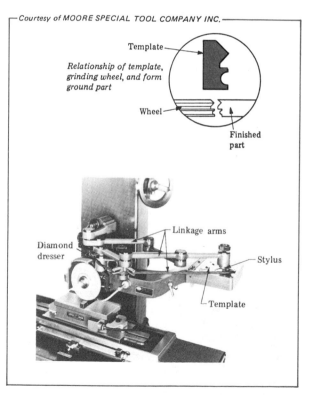

Template

Relationship of template, grinding wheel, and form ground part

Wheel

Finished part

Diamond dresser

Linkage arms

Stylus

Template

Figure 22-17 Pantograph Dresser Attached to Surface Grinder for Dressing Intricate Contour on a Grinding Wheel

a template of the required contour and linkage system that guides the diamond dresser to reproduce the form accurately (resulting from a maximum 10:1 reduction ratio).

FORMING CRUSH ROLLS ON A MICROFORM GRINDER

The *microform grinder* is a precision grinding machine. While it is used extensively to grind complex profiles on crush rolls, other circular and flat forms may be ground to shape. The microform grinder is used to grind form tools, templates, cams, gages, punches, dies, and other intricate forms. Such grinding is done on hardened materials, as well as on carbides. The principal design features of a microform grinder are the viewing screen, scope, pantograph, template, stylus, and drawing table (Figure 22–18A).

The pantograph feature provides for the stylus to follow the outline of a profile drawing. The pantograph is mounted on the drawing table. The arms permit the transfer, through a control mechanism, to form dress the crush roll to the required shape. The combination of viewing screen and scope permits the operator to follow the profile lines on the layout drawing and to control the form grinding process. The ratio between the enlarged drawing size and the reproduced profile on the crush roll is 50:1 on the model illustrated. This ratio helps to control the dimensional accuracy and the reproduced profile on a ground crush roll. An example of a complex profile that may be ground is shown in Figure 22–18B.

ABRASIVE CUTOFF WHEELS AND CUTTING-OFF PROCESSES

Cutting off (while not a widely used surface grinding process) means severing a material at a specific dimension. The cutting is done with a cutoff abrasive grinding wheel or abrasive saw. The process is particularly suited to cutting off hardened steel, ceramic, stainless steel, new high-temperature metals, and other difficult-to-cut materials. Unusual-shaped parts and extruded forms are held and positioned in fixtures for cutting-off processes.

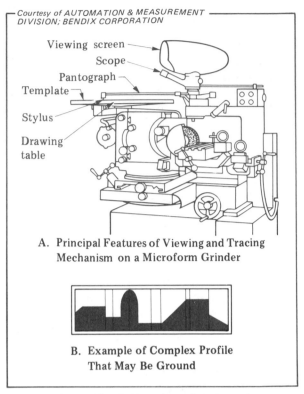

Courtesy of AUTOMATION & MEASUREMENT DIVISION; BENDIX CORPORATION

A. Principal Features of Viewing and Tracing Mechanism on a Microform Grinder

B. Example of Complex Profile That May Be Ground

Figure 22–18 Microform Grinder and Application in Grinding Intricate Forms

The process is fast cutting. The temperature may be controlled so that the parts are not work hardened. Cutting-off accuracy is better controlled than heat (flame) or arc cutting processes. The quality of surface finish is also far superior.

Cutoff wheels vary in thickness from 0.006″ to 0.187″ (0.15mm to 4.8mm) for general operations.

NONREINFORCED CUTOFF WHEELS

Cutoff wheels are available with or without reinforcing. The severity of the forces and the conditions under which parts are cut off determine whether a *reinforced* or *nonreinforced* wheel is used. The *nonreinforced cutoff wheel* is designed for machines where the work is clamped securely, the wheel operates in a controlled plane, and the wheel is adequately guarded. The surface grinder meets these requirements.

The nonreinforced wheel does not contain a strengthening fabric or filaments throughout the wheel diameter or reinforcement in special areas of the wheel. General cutoff wheels are made of aluminum oxide or silicon carbide for cutting ferrous metals, except titanium. Silicon carbide cutoff wheels are used for titanium and nonmetallics such as ceramics, plastics, and glass.

SPECIFIC BOND RECOMMENDATIONS

All cutoff wheels are type 1. They are furnished for both wet and dry grinding operations. Resinoid, rubber, and shellac cutoff wheel bonds are available for a wide variety of operations. For example, resinoid wheel bonds permit applications that range from the high production cutting of bar stock to the cool cutting action on nonproduction, heat-sensitive materials. Resinoid wheels are adapted to dry grinding.

Rubber wheel bonds range from free-cutting ferrous and nonferrous metals (either wet or dry) to the soft cutting action required for cutting glass and ceramics. Rubber-bonded wheels produce a good surface finish with a minimum of burrs and discoloration. Rubber wheels produce smooth side (vertical) surfaces and square corners. Thus, they are good to use for grooving and slotting operations.

Shellac wheel bonds are adaptable to highly critical applications where no burrs, checking from heat, or discoloration is permitted. Some shellac-bonded wheels are adaptable to extreme applications requiring fine grit sizes and soft grades. Shellac wheels are generally used in toolroom applications where the quality of cut and versatility are the main requirements.

Diamond cutoff wheels are recommended for cutting ceramic, stone, and cemented carbide materials. A *metal bond* is used because of its strength and ability to hold diamond abrasives securely in the wheel. Cutoff wheels are marked with letters to designate the abrasive: *A* designates aluminum oxide as recommended for metal cutoff operations; *C* represents silicon carbide for nonmetallic and masonry materials; *D* indicates a diamond wheel.

REINFORCED CUTOFF WHEELS

The *reinforced cutoff wheel* is termed *fully reinforced* when internal or external strengthening fabrics are incorporated throughout the entire diameter. When the strengthening material covers just the hole and flange area, the wheel is termed *zone reinforced.*

Reinforced wheels are designed for use where the grinding machine is hand guided or the workpiece is not clamped securely. Reinforced wheels resist breakage that normally results from severe cross bending. Reinforced cutoff wheels are manufactured in aluminum oxide or silicon carbide. These wheels may be used for surface speeds up to 16,000 sfpm for wheel diameters up to 16″ (406mm) and 14,200 sfpm for wheels over 16″ (406mm) diameter.

A. Safe Practices in Grinding Angular and Vertical Surfaces and Shoulders

- Check the parallel alignment of the solid jaw of a precision vise. A dial indicator is generally used to test for parallelsim.
- Tighten the clamping nuts on the knees of the compound vise to secure the tables from possible movement during grinding.
- Disengage the automatic transverse feed upon approaching a vertical, filleted, or angular shoulder. Disengagement permits hand control in order to take fine, precise cuts in forming a corner and the adjacent faces.
- Judgment is needed to avoid excessive side force on the grinding wheel. Such a force may cause distortion, uneven wheel rotation, and even fracturing.
- Avoid a sudden tap or hitting the side of the grinding wheel.
- Work from a side position and never directly in line with the revolving wheel. When a grinding wheel fractures, the spindle speed and the centrifugal force cause the broken pieces to fly at a terrific speed. These pieces can cause serious injury to the operator and damage to the machine.
- Check each setup to ensure that the work is held securely, the grinding wheel is free to rotate and travel without interference, and the trip dogs are tightly fastened.

B. Safe Practices in Using Form Dressers and in Crush Dressing and Form Grinding

• Remove all dust and abrasive particles during the form grinding of the wheel. The air exhaust system on the wet grinding system on the machine must be adequate to remove the volume of dust and particles produced when form grinding the wheel.

• Crush dress only on a surface grinder that is designed with a spindle that compensates for the severe forces required during the process.

• Select a grinding wheel with appropriate characteristics for crush grinding without damage to the wheel. Also, the wheel must have properties that permit it to reproduce the form accurately with maximum operator safety and wheel wear life.

• Select a cutoff wheel with adequate reinforcement to withstand the cutting and bending forces required for a specific set of conditions.

• Analyze the cutting action, burring, and surface finish produced with a cutoff wheel.

• Extend the workpiece (with a minimum overhang) so that the section to be cut off drops away from the grinding wheel onto a protective tray.

TERMS USED IN ANGULAR, VERTICAL SURFACE, AND FORM GRINDING AND CUTTING-OFF PROCESSES

Swivel-type magnetic chuck	An adaptation of the magnetic chuck. A magnetic chuck that may be adjusted from a horizontal plane.
Magnetic sine plate	A work-holding accessory having two hinged sections at right angles to each other. A sine plate with a magnetic top plate. An accessory that is set at precise angles by using gage block combinations.
Wheel Relieving	Reducing the area of grinding wheel contact. Recessing the wheel from near its periphery to the hub. Cutting away part of the wheel face to reduce the area of contact.
Form grinding (surface grinder)	The process of grinding an angle, radius or curved surface, or any combination of these forms. Form grinding requires a formed grinding wheel and a reciprocating horizontal table movement.
Forming a concave or convex wheel	A specific regular curve form grinding process. Setting up a radius dresser and positioning a diamond dressing tool above center to produce a concave arc; below center, a convex arc.
Sine dresser	An adjustable angle device that is set with the aid of gage blocks to produce a precise angle setting.
Combination angle and radius dresser	A form dressing accessory for a surface grinder. The dresser that may be set to form a continuous profile of straight-line and curved surfaces on a grinding wheel.
Cone-point, full-ball radius, and radius diamond dressing tools	Standard diamond-point holder designations. Terms used to identify the shape and size of the shank in the area surrounding the cutting edge of the diamond.

Reference roll	A duplicate formed roll of a work roll. A crush roll used to control the accuracy of the form on a worn work roll. A crush roll used in production work to precision dress the grinding wheel. In turn, the grinding wheel is used to regrind the original form on the worn work roll.
Microform grinder	A precision grinding machine for form grinding complex forms on precision tools and parts, including work and reference rolls. A form-generating machine that uses a stylus, tracer, and pantograph arrangement.
Reinforced grinding wheel	A cutoff wheel that is laminated with tough, flexible fibres. A grinding wheel capable of withstanding extreme forces in offhand processes.
Pantograph dresser	A surface grinder accessory incorporating a tracer linkage system for reproducing a template profile to form dress a grinding wheel.
Crush dressing	Rolling the form of a slowly turning work roll (with great force) into a grinding wheel. Using a hardened high-speed steel or carbide preformed work or reference roll to impress the required form (in reverse) in a grinding wheel.
Variable factors	A material, machine component, workpiece, or process that, if changed, results in a change in the quality of a surface finish (product) and dimensional accuracy.

SUMMARY

- Four groups of accessories are used as common work-holding devices:
  (1) vises; (2) angle plates; (3) magnetic blocks, chucks, and tables;
  and (4) fixtures for production work.
    - Workpieces held on a right angle block are often positioned at a required angle by using a sine bar and gage blocks or a combination of angle gage blocks.
- The magnetic sine table consists of two hinged plates. Each is set for a different angle. Gage blocks are used in each angle setting.
    - Formed grinding wheels are commercially available for regular radii, angles, and combinations of these forms.
- Grinding wheels are generally dressed to a required precision form by one of three wheel forming accessories: a radius dresser, an angular dresser, or a combination radius and angle dresser.
    - The diamond point on a radius dressing accessory is set above the axis of the cradle to form a concave arc in a wheel. A below-the-axis position is required to form a convex arc.
- Stops on a radius dresser control the arc over which the diamond is swung to produce the required radius.

- A feed from 0.002″ to 0.003″ (0.05mm to 0.08mm) per revolution is used in rough dressing to form a grinding wheel. This feed is reduced to 0.001″ to 0.0005″ (0.02mm to 0.001mm) for the final dressing cuts.

- Angle and radius dressers are designed to form wheels for grinding precision cutting, stamping, forming, shearing, and other tools and work parts. Such parts are usually made of hardened steels.

  - Three common shapes of diamond dressing tools include the cone-point, full-ball radius, and special point for concave and convex radii on small-wheel diameters.

- The pantograph dresser is a versatile form dressing accessory. The form on an enlarged template is duplicated at a reduced actual size on the face of the wheel.

  - Grinding wheels that are formed with a diamond dresser usually cut harder. However, the form produced is within closer dimensional and surface finish tolerances than form wheels that are crush dressed.

- The size of the abrasive grains in a form grinding wheel must be smaller than the finest radius that is to be form ground. The wheel grade should be harder than conventional grinding. Such a wheel maintains the required form to grind a number of workpieces.

  - Crush dressing requires a preformed work roll, a low-speed spindle (between 100 and 300 sfpm), a device that idles a work roll, and a surface grinder spindle that can withstand great force.

- A work roll and a reference roll are used to maintain quality form control on production runs. The reference roll serves as the master to correct the form of the work roll when it wears beyond the allowable tolerance.

  - The automatic feed rate for general crush grinding ranges from 0.0001″ to 0.001″ (0.002mm to 0.02mm) per revolution. Fine feeds are used for the final forming revolutions.

- Crush dressed wheels cut faster than wheels dressed with diamonds. The sharper cutting grains produce a rougher surface finish and a profile that is ground to a coarser dimensional tolerance.

  - Shellac wheels are suited to toolroom applications requiring a high-quality surface finish and accurate cut. Resinoid wheels are adpated to dry cutting-off processes in production. Rubber wheels are used extensively for production wet grinding.

- Shellac bonds are used for toolroom cutoff and grooving operations where precision dimensions and surface finish are important considerations.

  - A metal-bonded diamond cutoff wheel is recommended for cemented carbide, ceramic, stone, and other hard and/or abrasive materials.

- Crush rolls, and other parts requiring the grinding of a complex profile, are formed on a microform grinder.

UNIT 22 REVIEW AND SELF-TEST

A. GRINDING ANGULAR AND VERTICAL SURFACES

1. State three different work-positioning requirements for setting a workpiece to grind a compound angle. A toolmaker's universal vise is to be used.

2. Prepare a work production plan for grinding an angular surface on a workpiece. A standard angle plate setup is used. The workpiece is positioned by using a sine bar and gage blocks.

3. State two reasons why the use of a dish wheel is often preferred to the use of a type 1 wide-face abrasive wheel when grinding two surfaces of a shoulder simultaneously.

4. List the steps for form dressing a grinding wheel to a required cutting angle.

5. State three personal safety precautions the surface grinder operator must observe, particularly when grinding vertical shoulders and angular surfaces (with the side face of a grinding wheel).

B. FORM (PROFILE) GRINDING AND CUTTING-OFF PROCESSES

1. List the steps that must be taken to form grind hardened workpieces that have an internal corner radius of 0.75″ (18.8mm). The grinding wheel is to be formed with a cradle-type radius dresser. Stock is left on the part for a series of roughing cuts and a finish cut.

2. a. Cite two advantages to using a sine dresser for dressing a grinding wheel at angles up to 45° from either side of a 0° horizontal plane.
   b. Calculate the gage block height for setting the sine dresser at a 7°15″ angle above the 0° horizontal plane.
   c. Select a set of gage blocks that will produce the required height.

3. a. Explain why it is impractical to crush form rubber- or resinoid-bonded grinding wheels.
   b. Compare (1) the cutting action and (2) the surface finish of a crush formed wheel with a contour formed on an abrasive wheel when a diamond nib and wheel dresser are used.

4. Secure an abrasive wheel manufacturer's technical manual on cutoff wheels. Also, check the specifications of a surface grinder in the shop with respect to spindle features, size, and speed range.
   a. Select the wheel size and determine the specifications for a wheel that is adapted to cutting off 20 pieces 4mm thick from a 25mm square bar of SAE 1020 mild steel.
   b. Give the wheel size and specifications.
   c. Compute the spindle speed (RPM) and indicate the rate of table feed.

5. State two personal safety practices the grinder machine operator must follow (a) before crush dressing a wheel and (b) when mounting a cutoff wheel.

SECTION THREE

Cylindrical Grinding:
Machines, Accessories, and Processes

This section begins with a general description of basic types of cylindrical grinding machines, design features and their functions, work-holding devices, and accessories. Peripheral speeds, problems and corrective action, and basic cylindrical grinding process data are applied.

Procedures are given for setting up and operating the machine, grinding external straight and taper cylindrical surfaces, angle grinding, face grinding, the internal grinding of straight and slight and steep taper cylindrical surfaces, and using angle and radius truing and dressing devices. CNC grinding processes are introduced.

UNIT 23

Machine Components:
External and Internal Grinding

OBJECTIVES

After satisfactorily completing this unit, you will be able to:

- Understand major features of cylindrical grinders for straight, taper, form, and face grinding and internal grinding.
- Apply specifications to wheel selection for cylindrical grinding processes.
- Describe traverse feed and plunge grinding methods on plain and universal cylindrical grinders.
- Select work-holding chucks and fixtures, work centers and work supports, and wheel truing and face forming dressing devices.
- Discuss automated functions as wheel dressing, in-feed, table traverse dwell, size control, automatic cycling, and typical CNC cylindrical grinding processes.
- Determine cylindrical grinding problems, causes, and corrective steps.
- Perform each of the following processes.
    - Setting Up and Operating a Universal Cylindrical Grinder.
    - Grinding Straight Cylindrical Work (including the use of work supports).
    - Grinding Small-Angle and Steep-Angle Tapers.
    - Cylindrically Grinding Internal Surfaces.
    - Truing and Dressing Straight, Angle, and Form Grinding Wheels.
- Apply formulas and tables for work and wheel surface speeds, depths of cuts for roughing and finishing operations, wheel traverse rates, and wheel recommendations for cylindrical grinding.
- Follow *Safe Practices* for setting up and operating cylindrical grinders and correctly use related *Terms.*

GENERAL TYPES OF CYLINDRICAL GRINDING MACHINES

The *plain and universal cylindrical grinders* continue to be the basic types. The plain cylindrical grinder is primarily used for straight and taper cylindrical grinding. The machine may be adapted to plunge grinding and to the producing of contour-formed cylindrical surfaces.

The term *universal* indicates that multiple processes may be performed in addition to plain cylindrical grinder processes. Slight and steep tapers (angles), faces, and combined internal and external grinding processes are performed on the universal cylindrical grinder.

DESIGN FEATURES OF THE UNIVERSAL CYLINDRICAL GRINDER

The six principal units of the universal cylindrical grinder (Figure 23–1) are as follows:

- Base, which houses motors for the coolant system, lubrication system, and drive components for the table. The machined ways accommodate the saddle;

- Saddle, swivel table, and drive mechanism for longitudinal travel;

- Headstock;

- Footstock;

- Wheel head, carriage, and cross feed mechanism;

- Operator control panel.

Toolroom universal cylindrical grinders are capable of grinding perfectly round cylinders to less than 25 millionths of an inch (0.0006mm)

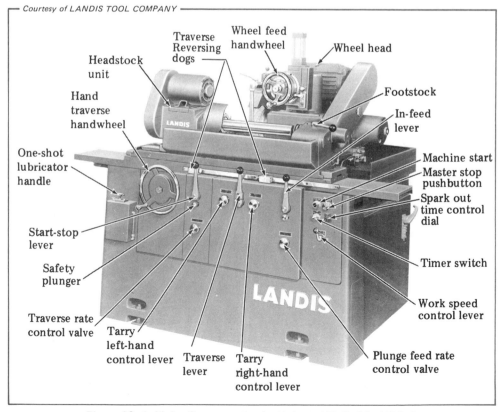

Figure 23–1 Major Components of a Universal Cylindrical Grinder (Hydraulic Traverse and In-Feed Model)

of roundness. Diametral sizes may be held to less than 50 millionths (0.0013mm). Surface finishes may be ground to less than 2 microinches and metric equivalent.

The universal cylindrical grinder is usually identified by size and whether the table traverse and/or the wheel feed is hand or hydraulically operated. Two popular toolroom sizes are 6″ × 12″ and 10″ × 20″ (152mm × 305mm and 254mm × 508mm). The grinder may be designed for:

• Hand traverse and hand wheel feed,
• Hydraulic traverse and hand wheel feed,
• Hand traverse and hydraulic in-feed,
• Hydraulic traverse and hydraulic in-feed.

SWIVEL TABLE

The table may be swiveled to position the work axis at a short angle to the grinding wheel face. The swivel table is adjustable from 0° to 20° included angle and up to 4″ per foot or metric equivalent.

The table may be moved by hand for positioning the workpiece in relation to the grinding wheel or for traversing the workpiece past the wheel face. Trip dogs are provided to control the length of table movement and to actuate the table reversing lever. Angle settings are made by turning an adjusting screw to position the table at the required angle indicated on the swivel table scale and tightening the table clamping screws (Figure 23-2). It should be noted that the taper angle and the table setting depend on the location of the taper on the workpiece, the work diameter, the length to be ground, and the grinding wheel diameter.

LONGITUDINAL TABLE TRAVEL

Table travel is controlled manually or automatically. The table travel rate is selected with a table speed selector knob. On some 10″ × 20″ (254mm × 508mm) models, the traverse speed is from 2″ to 240″ (50mm to 6,096mm) per minute. Models are designed so that the table may *dwell* from 0 to 2 1/2 seconds at the end of one or both table reversals.

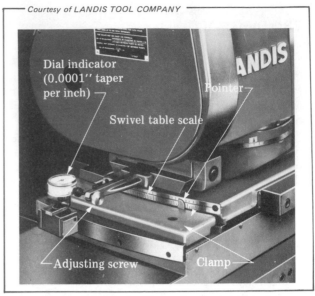

Courtesy of LANDIS TOOL COMPANY

Figure 23-2 Adjustments for Setting Swivel Table for Grinding a Taper

HEADSTOCK AND FOOTSTOCK

Figure 23-3 illustrates the general design features of a headstock. On the model illustrated, there are six work speeds of 60, 90, 150,

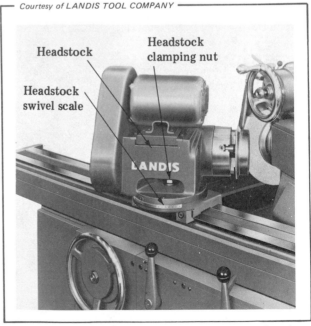

Courtesy of LANDIS TOOL COMPANY

Figure 23-3 General Design Features of a Headstock

225, 400, and 600 RPM. The work speeds are controlled by the headstock speed control knob.

The headstock is powered by its own motor. The headstock also includes a work-driving plate and arm that are mounted on the spindle. The base is graduated so that the headstock may be swiveled through a 180° arc. The swivel feature permits steep angle grinding and face grinding on workpieces that are held in a mechanical or magnetic chuck or collet or that are mounted in a fixture on the headstock spindle.

The rotation of the headstock spindle (and workpiece) is controlled by a lever or a handwheel control. The start, stop, jog, and brake of the headstock spindle (workpiece) is controlled by a spindle control lever. The headstock spindle for the model illustrated has a 1″ (25.4mm) diameter hole through the spindle. The headstock work center has a #10 Jarno taper; the footstock, a #6 Jarno taper.

The hardened steel footstock spindle is mounted in precision balls. The containers are hardened steel bushings. The design permits axial movement of the spindle with no radial clearance. The preloaded sleeve design eliminates clearance that is normally required for sliding fits. Also, the preloaded balls are packed and sealed for life to protect the balls against coolant and grit.

The work center is carbide tipped for wear life and to provide an efficient bearing surface for the workpiece and grinding processes. The force on the work center is adjustable. The center movement is controlled by the quick-acting control lever.

WHEEL HEAD AND CROSS FEED MECHANISM

The wheel head is a self-contained unit. It includes the spindle and a swivel base on which are mounted the drive motor, spindle, wheel feed, lubricating pump, oil reservoir, and pressure switch.

The handwheel is graduated for a work diameter reduction of 0.100″ (2.5mm) per revolution. The work diameter reduction per graduation of the regular feed knob is 0.001″ (0.025mm). The fine wheel feed graduations are in increments of 50 millionths (0.000050″) of an inch (0.00125mm) on the work diameter.

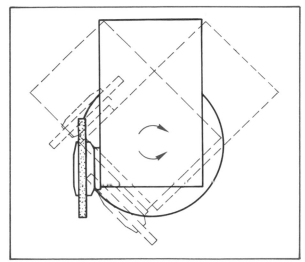

Figure 23–4 Universal Wheel Head Maximum Angular Settings at 45° Each Side of Center

Graduated Base. The wheel feed unit on the universal cylindrical grinder may be swiveled when the wheel head is swiveled. The wheel head may be swiveled 90° to position it at a required angle either side of center. Figure 23–4 shows universal wheel head settings at 45° each side of center. The large, circular graduated swivel base is graduated in degrees. Graduated machine scales are used only for general reference. Other precision instruments and gages are employed to measure within highly precise limits of accuracy.

CROSS FEED MECHANISM

The cross feed mechanism provides for feeding the grinding wheel into the workpiece and automatically stopping the in-feed movement when the required diameter is produced. In-feeding is controlled manually or by power feed.

The cross feed mechanism is started by the cross feed control lever. Positive stop mechanisms differ among machine-tool builders. However, the function in each case is to stop cross feeding when the workpiece is ground to a required diameter. The positive stop is set for the desired diameter by in-feeding the grinding wheel until it grazes the work surface. The handwheel disengaging knob is pulled out to free the handwheel. The handwheel is then turned until the fixed stop pin comes in contact with the right side of the switch operating slide. The cross feed is engaged so that the work is ground to the size to which the stop is set.

When the grinding has sparked out at the footstock end, the cross feed is disengaged. The work diameter is then measured. The index dial is set for any material that still is to be removed. The cross feed is reengaged to take successive cuts until the cross feed is stopped by the positive stop. The cylindrical grinder operator usually sets the stop to grind from 0.001″ to 0.002″ (0.025mm to 0.05mm) oversize. Further adjustments are made in a second setting to bring the workpiece to the required size and within surface texture limits.

BASIC CYLINDRICAL GRINDER ACCESSORIES

Standard jaw and magnetic chucks and collets are accessories that are mounted on and secured to the headstock spindle. Although chucks are primarily used to hold parts for internal grinding. such chucking permits external grinding. Face grinding operations require holding the workpiece in a chuck or on a magnetic faceplate or fixture.

Other accessories include work-supporting *universal spring back rests* (Figure 23–5) and a series of wheel truing, dressing, and forming devices with functions similar to surface grinding devices. Main differences relate to design modifications of the device for mounting on the table and wheel forming from a different position.

OPERATIONAL DATA FOR CYLINDRICAL GRINDING

The following *operational data* is commonly found in handbook tables and other technical manuals. For cylindrical grinding purposes, basic process data must be known for:

- The work surface speed in feet or meters per minute,
- The in-feed in fractional parts of an inch or millimeter per pass for roughing and finish grinding,
- The fractional part of the wheel width that traverses the workpiece for each revolution.

Table 23–1 provides operational data for traverse and plunge grinding of common work-

Figure 23–5 Universal Spring Back Rest for Work Support

piece materials on a cylindrical grinder. Using the same work surface speed for a specific material, the information needed for plunge grinding deals with the amount of in-feed.

AUTOMATED CYLINDRICAL GRINDING MACHINE FUNCTIONS

Automation is possible in at least seven different categories, briefly described as follows:

- *Work loading and positioning*, which involves work-feeding devices, loading and unloading, and work positioning;
- *In-feed*, which involves automatic presetting of advance feed rates, cutoff points, and duration of time-process functions. In-feed functions relate to rapid approach rates, feeds for roughing and finishing cuts, and the sparkout period that follows;
- *Table traverse tarry (dwells)*, which involves continuation of wheel and workpiece contact (exposure) at the ends of a cut for a preset time to ensure uniform cutting conditions;
- *Size control*, which provides for automatically adjusting the cutoff points of the in-feed to control the advance movement of the grinding wheel. Provision is also made to automatically compensate for wheel size variations through signals originating with a master gage;
- *Automatic cycling*, which relates to the actuating and shutting down of all mechanisms dealing with work rotation, in-feed,

Table 23-1 Operational Data for Traverse and Plunge Grinding of Common Workpiece Materials (Partial Table)

Traverse Grinding									
Workpiece Material		Surface Speed of Work		In-Feed per Pass				Fractional Wheel Width Traverse per Revolution of Workpiece	
Kind	Machinability (Hardness) Characteristics			Roughing		Finishing			
		(fpm)	(m/min)*	(0.001″)	(0.01mm)	(0.0001″)	(0.001mm)	Roughing	Finishing
Carbon steel (plain)	Annealed	100	30	0.002	0.05	0.0005	0.013	1/2	1/6
	Hardened	70	21	0.002	0.05	0.0003 to 0.0005	0.008 to 0.013	1/4	1/8
Tool steel	Annealed	60	18	0.002	0.05	0.0005 max.	0.013 max.	1/2	1/6
	Hardened	50	15	0.002	0.05	0.0001 to 0.0005	0.003 to 0.013	1/4	1/8

Plunge Grinding					
Workpiece Material		In-Feed of Grinding Wheel per Revolution of Workpiece			
Kind	Machinability (Hardness) Characteristics	Roughing Cuts		Finishing Cuts	
		(0.0001″)	(0.001mm)**	(0.0001″)	(0.001mm)
Steel	Soft	0.0005	0.013	0.0002	0.005
Alloy and tool steel	Hardened	0.0001	0.003	0.000025	0.0006

*Values rounded to closest meter.
**Values rounded to closest 0.001mm.

table traverse, and coolant supply as required through a complete machining cycle. At the completion of the operation, cycling refers to the retracting of the wheel slide and grinding wheel and the stopping of the table movement, work rotation, and coolant supply;

• *Wheel dressing*, which specifies wheel dressing at a preset frequency. Compensation is made for wheel size reduction through automatic in-feed movement;

• *Computer numerical control operations*, which involve advanced levels of automating cylindrical processes and are designed to improve the speed, accuracy, and efficiency of the setup and to check out the program and troubleshoot problems.

The five line drawings (Figure 23-6) illustrate a variety of cylindrical grinding process cycles that are programmable on CNC machines. Figures 23-6A and -6B provide examples of grinding with a *straight wheelhead* model. The process at (A) shows a wide grinding wheel that spans two or more diameters and grinds them simultaneously using plunge feed.

By contrast (Figure 23-6B), a narrow wheel plunge grinds the shorter, smaller diameter. The straight wheelhead is then moved back and fed into the workpiece by incremental infeed and multiple passes to grind the second diameter.

Figures 23-6C and -6D show operations performed on an *angular wheelhead* model using plunge feed. The workpiece at (C) requires a wheel dressed to the required shoulder radius. The radius, shoulder, and diameter are ground in one plunge feed pass. Another common example of grinding two diameters with the same radius shoulder is shown at (D). The surfaces are ground in two plunge cut passes.

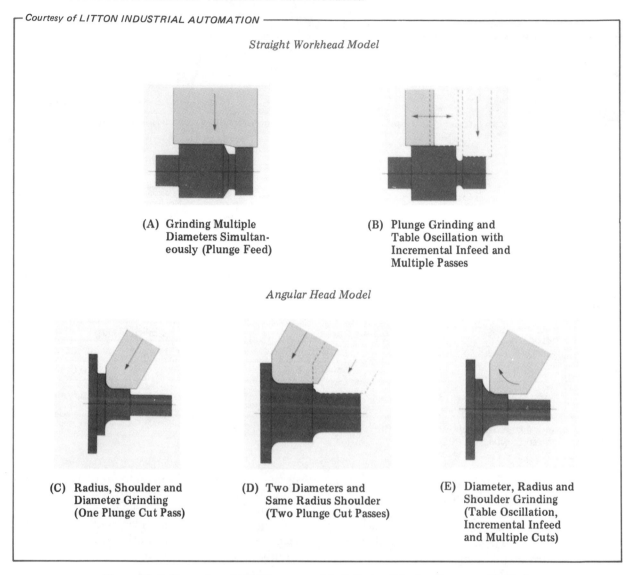

Straight Workhead Model

(A) Grinding Multiple Diameters Simultaneously (Plunge Feed)

(B) Plunge Grinding and Table Oscillation with Incremental Infeed and Multiple Passes

Angular Head Model

(C) Radius, Shoulder and Diameter Grinding (One Plunge Cut Pass)

(D) Two Diameters and Same Radius Shoulder (Two Plunge Cut Passes)

(E) Diameter, Radius and Shoulder Grinding (Table Oscillation, Incremental Infeed and Multiple Cuts)

Figure 23–6 Examples of CNC Programmable Cylinder Grinding Processes (Cycles)

The fifth application, using an angular wheelhead model (Figure 23–6E), requires CNC input to grind the large radius using programmed *circular interpolation*. The shoulder and diameter require table oscillation and incremental infeed for multiple wheel cuts. CNC models also include *linear interpolation* wheelhead control capability to grind tapered surfaces using *incremental infeed*, *table oscillation*, and multiple cuts.

CYLINDRICAL GRINDING PROBLEMS AND CORRECTIVE ACTION

The four fundamental cylindrical grinding problems that confront the machine operator are as follows:

- Chatter as related to the wheel, workpiece, and machine units;
- Work defects;
- Improperly operating machine;
- Grinding wheel conditions.

Examples of the probable causes of these major problems and recommended corrective action are provided in Table 23–2.

WHEEL RECOMMENDATIONS FOR CYLINDRICAL GRINDING (COMMON MATERIALS)

The sequence of markings on a typical cylindrical grinding wheel (according to ANSI designation) is the same as for markings described in earlier units. Letters and numbers appear in sequence to designate six categories of information: (1) *abrasive type* (letter), (2) *grain size* (number), (3) *grade* (letter), (4) *structure* (number), (5) *bond type* (letter), and (6) the optimal manufacturer's record (number).

In addition, the shape of the grinding wheel is identified according to another set of standards (ANSI Standard B74.2). Within this standard are thirteen different shapes of grinding

Table 23–2 Selected Common Cylindrical Grinding Problems, Probable Causes, and Corrective Action

Cylindrical Grinding Problem	Probable Cause	Corrective Action
Chatter Wheel	—Out-of-balance	—Run the wheel without coolant to remove the excess or unevenly distributed fluid in the wheel
		—Rebalance the wheel before and after truing and dressing
	—Out-of-round	—True the wheel face and sides before and after balancing
	—Too hard	—Change to coarser grit, finer grade, and a more open bond
	—Improperly dressed	—Check the mounting position and rigidity of the diamond dressing tool and its sharpness
Work Defects Check marks	—Improper cutting action	—Dress the grinding wheel to act softer
		—Increase the coolant flow and position the nozzle for maximum cooling and cutting efficiency
	—Incorrect wheel	—Use a softer-grade wheel and check the correctness of grain size, abrasive, and bond
Improperly Operating Machine Dimensional inaccuracies	—Out-of-roundness	—Relap the work centers
		—Check the condition of the machine centers, accuracy of mounting, and the footstock force against the workpiece
Grinding Wheel Conditions Wheel defects	—Faulty wheel dressing	—Incline the dressing tool at least three degrees from the horizontal plane
		—Reduce the depth of the dressing cuts
		—Round off the wheel edges
	—Wheel acting too hard (problems of loading, glazing, chatter, burning, and so on)	—Increase the rate of in-feed, the RPM of the work, and the traverse rate
		—Decrease the wheel speed or width
		—Select a softer wheel grade and coarser grain size or dress the wheel to cut sharper

wheel faces. The range of shapes and sizes are specified in handbooks and manufacturers' technical literature. Examples of other standard types of regular and diamond grinding wheels are generally used for cylindrical grinding:

- Type 1 (straight wheel),

- Type 5 (straight wheel with one side recessed),

- Type 7 (straight wheel recessed on two sides),

- Types 20 through 26 (straight wheels relieved and/or recessed on one or both sides).

Variables of machines, abrasives, materials, processes, and other conditions make it impractical to recommend a precise grinding wheel. Therefore, manufacturers' and handbook tables are based on typical applications with conditions assumed for common practices. Each table provides for a starting selection. Under actual conditions, the selection is subsequently refined. Table 23–3 provides an example of wheel recommendations of one manufacturer for the cylindrical grinding of common metals. In some instances, grinding processes are identified.

BASIC EXTERNAL AND INTERNAL CYLINDRICAL GRINDING PROCESSES

Once the cylindrical grinder is set up, five basic processes are performed in toolroom work or for small production runs. These processes include:

- Grinding parallel surfaces,

- Grinding tapered surfaces,

- Grinding contour-shaped (curvilinear) surfaces,

- Grinding at a steep angle to form beveled surfaces and shoulders up to 90° to the work axis.

- Face grinding.

While most surface forming on a cylindrical grinder is on external surfaces, the same applications are made to internal grinding.

Table 23–3 Wheel Recommendations for Cylindrical Grinding of Materials Generally Used in the Custom Shop

Material	Wheel Marking*
Aluminum parts	SFA 46–18 V
Brass	C 36–K V
Bronze	
Soft	C 36–K V
Hard	A 46–M 5 V
Cast iron (general, bushings, pistons)	C 36–J V
Cast alloy (cam lobes)	
Soft	
Roughing	BFA 54–N 5 V
Finishing	A 70–P 6 B
Steel	
Forgings	A 46–M 5 V
Soft	
1″ (25.4mm) diameter and under	SFA 60–M 5 V
Over 1″ (25.4mm) diameter	SFA 46–L 5 V
Hardened parts	
1″ (25.4mm) diameter and under	SFA 80–L 8 V
Over 1″ (25.4mm) diameter	SFA 60–K 5 V
Stainless (300 series)	SFA 46–K 8 V
General-purpose grinding	SFA 54–L 5 V

The prefix BF with the aluminum oxide designation (A) indicates *blended friable* (a blend of regular and friable). SFA indicates *semifriable*.
*The grain size, grade, and structure are the recommendation of one leading manufacturer.

How to Set Up and Operate a Universal Cylindrical Grinder

Setting Up the Wheel and Wheel Head

STEP 1 Determine the wheel specifications for the material to be ground.

STEP 2 Mount and secure the wheel on the spindle.

STEP 3 Start and stop the spindle several times if the machine has been idle to lubricate the spindle bearings.

STEP 4 Check the graduation on the base of the swivel head. If necessary, adjust the position to a zero reading.

STEP 5 Mount and position a diamond dressing tool on the table. True and dress the wheel according to job requirements.

Setting the Swivel Table

STEP 1 Determine the nature of the table setting.

Note: Parallel cylindrical surfaces require a zero setting. Small-angle surfaces may be produced by setting the table at the required angle or taper. Steep-angle grinding may require both the table and wheel head to be set at complementary angles.

STEP 2 Loosen the clamping bolts at each end to set the table at the required angle.

STEP 3 Turn the table adjustment nut until the table is set at the required angle. Tighten the clamping bolts.

STEP 4 Check the accuracy of the table setting.

Note: The test bar and dial indicator method may be used for checking parallelism. Angle table settings are usually checked by taking a trial cut on the workpiece and measuring or gaging the included angle.

Positioning the Headstock and Footstock

STEP 1 Clean the table face and the bases of the headstock and footstock.

STEP 2 Position the headstock and footstock so that they are centered on the table.

STEP 3 Check the spindle bore and the condition of the centers. If worn or damaged, the centers should be reground.

Note: Carbide-tipped centers are commonly used to provide greater wear life and an accurate bearing surface for center work.

Checking the Coolant System

STEP 1 Determine what cutting fluid (coolant) is required for the job.

STEP 2 Check the cutting fluid in the system to meet the job grinding requirements.

STEP 3 Position the nozzle to deliver the fluid for maximum cutting and cooling efficiency.

STEP 4 Attach the cutting fluid guards to the table.

Mounting the Workpiece

STEP 1 Check the angle and bearing condition of the center holes. Redrill, grind, or lap, if required.

STEP 2 Select a driving dog. Secure it on one end of the workpiece.

STEP 3 Move the wheel head so that the workpiece may be placed between centers without interference from the grinding wheel.

STEP 4 Make sure the centers are free of particles and clean.

STEP 5 Move the footstock lever to permit mounting the workpiece between centers.

STEP 6 Place the driving dog and the drive end of the workpiece on the headstock center.

STEP 7 Move the spindle lever of the footstock to permit inserting the center in the workpiece.

Setting the Table Position and Travel

STEP 1 Move the table to the left until the grinding wheel extends beyond the footstock end of the workpiece. The distance should equal the width of the traverse cut of the grinding wheel (from 1/6 to 1/2 wheel width, depending on the nature of the cut).

STEP 2 Move the table reversing lever as far to the left as it will go.

STEP 3 Move the right-side table trip dog until it touches the table reversing lever. Secure the trip dog in position.

STEP 4 Move the table to the right-end location where the cut is to end.

STEP 5 Move the table reversing lever to its extreme right position.

STEP 6 Bring the left-side table trip dog against the reversing lever. Secure the trip dog.

STEP 7 Recheck for operational clearance between the driving dog, headstock, workpiece, and grinding wheel and wheel head and tailstock.

STEP 8 Set the dwell (sparkout timer) control to permit adequate time at each reversal for grinding to depth.

Cross Feeding Manually and Automatically

STEP 1 Position the switch-operating cross slide throwout lever to disengage the automatic cross feed.

STEP 2 Start the grinding wheel, coolant, system, and headstock spindle. Engage the table traverse.

STEP 3 Feed the grinding wheel until it grazes the workpiece. Set the cross slide handwheel at zero.

STEP 4 Move the cross slide handwheel to advance the wheel to the depth of the first roughing cut. Take the cut until the wheel sparks out at the footstock end of the cut.

STEP 5 Stop the work rotation and table traverse at the end of the cut. Measure the workpiece. Determine what the final depth reading should be on the graduated handwheel.

STEP 6 Set the cross feed positive-stop mechanism to permit feeding to within 0.001" (0.025mm) of the diameter to which the workpiece is to be ground.

STEP 7 Continue to in-feed for the roughing cuts until the positive-stop mechanism stops further in-feeding.

STEP 8 Spark out the final cut at the footstock end of the workpiece. Then, stop the table travel and headstock spindle (work rotation).

STEP 9 Measure the work diameter. Determine precisely how much additional stock is to be removed.

STEP 10 Start the work rotating. Feed the wheel for the first of several light finish cuts.

STEP 11 Engage the table travel. Take successive cuts and permit the final cut to spark out

STEP 12 Make a final check for dimensional accuracy and quality of surface texture.

Note: The quality of surface texture may be established by checking with a portable surface analyzer. An optical check may be made using a comparator scale and optical viewer.

Note: The wheel itself or the speed and sharpness to which the wheel is dressed may need to be changed to produce the required surface texture.

STEP 13 Back the wheel away from the workpiece when the grinding is completed. Stop the table travel and spindle (workpiece) rotation.

STEP 14 Move the footstock lever to release the workpiece.

How to Grind Straight Cylindrical Work

STEP 1 Select the appropriate cylindrical grinding wheel. Check it for balance. Mount the wheel on the spindle and true and dress it.

STEP 2 Set the wheel head at zero.

STEP 3 Attach a driving dog on the workpiece so that there is clearance for grinding.

STEP 4 Check the center holes (on center work) for quality of surface and accuracy of the included angle.

Note: Some workpieces may be held in a chuck when there is limited overhang. Longer chucked pieces may be supported by the footstock center. Collets and fixtures may also be used in the headstock.

STEP 5 Position the grinding wheel. Set the table trip dogs for the length to be ground. Set the dwell control to spark out each cut. Adjust the traverse rate. Set the spindle speed at the RPM required for rotation of the workpiece.

STEP 6 Feed the grinding wheel into the revolving workpiece until it just grazes the surface. Set the cross feed handwheel at zero.

STEP 7 Set the cross feed in-feed trip to permit grinding to the required finished dimension.

STEP 8 Take successive roughing and finishing cuts.

SUPPORTING LONG WORKPIECES

Long, slender workpieces need to be supported by *spring back rests* or *steady rests*. The number and placement of the back rests depend on the work diameter and the length and nature of the grinding process. Universal spring back rests are designed to be positioned along and secured to the table. The *shoes* (bearing surfaces) are usually made of bronze to protect the work surfaces. Shoes are available in increments of 1/16″ (1.5mm) for work sizes beginning at 1/8″ (3mm) through 1″ (25.4mm). Each shoe remains in contact with the workpiece automatically, giving constant support.

In general practice, long workpieces that are 1/2″ (12.7mm) in diameter are supported at 4″ to 5″ intervals along the length. Fewer spring back rests are needed to support stronger, larger-diameter workpieces. Steady rests are used primarily as an intermediate work support.

How to Grind Small-Angle and Steep-Angle Tapers

Grinding Small-Angle Tapers up to 8°

STEP 1 Release the swivel table clamping bolts.

STEP 2 Swivel the table to the required angle or taper per foot (or metric unit). Read the setting on the graduated scale on the end of the table.

STEP 3 Tighten the table clamping bolts to secure the table at the angle setting.

STEP 4 Set the headstock spindle RPM, the rate of table traverse, in-feed rate, and dwell time. Adjust the flow rate of the cutting fluid and the position of the nozzle. Secure the wheel and splash guards.

Caution Use safety goggles or a protective shield.

STEP 5 Bring the grinding wheel into position to take a first cut. Set the wheel head positive-stop mechanism to automatically trip the in-feed at the required position.

STEP 6 Take several cuts until the ground taper length is sufficient to be measured.

STEP 7 Adjust the table setting to compensate for any variation between the taper being ground and the required one.

STEP 8 Take successive roughing cuts by using power table traverse. Leave the workpiece from 0.001″ to 0.002″ (0.025mm to 0.050mm) oversize.

STEP 9 Set the feed rate on the fine-feed dial to remove the remaining material. Take several light finish cuts.

STEP 10 Spark out at the footstock end so that the grinding wheel clears the workpiece. Then, stop the table travel and rotation of the part.

Grinding Steep-Angle Tapers

Steep tapers (Figure 23–7) may be ground by three fundamental methods:

- Swiveling the universal wheel head on its swivel (angle-graduated) base to the required angle;
- Swiveling the headstock for workpieces that are held directly in a chuck, fixture, collet,

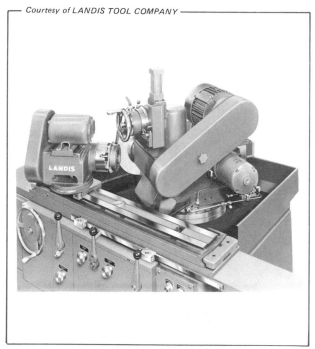

Courtesy of LANDIS TOOL COMPANY

Figure 23–7 Wheel Head and Headstock Swiveled to Grind a Steep Angle

or the tapered spindle. The angle setting of the headstock is read directly from graduations on the swivel base;

• Swiveling the universal wheel head and headstock spindle axes (as shown in Figure 23–7).

FACE GRINDING ON A UNIVERSAL CYLINDRICAL GRINDER

Face grinding refers to the grinding of a plane surface at a right angle to the axis of the part. Face grinding on a universal cylindrical grinder requires the workpiece to be set in a plane that is parallel to the axis of the grinding wheel spindle. The headstock is swiveled 90° for face grinding. The grinding wheel is hand fed to the depth of cut against the revolving part. The part is fed across the face of the grinding wheel either by hand or automatically by using the table traverse feed. The wheel is fed into the part for successive cuts.

GRINDING INTERNAL SURFACES ON A UNIVERSAL CYLINDRICAL GRINDER

STRAIGHT (PARALLEL) GRINDING

An internal grinding attachment permits grinding outside diameters and inside diameters without removing the part, grinding wheel, or spindle. The attachment design permits it to be mounted on the wheel head. It may be swung to a down position for internal grinding (Figure 23–8).

The spindle setup in Figure 23–8 permits grinding a hole to 4 3/8″ (110mm) diameter × 2 1/2″ (63mm) deep without removing a standard 12″ (304mm) diameter external grinding wheel. Spindle quills extend the depth to which an internal surface may be ground.

On the model illustrated, the belted motor drive produces spindle speeds up to 16,500 RPM. Smaller machines are designed to deliver infinitely variable spindle speeds to 45,000 RPM through a dial control on an electrical panel.

INTERNAL GRINDING PROBLEMS

Internal grinding spindles, by the nature of their size and other design limitations, are not as strong or rigid as spindle assemblies for external grinding. As a result, three common grinding problems may occur. Unless the wheel is dressed correctly and run at the required cutting speed with proper in-feed and traverse feed rate, the internally ground hole may have one or more of the following dimensional defects:

• The hole may be bell-mouthed. This defect is caused by the wheel overlapping the hole too much and may be corrected by reducing the overlap at each end of the hole. The bore length may also be too long and outside the capacity of the spindle to accurately machine within the required tolerance limits.

• The hole may be out-of-round. This defect may be caused by overheating during the grinding process or through distortion in chucking or holding the workpiece.

• The hole may be tapered. This defect may be caused by using a wheel that is too soft to hold its size, too fast a feed, or the setting of the workpiece axis at a slight angle.

Other general grinding problems, probable causes, and recommended corrective action were summarized earlier in Table 23–2.

Courtesy of LANDIS TOOL COMPANY

Figure 23–8 Internal Grinding Attachment Locked in Operable Position (Setup for Internal Grinding)

How to Set Up for and Cylindrically Grind Internal Surfaces

Grinding Straight Internal Surfaces

STEP 1 Move the internal grinding attachment into operating position. Lock the attachment.

STEP 2 Determine the spindle RPM required to grind at the specified fpm (m/min). Set the wheel speed controls accordingly.

STEP 3 Select an appropriate diamond dressing tool. Mount and position to dress the wheel.

STEP 4 Bring the internal grinding wheel spindle to operating speed. Dress the wheel and use a cutting fluid (coolant).

STEP 5 Turn the cross feed selector knob to the internal grinding position.

STEP 6 Turn the cross feed handwheel counterclockwise to advance the wheel into the workpiece.

Note: The setting of the selector knob for internal grinding permits using the wheel head positive-stop mechanism, power traverse feeds, and graduations on the handwheel in the same manner as for external grinding.

STEP 7 Mount the workpiece in a suitable chucking device on the headstock spindle.

STEP 8 Calculate the required speed (RPM) for the workpiece and for the grinding wheel and determine the rate of table travel.

STEP 9 Adjust the headstock, internal grinding spindle controls, and the table traverse feed rate control.

STEP 10 Disengage the power cross feed control lever.

Caution: Make sure the in-feed safety plunger is pushed in at all times on grinders equipped for hydraulic in-feed to prevent accidental in-feed of the wheel head.

STEP 11 Adjust the table trip dogs.

STEP 12 Position the wheel partly into the revolving workpiece. Bring the wheel into contact with the hole surface. At the same time, feed the workpiece so that the grinding wheel grinds toward the back of the hole.

STEP 13 Withdraw the workpiece to clear the wheel at the end of the cut.

STEP 14 Measure the inside diameter. Set the graduated dial at zero.

STEP 15 Take repeated cuts until the hole is ground to the correct diameter.

STEP 16 Clear the work and the wheel at the end of the grinding process by moving the right-hand trip dog.

STEP 17 Recheck the finished diameter. Clean the machine, tools, and workpiece. Leave the machine in a safe, operating condition.

Grinding Small-Angle Internal Tapers

STEP 1 Check the wheel head to be sure it is set at 0°.

STEP 2 Check the 0° setting of the headstock.

STEP 3 Set the swivel table at the required taper angle, taper per foot, or metric equivalent.

STEP 4 Proceed to grind internally following the same steps as for grinding a straight internal cylinder.

Note: The correctness of taper is checked with a taper plug gage or the mating part.

Grinding Steep-Angle Internal Tapers

STEP 1 Check the wheel head to be sure it is set at 0°.

STEP 2 Check that the swivel table is set at the zero position.

STEP 3 Loosen the clamping bolts on the headstock base.

STEP 4 Swivel the headstock to the required angle. Secure the headstock in position.

STEP 5 Proceed to grind the steep taper. Follow the same steps as were used for small-angle internal taper grinding and measuring.

CYLINDRICAL GRINDER WHEEL TRUING AND DRESSING DEVICES

There are three basic types of wheel truing and dressing devices for cylindrical grinders. The most common are the wheel truing and dressing fixtures for generating a flat wheel face. These fixtures are designed for mounting either on the footstock or on the table.

Grinding wheels may be dressed at an angle by using a table-mounted *angle truing attachment.* Concave and convex radii are formed by using a *radius wheel truing attachment* (Figure 23–9). The *combination angle and radius wheel truing attachment* is used to accurately form wheels to different radius and angle combinations.

The principles for setting these attachments are the same as described for similar surface grinding attachments.

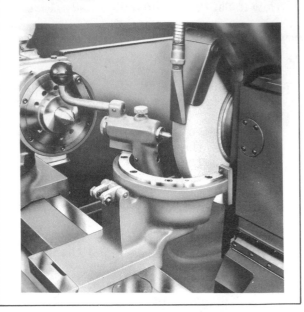

Courtesy of LANDIS TOOL COMPANY

Figure 23–9 Radius Wheel Truing Attachment

How to True and Dress Cylindrical Grinding Wheels

Producing a Straight Grinding Wheel Face

STEP 1 Select and mount a wheel truing fixture on the footstock or secure it to the table.

STEP 2 Position the footstock and grinding wheel with adequate, safe working clearance.

STEP 3 Position the diamond dresser. The diamond point is set just below the axis of the wheel and at a slight downward angle.

STEP 4 Position the reversing trip lever to set the right-side and left-side dogs.

STEP 5 Set the rate of table travel to a fairly fast one.

STEP 6 Set the grind-true switch in the *true* position. This setting controls the constant flow of cutting fluid.

STEP 7 Bring the wheel spindle to operating speed.

STEP 8 Move the diamond dresser until it is in contact with one edge. Round this edge of the grinding wheel slightly by manipulating the longitudinal (traverse) and cross feed movement handwheels.

STEP 9 Repeat the process of rounding the other edge of the grinding wheel.

STEP 10 Start the table movement with the start-stop lever.

STEP 11 Advance the wheel to start the first cut.

STEP 12 Move the diamond past the edge of the wheel.

STEP 13 Take the first dressing cut.

STEP 14 Continue to in-feed the wheel and take successive cuts until the wheel is trued.

STEP 15 Finish dressing the wheel. The rate of table travel and the depth of the finishing cuts are reduced.

STEP 16 Remove the truing and dressing device.

STEP 17 Clean the machine thoroughly.

Forming a Corner Radius on a Grinding Wheel

STEP 1 Select a radius wheel truing attachment. Position it to shape the right- or left-hand corner of the wheel.

STEP 2 Determine the radius setting of the dressing tool. The setting for an external radius is established by subtracting the required radius from the measurement shown on the dressing-tool holder.

STEP 3 Set the diamond dressing tool at the calculated setting.

STEP 4 Position the diamond-tool holder at a right angle to the wheel face.

STEP 5 Move the truing attachment forward until the diamond tool touches the face of the wheel. Tighten the clamping bolt nut in order to secure the slide in position.

STEP 6 Move the table to clear the diamond tool and the grinding wheel. Swivel the diamond-tool holder again. This time the diamond tool is at a right angle to the side of the wheel.

STEP 7 Move the table until the diamond tool just grazes the wheel. This second setting correctly positions the diamond tool in relation to the face and side of the wheel.

STEP 8 Move the table away from the wheel to permit the diamond tool to swing without hitting the corner (edge) of the wheel.

STEP 9 Feed the table carefully by hand and swing the holder back and forth in a 90° arc.

STEP 10 Keep advancing the diamond tool by moving the table until the complete radius is formed.

Forming an Angular Face on a Grinding Wheel

STEP 1 Select an appropriate angle wheel truing attachment. Position and secure the attachment to the table.

STEP 2 Swivel the attachment to the required angle. Read this angle directly on the graduated base of the swivel section.

STEP 3 Position the wheel by turning the longitudinal table handwheel and the cross feed handwheel.

STEP 4 Move the table so that the diamond tool is fed into position for the first light cut of from 0.0005″ to 0.001″ (0.0125mm to 0.025mm).

STEP 5 Continue to feed the wheel and to traverse the diamond across the wheel face for each pass.

Forming a Concave or Convex Face on a Grinding Wheel

STEP 1 Select an appropriate radius wheel truing attachment. Position the attachment so that the axis of the diamond cutting tool is aligned with the centerline of the grinding wheel. Secure the attachment to the table.

STEP 2 Postion the diamond tool in the holder so that contact is made when the diamond is positioned slightly below the axis of the wheel (and spindle) and at a small downward angle.

STEP 3 Position the diamond point. Use the setting gage on the front of the toolholder. Secure the tool in this starting position.

STEP 4 Set the slide on which the diamond tool is mounted to the required radius.

STEP 5 Check the setup for the alignment of the radius truing attachment with the wheel centerline and the position of the diamond and check the position of the slide for forming the convex or concave radius.

STEP 6 Bring the spindle up to operating speed. Advance the wheel to the diamond tool for the first cut either at the center of the wheel (for a concave form) or across the ends of the wheel (for forming a convex face).

Safe Practices in Setting Up and Operating Cylindrical Grinders

---

- Observe the ANSI "Safety Code for the Use, Storage, and Inspection of Abrasive Wheels." Visual inspection is accompanied by ring testing for possible cracks.

- Examine the construction and clearance between the guard and the grinding wheel. The clearance should be adequate to avoid interference with the grinding operation.

- Check the balance of the grinding wheel.

- Establish from the manufacturer's specifications what the safe operating (fpm or m/min) speed is for the grinding wheel. Convert the surface speed into the RPM at which the grinding wheel is to revolve.

- Position splash guards to contain the cutting fluid. Immediately wipe up spilled cutting fluid and use a floor surface compound to keep the machine area clean.

- Control the in-feed of the grinding wheel (wheel head) by carefully adjusting with the in-feed handwheel. Power in-feed is not recommended until the first cut is set by hand, clearances are checked, and everything is operating correctly.

- Recheck the settings of the trip dogs by moving the table longitudinally by hand. Finer adjustments are made before the power feed is engaged.

- Lock the internal grinding attachment when it is swung out of operating position.

- Set the wheel head travel movement so that it is automatically stopped when the required ground diameter is reached.

- Review the table of common cylindrical grinding problems, probable causes, and corrective action (Table 23–2). Study the potential machining problems and surface and dimensional defects that may result from unsafe practices.

- Use eye protection devices against flying chips and abrasive particles.

- Leave the machine tool in a safe, operating condition.

TERMS USED IN CYLINDRICAL GRINDING TECHNOLOGY AND PROCESSES

Traverse grinding	A form of cylindrical grinding where the revolving workpiece is moved across the grinding wheel face. Generally related to the grinding of a straight or taper cylindrical surface by reversing the direction of table travel feed for each cut across a workpiece.
Plunge grinding (cylindrical grinding)	Forming a flat or other contour cylindrical surface by in-feeding the wheel head and grinding wheel into a workpiece without moving the table longitudinally.
In-feeding	Advancing the grinding wheel into the workpiece for purposes of positioning the wheel at a cutting depth.
Cross feed control	A positive stop (fixed pin) that actuates a switch to stop the cross feeding when the required diameter to which the cross feed is set is reached.
Headstock	A unit designed for mounting on the swivel table. The cylindrical grinder unit for holding, mounting, and positioning a workpiece and rotating it in a fixed relationship (fpm or m/min) to the revolving wheel.

Footstock   A machine device adjustable along the table to support workpieces. A base, movable spindle, and center for supporting center work.

Wheel head   The entire mechanism for housing, driving, and controlling the grinding wheels for external and internal grinding. The grinding spindle head that is mounted on a carriage to permit swiveling the wheel at an angle to the axis of the workpiece.

Spring back rests or steady rests   Work-supporting devices. Supports that prevent small-diameter or proportionally long, thin workpieces from becoming distorted during grinding.

Automated functions   Machine setups, processes, and systems that when once programmed are cycled automatically. Seven basic categories of automatic cylindrical grinding activities related to work, feed, size, cycling, wheel dressing, checkout, and troubleshooting.

SUMMARY

- The principal design components of a universal cylindrical grinder include the base; saddle, swivel table, and drive mechanism; headstock and footstock; wheel head, carriage, and cross feed mechanism; cutting fluid system; and operator control panel.

  - Controls are provided for revolving the workpiece at the required RPM. The controls permit starting, stopping, jogging, and braking.

- Carbide-tipped centers are widely used to provide a smooth, precision bearing surface for center work.

  - Handwheel graduations permit work diameter reductions of 0.100″ (2.5mm) per revolution. The fine wheel feed graduations are in increments on the work diameter of 50 millionths (0.000050″) of an inch (0.00125mm).

- The wheel head is adjustable on its swivel base. Angle settings are made directly according to the angle graduations.

  - Power in-feed is controlled by a positive-stop mechanism. When set to grind a workpiece to a required diameter, movement of the wheel head (grinding wheel) is automatically tripped (stopped).

- The basic accessories for cylindrical grinders include work-supporting devices (back rests); standard chucking and work-holding devices; and wheel truing, dressing, and forming devices.

  - Cutting speeds may be given in feet or meters per minute or may be calculated by standard formula when the spindle RPM and wheel diameter are known. Similarly, the RPM may be computed.

- High-speed cylindrical grinding is performed on machines that are designed to operate at spindle speeds in excess of 6,500 fpm (1,980 m/min). High-speed machines are more ruggedly constructed to withstand the increased cutting forces.

■ Workpieces may be held between centers or in a chuck or collet or may be mounted on a faceplate or fixture. Work supports are used for long, small-diameter workpieces or where heavy cuts may cause distortion.

■ Traverse grinding provides for a revolving workpiece to be moved at cutting depth across the face of a grinding wheel. Traverse grinding cuts may be taken parallel to or at an angle from 0° to 90° to the work axis.

■ Steep angles may require a combination of angle settings of the table, swivel head for the grinding wheel, or swivel of the workpiece in the headstock.

■ Face grinding requires the angle relationship between the grinding wheel and workpiece axis to be 90°.

■ The internal grinding spindle is moved into position and locked. The spindle RPM is set to the recommended spindle speed requirement. The distance of travel (depth) inside the workpiece is set by using the table trip dogs.

■ Improper wheel dressing, the rate, and depth of feed may result in a bell-mouthed, out-of-round, or tapered hole. These inaccuracies may be partially corrected by a sharp-cutting wheel and by changing the feed and speed rates.

■ Universal cylindrical grinders are designed with an internal grinding head. This head is swung into position and locked. The internal grinding head is an integral unit on the wheel head. The use of smaller-diameter wheels requires the internal grinding spindle to rotate at higher speeds than for external grinding.

■ Manufacturers' tables are used for the suggested fractional width of grinding wheel to use for roughing and finishing cuts and traverse or plunge cutting rates.

■ Operational data for controlling machine speeds, feeds, and the grinding process are based on factors such as design features and characteristics of the workpiece; machine stability; conditions of grinding; and the grinding wheel.

■ The grinder operator must establish the surface speed and RPM of the workpiece, in-feed, and rate of traverse.

■ Longitudinal table travel controls permit the table to dwell up to a few seconds at each reversal of the table.

■ In addition to the automatic controls for feed, speed, and traverse on standard machines, other cylindrical grinder functions may be automated. These functions may be grouped to include the loading and positioning of the workpiece; presetting, cutoff, and duration of infeed; table traverse for dwell and sparking out; size control with compensation for variations in wheel size; cycling for setup, machining, and shutdown; frequency dressing of the grinding wheel; and computerized numerical control.

■ Cylindrical grinding problems fall into four groups: (1) chatter, (2) work defects, (3) machine operation, and (4) grinding wheel condition.

■ ANSI wheel markings are used for abrasive type, grain size, grade, structure, and bond type. Following the ANSI grinding wheel and machine process code of safety is an industry requirement.

■ Tables provide feeding rate information for roughing and finishing cuts. In general nonproduction grinding, the in-feed for roughing cuts is 0.002″ (0.05mm). Finish cuts are reduced to 0.0001″ to 0.0005″ (0.0025mm to 0.0125mm) per pass.

■ The five basic cylindrical grinding processes are parallel, taper, form, steep-angle, and face grinding.

UNIT 23 REVIEW AND SELF-TEST

1. a. State when (1) regular feed knob graduations and (2) fine wheel feed graduations are used in cylindrical grinding.
   b. Indicate the customary inch and metric graduation increments for reducing the work diameter on (1) a regular feed knob and (2) the fine-feed graduated handwheel.

2. Identify the function of each of the following: (a) sparkout timer dials, (b) positive-stop cross feed mechanism, and (c) universal spring back rests.

3. Cast steel parts are to be rough and finish ground using the table traverse method.
   a. Refer to a table of operational data for traverse and plunge grinding.
   b. Establish (1) work surface speed in m/min, (2) roughing in-feed per pass in mm, (3) finishing cut in-feed per pass, and (4) fractional wheel width traverse per revolution of the work for roughing cuts, and (5) fractional wheel width traverse for finishing cuts.

4. Calculate the headstock spindle RPM at which a 3.625″ diameter hardened tool steel punch may be ground. Round off the work speed at which the headstock speed control is set.

5. List the steps or set up a cylindrical grinder and grind the mandrel to the required taper and surface finish.

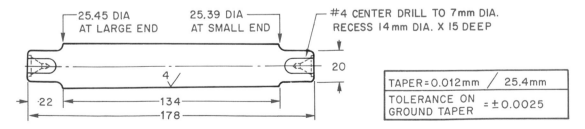

6. List the steps or dress a type 1 wheel for parallel (straight) grinding on a cylindrical grinder.

7. State three personal safety precautions the cylindrical grinder operator must observe with respect to the grinding wheel.

Cutter and Tool Grinding:
Machines, Accessories, and Processes

This section covers the functions of the major components of the cutter and tool grinder, attachments and accessories, and basic machine setups in preparation for grinding straight and helical teeth on a plain milling cutter; slitting saws; staggered-tooth side milling cutters; shell and small end mills; inserted-tooth face milling cutters; angle cutters; and form relieved cutters. Calculations are required to establish the offset for grinding relief, clearance, and rake angles. Considerations for grinding solid and adjustable machine reamers and starting teeth on taps are also included. The designation tool and cutter grinder and cutter and tool grinder are used interchangeably.

UNIT 24

Cutter and Tool Grinder
Features and Setups

OBJECTIVES

After satisfactorily completing this unit, you will be able to:

- Identify cutter and tool grinding machine features and functions.
- Select attachments for surface and internal and external cylindrical grinding, gear cutter, and small end mill grinding.
- Illustrate primary relief and secondary clearance angles and check these angles with a dial indicator or angle gage.
- Make tooth rest setups for straight wheel and flaring-cup wheel applications.
- Apply shop formulas for calculating tooth rest or wheel head offset, spindle speeds, and dial indicator drop.
- Perform or simulate the following processes.
  - Setting up a Cutter and Tool Grinder for Sharpening Plain Milling Cutters (Hollow and Flat Grinding Methods).
- Follow *Safe Practices* for cutter and tool grinder setups and grinding processes and correctly use related *Terms.*

MACHINE DESIGN FEATURES OF THE UNIVERSAL CUTTER AND TOOL GRINDER

One popular size of toolroom universal cutter and tool grinder has a 10″ (254mm) swing capacity over the table and a 16″ (406mm) longitudinal table movement. The major components and controls of a cutter and tool grinder are illustrated in Figure 24–1.

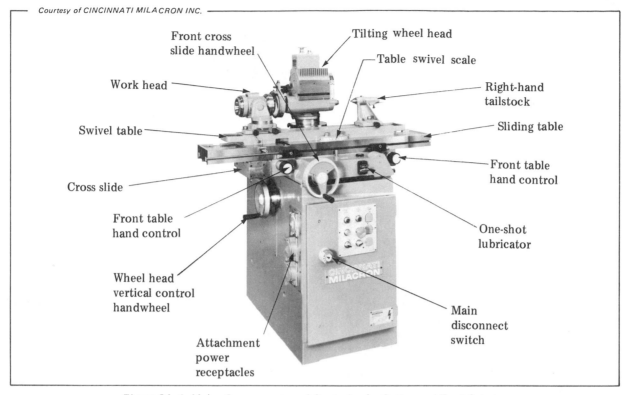

Figure 24-1 Major Components and Controls of a Cutter and Tool Grinder

TABLE

The cross (traverse) movement of the table is 10″ (254mm). The table swivels 90°. A taper setting device permits taper settings toward or away from the wheel head up to 5″ (127mm) per foot on the work diameter. The swivel table is graduated in degrees from either side of center.

The table is fitted with spring-cushioned table dogs (Figure 56-2). The dogs govern the length of table traverse, absorb the shock of table reversal, and provide positive table stop, where required.

"Tange Bar" Taper Setting Device. The "tange bar" taper setting device (Figure 24-3) uses a standard setting block and a gage block. The center distance from the pivot point of the table to the center of the block setting combination is fixed at 12″ (304.8mm). The gage block size is determined by the height measurement of 1/2 the included angle (Figure 24-3A). This method is used to set the swivel table to the required angle for precision taper grinding (Figure 24-3B).

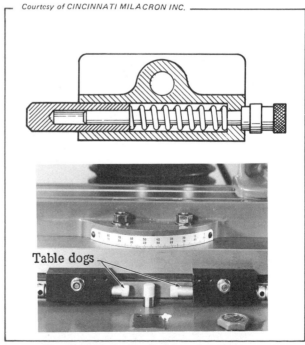

Table dogs

Figure 24-2 Spring-Cushioned Table Dogs

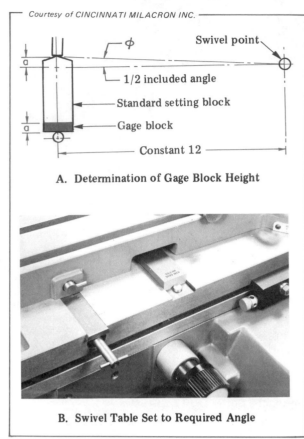

A. Determination of Gage Block Height

B. Swivel Table Set to Required Angle

Figure 24-3 "Tange Bar" Taper Setting Device

WORK HEAD

The standard universal work head (Figure 24-4) is designed with two tapered holes running through the spindle. One end accommodates a Morse or Brown & Sharpe taper. The other end is designed for a standard—for example, a #50—National Series taper.

WHEEL HEAD

Wheel heads are designed with a graduated base that permits settings through 360°. Wheel heads may also be swiveled to an angle in the vertical plane. An added feature on some machines is called an *eccentric column*. The eccentric column is also graduated through 360° of arc.

Since the table saddle may also be swung through 180°, the traverse range of the swivel table on a 10" × 16" (254mm × 406mm) universal model is extended by 3-1/2" (89mm). Thus, the offset table, wheel head, and eccentric column provide for machining over an extended cross (traverse) range. The normal 10" (254mm) cross range is extended to 13-1/2" (343mm) by swiveling the table 180°. The cross range is increased to 17" (431mm) by swiveling the eccentric column to the rear position.

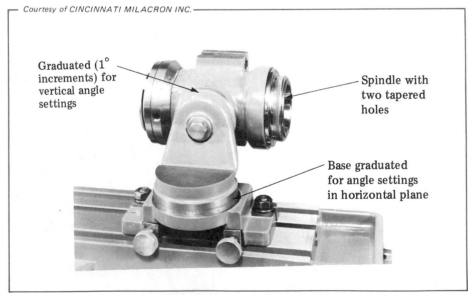

Figure 24-4 Universal Work Head

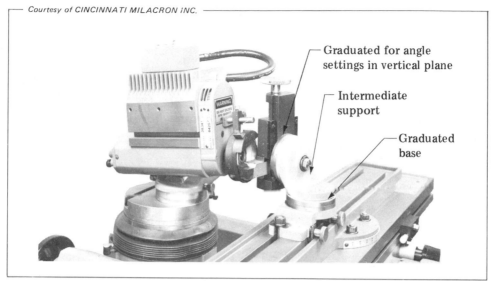

Figure 24-5 Surface Grinding Attachment Positioned for Grinding
a Clearance Angle on a Flat Cutting Tool

Standard toolroom model wheel heads are operable at spindle speeds of 3,834 RPM for 6″ (152mm) diameter wheels having a maximum wheel surface speed of 6,022 fpm (1,835 m/min). The spindle speed of 6,425 RPM is used for 3-1/2″ (89mm) diameter grinding wheels at a maximum surface speed of 5,887 fpm (1,794 m/min).

FORCED LUBRICATION SYSTEM

The cutter and tool grinder is equipped with a one-shot forced lubrication system. A single *actuating plunger* is located on the front of the machine on the cross slide. Lubrication is provided to the moving parts on the cross feed screw and the cross slide bearings.

CUTTER AND TOOL GRINDER ATTACHMENTS

SURFACE GRINDING ATTACHMENT

The *surface grinding attachment* is applied to the grinding of flat forming tools for turning and other machine tool processes. This attachment consists of a swivel base, a vise, and an intermediate support between these two units. The base and intermediate support are graduated

and are designed to be swiveled 360° in both the horizontal and vertical planes.

CYLINDRICAL GRINDING ATTACHMENT

The *cylindrical grinding attachment* provides for the grinding of outside diameters on work that is held in a chuck or fixture or between live or dead centers. The attachment is adapted to straight and taper cylindrical grinding and for facing operations.

INTERNAL CYLINDRICAL GRINDING ATTACHMENT

Internal diameters may be ground by the addition of an *internal grinding spindle*. The spindle is used with the work head drive unit of the cylindrical grinding attachment (Figure 24-6). Spindle RPM up to 23,000 is within the range of this attachment. This speed provides for the use of small-diameter grinding wheels to grind smaller-diameter holes than are normally produced.

GEAR CUTTER SHARPENING ATTACHMENT

The *gear cutter sharpening attachment* (Figure 24-7) is designed for sharpening form relieved cutters by grinding the face of the teeth. The cutter is supported on a bracket. The bracket

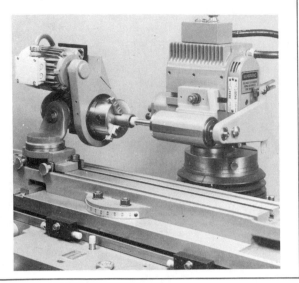

Figure 24–6 Application of Work Head Drive Spindle and Internal Grinding Spindle to Internal Cylindrical Grinding

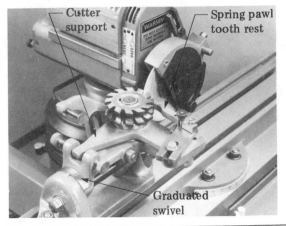

Figure 24–7 Application of Gear Cutter Sharpening Attachment for Grinding the Tooth Face on Form Relieved Cutters

may be swiveled to accommodate the angle of the tooth face. The spring pawl tooth rest lies in a horizontal position so that the edge of the spring locates against the back of the tooth.

SMALL END MILL GRINDING ATTACHMENT

For cylindrical grinding purposes, small end mills are held in collets. The *small end mill grinding attachment* (Figure 24–8) includes an intermediate support, a 24-division master index plate, and a plunger-type indexing mechanism. The spindle is designed to take straight cylindrical and taper collets.

RADIUS GRINDING ATTACHMENT

Ball-shaped end mills with straight or helical flutes and cutters that are to be ground to an accurate 90° radius require the use of a *radius grinding attachment* (Figure 24–9). The design features include a base plate with two mounted table slides and a swivel plate. Movement of each slide is controlled by a micrometer adjustment knob (or graduated collar). A 24-notch index plate may be mounted at the back of the work head to permit direct indexing of straight-fluted cutters.

The starting position is established by a micrometer gage. Radius grinding attachments are available with capacity to position cutters for grinding radii from 0″ to 2″ (0mm to 50mm), cutter diameters up to 12″ (305mm), and to a maximum width of cutter face of 3″ (76mm).

Figure 24–8 Small End Mill Grinding Attachment

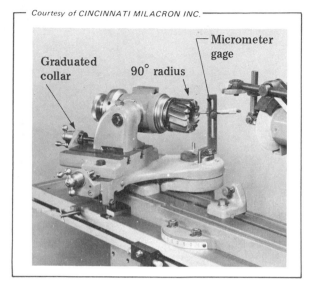

Figure 24-9 Radius Grinding Attachment for a Universal Cutter and Tool Grinder

ADDITIONAL ATTACHMENTS

A number of other attachments are optional, depending on the variety of toolroom or small manufacturing cutter and tool grinding operations. For example, the standard work head may be equipped with an *indexing attachment.* This attachment eliminates the tooth rest as the teeth may be indexed according to the notches on an index plate.

A *micrometer table positioning attachment* adds the feature of a precision lead screw for the machine table. A *sine bar attachment* permits the work head to revolve at a predetermined lead without a tooth rest. A *heavy-duty tailstock* increases the swing capacity over the table. A special *extended grinding wheel spindle* is interchangeable with the conventional spindle. The extended spindle provides added range. A *draw-in collet attachment*, having similar design features to draw-in attachments for turning machines, may be used with the work head, Collets are available in sizes from 0.125″ to 1.125″ in increments of 1/64″. Metric sizes range from 3mm to 28mm in increments of 1mm. Small, straight-shank cutters are conveniently held in draw-in collets.

CENTERING GAGE, MANDREL, AND ARBOR ACCESSORIES

CENTERING GAGE

The *centering gage* consists of a base, arm, and center. As the name indicates, it is used to position the wheel head (and spindle axis) at the center height of the work head and tailstock centers. It is also used to locate the cutter teeth to coincide with the center axis (when this setup is required).

GRINDING MANDREL

The *grinding mandrel* serves to hold a cutter so that it may be accurately mounted between centers. The cutter grinding mandrel has a slight taper for about one-third of its length. When pressed against this slight taper, the cutter is held securely for grinding.

CUTTER GRINDING ARBOR

A *cutter grinding arbor* is used when there is considerable grinding to be done. The general design of the arbor includes a centered, ground shaft to accommodate the bore diameter of the cutter and a series of spacing collars, washer, and nut.

FUNCTIONS AND TYPES OF TOOTH RESTS AND BLADES

FUNCTIONS OF TOOTH RESTS

A tooth rest serves three functions:

• To position the tooth or surface to be ground in a fixed relationship to the grinding wheel,
• To provide a support (the top face of the tooth rest blade) for the tooth during the grinding process,
• To permit the tool or cutter to be indexed to the next tooth. The tooth rest springs back into working position after each index.

TYPES OF TOOTH RESTS

Tooth rests are of two general types: (1) stationary and (2) adjustable. The adjustable type is designed with a micrometer adjustment. The blade (and tooth rest) may be adjusted to more

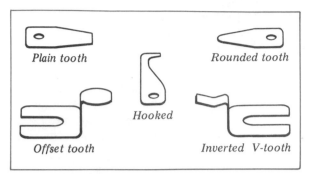

Figure 24–10 Five Common Shapes of Tooth Rest Blades

accurately position the cutter for grinding the relief angle. The two types of cutter tooth rests may be fastened to the wheel head or the table.

FORMS OF TOOTH REST BLADES

Different forms of tooth rest blades are available. The blade form depends on the shape of the cutter teeth that rest against the blade. Five common shapes of tooth rest blades are shown in Figure 24–10.

Straight-tooth milling cutters are supported by a *plain tooth rest blade*. Shell end mills, small end mills, taps, and reamers are ground with the support of a *rounded tooth rest blade*. An *offset tooth rest blade* is applied in the grinding of coarse-pitch helical milling cutters and large face mills with inserted cutter blades. A *hooked blade* is used with straight-tooth plain milling cutters having closely spaced teeth, end mills, and slitting saws. An *inverted V-tooth blade* is used for grinding staggered-tooth cutters.

TOOTH REST APPLICATIONS AND SETUPS

Cutters are divided for sharpening purposes into two general groups. The first group includes cutters that are sharpened by grinding primary relief and/or secondary clearance angles behind the cutting edge of each tooth. Examples of cutters that are ground on the periphery are plain and helical milling cutters, cutoff saws, and reamers. Cutters that are ground on the sides or ends are also included.

The second group of form-relieved cutters are sharpened by grinding the cutting faces of the teeth. Such cutters have a definite profile for form machining. Gear tooth cutters, combination radius and angle form cutters, and taps are examples.

CUTTER GRINDING WITH TOOTH REST MOUNTED ON TABLE

The height of the tooth rest depends on the type of grinding wheel and its direction of rotation in relation to the cutting edge of the cutter and the location of the cutter and machine centerline. One typical setup of a tooth rest for hollow grinding a plain milling cutter with a straight-type wheel is illustrated in Figure 24–11. The direction of the grinding force is clockwise. The grinding process produces a counterclockwise force on the cutter to hold it against the tooth rest.

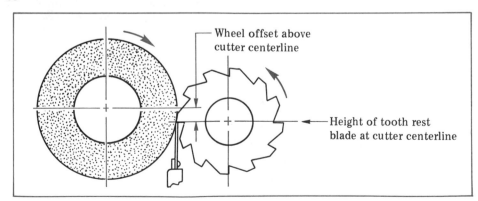

Figure 24–11 Typical Setup of Tooth Rest for Hollow Grinding Plain Milling Cutter with Straight Wheel

While there is no tendency for the tooth to dig into the wheel during grinding, a slightly burred cutting edge is produced. In addition, the heat generated in grinding moves toward the smallest area, which is the cutting edge. If too much heat is generated, the temper of the teeth at the cutting edge may be drawn.

CUTTER GRINDING WITH TOOTH REST MOUNTED ON WHEEL HEAD

When the tooth rest blade is positioned above the cutter, the centers of the wheel and cutter are offset to produce the required clearance angle (Figure 24-12). Here, the wheel is offset below the center line of the cutter. A caution is in order when using this setup. The cutter must be held securely against the tooth rest. The cutting action tends to move the cutter away from the tooth rest. Thus, there is the possibility of personal, wheel, and/or cutter damage.

With the wheel and cutter properly set up, there are two advantages of using this method. First, burr-free cutting edges are produced. Second, the possibility of overheating the cutting edges is reduced.

POSITIONING OF FLARING-CUP WHEELS

Flaring-cup wheels are widely used for cutter and tool grinding. The two general methods of positioning the tooth rest are similar to the setups used with standard straight grinding wheels. The main difference is that the axes of the cutter

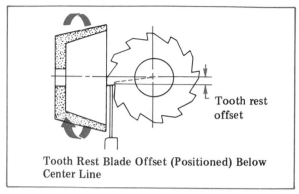

Tooth Rest Blade Offset (Positioned) Below Center Line

Figure 24-13 Setup for Grinding Primary Relief and/or Secondary Clearance Angles with Flaring-Cup Wheel

and wheel fall on the same center line. The clearance angle for grinding the cutter teeth is set by adjusting the tooth rest.

The setups for grinding the primary relief and/or secondary clearance angles with a flaring-cup wheel are shown in Figure 24-13. The positioning of the tooth rest below or above center at the required angle depends on whether the tooth rest is mounted on the table or the wheel head.

There usually is a greater area of contact when grinding with cup wheels. Therefore, the cuts are lighter than cuts taken with a straight type 1 wheel.

GRINDING CONSIDERATIONS AND CLEARANCE ANGLE TABLES

The clearance angle produced by a straight grinding wheel depends on the diameter of the *wheel*. When a cup wheel is used, the diameter of the *cutter* determines the cutting angle to be ground. The distance the tooth rest is set above or below center determines the clearance angle.

PRIMARY RELIEF AND SECONDARY CLEARANCE ANGLE CONSIDERATIONS

Summary tables provide a guide for grinding clearance angles. Table 24-1 gives recommended clearance angles for high-speed steel and cemented carbide cutters. Note that different cutting angles depend on the type of cutter and whether the cutting edge is on the periphery, corner, face, or a combination.

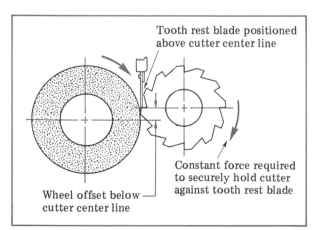

Figure 24-12 Setup for Cutter Grinding with Tooth Rest Blade Positioned Above Cutter

Table 24-1 Recommended Clearance Angles for High Speed Steel and Carbide Milling Cutters (Partial Table)

Recommended Relief and Clearance Angles for High-Speed Steel Milling Cutters

Material to Be Machined	Primary Relief Angle	Secondary Clearance Angle
Gray cast iron Malleable cast iron	4° to 7°	9° to 12°
Plain carbon steel Cast steel Tool steel High-speed steel Alloy steel	3° to 5°	8° to 10°

Type of Milling Cutter	Periphery Angle			Corner Angle			Face Angle		
	Steel	Cast Iron	Aluminum	Steel	Cast Iron	Aluminum	Steel	Cast Iron	Aluminum
				Primary Clearance Angle					
Face or side	4° to 5°	7°	10°	4° to 5°	7°	10°	3° to 4°	5°	10°
Saw and slotting	5° to 6°	7°	10°	5° to 6°	7°	10°	3°	5°	10°

ESTABLISHING THE OFFSET

Table 24-2 lists the offset for sharpening milling cutters with straight (type 1) wheels. The amount of offset shown is for standard clearance angles from 3° to 7° for wheel sizes ranging from 3″ (76.2mm) to 6″ (152.4mm).

Table 24-3 gives the offset for sharpening milling cutters with flaring-cup wheels. Here, the cutter diameter influences the offset. While the table covers cutter diameters to 3 1/2″ (88.9mm), the values given in Table 23-2 for larger straight-wheel diameters may be substituted.

AMOUNT TO RAISE OR LOWER THE WHEEL HEAD

When the wheel head is raised or lowered (depending on the grinding method) to produce the required primary relief angle, the amount may be calculated by formula. The cutter tooth level must be on center. The wheel head offset measurements originate from the same spindle and work center axes. The formula is as follows:

Table 24-2 Examples of Offsets for Sharpening Milling Cutters with Straight (Type 1) Wheels

Grinding Wheel Diameter		Cutter Clearance Angle									
		3°		4°		5°		6°		7°	
		Offset Required to Grind Cutter Teeth to Clearance Angle									
(Inches)	(mm)	(0.001″)	(mm)	(0.001″)	(mm)	(0.001″)	(mm)	(0.001″)	(mm)	(0.001″)	(mm)
3	76.2	0.079	2.01	0.105	2.67	0.131	3.33	0.158	4.01	0.184	4.67
4	101.6	0.105	2.67	0.140	3.56	0.175	4.45	0.210	5.33	0.245	6.22
6	152.4	0.157	3.99	0.210	5.33	0.262	6.65	0.315	8.00	0.368	9.35

Table 24–3 Offset for Sharpening Milling Cutters with Flaring-Cup Wheels

Cutter Diameter		Cutter Clearance Angle								
		3°		4°		5°		6°		7°
		Offset Required to Grind Cutter Teeth to Clearance Angle								
(Inches)	(mm)	(0.001″) (mm)		(0.001″) (mm)		(0.001″) (mm)		(0.001″) (mm)		(0.001″) (mm)
1/2	12.7	0.013	0.33	0.017	0.43	0.022	0.56	0.026	0.66	0.031	0.79
3/4	19.1	0.019	0.48	0.026	0.66	0.033	0.84	0.040	1.02	0.046	1.17
1	25.4	0.026	0.66	0.035	0.89	0.044	1.12	0.053	1.35	0.061	1.55
1 1/2	38.1	0.039	0.99	0.053	1.35	0.066	1.68	0.079	2.01	0.092	2.34
2	50.8	0.052	1.32	0.070	1.78	0.087	2.21	0.105	2.67	0.123	3.12
2 1/2	63.5	0.065	1.65	0.087	2.21	0.109	2.77	0.134	3.40	0.153	3.89
3	76.2	0.079	2.01	0.105	2.67	0.131	3.33	0.158	4.01	0.184	4.67
3 1/2	88.9	0.092	2.34	0.122	3.10	0.153	3.89	0.184	4.67	0.215	5.46

$$\text{Wheel Head Offset} = \text{Sine of Required Primary Relief Angle} \times \text{Grinding Wheel Radius}$$

AMOUNT TO RAISE OR LOWER THE TOOTH REST

The distance the tooth rest is lowered or raised to position the edge of the cutter in relation to the center line of a flaring-cup wheel used to grind the required primary relief angle may be calculated by formula:

$$\text{Tooth Rest Offset} = \text{Sine of Required Primary Relief Angle} \times \text{Cutter Radius}$$

METHODS OF CHECKING ACCURACY OF CUTTER CLEARANCE ANGLES

DIAL INDICATOR "DROP" METHOD

Cutter clearance angles may be checked by using the *dial indicator "drop" method*. The term *drop* is derived from the fact that when a dial indicator is used to measure the ground clearance angle, the pointer end "drops" (moves downward) as the cutter is revolved. The pointer measures the movement from the front to the back of the land. Tables are available that give the dial indicator movement for different cutter diameters, land widths, and relief angles.

Figure 24–14 illustrates how the dial indicator "drop" method is used to check a primary relief angle. As a general rule of thumb, for each degree of relief on a 1/16″ (1.5mm) land, there is a 0.001″ (0.025mm) movement (drop) on a dial indicator.

CUTTER CLEARANCE GAGE METHOD

Cutter clearance angles may also be checked by using the *cutter clearance gage method*. There are two common designs of cutter clearance angle gages. One design consists of two hardened steel arms that are at right angles to each other. The arms are placed on top of two teeth on the cutter. A hardened sliding center blade, ground on the end with the required angle, is brought into contact with the face of a ground tooth. The tooth is ground to the required angle when the ground angle and the blade angle coincide.

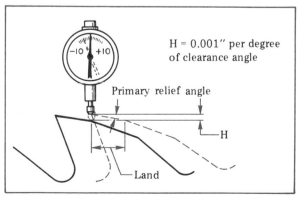

H = 0.001″ per degree of clearance angle

Figure 24–14 Checking Primary Relief Angle by Dial Indicator "Drop" Method

The second design of cutter clearance angle gage consists of a graduated frame with a fixed foot and beam. The gage is set by positioning the feet (fixed and adjustable feet) on two alternate teeth of the cutter and at a right angle to the tooth face. The blade is set to the clearance angle of the cutter. The reading is taken directly from the graduated scale on the frame. The gage may be used to read angles from 0° to 30° on straight, helical, side, or inserted-tooth milling cutters; straight or helical end mills; saws and slitting cutters; and T-slot cutters. The gage is practical for measuring or checking cutter diameters from 2" (50mm) to 30" (762mm). Angles may be measured on end mills (if the teeth are evenly spaced) in diameters that range from 1/2" (12.7mm) to 2" (50mm).

How to Set Up a Cutter and Tool Grinder for Grinding Plain Milling Cutters

Using a Straight Wheel (Hollow Grinding Method)

STEP 1   Select the largest possible straight wheel.

> Note: The larger the diameter, the smaller the radius (hollow) ground behind the cutting edge.

STEP 2   Run the standard grinding wheel checks. Mount, true, and dress the wheel.

STEP 3   Center the tailstock (or a work head and tailstock) on the table. Secure them at positions that will accommodate the mandrel or arbor assembly.

STEP 4   Use a test bar between centers to check the alignment of the table with respect to the wheel head.

STEP 5   Center the wheel head at the height of the tailstock (work head) center. Use a centering gage.

STEP 6   Set the wheel head at the zero graduation.

STEP 7   Position the height of the tooth rest blade edge on center by using the same centering gage.

> Note: The tooth rest attachment is secured to the table for straight-tooth cutters. The tooth rest is attached to the wheel head post for grinding spiral-tooth (helical) plain milling cutters.

> Note: The angle of the tooth rest blade must be the same as the helix angle of the spiral-tooth cutter. The angular face of the cutter rests on the angle slope of the tooth rest blade.

STEP 8   Set the wheel guard in position.

STEP 9   Set the vertical wheel head control handwheel at zero.

STEP 10   Calculate the wheel head offset to produce the required primary relief angle.

STEP 11   Offset the wheel head.

STEP 12   Set the arbor and cutter between centers. Position the tooth rest blade on the cutter face just behind the cutting edge.

STEP 13   Start the spindle. Allow for a short start-up time at spindle speed.

STEP 14   Move the workpiece for the start of the first cut on the first tooth.

STEP 15   Traverse the cutter past the straight grinding wheel.

> Caution: A continuous force must be applied to keep the cutter face against the tooth rest blade.

STEP 16   Clear the wheel from the workpiece at the end of the cut. Rotate the cutter 180°. Position the face of the tooth opposite the first ground tooth against the tooth rest.

STEP 17   Grind this tooth at the same cross feed handwheel setting as the first tooth.

STEP 18   Stop the spindle. Clean the workpiece and wheel. Measure the outside diameter at both ends of the cutter. The table is adjusted (aligned) for any taper.

STEP 19   Continue to grind each tooth in succession to the same depth.

STEP 20   Make a final sparking-out pass without infeeding.

STEP 21   Measure the primary relief angle by using either the dial indicator drop method or a cutter clearance angle gage.

Note: The angle is usually measured after the first cut is taken across three teeth.

Note: The cutting teeth are stoned to remove the burrs produced by grinding.

STEP 22   Determine the required offset to grind the secondary clearance angle.

Note: If the width of the land is greater than is recommended, it is necessary to grind the secondary clearance until the required width is reached. The same steps are followed to reposition the grinding wheel in relation to the axis of the workpiece.

STEP 23   Reset the grinding wheel offset for the secondary clearance angle.

STEP 24   Proceed to grind the secondary clearance on all teeth.

STEP 25   Stop the machine. Remove the cutter and arbor or mandrel assembly.

Caution: Protect the hands and fingers from rubbing against the razor-sharp cutter edges.

Using a Flaring-Cup Wheel (Flat Grinding Method)

STEP 1   Select an appropriate flaring-cup wheel.

STEP 2   Check the wheel for proper fitting. Mount on the wheel spindle. Lock the guard in place. True and dress.

Note: Many operators dress the wheel grinding face to a narrow land by using a dressing stick.

STEP 3   Set the spindle axis at center height by using a center gage.

STEP 4   Select the tooth rest blade form that is suitable for the cutter (straight or helical teeth). Set the tooth rest attachment so that the blade is at the right height.

STEP 5   Set the graduated ring on the vertical control handwheel at zero.

STEP 6   Calculate or check tables for the offset required to produce the primary relief clearance angle.

STEP 7   Position and set the tooth rest blade at the offset distance.

Note: If the tooth rest is attached to the table or is on the wheel head just below the wheel, the blade is positioned below the center line. The blade is positioned the offset distance above the center line when the tooth rest is mounted so that the cutting takes place starting at the face and continuing back to the secondary clearance angle.

STEP 8   Proceed to take a first cut on two opposite teeth. Measure the outside diameter for parallelism.

STEP 9   Proceed to grind successive teeth by taking one or more cuts.

Note: The primary relief angle is measured after a few teeth are ground. Then, offset adjustments are made to grind the correct angle.

STEP 10   Finish grind by sparking out the last cut.

Note: If, as a result of grinding, the land width becomes too wide, the offset for the secondary clearance is reset. The secondary clearance angle is ground the required land width.

STEP 11   Stop the machine. Remove the cutter and arbor assembly.

Safe Practices for Setting Up and Operating a Cutter and Tool Grinder

- Make standard wheel checks for soundness, balance, truing, and dressing.
- Test the wheel fit on the spindle, correct mounting, and condition of the wheel for grinding.
- Check the grinder stops and tripping devices.
- See that the blade and tooth rests are securely attached to the table or wheel head column.
- Examine the work-holding device and the setup. Adequate, safe working space must be provided. The operator must be sure the cutter is held securely against the work rest during grinding.
- Test the condition of the tailstock center and the tension (force) of the adjustable center against the arbor. The force exerted must be sufficient to hold the work securely.

- Check the diameter of the grinding wheel and the spindle speed before starting the spindle. The fpm or m/min must be within the maximum recommended wheel speed. In the case of small-diameter wheels, it is equally important to bring the fpm up to maximum wheel speed in order to grind efficiently.

- Regulate the amount of in-feed per cut. Avoid undue force on the grinding wheel, particularly when flaring-cup, saucer, and relieved types of wheels are used.

- Feed the cutter carefully so that the wheel is not brought into sharp contact with the workpiece.

- Protect machine surfaces when dressing the wheel.

TERMS USED WITH CUTTER AND TOOL GRINDER MACHINES AND SETUPS

Cutter and tool grinder	A machine tool primarily designed to generate primary relief and secondary clearance angles on hand and machine single- and multiple-point cutting tools. A grinder that may be adapted to internal and external cylindrical grinding and special tool grinding by adding special accessories.
Tange bar	A taper setting device for positioning the swivel table of a cutter and tool grinder to a precise angle.
Work head (cutter and tool grinder)	A work head with graduated base and upright arm for housing the spindle. A work-holding device that may be swiveled through 360° in horizontal and vertical planes.
Cutter grinder attachments	Mechanisms that when added to the basic machine, increase the capacity to perform additional processes. Examples include micrometer table positioning, sine bar, and heavy-duty tailstock.
Centering gage	A device having a base and upright mounting for a pointer (center). A gage used for positioning the wheel head and tool rest blade at center height.
Tooth rest blade forms	Plain, rounded, offset, hooked, and inverted V-tooth formed blades. A variation of blade shapes to accommodate different cutter teeth forms.

Offset (wheel head or tooth rest)	The distance a wheel head or tooth rest blade is moved in relation to the axis of the cutter or the grinding wheel. The distance between axes needed to grind a primary relief or secondary clearance angle.
Cutter clearance angle measurement	Measuring the angle at which tool or cutter relief or clearance angles are ground. The use of a dial indicator or a cutter clearance angle gage.
Hollow ground angle	The radial contour generated by grinding the primary relief or secondary clearance angle on a cutter tooth with a straight-type grinding wheel.
Flat grinding method	Use of a flaring-cup wheel and offsetting the cutter to grind a flat primary or secondary angle surface.

SUMMARY

- The capacity of the cutter and tool grinder is increased by using attachments to perform internal and external cylindrical grinding, surface grinding, and other special grinding processes.

  - The swivel table with fine adjustment settings provides for accurately grinding straight and taper surfaces.

- The work head spindle design permits direct mounting on one end of parts with Morse or Brown & Sharpe tapers. Accessories and other parts with large standard tapers in the National Series are mounted on the opposite end.

  - The gear cutter sharpening attachment permits grinding tooth faces on form cutters.

- A radius grinding attachment is used to accurately blend a 90° radius edge with the periphery and face of a cutter.

  - The wheel head (spindle) and tooth rest blade heights are set on center with the aid of a centering gage.

- Tooth rests may be stationary or adjustable. They are fitted for table, workhead, or wheel head mounting.

  - Five basic forms of blades accommodate the standard forms of cutters and tools used in the jobbing shop. The blade forms are: plain, rounded, offset, hooked, and inserted V-tooth.

- Four principal design features of multiple-tooth cutters to which the grinder operator relates include: primary relief (clearance) angle, secondary clearance angle, cutting edge, and land.

  - Grinding from the cutting edge back toward the heel of the land produces a burr-free cutting edge. Grinding heat is distributed away from the cutting edge.

- Straight grinding wheels produce an angle surface ground to a slight radius (hollow ground). The largest-diameter wheel is used to reduce the curvature.

    - Clearance angle tables provide manufacturers' recommendations of primary relief and secondary clearance angles according to cutter material, size, and work processes.

- The amount of offset of the wheel head (spindle) or cutter may be computed as the product of the sine of the primary relief angle and the radius of the grinding wheel or the cutter.

    - Cutter angles may be measured by the dial indicator drop method. A more functional shop practice is to use a cutter clearance angle gage.

- Flaring-cup wheels are used for flat grinding relief and clearance angles. The distance the tooth rest blade (and position of the tooth) is offset from the center determines the angle to which cutter teeth are ground.

    - Light first cuts are taken on opposite teeth. The cutter is then measured for diametral accuracy. Any taper is compensated for by adjusting the swivel table.

- Standard safety checks are made for grinding wheels and cutting speed limits, machine parts and controls, accessories, and work-holding setups.

UNIT 24 REVIEW AND SELF-TEST

CUTTER AND TOOL GRINDER FEATURES AND SETUPS

1. State three functions of tooth rests.
2. Make a simple freehand sketch of a setup to flat grind a primary relief angle on a cutter. A flaring-cup wheel is to be used. The cutting action is from the face of the cutting edge back across the land.
3. Calculate the tooth rest offset required to grind (a) a 6° primary relief angle and (b) an 11° secondary clearance angle on a 100mm diameter helical-tooth milling cutter.
4. List the steps for measuring a ground secondary clearance angle on a helical-tooth milling cutter by using a cutter clearance angle gage.
5. a. Select a straight face milling cutter from the toolroom.
   b. Use a cutter clearance angle table. Establish the smallest primary relief angle required for the machining of annealed gray cast iron workpieces.
   c. Calculate the offset between the centers of the cutter and a straight (type 1) grinding wheel.
   d. Set up the cutter grinder to use a type 1 wheel to hollow grind the primary angle. The grinding wheel is to rotate in a direction toward the cutting edge of the cutter teeth.
6. Give two reasons why it is important to maintain a constant force to hold the cutter securely against a tooth rest blade when grinding.

Grinding Milling Cutters and Other Cutting Tools

Cutter grinding and tool sharpening as carried on within jobbing shops and toolrooms require setups and dimensional requirements that vary from one cutter to another. These variations are produced by factors such as: type of cutter, shape of tooth face, cutting angles for particular machining processes, cutter size, and design of the part. There are, however, fundamentals that apply in general.

The types of cutters may be grouped to include the grinding of peripheral, side, face, and end cutting teeth. Therefore, the underlying technology and processes described in this unit relate to sharpening side milling cutters, face

mills, end mills of shell or solid-shank form, angle cutters, and form relieved milling cutters—cutters the machine operator usually handles. In addition, there are conventional machine reamers, taps, and other fluted cutters. Tool grinding also includes the sharpening of flat forms of cutters used for lathe, planer, and shaper work.

Cutter grinding operations deal with tooth cutting angles; primary relief, secondary clearance, and radial rake angles; axial movements; and straight or helical-angle teeth. The setups in this unit deal with hollow and/or flat ground teeth.

OBJECTIVES

After satisfactorily completing this unit, you will be able to:

- Describe slitting saws and the sharpening of peripheral and side teeth; staggered-tooth milling cutters, end mills, carbide-tooth shell mills, single- and double-angle milling cutters, and form relieved cutters.

- Give features of machine reamers, adjustable hand reamers, and taps, and suggest methods of sharpening.

- Establish grinding requirements for flat cutting tools and other cutter and tool sharpening.

- Use manufacturers handbook tables for grinding wheel specifications, primary and secondary clearance angles, radial relief angles for end mills, and concentric relief.

- Compute wheel head and work head offsets.

- Perform or simulate the following processes.
  - Sharpening Peripheral Teeth on a Slitting Saw.
  - Sharpening Staggered-Tooth and Shell Milling Cutters (including carbide-insert shell mills) and End Mills.
  - Grinding the Teeth of a Form Relieved Cutter.
  - Sharpening the Chamfer of a Machine Reamer and an adjustable Hand Reamer.
  - Grinding Cutting, Clearance, and Rake Angles on Flat Cutting Tools.

- Follow *Safe Practices* dealing with personal, machine tool, and cutter protection when grinding milling cutters and other cutting tools and correctly use related *Terms*.

CHARACTERISTICS OF SLITTING SAWS

Slitting saws are used for medium-width slotting and cutoff operations. The saws may be plain or designed with side chip clearance for deeper slotting and sawing applications. An example of a metal-slitting saw with staggered teeth and side chip clearance is shown in Figure 25–1. This cutter has alternate right- and left-hand, side-cutting teeth. These teeth provide chatter-free cutting action and minimize chip removal problems. The narrow lands of the alternate side teeth serve to maintain the cutter width when slotting and to reduce rubbing between the workpiece and the cutter.

As with all other cutters, the tooth form, numbers of teeth, cutting edge angle, and primary relief and secondary clearance angles vary according to the job requirements. Slitting saws are made of high-speed steel or have teeth that are carbide tipped. The basic procedures for setting up and using a straight (type 1) grinding wheel to grind a plain milling cutter are used for sharpening slitting saws.

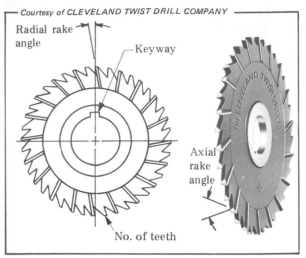

Courtesy of CLEVELAND TWIST DRILL COMPANY

Figure 25–1 Metal-Slitting Saw with Staggered Teeth and Side Chip Clearance

How to Sharpen Peripheral Teeth on a Slitting Saw

STEP 1 Set the edge of the tooth rest blade and the wheel head at center height with the centering gage.

STEP 2 Position the micrometer collar on the wheel head at zero.

STEP 3 Calculate the wheel head offset required to produce the primary relief angle.

Note: The following formula is used:

$$\text{wheel head offset} = \text{sine of required relief or clearance angle} \times \text{grinding wheel radius}$$

STEP 4 Raise the wheel head the amount of offset.

STEP 5 Mount the arbor and cutter between centers.

STEP 6 Position the tooth rest on the table so that the blade supports the face of the tooth to be ground. Secure the tooth rest to the table.

Note: Hook or L-type rest blades are adapted to the grinding of slitting saws. A flicker-type tooth rest support permits each tooth to be ratcheted into position for grinding.

STEP 7 Start the spindle and run it at operating speed for a minute.

Caution: Stand clear and work from the side of the revolving grinding wheel.

STEP 8 True and dress the wheel face.

STEP 9 Bring the grinding wheel into position for the first cut.

Caution: The cutter must be held securely against the tooth rest blade throughout the grinding of each tooth.

STEP 10 Grind each tooth by traversing the table and cutter past the grinding wheel. Use the back machine controls for table traverse.

Grinding the Secondary Clearance Angle

STEP 1   Calculate the height needed to produce the secondary clearance angle.

STEP 2   Raise the wheel head the additional amount.

> Note: A smaller-diameter wheel may be needed to grind the secondary clearance angle clear of the primary relief angle.
>
> Note: Many times, just the primary relief angle is ground.

STAGGERED-TOOTH MILLING CUTTERS

The functions of the alternate right- and left-hand helical teeth on staggered-tooth milling cutters and on staggered-tooth slitting saws are similar. The design features of peripheral and side teeth on a staggered-tooth side milling cutter (Figure 25–2) provide for smooth cutting action, minimize chip removal problems, help to maintain cutter width, and reduce rubbing. Each successive tooth is sharpened on a different helix. Staggered-tooth milling cutters are sharpened by using a type 1 straight wheel. The wheel face is trued and formed to a narrow width.

An inverted V-tooth rest blade, with a slightly smaller included angle, is used to support the cutter teeth. The blade requires accurate positioning at the center of the narrow, flat area on the grinding wheel. Unless the V is centered, every other tooth will be ground slightly higher or lower than the adjacent tooth. The tooth rest support is mounted on the wheel head assembly. The relief angle and the secondary clearance angle are produced by lowering the wheel head. The wheel head offset equals the sine of the primary relief angle multiplied by the cutter radius.

CHECKING THE TEETH FOR CONCENTRICITY

The teeth must be checked for concentricity on both helixes of the cutter. The position of the apex of the V-shaped tooth rest blade determines concentricity. Concentricity is checked with a tenth dial indicator (0.0001″ or 0.0025mm). A tolerance of 0.0003″ (0.0075mm) is accepted for general milling purposes. When the difference between two teeth exceeds this amount, it is necessary to adjust the location of the blade apex. The apex (blade) is moved in the direction of the helix with the higher tooth. The blade adjustment moves the tooth slightly higher. The result is that a slight amount of additional material is ground away.

GRINDING SECONDARY CLEARANCE

The secondary clearance may be ground by following steps similar to the steps used for grinding the primary relief angle. The wheel head is lowered to permit grinding at the required secondary clearance angle.

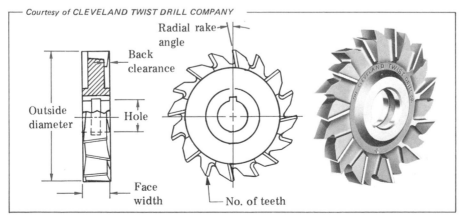

Courtesy of CLEVELAND TWIST DRILL COMPANY

Figure 25–2 Design Features of Peripheral and Side Teeth on a Staggered-Tooth Side Milling Cutter

Another method is to use a tooth rest with a micrometer adjustment. The blade is lowered for the required offset. The table is swiveled for grinding the peripheral teeth at the right- and then the left-hand helix. Grinding the secondary clearance at the helix angle also produces a uniform land width for the primary relief angle.

CONSIDERING SIDE TEETH AND LAND WIDTH

Generally, the smaller the side tooth land width, the less heat is generated through contact. A slight (about $1/2°$) back clearance further reduces the contact area of the land. The design of land width, small back clearance angle, and minimal relief angle produces a high-quality surface finish on the sides of deeply milled slots. A flaring-cup wheel is used for grinding the side teeth. The wheel face is narrowed to a small, flat width.

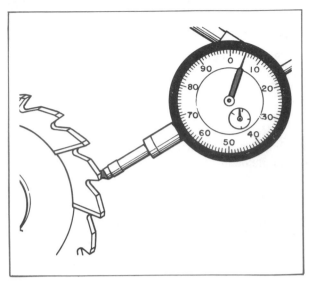

Figure 25–3 Checking Concentricity of Teeth with Right- and Left-Hand Helix on a Staggered-Tooth Milling Cutter

---

How to Sharpen a Staggered-Tooth Milling Cutter

Grinding the Teeth Peripherally

STEP 1 Mount, true, and dress a straight (type 1) wheel to a narrow-width grinding face.

STEP 2 Use a tooth rest with an inverted V-tooth rest blade. Secure the unit to the wheel head.

STEP 3 Set the wheel head and the tooth rest blade edge at center height.

STEP 4 Position the handwheel graduated collar at zero.

STEP 5 Calculate the offset required to produce the primary relief angle.

STEP 6 Lower the wheel head the required offset.

STEP 7 Mount the cutter carefully so that each tooth is supported in correct relation to the narrow-width grinding area of the wheel.

STEP 8 Start the spindle and run it at operating speed for part of a minute.

STEP 9 Take a fine cut across two successive teeth.

STEP 10 Stop the spindle. Set up a dial indicator and check for grinding concentricity on two teeth of the right- and left-hand helix (Figure 25–3).

Note: If adjustment is needed, the V-blade apex is moved slightly in the direction of the helix with the high (ground surface) tooth.

STEP 11 Take one or more cuts on all teeth.

STEP 12 Finish grind with no additional in-feed of the cross slide on the last pass.

Sharpening the Side Teeth

STEP 1 Mount the milling cutter on a stub arbor. Secure the assembly in a universal work head.

STEP 2 Set one of the side teeth so that it is positioned in a horizontal plane. Lock the work head spindle.

STEP 3 Use a hook or L-shaped tooth rest blade. Mount the tooth rest with micrometer adjustment on the work head.

STEP 4 Set the work head at the required relief angle for the side teeth.

Note: The relief on the land of the side teeth of staggered-tooth milling cutters is usually minimal. On other cutters, a narrow uncleared land on the side teeth maintains the original cutter width and reduces rubbing.

STEP 5 Swivel the wheel head until the spindle axis is parallel with the work head axis. Center the flaring-cup wheel with the centering gage.

STEP 6 Move the cutter carefully up to the revolving grinding wheel for the first cut.

STEP 7 Hold the cutter securely against the tooth rest blade. Traverse the cutter across the wheel face. Continue to index and grind the primary relief angle on all teeth.

STEP 8 Make a final sparking-out pass without additional in-feed.

STEP 9 Clear the cutter and grinding wheel.

STEP 10 Swing the tool head to the setting for the secondary clearance angle.

STEP 11 Take one or more cuts on all side teeth until the required width of land is reached.

STEP 12 Remove the cutter from the stub arbor. Stone any burrs on the side teeth.

STEP 13 Measure the width of the cutter to determine how much the teeth on the second side are to be ground in order to keep both sides even.

STEP 14 Remount the cutter and secure the arbor.

STEP 15 Swivel the work head downward for the primary relief angle. Reposition the tooth rest so that the tooth face of the cutter is held against the blade.

STEP 16 Proceed to take a series of cuts until the required amount is ground from the face of the side teeth. Spark out the final cut.

STEP 17 Reposition the work head to the secondary clearance angle. Continue grinding each tooth. Take a series of cuts until the land is ground to the required width.

STEP 18 Remove the cutter. Stone burrs on the teeth edges.

STEP 19 Clean the machine and leave it in a safe operable condition.

DESIGN FEATURES OF SHELL MILLS

Shell mills (Figure 25–4) are designed for both face and end mill operations and slabbing or surfacing cuts. These cutters are available with square, chamfered, or round corners. Right-hand cutters with right-hand helix are standard. Left-hand cutters with left-hand helix are available on special order. Shell mills are held and driven by arbors and adapters. The cutter material is usually high-speed steel. Shell mills are also furnished with carbide teeth.

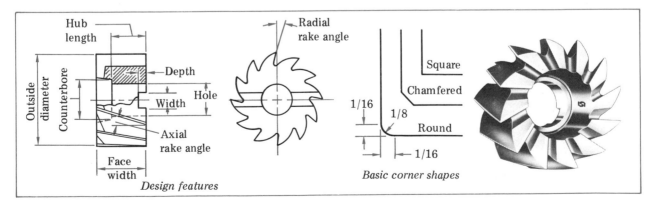

Figure 25–4 Shell Mill Designed for Face or End Milling Operations

How to Grind a Shell Mill

Grinding the Peripheral Teeth

STEP 1 Swivel the wheel head to 90°. Mount, true, and dress a flaring-cup wheel.

STEP 2 Set the table at 0°. Mount the cutters on a mandrel between the tailstock centers.

STEP 3 Attach the tooth rest with an offset blade to the wheel head. Center the high point of the tooth rest blade with the center of the narrow grinding area on the wheel.

STEP 4 Center the wheel head and the tooth rest blade height with the centering gage.

STEP 5 Lower the wheel head and tooth rest to the offset required to produce the primary relief angle.

STEP 6 Hold the cutter firmly against the tooth rest blade.

STEP 7 Bring the cutter and grinding wheel carefully into contact for the first cut.

Note: The depth of cut is checked at several places along the cutter tooth.

STEP 8 Traverse the table in the direction of the tooth angle. Position each successive tooth.

STEP 9 Spark out the final cut.

STEP 10 Reposition the wheel head and tooth rest. Lower them to the required offset to produce the secondary clearance angle.

STEP 11 Start the first cut. Then, continue to take successive cuts until the primary land width is ground to size.

STEP 12 Remove any grinding burrs from the cutting edge.

Caution: Grinding burrs are razor sharp. Use a cloth or other protective covering when handling the cutter.

Grinding the End Teeth

STEP 1 Mount the shell mill on a stub arbor. Secure this assembly in the work head spindle.

STEP 2 Position a micrometer adjustable tooth rest and secure it to the machine table. The blade is located near the periphery of one of the side teeth. The blade is positioned so that the first face tooth (and all other teeth) is held in a horizontal plane.

STEP 3 Lower the wheel head until the grinding wheel clears the tooth above the one being ground.

STEP 4 Swivel the work head to the required angle setting for the primary clearance angle.

STEP 5 Take a series of cuts until all teeth are ground to a sharp cutting edge and the land is ground to the correct width or wider.

STEP 6 Reset the work head to the secondary clearance angle.

STEP 7 Proceed to take one or more cuts at the secondary clearance angle until the land is ground to width.

Grinding a 45° Beveled Corner

STEP 1 Swivel the work head to the 45° (bevel angle) graduation on the base.

STEP 2 Set the work head at the 0° (horizontal plane) graduation.

STEP 3 Use the centering gage to bring the face tooth level horizontally and on center.

STEP 4 Rotate the cutter the number of degrees required for the clearance angle. Lock the work head spindle in position.

STEP 5 Swivel (tilt) the work head to the same angle.

Note: The two angle settings (work head and shell mill) are necessary. The clearance angle produced by grinding must be greater than the single angle setting in order to prevent heel drag.

STEP 6   Hold each tooth firmly against the tooth rest blade. Bring the cutter carefully into position for the first sharpening cut.

> Note: The wheel head and the grinding wheel must be lowered to clear the tooth above the one being ground.

STEP 7   Continue to grind the beveled edge on each tooth.

CONSIDERATIONS FOR GRINDING CARBIDE-TOOTH SHELL MILLS

Similar setups and procedures are used for grinding primary relief and clearance angles on the side, bevel, and face on a carbide-tipped (inserts) shell mill. A diamond-grit cup wheel is generally used for sharpening purposes. Carbide cutters may be ground to positive or negative radial rake. A coolant is used to prevent overheating and checking at the cutting edge. The depth of each cut is usually restricted to 0.001″ (0.025mm) or less.

How to Grind Face Teeth and Chamfer Edges on a Carbide-Insert Shell Mill

STEP 1   Mount and true up the face of a diamond-grit cup wheel. Set the wheel head at 0°.

STEP 2   Mount the cutter on an arbor. Secure the assembly in the work head spindle.

STEP 3   Proceed to position the tooth rest, blade, and tooth face in the same manner as for grinding a high-speed steel shell mill.

DESIGN FEATURES OF END MILLS

The grinding of peripheral and end teeth on an end mill is one of the most common cutter grinding processes. On small-size end mills, no outside diameter regrinds are considered. Design features provide for common two-, three-, four-, and six-flute end mills; single-end general-purpose and double-end end mills; regular, long, and extra-long end mills; standard and high-helix end mills; and straight and ball-end end mills. Heavy-duty premium cobalt high-speed steel end mills are used for machining typical exotic metals such as high-temperature alloys and high-tensile steels.

RADIAL RELIEF ANGLES

The general range of outside diameters of end mills is from 1/8″ to 2″ (3.2mm to 50.8mm). The primary relief angle range for these diameters is from 16° for a 1/8″ (3.2mm) diameter high-speed steel end mill for machining carbon steels to 6° for a 2″ (50.8mm) diameter mill.

It is important that the runout of the diamond wheel must be checked and corrected. The diamond wheel must cut as it rotates in the cutting edge. Indexing methods are used, if required, to control the equal spacing of the cutter teeth. Cutter teeth are usually honed with a 400-500 grit hand hone after they have been stoned with an abrasive stick.

Table 25-1 provides recommended radial relief angles for high-speed steel end mills (for machining mild carbon steels, tool steels, and nonferrous metals). The angles are for the

Table 25-1 Recommended Radial Relief Angles for High-Speed Steel End Mills

Material to Be End Milled	End Mill Diameter							
	1/8″ (3.2mm)	1/4″ (6.4mm)	3/8″ (9.5mm)	1/2″ (12.7mm)	3/4″ (19.0mm)	1″ (25.4mm)	1-1/2″ (38.1mm)	2″ (50.8mm)
	Radial Relief Angle* (in Degrees)							
Carbon steel	16	12	11	10	9	8	7	6
Tool steels	13	10	9	8	7	6	6	5
Nonferrous metals	19	15	13	13	12	10	8	7

*For secondary clearance angles, multiply the radial relief angle by 1.33.

conventional grinding of radial relief and accompanying secondary clearance angles.

How to Sharpen an End Mill

Grinding the Side Cutting Edges (Conventional Method)

STEP 1 Select, mount, and true a straight grinding wheel. Dress the wheel and narrow the face to a width of about 1/16″ (1.5mm).

STEP 2 Mount the small end mill grinding attachment.

STEP 3 Select a narrow-width blade to match the helix of the end mill. Install the tooth rest and blade assembly on the wheel head.

STEP 4 Secure the end mill in a straight or taper collet, depending on the body shape.

STEP 5 Adjust the wheel head, tooth rest blade, and mounted end mill to the same center height.

Note: Each end tooth must be in a horizontal plane.

STEP 6 Set the micrometer collar on the wheel head column at zero.

STEP 7 Calculate the required offset for lowering the wheel head:

$$\begin{array}{l} \text{wheel} \\ \text{head} \\ \text{drop} \end{array} = \begin{array}{l} \text{sine of} \\ \text{primary} \\ \text{relief angle} \end{array} \times \begin{array}{l} \text{cutter} \\ \text{radius} \end{array}$$

STEP 8 Lower the wheel head and tooth rest to grind the required primary relief angle.

STEP 9 Move the fixture away from the grinding wheel to clear the end mill.

STEP 10 Advance the cutter forward along the tooth rest blade until the cutting edge at the shank end is in position for the start of the cut.

Note: While an experienced operator may start the cut from the end of the shank area, it is safer (in terms of possible damage to the face end of the end mill if there is accidental contact) to start at the shank end.

STEP 11 Take a light first cut on each flute. Increase the feed increments until the primary relief area of the land is ground to correct any cutter wear or damage.

STEP 12 Spark out the final cut.

STEP 13 Determine the amount of wheel head offset required to grind the secondary clearance angle.

STEP 14 Lower the wheel head the additional amount.

STEP 15 Grind the secondary clearance until the land width is ground to the required size.

Grinding the End Teeth (Universal Work Head Method)

STEP 1 Select a small-diameter flaring-cup wheel. True the wheel and dress it to produce a narrow 1/8″ to 3/16″ (3mm to 5mm) grinding area. Stop the spindle.

STEP 2 Swivel the workhead counterclockwise past the 90° graduation on the base to 88°.

Note: An additional 2° to 3° is recommended so that the teeth are ground slightly lower at the center than at the outside edge.

STEP 3 Tilt the work head spindle to the number of degrees specified for the relief angle of the axial end teeth.

STEP 4 Attach a flicker-type tooth rest support with micrometer adjustment to the work head to provide ratchet indexing.

STEP 5 Insert the end mill in a collet and level the tooth. Adjust the tooth rest support in relation to the master index plate.

STEP 6 Traverse the first end tooth so that the grinding wheel completes the cut at the center of the end mill.

Note: The table stop is set to prevent feeding with possible damage to the opposite tooth face.

STEP 7 Return the grinding wheel to clear the cutter. In-feed about 0.003″ (0.08mm). Traverse the cutter across the wheel face.

STEP 8 Index (ratchet) the spindle to position the next tooth. Continue the grinding and indexing until all teeth have been ground to correct cutter wear or damage.

STEP 9 Spark out the final cut without additional in-feed.

STEP 10 Calculate the additional offset to which the work head is to be set to grind the secondary clearance angle.

STEP 11 Adjust the work head angle and wheel to permit grinding the secondary clearance.

STEP 12 Take successive cuts until the required land width is reached for the primary relief angle.

ECCENTRIC RELIEF AND METHOD OF SHARPENING

Side cutting edges of end mill flutes are also designed with eccentric relief. Figure 25–5 shows the characteristics of eccentric relief on a two-flute end mill. The cutter relief starts at the cutting edge and continues in an arc until it joins the body of the end mill. Generally, there is no secondary clearance. End mills that are ground with eccentric relief are especially adapted to milling low tensile strength materials requiring high surface finishes and to numerically controlled machining where the end mill is not reused for a particular process once it wears.

The eccentric relief is generated by swiveling the wheel head at a specified number of degrees. The wheel head angles for eccentric relief grinding of end mill cutter diameters from 1/8″ (3.18mm) to 1 1/4″ (31.7mm) are given in Table 25–2. An *air-spindle grinding device* with a support finger is attached to the table. The air spindle provides both rotary and axial movement of the end mill. The face of the grinding wheel is trued parallel to the table. The face is then dressed to narrow it to the width across the end mill land.

The wheel head is rotated counterclockwise according to the diameter of the end mill and the cutter helix angle (20°, 30°, or 40°). The cutter tooth and wheel head are set at center height. The tooth rest is set to make contact at the right edge of the grinding wheel. Unlike conventional cutter grinding, the cut is started from the end of the cutter. The air spindle turns

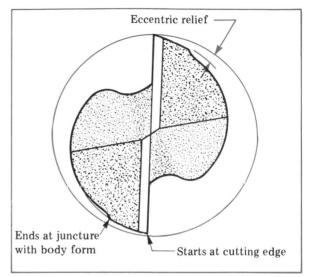

Figure 25–5 Eccentric Relief on a Two-Flute End Mill

the end mill while advancing it axially along the helical flute. The cutting edge is generated by the smaller right side of the wheel (due to the wheel head angle position). A greater amount of clearance is generated from the cutting edge to the heel of the land. The relatively larger left side of the grinding wheel produces the greater clearance as the cutter (supported on the tooth rest) axially traverses the grinding wheel during the rotation of the end mill. The quality of the eccentric relief grind depends on the correct dressing and angle setting of the grinding wheel, narrowing the cutting area of the wheel, the speed of radial and axial movement of the air spindle and mounted cutter, and the shape and correct positioning of the tooth rest.

Table 25–2 Radial Relief and Wheel Head Angles for Eccentric Relief Grinding of End Mills

Diameter of Cutter		Recommended Radial Relief Angle	Cutter Helix Angle		
			20°	30°	40°
(in.)	(mm)		Required Wheel Head Angle		
1/8	3.18	13°	4°30′	7°	11°
1/4	6.35	10°30′	3°45′	6°	10°
1/2	12.70	10°	3°15′	5°15′	9°15′
3/4	19.05	9°	3°	4°40′	8°30′
1-1/4	31.75	8°	2°30′	4°	7°30′

DESIGN FEATURES AND GRINDING OF FORM RELIEVED CUTTERS

Form cutters are usually marked with the radial rake. Some of the most frequently used form cutters that require grinding on the tooth face include gear tooth milling cutters, convex and concave cutters, and multiple-form gang milling cutters. Before the faces of the cutter teeth are ground, it is necessary to check the uniformity of each tooth from the cutting face to the back of the tooth form. If there is a variation in micrometer measurement, it is necessary to accurately grind a reference area on the back of each tooth for indexing. Dish (type 12) wheels (Figure 25–6) are used to grind the radial face of a cutter tooth down to and including part of the tooth gullet.

The two general methods of grinding the face of the teeth include (1) mounting the cutter between centers and (2) positioning the cutter horizontally in a gear cutter attachment or form relieved cutter holder.

Courtesy of AMERICAN TOOL & GRINDING COMPANY INC.

Roughing depth

Figure 25–6 Wheel Head (and Wheel) Set for Depth of Down-Feed Roughing Cut

How to Grind the Teeth on a Form Relieved Cutter

STEP 1 Select, test, and mount a type 12 dish grinding wheel on the spindle.

STEP 2 Use an abrasive dressing stick to shape the curved side of the wheel to clear the back side of the cutter teeth.

STEP 3 Shape the peripheral face of the wheel to conform to the shape of the tooth gullet. Dress the side cutting face of the wheel.

STEP 4 Mount the form relieved cutter on an arbor between centers.

Note: The cutter is positioned to grind the reference area on the back of the teeth if there are variations of more than 0.002″ (0.05mm) in the width of the teeth. A similar grinding wheel setup and indexing are used as for face grinding the teeth.

STEP 5 Position the radial face of one tooth face so that it is in a vertical plane.

Note: The tooth rest blade is adjusted to support the cutter in this position.

STEP 6 Position the dish wheel so that the face just clears the radial face of the cutter at the gullet depth.

STEP 7 Set the graduated micrometer collar on the cross slide handwheel at zero.

STEP 8 Clear the setup. In-feed the cutter by turning the micrometer screw on the tooth rest support.

STEP 9 Set the wheel head (and wheel) for the depth of the down-feed roughing cut (Figure 25–6).

STEP 10 Traverse the cutter for each increment of down-feed.

STEP 11 Dress the wheel and relieve the side face in preparation for finish cuts.

STEP 12 Take additional finish cuts by making in-feed increments with the micrometer tooth rest adjustment.

DESIGN FEATURES AND GRINDING SINGLE- AND DOUBLE-ANGLE MILLING CUTTERS

The two general forms of angle milling cutters are single angle and double angle. The design features of single- and double-angle milling cutters are shown in Figure 25–7. Single-angle milling cutters are commercially available with 45° and 60° included angles and are either right hand or left hand.

The teeth of a single-angle cutter are on the angular face and on the vertical face (Figure 25–7A). The teeth on the vertical face are ground with a slight back clearance. Single-angle milling cutters are used primarily for milling dovetails and angular surfaces. Single-angle cutters are ground with the work head set at 0° while the table is swiveled to the included angle.

Double-angle milling cutters (Figure 25–7B) are designed with included angles of 45°, 60°, and 90°. These cutters are used for milling bevel edges, angular notches, serrations, and so on.

Angle cutters are ground with a flaring-cup wheel. The teeth are set parallel to the table. Usually, the cutter is mounted on an arbor and held in the work head. The wheel head and grinding wheel spindle are set at the back clearance angle for grinding the vertical face of a single-angle milling cutter. The work head is tilted to the primary relief or secondary clearance angle for the vertical face teeth.

The teeth on the angular surfaces of single- and double-angle cutters are ground by setting the work head at 0° and swiveling the table to each required angle setting. The blade of the tooth rest is aligned at center with the angular cutter. The work head is tilted to the required primary relief or secondary clearance angle when the cutter is mounted on the work head. On cutters that are mounted between centers, either the spindle is tilted or the cutter is offset to produce the relief and clearance angles.

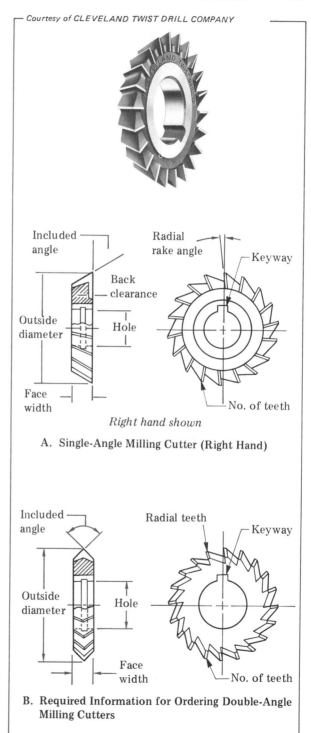

Courtesy of CLEVELAND TWIST DRILL COMPANY

Right hand shown

A. Single-Angle Milling Cutter (Right Hand)

B. Required Information for Ordering Double-Angle Milling Cutters

Figure 25–7 Design Features of Single- and Double-Angle Milling Cutters

DESIGN CONSIDERATIONS FOR GRINDING MACHINE REAMERS

When the lands on solid reamers become worn, they are cylindrically ground to a smaller size. The lands are ground to a slight taper of 0.0002″ (0.005mm) per inch to provide clearance (longitudinal relief), especially on deep-hole reaming.

One caution must be observed in cylindrically grinding of the lands. The direction the reamer turns must be opposite to the direction of normal rotation. The back of the land margin contacts the grinding wheel first. A second caution requires that the design of the cutter grinder must permit the direction of spindle rotation to be reversed and operated safely without loosening the wheel setup.

Secondary clearance is primarily provided to reduce the margin area along the flutes of a reamer. The procedure is basically the same as for grinding secondary clearance on a plain milling cutter or end mill. Either a straight wheel or a flaring-cup wheel may be used. The amount of offset (using the type 1 straight wheel method) is equal to the sine of the secondary clearance angle multipled by the radius of the grinding wheel. In the case of solid-type machine reamers, there is no primary clearance angle to the ground narrow band margin.

How to Sharpen the Chamfer of a Machine Reamer

Sharpening the Chamfer Edge by Offsetting the Wheel Head

STEP 1 Set the wheel spindle parallel with the table ways. Position the wheel head at center height.

STEP 2 Select and mount a type 12 dish wheel. Secure the guard.

STEP 3 True and dress the wheel to a narrow margin width and slightly relieve the vertical face.

STEP 4 Swivel the table to the 45° (or other) chamfer angle. Secure at this angle position.

STEP 5 Locate the tailstock centers to accommodate the reamer length.

STEP 6 Mount a micrometer-type tooth rest support on the machine table.

STEP 7 Adjust the chamfer cutting edge with the tooth rest blade to center height.

STEP 8 Feed the chamfer of the reamer almost to the outer edge of the grinding wheel.

Caution: The table stop is set to avoid possible damage to the cutter. The stop prevents additional movement toward the tailstock center.

STEP 9 Raise the wheel head according to the offset as calculated by the formula:

$$\frac{\text{wheel}}{\text{head rise}} = \text{sine of clearance angle} \times \text{grinding wheel radius}$$

STEP 10 Start the wheel spindle. Feed the reamer to the grinding wheel by using the back cross slide handwheel.

STEP 11 Take roughing cuts until all teeth are ground sharp clear to the outside diameter of the machine reamer.

Note: The movement of the table is coordinated with the indexing of each individual chamfer.

STEP 12 Dress the wheel and take the first finish cut.

STEP 13 Spark out the final finish cut. Stop the machine. Remove the reamer.

STEP 14 Stone to remove grinding burrs and to produce finer cutting edges. Hone the chamfer edges and the faces of each land.

STEP 15 Check the quality of the surface finish against a comparator specimen, if required.

STEP 16 Clean the machine and accessories. Return all items to appropriate storage areas.

Table 25–3 Recommended Primary Relief and Secondary Clearance Angles for Hand Reamers

Diameter of Reamer		Material to Be Reamed			
		Steel		Cast Iron and Bronze	
		Width of Margin			
		0.005″ to 0.007″ 0.12mm to 0.17mm		0.025″ to 0.030″ 0.61mm to 0.75mm	
		Grinding Angle			
(in.)	(mm)	Primary Relief	Secondary Clearance	Primary Relief	Secondary Clearance
1/2	12.7	3°	12°	7°30′	17°
1	25.4	1°30′	10°30′	4°30′	14°
1 1/2	38.1	1°	10°	3°30′	13°
2	50.8	45′	10°	3°15′	12°30′

DESIGN FEATURES OF ADJUSTABLE HAND REAMERS

Table 25–3 gives the primary relief and secondary clearance angles for reamer diameters ranging from 1/2″ (12.7mm) to 2″ (50.8mm). It shows a range of margins from 0.005″ (0.12mm) for reaming steel to a maximum width of 0.030″ (0.75mm) for bronze and cast iron. It is important to grind the land width to the recommended margin. Otherwise, the primary relief may be insufficient to prevent land interference when reaming.

To grind an adjustable machine reamer to ream accurately, the reamer blades are adjusted oversize. Enough stock is allowed to grind the blades to the correct diameter, starting taper, and correct margin (width) of the land.

How to Sharpen an Adjustable Hand Reamer

STEP 1 Mount a type 1 straight or type 12 dish wheel. Secure the guard.

STEP 2 Dress the outside diameter of either wheel to produce a narrow-width (about 1/16″ or 1.5mm) cutting area.

STEP 3 Attach a tooth rest with a narrow blade to the wheel head.

Note: The tooth rest blade is positioned directly in line with the narrow grinding area of the wheel.

STEP 4 Set the wheel head and tooth rest blade at center height.

STEP 5 Prepare the adjustable blades on the reamer. The blades are checked for proper seating. The reamer is adjusted oversize from 0.008″ to 0.010″ (0.2mm to 0.25mm).

STEP 6 Refer to Table 25–3 for recommended primary relief and secondary clearance angles. Compute the distance the wheel head is to be lowered to generate each required angle.

STEP 7 Mount the reamer on centers. Hold the reamer blade securely against the tooth rest blade.

STEP 8 Take a first cut by grinding a narrow margin across the length of one blade.

STEP 9 Take the same cut acorss the opposite blade.

STEP 10 Stop the wheel. Check the diameter for size and adjust to correct for any taper.

STEP 11 Continue to finish grind the primary relief angle. Spark out the final cut.

Note: Spark out the final cut without additional in-feed.

STEP 12 Lower the wheel head the additional distance to produce the secondary clearance angle.

STEP 13 Continue to take successive cuts across each reamer blade until the correct margin is reached.

STEP 14 Reposition the reamer at center height in order to grind the taper portion of the reamer in correct relation to the margin and primary relief angle on each blade.

STEP 15 Loosen the table swivel locking nuts. Swivel the table to a taper of 1/4″ (6.25mm) per foot.

Note: This taper is the customary taper at the starting end of hand reamers.

STEP 16 Offset the wheel head to again produce the primary relief angle.

STEP 17 In-feed for the first cut. Start at the end and traverse the blade until the primary starting taper extends approximately one-fourth the length of the blade (flute).

STEP 18 Spark out the final finish cut to ensure concentricity.

STEP 19 Change the wheel head offset (if necessary) to grind the secondary clearance angle for the starting taper section for reaming steel and where the primary relief angle is small (Table 25–3).

STEP 20 Stop the machine. Remove the reamer. Stone and then hone the face of the blades to produce a fine cutting edge.

STEP 21 Clean the machine. Return all items to proper storage areas.

GRINDING CUTTING, CLEARANCE, AND RAKE ANGLES ON FLAT CUTTING TOOLS

Flat cutting tools such as thread chasers often require the grinding of a chamfer or face, top rake, lip rake, and clearance angles. While these processes are often performed on surface grinders, the use of a surface grinding accessory extends the versatility of the cutter and tool grinder to cover the grinding of flat cutting tool forms.

CONSIDERATIONS FOR SHARPENING TAPS

Where precision tapping is required, using either hand or machine taps, the taper cutting portion is sharpened by machine grinding. Usually a grinding fixture is used. The tap is mounted between centers. The finger rest is set behind each flute. Small taps are held in a chuck and are positioned with a ratchet stop. The taper at the front end of the tap and the clearance are produced by adjusting the attachment and turning the tap by a handle on the tap grinding attachment.

A type 1 grinding wheel is used. The wheel head is set at 1° more than the angle of the attachment setting to reduce the width of the grinding face. The taper section of each flute is indexed and secured in position by the finger rest.

How to Grind Cutting, Clearance, and Rake Angles on Flat Cutting Tools

STEP 1 Use a standard type 1 straight wheel with a spindle extension, if needed.

STEP 2 Swivel the wheel head 90° and up so that it clears the work height.

STEP 3 Secure a thread chaser grinder attachment to the machine table.

STEP 4 Set the thread chaser in the fixture. Then, swing the fixture to the required angle.

STEP 5 Lower the grinding wheel for the first cut. Traverse the chaser until the chamfer angle is ground across the required number of teeth. Note the handwheel graduation reading for the final depth of cut.

STEP 6 Remove the first chaser. Replace with the second chaser.

STEP 7 Grind to the same depth as the first chamfer.

STEP 8 Continue to set and grind each of the remaining chasers.

STEP 9 Change the fixture setup to permit each chaser to be positioned horizontally so that the face (where required) may be ground to a particular rake angle.

STEP 10 Dress the wheel to produce a corner radius.

STEP 11 Feed the chaser for the first cut while traversing the cutter face past the grinding wheel.

STEP 12 Take additional cuts until the teeth along the face are sharp.

STEP 13 Remove the first chaser. Proceed to follow the same steps to sharpen the other chasers in the set.

STEP 14 Follow standard procedures for machine cleanup.

OTHER APPLICATIONS OF CUTTER AND TOOL GRINDING

The applications in this unit relate to basic set-ups and cutter grinding (sharpening) processes. The cutters and tools described are among the most widely used in custom jobbing shops. In addition, there are many special fixtures and setups for form relief and clearance grinding. For example, combination drills and counter-sinks may be ground to required cutting angles. Step drills require special positioning and sharp-ening techniques.

Other tool cutting-off processes are per-formed to remove sections of broken or damaged cutters. Cutoff wheels are used in such applica-tions. The face end on end mills is often gashed to sharpen the end teeth.

Safe Practices for Machine Setups and Cutter Grinding Processes

---

- Allow the wheel spindle to reach its operating speed when starting up for a grinding operation. Stand clear of the revolving grinding wheel at all times.

- Apply a continuous force against the cutter during grinding to hold it securely against the tooth rest blade.

- Dress wheels for cutter grinding to a small area of contact on the peripheral and side faces (as applicable) to reduce the cutting forces and the amount of heat generated.
- Stop the wheel spindle when measuring the ground angle areas or making adjustments to the tooth rest setup.
- Make a test cut on wide cutters to correct for unusual wear. It may be necessary to reduce the depth of the first cut.
- Stone and hone the razor-sharp grinding burrs produced by grinding.
- Use a cloth as protection against burrs when removing a cutter from the setup or during burring or other handling.
- Grind toward the cutting edge of carbide-tipped cutting tools. Use a coolant to avoid overheat-ing and checking.
- Check the safety features of the spindle head when it is necessary to change the direction of wheel rotation. The direction of rotation of a reamer (when grinding the margin) must be op-posite to its normal rotation.
- Replace guards after a wheel is mounted and before the spindle is started.
- Use safety goggles and/or a protective shield at all times.
- Clean the work station so that all machine sur-faces are clear of abrasive and foreign particles. Leave the machine in a safe operable condition.
- Use safety goggles and/or protective shield at all times.

TERMS USED IN MACHINE SETUPS AND CUTTER GRINDING PROCESSES

Alternate side teeth A design feature of staggered-tooth side mills, slitting saws, and other milling cutters. An alternate tooth pattern where the narrow land area reduces friction and permits machining to a controlled width.

Inverted V-tooth rest blade A special-form blade for use with staggered-tooth, angle, and other cutters that require precise location for grinding. A blade that is positioned to support each tooth in relation to the center of the wheel grinding area.

Stub arbor A work-holding arbor that centers and holds the cutter to be ground. The taper shank is secured in the work head spindle.

Sparking-out pass	A final pass across the cutter teeth without additional in-feed. Continuation of a grinding process until the wheel traverses the cutter almost without cutting (sparking).
Chamfer cutting edge	The angular surface formed between the face and sides of a cutter. An angular-formed cutting edge on a milling cutter.
Eccentric relief	Relief ground as a continuous arc from the cutting edge of an end mill to the flute form.
Air-spindle grinding device	A machine attachment with a cutter support arm and finger. A work-holding device that feeds and advances the cutter axially along the helical flute past the grinding wheel. A holding, indexing, and feeding device usually used when grinding a quantity of end mills.
Permanent form relief	A design characteristic for relieving form teeth so that the tooth profile remains unchanged after numerous sharpenings. Grinding the face of a form cutter at a constant radial rake and maintaining the same tooth form.
Type 11 flaring-cup wheel	A side grinding (hollow-cup) wheel with shell tapered from the back outward. Shapes and sizes are based on ANSI standards.
Type 12 dish wheel	A dish-shaped wheel designed for grinding on the side or face. Features conform to ANSI standards.
Longitudinal (back taper) relief (machine reamers)	Grinding the margins of a machine reamer to a slight taper of 0.0002″ (0.005mm) per inch. A slight clearance from the cutting end to the shank end of the reamer margin (land).
Back cross slide handwheel	The handwheel position for actuating the cross slide from the back of the cutter grinder.
Land interference	A condition where the heel of the land rubs against the wall of a workpiece during grinding. Insufficient relief at the heel of the land.
Milled thread chaser	A flat thread formed cutting tool used in sets for machine cutting a thread. A formed cutter requiring sharpening on the starting chamfer, along the edge, and across the face.

SUMMARY

- Primary relief, eccentric relief, secondary clearance, and rake angles are ground on milling cutters, formed chasers, and other flat cutting tools.

  - The cutters usually ground (sharpened) include slitting saws, staggered-tooth milling cutters, shell mills, carbide-insert end mills, form relieved cutters, single- and double-angle cutters, machine reamers, adjustable hand reamers, and flat cutting tools.

- The alternate narrow lands on staggered-tooth cutters, particularly slitting saws, serve to maintain the cutter width and reduce rubbing on deep slotting and sawing operations.

- ■ The cutting edge of each slitting saw tooth and the wheel head are set at center height. Primary relief and secondary clearance are produced by offsetting (raising) the wheel head.

■ Each tooth on a staggered-tooth milling cutter is alternately ground to a right- or left-hand helix. Accurate tooth grinding requires the inverted V-tooth rest blade to be positioned at the exact center of the narrow, flat grinding area on the wheel face.

- ■ Relief and clearance angles for staggered-tooth milling cutters are generated by lowering the wheel head to the required offset.

■ The wheel head offset equals the product of the sine of the relief angle and the radius of the cutter.

- ■ Side teeth are positioned at a horizontal plane. The work head is set at the required relief angle and with a slight back clearance. A flaring-cup wheel is used for grinding the side teeth.

■ Shell mills require sharpening on the periphery. The wheel head and cutter tooth height are set with a centering gage.

- ■ The end teeth of a shell mill are ground by mounting the cutter on a stub arbor secured in the work head spindle.

■ Beveled corners (cutter teeth chamfer) are ground by swiveling the work head to the required chamfer angle.

- ■ Clearance angles for chamfers on shell and face mills require two angle settings: tilting the work head, and rotating the cutter to the clearance angle. The double angle permits grinding to prevent heel drag.

■ Diamond-grit cup wheels are used for grinding carbide-tooth shell mills with positive or negative rake.

- ■ Form cutter faces are ground at the radial rake angle indicated on the cutter. Dish (type 12) wheels are formed to the gullet radius. The grinding wheel face is dressed to provide a small cutting area on the wheel.

■ A form cutter is positioned against and indexed by using a pawl on the gear attachment. The cutter is in-fed by micrometer adjustment.

- ■ Angle cutters are ground with flaring-cup wheels. On single-angle cutters, the face teeth are set in a horizontal plane. The work head is adjusted to the primary or secondary clearance angle. The angle teeth require setting the table to the specified cutter angle. The teeth for the second angle of a double-angle cutter are set by swiveling the table to the second angle setting.

■ The side cutting edges of the adjustable blades on hand reamers do the cutting. A starting taper section is ground for about one-fourth the length of the blade.

- ■ Thread chasers provide an example of flat cutting tools that require grinding of a chamfer or face, top rake, lip rake, and clearance angles.

■ Hand and machine taps are generally ground by sharpening the starting (cutting) threads. Sharpening by grinding longitudinally along the flutes reduces the dimensional accuracy of the teeth.

■ Gashing; cutting off damaged ends of cutters; reforming; and grinding drills, countersinks, and other flat and circular cutters are performed on cutter and tool grinders.

UNIT 25 REVIEW AND SELF-TEST

1. Give the formulas for calculating the offset of the wheel head to grind the primary relief (or secondary clearance angle) on the peripheral teeth of (a) slottings saws and (b) staggered-tooth milling cutters.

2. a. Describe briefly what effect the positioning of the tooth rest blade apex has on the concentric grinding of the peripheral cutting teeth edges on a staggered-tooth milling cutter.
   b. Tell how nonconcentric grinding of the cutter teeth may be corrected.

3. Refer to a manufacturer's table of wheel recommendations for grinding selected materials.
   a. Look up and record the recommendations for sharpening high-speed steel milling cutters with the following wheels: (1) type 1 straight wheel (dry), (2) type 11 flaring-cup wheel (wet), and (3) type 12 dish wheel (dry).
   b. Give the recommendations for sharpening cemented carbide cutting tools with (1) a type 11 cup wheel for offhand grinding (roughing cuts) single-point cutting tools and (2) a cup wheel for backing-off (finishing cuts) cutters.

4. a. Select a staggered-tooth side milling cutter that requires sharpening of the peripheral and side teeth.
   b. Determine the most suitable types, sizes, and specifications of the required grinding wheels.
   c. Establish the primary relief and secondary clearance angles for the cutter and the amount of offset required.
   d. Set up the workpiece, machine, and attachment to sharpen the peripheral and side teeth. Grind the lands to the specified size.
   e. Sharpen the cutter. Test for concentricity. Stone and hone the cutting edges. Check the angles.

5. a. Select a dull adjustable hand reamer or one with unevenly worn or out-of-parallel blades.
   b. Grind the outside diameter.
   c. Regrind the primary relief and secondary clearance on each blade. The reamer is to be used for reaming holes in cast iron plates.
   d. Regrind the starting taper.
   e. Check the concentricity and parallelism of the blades, margin width, and starting taper.

6. Identify two personal safety factors the cutter grinder operator observes in relation to the use of tooth rests and blades.

SECTION FIVE

Superabrasives: Technology and Processes

Characteristics, properties, and applications of superabrasive cubic boron nitride (CBN) and manufactured diamond grinding wheels, cutting tool blanks, and inserts are covered in this unit. This new technical content builds upon information and processes covered in earlier units on grinding wheel materials, systems of coding, preparation, and applications to surface, cylindrical, and tool and cutter grinding.

UNIT 26

Cutting Tools for Grinding and Other Machining Applications

OBJECTIVES

After satisfactorily completing this unit you will be able to:

- Describe properties and characteristics of regular and coated carbide, ceramic, and polycrystalline diamond (PCD) and cubic boron nitride (PCBN) superabrasives, and identify applications for grinding and other machining processes.

- Understand and apply background information on hardness, comprehensive, strength, abrasion resistance, and thermal conductivity.

- Identify basic types of CBN crystals.

- Relate resin, vitreous, metal, and electroplating bond systems, crystal concentration, and micron powders to grinding, honing, lapping and polishing and to milling, turning, shaping, and other machining processes.

- Interpret the coding system by specifying polycrystalline CBN (PCBN) cutting tools.

- Apply tool geometry and technology about rake, clearance, and cutting angles, and cutting speeds and feeds, to full-face inserts, tipped inserts, and brazed-shank PCBN inserts.

- Understand the effects of sulphurized and sulfo-chlorinated cutting oils on superabrasive cutting tools and production.

- Perform the necessary processes to prepare for machining with PCBN cutting tools.

- Follow recommended industry codes, manufacturer's specifications, and personal operator, machine tool, and tooling *Safe Practices* and correctly apply new *Terms*.

- Solve the assigned problems from the *Review and Self-Test Items*.

IMPACT OF SUPERABRASIVE GRINDING AND MACHINING

The use of superabrasive cutting materials is moving rapidly from toolroom applications into production grinding and machining of hard alloy steels, metal matrix composites, and extremely tough and abrasive materials.

This growth in applications has been made possible by parallel and accelerating developments in heavy-duty, high-speed, multi-axis machine tools, machining centers, and the impact of computer-aided design and factory integrated manufacturing systems.

Superabrasives are being widely used in grinding, honing, lapping, and polishing and for standard turning, shaping, milling, slotting, threading, and other machining processes. It is possible to grind at cutting speeds above 25,000 sfpm and to machine at rates of 10,000 sfpm and still meet high statistical process control (SPC), surface texture standards, and geometric tolerancing requirements.

Standard cutting tools with improved properties and superabrasive cutting tools are also influencing cost-effective, just-in-time (JIT), zero stock inventory developments. It is possible to exactly match production reliability and predictability (SPC) with parts inventory requirements. JIT contrasts sharply with costly, long-time stocking of parts and components.

MAJOR CATEGORIES OF METAL CUTTING MATERIALS

CARBIDE CUTTING TOOLS

Tungsten carbide is known for its high abrasion resistance, high shock resistance, and high mechanical properties. As a commercially, readily available material, coated and uncoated carbide cutting tools have had the potential of meeting 90% of the requirements for metal removal.

Tungsten carbide cutting tools are coated to increase tool wear life and to improve abrasion resistance. Combinations of titanium carbide (TiC), titanium nitride (TiN), and alumina coatings are being used with a tough carbide substrate. Additional chip control designs are being developed for ever-wider ranges of feed rates and depth of cuts ranging from heavy roughing to fine finishing cuts.

Coated carbides have high abrasion resistance, chemical stability, and lubricity properties that make these cutters adaptable for high speed machining.

CERAMIC CUTTING TOOLS

Alumina-based ceramic cutting tools, with additions of titanium oxide (TiO) or titanium carbide (TiC) are chemically stable. Ceramic cutting tools may be used at higher cutting speeds than carbide cutting tools. Also, ceramic cutting tools have increased tool life and produce a higher quality and accuracy of surface texture.

CUBIC BORON NITRIDE (CBN) CUTTING TOOLS

Cubic boron nitride (CBN) has excellent thermal shock resistance and the ability to retain its hardness at high temperatures. It combines the properties of hot-hardness and remains chemically inert (stable). CBN cutting tools are adaptable to machining hard materials in a Rockwell hardness range of Rc 45 to 70.

CBN crystals permit efficient grinding, honing, lapping, and polishing, and the machining of hardened carbon and alloy steels, tool and die steels, cast irons, and nickel or cobalt-base *superalloys*. While superalloys may not be as hard as hardened steel, they are hard-to-machine due to a combination of high strength and the tendency to deform plastically as a result of cutting forces. When grinding, CBN crystals resharpen themselves by *cleaving* or *microfracturing* to create new, sharp, cutting edges.

CUBIC BORON NITRIDE (CBN) INSERTS

CBN inserts are brazed onto standard carbide inserts and are adapted to machining hard, chilled, cast iron, hardened steels (Rc 45 and harder), nickel-cobalt steels (Rc 35 and harder), exotic alloys, fiber-reinforced polymers, composites, and laminates.

One of the newer CBN inserts consists of a polycrystalline CBN layer bonded at high-pressure and high-temperature onto a tungsten carbide substrate (Figure 26–1). This insert is functional for interrupted cuts and high cutting rates. The cutting tool (insert) is not subject to chemical attack, oxidation, or softening at high temperature.

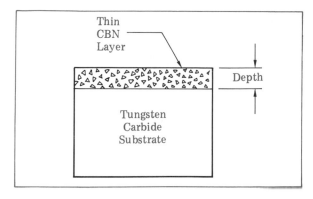

Figure 26-1 Composition of PCBN Blanks and Inserts

COMPARISON OF CBN WITH ALUMINUM OXIDE AND SILICON CARBIDE CUTTING TOOLS

CBN ranks next to the diamond in hardness and is over 2 1/2 times harder than aluminum oxide. CBN crystals also have sharp, long-lasting cutting edges that microfracture in cutting. Thus, new, sharp edges are produced continuously.

The hardness and sharpness of CBN abrasives make it possible to achieve higher material removal rates that exceed conventional aluminum oxide and silicon carbide abrasive wheels. Free-cutting CBN wheels are capable of grinding without metallurgical damage to the workpiece. By comparison, a photomicrograph of a part ground with a conventional wheel often shows a damaged surface structure.

Wheel wear resulting from high feed rates is lower in a CBN wheel as compared with rapid wheel wear of aluminum oxide and silicon carbide wheels. Lower CBN wheel wear and fewer adjustments to compensate for such wear means less machine down time and more cost-effective machining. Initial higher CBN wheel costs are offset by increased productivity.

Finish dressing and spark out passes may also be eliminated in CBN grinding. Whenever a conventional wheel is replaced by a CBN wheel, a grinding wheel of the same diameter and width and the largest abrasive grain size (that will produce the required surface finish) is selected.

POLYCHRYSTALLINE DIAMOND CUTTING TOOLS (PCD)

The polycrystalline diamond (PCD) is a synthetic cutting tool. PCD is produced by subjecting a combination of very fine diamond powder and tungsten carbide to extremely high pressure (one million psi) and temperature (around 3000°F). The diamond crystals are sintered and bonded to a carbide substrate.

Polycrystalline diamond cutting tools are used to machine *nonferrous* metals and abrasive *nonmetallic materials*. PCD cutting tools have excellent shock resistance and considerably greater tool wear life than carbide cutting tools. However, there are limitations due to the fact that polycrystalline diamonds are affected by their reaction with carbon in steels and other metals at machining temperatures above 1200°F. Polycrystalline diamond crystals react with carbon to form carbides.

In other words, polycrystalline diamonds are incompatible chemically with materials like iron, cobalt, tungsten, tantalum, and other metals. The *solubility* of the diamond in the material being machined greatly affects the cutting edges and tool wear life.

PCD tooling outperforms carbide cutting tools on alloys of brass and bronze, magnesium, and other metals that do not react to carbon; nonabrasive plastics, fiberglass epoxies, and pre-sintered ceramics. A variety of standard, commercially available, superabrasive CBN and diamond (PCD) products are illustrated in Figure 26-2.

Courtesy of NORTON COMPANY

Figure 26-2 Examples of Polycrystalline CBN and Diamond (PCD) Superabrasive Products

Table 26–1 Characteristics and Applications of Basic Cutting Materials

Cutting Tool Material	Characteristics		General Machining Applications
	Strengths	Weaknesses	
High Speed Steel	• Readily available • Versatility in multi-machining processes • Excellent shock resistance	• Limited tool wear resistance • Limited speed range capability	• General toolroom processes • Comparatively low speed operations and low horsepower requirements • Interrupted cuts
Carbide	• Currently, predominant (most versatile) cutting tool material • Excellent shock resistance	• Limited to moderate range cutting speeds	• Roughing to finishing cuts on cast irons, steels, plastics, and other exotic materials
Coated Carbide	• Operate at higher cutting speeds than uncoated carbides • Excellent versatility • High shock resistance • Good performance at moderate speeds	• Moderate speed limitation	• Identical to carbide cutting tools, but at higher cutting speed rates
Ceramic (Hot/Cold) Pressed	• High versatility • High abrasion resistance • High speed range capabilities	• Low resistance to thermal shock • Low resistance to mechanical shock	• Heavy duty resurfacing and finish machining operations on cast irons and steels
Ceramic-Silicon Nitride	• High shock resistance • Excellent abrasion resistance	• Very limited cutting tool applications	• Roughing to finishing operations on cast irons
Cubic Boron Nitride (CBN) — Grinding	• High comprehensive strength (approx. 940,000 psi) • High hot hardness • High thermal shock resistance • High thermal conductivity • High abrasion resistance • Consistent machining quality	• Limited applications and machining performance on materials below Rc 38 • Initial cost factor	• Machining processes on hardened workpieces in the Rc 45–70 range at high material removal rates • Machine hard, tough, abrasive, difficult-to-grind materials
Polycrystalline Cubic Nitride (CBN) — Tool Blanks and Tipped Inserts	• High tool wear life • Resistance to chipping and cracking • High resistance to impact • High material removal rates • Produces uniformly good surface finishes • Uniform hardness and abrasion resistance	• Low shock resistance • Solubility at high temperature machining with carbon in ferrous materials	• Cutting hard, tough, abrasive materials and hardened machine steels (Rc 45 and above) cobalt-base, high machining temperature alloys (Rc 35 and above) • Materials with tough, abrasive scale • Interrupted cuts
Diamond (Grinding Wheels)	• Excellent wear life • Produces fine quality surface finish • High cutting speeds • Fast, cool cutting		• Machining nonferrous materials and abrasive non-metallic materials • Grinding processes requiring a high quality surface finish on difficult to finish materials
Polycrystalline Diamond (PCD) Tools	• Availability in three crystal microstructures for various applications • Superior impact resistance of coarse grains • Hardness • Comprehensive strength and thermal conductivity (stability); highest of all cutting tools • Diamond layer is uniformly distributed among the matrix and wheel mass (periphery) • Machining of high quality parts and surface finish • Rate of tool wear to increased speed is lower in comparison to carbide. • Longer production runs.	• Initial higher tool costs (offset by greater production) • Machine tool capacity requirements of operating at high speeds and under severe cutting forces • Instability of cutting tool to efficiently cut ferrous materials	• Machining nonferrous and nonmetallic materials • Fine grain: exceptional fine surface finish • Medium grain: nonferrous metals and general-purpose machining • Coarse grain: interrupted cuts and milling processes

SUMMARY OF BASIC CUTTING TOOL MATERIALS

Table 26-1 summarizes the characteristics and common applications of conventional and superabrasive cutting tool materials.

PROPERTIES OF DIAMONDS AND CBN SUPERABRASIVES

Superabrasives have the ability to cut the hardest known materials. This cutting ability depends upon four main physical proporties.

HARDNESS

Hardness is the measured resistance of a material to penetration. The hardness of a superabrasive is designated according to either a *Brinell Hardness Number (BHN)*, a *Rockwell Hardness Scale* (like *Ra* or *Rc*) reading, a *Knoop Hardness Test Number (KHN)*, or a value converted from one system to another. The Knoop Hardness Test is used for extremely hard, brittle materials.

Chart 26-1A provides a visual comparison of two superabrasives (diamond and cubic boron nitride (CBN) with silicon carbide (SiC) and aluminum oxide (Al_2O_3) abrasives. Note the extreme hardness of the diamond (7,000 to 10,000 kg/m^3 on the Knoop Hardness Scale). Second in hardness is CBN with a hardness of 4,500 kg/mm^2. Third in hardness is silicon carbide at 2,500 kg/mm^2. Aluminum oxide is fourth in hardness at 2,000 kg/mm^2.

COMPRESSIVE STRENGTH

Compressive strength relates to the ability of a cutting tool to withstand cutting forces. Compressive strength is the maximum compressive force that a material can withstand before it fractures.

Chart 26-1B compares the tremendous compressive strength of the diamond (at 1065 kg/mm^2) with cubic boron nitride (720 kg/mm^2), aluminum oxide (300 kg/mm^2), and silicon carbide (58 kg/mm^2).

Chart 26-1 Comparison of Physical Properties of Superabrasives, Silicon Carbide, and Aluminum Oxide

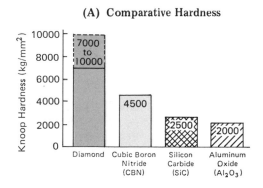

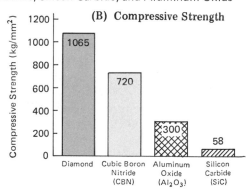

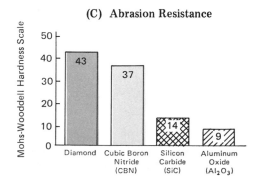

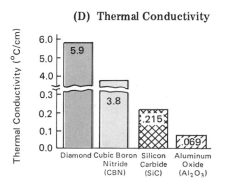

ABRASION RESISTANCE

The ability of superabrasives to maintain sharp cutting edges, to manufacture dimensionally accurate parts, and to produce high quality surfaces relates to the physical property called *abrasion resistance.*

Chart 26–1C shows the diamond as having the greatest relative abrasion resistance of (43) on a *Mohs-Woodell Hardness Scale.* This value compares with CBN (37), silicon carbide (14), and aluminum oxide (9). In other words, CBN superabrasive cutting tools have approximately 2 1/2 times greater abrasion resistance than silicon carbide cutting tools, and four times more abrasion resistance than aluminum oxide cutting tools.

THERMAL CONDUCTIVITY

Thermal conductivity deals with the transfer of heat during high cutting temperatures. Heat produced at high removal rates and under tremendous forces is dissipated rapidly through a superabrasive cutting tool material into a grinding wheel or cutting tool. A high thermal conductivity prevents the weakening of the cutting tool material and possible thermal damage to a workpiece.

Chart 26–1D shows that diamond cutting tools have approximately 1 1/2 times more thermal conductivity as compared with CBN, 27 times more thermal conductivity than silicon carbide, and 85 times more thermal conductivity than aluminum oxide cutting tools.

MANUFACTURED DIAMOND ABRASIVES

Diamond abrasives, when supported by a proper bond in a grinding wheel, are an effective cutting tool for grinding and reconditioning cemented carbide cutters. Diamond abrasive wheels are particularly adapted to the grinding of extremely hard materials due to the fact that the abrasive grains are from three to four times harder than silicon carbide grains. Also, it is possible to control the characteristics of diamonds during manufacture to meet very rigid specifications.

A brief description of the four most widely used *manufactured diamonds*, with applications, follows.

- **Regular Friable Diamond (Type RVG).** Wheels with this type of rough surface, irregular shape, free-cutting crystal are commonly used to regrind tungsten carbide cutting tools. Light feeds are used. Grinding may be done wet or dry.
- **Nickel-Coated Friable Diamond (Type RVG-W).** As the name implies, the friable diamond crystal has an added nickel coating. This coating provides a good holding surface around the crystals for a resin bond. The resin-bond, nickel-coated, friable wheel is especially suited for grinding cemented carbides under medium to severe grinding conditions.
- **Copper-Coated Friable Diamond (Type RVG-D).** This type of manufactured diamond crystal has harder properties than the regular friable diamond. The copper coating controls crystal fracturing. RVG-D abrasives are used for dry grinding tungsten carbide.
- **Blocky, Nickel-Coated Diamond (Type CSG II).** These tough crystals are designed for resin-bond wheels; wet or dry grinding. This wheel is used to grind both the extremely hard, brazed, tungsten carbide insert and an area of the steel shank of the holder surrounding the insert.

CUBIC BORON NITRIDE (CBN) SUPERABRASIVE PRODUCTS

There are at least seven basic types of cubic boron nitride (CBN) abrasives that are commercially available. Each type is manufactured for a particular bond system, machining process, rate of removal, or combination. In addition, there are *micron powders* in a variety of grades. CBN micron powders are used as loose abrasives, primarily for lapping and polishing operations.

TYPE I CBN CRYSTALS

This is a medium-toughness, *uncoated*, *monocrystalline*, CBN abrasive that is widely used in metal, vitreous, and electroplated bond systems. The term *monocrystalline* refers to a single crystal.

TYPE II CBN CRYSTALS

This CBN monocrystalline abrasive is *metal coated.* The metal coating promotes good bond adhesion and thermal dissipation during grinding. Type II crystals are used in resin-bond systems.

TYPE 500 CBN CRYSTALS

This type of monocrystalline CBN abrasive is used in electroplated-bond systems. It is engineered specially for severe production machining operations requiring high material removal rates.

TYPE 510 CBN CRYSTALS

These crystals are manufactured with a fine coating of titanium to produce a strong matrix/crystal bond. This cutting material has high metal removal rate capability.

CBN 510 is designed for metal- and vitreous-bond systems and applications requiring maximum resistance to abrasive pull-out. These qualities make this abrasive ideally suited for grinding soft steels (HRc 50 and less) and in honing processes.

Type I, II, 500, and 510 crystals have a monocrystalline structure. The monocrystals fracture along a great number of cleavage planes. In other words, monocrystals *macrofracture* a number of times during the life of a crystal. Macrofracturing is needed to keep CBN crystals sharp and free-cutting.

TYPE 550 CBN CRYSTALS

The availability of coarse size crystals up to 20/30 makes Type 550 abrasives suitable for high material removal production jobs. This tough *microcrystalline* abrasive has a higher thermal stability than other types of CBN abrasives. Type 550 crystals are used in vitrified-bond and metal-bond grinding wheels. Such wheels have excellent form holding ability and long wheel life.

TYPE 560 CBN CRYSTALS

The Type 560 crystal is metal coated (with a nickel coating) for use in resin-bond grinding wheels. The high toughness accounts for long wheel wear (as much as ten times more than other CBN products) and improved productivity.

TYPE 570 CBN CRYSTALS

This high toughness superabrasive is particularly adapted to machining operations on workpieces that are covered with heat treating processes scale or contain hard spot inclusions. Type 570 is produced by electroplating Type 550 crystals.

With the availability of coarse mesh crystal sizes, it is possible to achieve high metal removal rates and high productivity in grinding. Low wear rates makes this superabrasive valuable to form grinding. CBN wheels may be operated at ultra-high speeds (20,000 sfpm or 100 m/s).

RESHARPENING CBN 550, 560, AND 570 PARTICLES

Each dense superabrasive particle consists of thousands of micron-size crystalline regions. As the sharp edges of an abrasive particle begin to dull, the increased force exerted on the particle causes a cleavage in a submicron CBN crystalline region. This continuous process of microfracturing produces a resharpened superabrasive particle.

MACHINING APPLICATIONS OF SUPERABRASIVES

Six major classifications of metals and hardness scale values are given in Table 26–2. The appropriate polycrystalline cubic boron nitride (PCBN) or diamond (PCD) to use for either grinding or other machining processes is indicated for each metal.

CUTTING FLUIDS FOR CBN GRINDING

Dry Grinding. CBN grinding wheels (3 3/4" to 5" (95mm to 125mm) diameter wheels at 3500 rpm) operate efficiently for tool and cutter grinding and other dry grinding operations within the 3500 to 5,000 sfpm (17–25 m/s) range.

Wet Grinding. There are two basic types of grinding fluid (coolant) for CBN wheels; *water soluble* and *straight grinding oil*. Water soluble oils are used in concentrations of from 2 to 10%. Chart 26–2 shows the effect of richer concentrations of water soluble oil on CBN grinding wheel performance.

In general, the richer concentration improves the CBN material removal capability and wheel life. Heavier duty grinding operations require the use of sulphurized or sulfo-chlorinated straight grinding oils. It is important in wet grinding to *flood* the interface area between the workpiece and the grinding wheel.

Table 26-2 Applications of Superabrasives in Grinding and Machining Metals

Metals Classification		Hardness (Brinell or Rockwell)	Grinding Processes		Machining Processes	
					Polycrystalline	
			Cubic Boron Nitride (CBN)	Diamond (PCD)	Cubic Boron Nitride (PCBN)	Diamond (PCD)
Cast Iron	Gray	180 BHN +	X	–	X	–
	Ductile	200 BHN +				
	White	500 BHN +				
Hardened Steel	Tool and Die and High Speed	HRc 50 +	X	–	X	–
	Carbon					
	Stainless					
	Alloy					
Superalloys (High Temperature)	Nickel Base	HRc 35 +	X	–	X	–
	Cobalt Base					
	Iron Base					
Hard Facing Materials	Metal Base	HRc 60 +	X	–	X	–
	Carbide-Oxide Base	HRc 50 +	–	X	–	X
Aluminum Alloys	Mold Cast (Sand or Permanent)		–	–	–	X
	Die Cast					
	Wrought					
Cemented Tungsten Carbide	Tool and Die (All Grades)	HRa 84–95	–	X	–	X

Symbols: (+), Greater than the number given; (X), superabrasive used; (–), superabrasive is not used.

CONCENTRATION OF SUPERABRASIVE

CBN grinding wheels are identified according to the *concentration of contained abrasive.* This means: the *amount (percent) of superabrasive by unit volume of bond material.*

The *weight* of the superabrasive volume is expressed in *carats per cubic inch or carats per cubic centimeter.*

For example, an impregnated-bond superabrasive wheel with a concentration number of 100 weighs 72 carats per cubic inch (or the equivalent of 4.4 carats/cm^2). By volume, this CBN wheel has 25% superabrasive. Additional examples of concentration number, volume in percent and by weight, and wheel bond types are given in Table 26–3.

A high concentration of contained abrasive in an impregnated-bond superabrasive wheel has the following effects on grinding.

• The higher the concentration the greater the tool life.
• Less force is required per abrasive particle.
• The quality of surface finish is improved.

Chart 26-2 Effect of Grinding Fluids on Wheel Life and Material Removal Rates

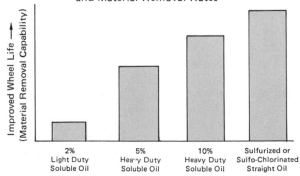

| 2% Light Duty Soluble Oil | 5% Heavy Duty Soluble Oil | 10% Heavy Duty Soluble Oil | Sulfurized or Sulfo-Chlorinated Straight Oil |

y-axis: Improved Wheel Life → (Material Removal Capability)

Table 26-3 Concentrations and Bonds of CBN Impregnated-Bond Wheels

Concentration Number	Percent of Superabrasive by Volume	Weight of Superabrasives		Wheel Bond
		Carats/in.^{3}	Carats/cm^3	
200	50	144	8.8	• Vitreous
150	37.5	108	6.6	• Vitreous
100	25	72	4.4	• Vitreous • Resinoid • Metal
75	18.75	54	3.3	• Resinoid • Metal
50	12.5	36	2.2	• Metal
25	6.2	18	1.1	• Metal (used only rarely)

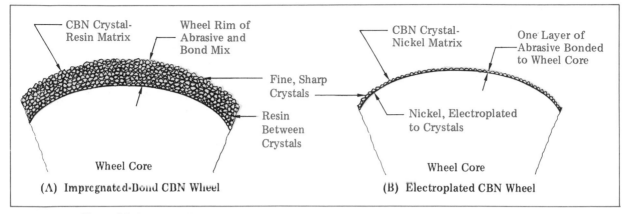

Figure 26-3 Design Features of Impregnated-Bond and Electroplated CBN Grinding Wheels

WHEEL BONDING SYSTEMS

The four basic bonding systems used to hold superabrasive grains to the periphery of the wheel core include: (1) resin, (2) vitreous, (3) metal, and (4) electroplating. A description follows of the characteristics, design features, and functions of each bond system.

CBN RESIN-BONDED WHEELS

A mixture of resin and filling agents and CBN abrasives of a selected grain size and concentration is flowed into a mold cavity around the wheel core (Figure 26-3). The mold and contents are then put under pressure and heated to 750°F (400°C). Under this pressure and temperature, the CBM abrasive, bonding agent, and wheel are bonded together. The bonding process is followed by a temperature curing cycle that strengthens the resin bonding.

Concentrations in resin-bonded wheels usually range from 125 abrasive grains to around 75. These wheels have a broad range of applications as they have free-cutting qualities and can remove metal fast. For these reasons CBN resin-bonded wheels are used for general purposes in toolrooms and for production grinding of tool steels, other steels, and superalloys.

VITREOUS- (CERAMIC-) BONDED WHEELS

These wheels have a greater range of concentration of CBN abrasive grains (from 200 down to 50) and a higher bonding strength than resin-

bonded wheels. Other characteristics of vitrified-bond CBN abrasive wheels are as follows.
- Ability to produce a wide range of good surface finishes.
- Excellent wear resistance.
- Ability to retain form and straightness.
- Porosity (open structure) of wheels to provide adequate chip clearance can be controlled.
- Permit use of coolants to prevent wheel loading.
- Wheels may be formed by crush dressing or with a rotary diamond dresser.

METAL-BONDED WHEELS

While the least used of CBN wheels, metal-bonded wheels are the toughest bonded. These wheels are functional for form, creep-feed, and internal stock removal for steels and superalloys. Metal-bonded wheels have the following additional characteristics and applications.
- Ability to hold the wheel form.
- Long cutting tool wear life.
- High heat and abrasion resistance.

Metal-bonded wheels are produced by mixing powdered metals (usually bronze) with measured amounts by type, size, and weight of CBN abrasive. This mixture is flowed into the mold cavity around the periphery of the wheel core.

The metal bond is formed as the mass is subjected to a controlled temperature and atmosphere. The melting and fusing of the powdered metals provide the high, tough, bond strength in metal-bonded wheels.

ELECTROPLATED WHEELS

Electroplated CBN wheels may be produced to any complex contour to follow the shape and size of the steel wheel core. Superabrasives are bonded to the surfaces of the grinding wheel through a nickel plating bath.

Electroplated wheels are important in grinding deep formed surfaces like slots, grooves, gear teeth, and splines. While these wheels cut at high stock-removal rates and have maximum abrasive particle exposure, the wheels have shorter wear life and produce a limited quality surface finish. Although wet grinding is preferred, electroplated wheels can be used dry. The rim depth of the CBN superabrasive is only one abrasive layer (Figure 26–3B) as contrasted with an abrasive depth of 0.08″ (2mm) with resin, metal, and vitreous bonded wheels.

ANSI SUPERABRASIVE GRINDING WHEEL SHAPES AND CODING SYSTEM

There are four characteristics of a superabrasive grinding wheel that are used in the ANSI (American National Standards Institute) identification code. The designations of ① *core shape*, ② *shape of the abrasive section*, ③ *location of the abrasive section*, and ④ *wheel modifications* are illustrated in Figure 26–4.

Figure 26–5 provides examples of the grinding face and the abrasive section for a number of CBN grinding wheels.

MICRON CBN POWDERS FOR POLISHING AND LAPPING

Loose diamond and CBN micron powders (of 50 microns or less) are used for metal polishing and lapping tool and die steels, hardened ferrous alloys, and superalloys.

CBN superabrasive powders have up to 10 times more productivity than conventional abrasive powders in lapping operations at normal operating pressures. Under extremely heavy machining conditions, aluminum oxide abrasive particles normally break down due to high temperatures up to 600°F (315°C) and pressures of 50,000 psi/3,515 kg/cm² in the cutting tool-workpiece interface zone.

Recommended applications of different superabrasive bond systems, wheel types, and micron

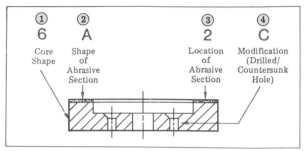

Figure 26–4 ANSI Superabrasive Wheel Shape Coding System

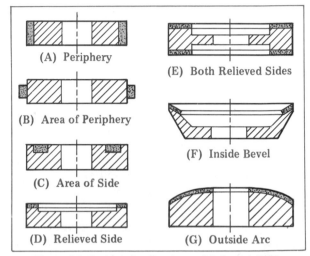

Figure 26–5 Abrasive Sections of Selected CBN Grinding-Wheel Shapes

powders are given in Table 26–4. Machine tools or representative grinding or other machining processes for each abrasive product are identified, including a number of metals that may be machined.

HONING WITH SUPERABRASIVES: ADVANTAGES

The following are the major advantages of using PCBN superabrasives in honing stones over such conventional abrasives as aluminum oxide and silicon carbide.

- CBN honing stones have very high honing and very slow wear rates (up to 100 times longer than regular abrasives).
- CBN stones hone quietly without high-pitched noises.
- The close match between bond and superabrasive wear rates prevents the premature loss of the superabrasive.

Table 26–4 Grinding and Machining Applications of CBN Superabrasives According to Type and Bond Systems

Applications (Machine Tools/Processes)	Bond System	Recommended Product	Workpiece Material
Tool and cutter grinding	Resin	Type II, 560	• Hardened steels • Difficult to grind steels • Wear-resistant alloys • Superalloys
Surface grinding	Resin	Type II, 560	
Jig grinding	Electroplated	Type I, 500, 570	
Form grinding	Vitrified, Metal	Type I, 510, 550	
I.D. grinding	Resin	Type II, 560	
	Electroplated	Type I, 500, 570	
	Vitrified, Metal	Type I, 510, 550	
Hob grinding	Resin, Metal	Type II, 510, 550, 560	
Vertical spindle rotary table surface grinding	Metal	510, 550	
Creep feed grinding	Vitrified Metal, Resin, Electroplated	Type I, Type II, 500, 510, 550, 560, 570	• Mild to hardened steels • Superalloys
Superabrasive machining	Vitrified, Metal, Resin, Electroplated	550, 560, 570	• Mild and hard steels • Nickel- and cobalt-base alloys
Honing	Metal	510, 550	• Mild steel • Hardened steel
Lapping and polishing		Micron powder	• Hardened steels

Adapted from *General Electric Company, Specialty Materials Department.*

- Faster honing due to increased metal removal rates (almost double the rate of conventional honing stones).
- CBN honing stones remain free cutting at lower machining pressures.
- Generally, less heat is generated with CBN honing stones.
- Uniformity of CBN stone quality ensures consistent production time from stone to stone.

POLYCRYSTALLINE CBN CUTTING TOOLS

Cutting tools of superhard polycrystalline CBN (PCBN), when supported by the strength and toughness of a tungsten carbide *substrate*, are excellent for the machining of hardened steels, abrasive cast irons, and tough superalloys. The term *substrate* is used to identify the material below a coating. The substrate provides the toughness and resistance to tool deformation and thermal shock.

Superalloys refers to materials that are not as hard as hardened steel but are hard-to-machine due to a combination of high strength and the tendency to deform plastically as a result of extreme cutting forces and heating.

Polycrystalline CBN (PCBN) cutting tools have high metal removal rates, long tool wear life, and may be used for heavy duty and interrupted cuts.

DESIGN FEATURES OF PCBN TOOL BLANKS AND INSERTS

As stated earlier, a *PCBN tool blank and insert* consists of a layer of CBN bonded *(sintered)* to a cemented tungsten carbide *substrate.*

Tool blanks are commercially available for specific operations. Three basic classifications of PCBN cutting tools are: (1) full-face inserts, (2) tipped inserts, and (3) brazed shank inserts.

Tool blanks and inserts are available in different sizes and shapes, including full circle and circle quadrants, triangular, and square. Blanks and inserts are produced in varying thicknesses of PCBN layer and depth of carbide substrate. The coding system that is used for specifying PCBN inserts is shown in Table 26–5.

Table 26-5 Coding System for Polycrystalline Cubic Boron Nitride Cutting Tools

① ② ③ ④   ⑤ ⑥ ⑦
B S P U - 4 3 2

Position in Code	Feature or Dimension	Code
①	Product Identification	**B** – PCBN Blank
②	Shape	**S** – Square **R** – Round **T** – Triangular
③	Rake (6° standard; other angles to be specified)	**N** – Negative **P** – Positive Rake
④	Insert Tolerances	**G** – Precision **U** – Utility
⑤	Blank Size (Outside)	Number of eights (1/8″)
⑥	Blank Thickness	Number of sixteenths (1/16″)
⑦	Nose Radius (applies only to square and triangular inserts)	Number of sixty-fourths (1/64″)

Note: In the example, the code indicates the following. A PCBN blank (**B**) that is square (**S**), has a positive 6° standard rake angle (**P**), and meets utility quality insert tolerances (**U**) is required. The PCBN blank is 4/8 or 1/2″ square (**4**), 3/16″ thick (**3**), and has a nose radius of 2/64 or 1/32″ (**2**).

FULL-FACE INSERT

The *full-face insert* is designed with a layer of PCBN bonded to a substrate (cemented carbide). As the cutting edges dull, new, sharp, cutting edges are produced by *downsizing*, (regrinding each cutting edge of the insert).

TIPPED INSERTS

The *tipped insert* fits into a pocket that is machined in the carbide substrate. The insert is brazed in place. These inserts have only one cutting face (edge) that when dulled may be reground any number of times.

BRAZED-SHANK CUTTING TOOLS

This type is similar in design to standard brazed insert cutting tools, except that a CBN blank is used instead of any other cutting tool material. A CBN blank of a particular shape and size is brazed into a machined pocket in the shank of the toolholder.

ADVANTAGES OF PCBN SUPERABRASIVE CUTTING TOOLS

Chart 26-3 provides a summary of the advantages of machining with PCBN cutting tools. The top row of blocks provides a list of the major properties of PCBN superabrasive cutting tools. The second row identifies advantages. Note that all blocks are interlocked. This means every PCBN property is a factor in establishing the working characteristics of a superabrasive material cutting tool. At the same time each factor enters into the determination of the total advantages of machining with PCBN cutting tools in contrast with other standard cutting tools.

Chart 26-3 Properties and Advantages of Polycrystalline CBN Cutting Tools

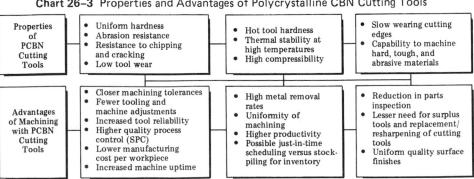

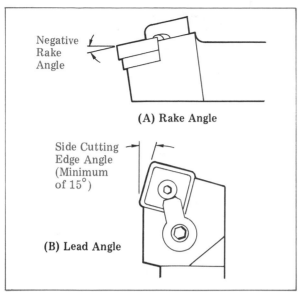

(A) Rake Angle

(B) Lead Angle

Figure 26-6 Recommended Rake and Lead Angles for PCBN Cutting Tools

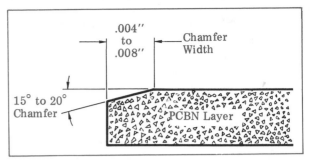

Figure 26-7 Recommended Preparation of Cutting Edge for PCBN Cutting Tools

PCBN CUTTING TOOL GEOMETRY

Cutting characteristics, quality of surface finish, dimensional accuracy, cutting speeds and feeds, and tool life, depend on correct rake, clearance, and cutting tool angles. The following tool geometry recommendations apply to PCBN cutting tools.

- Select negative-rake cutting tools (Figure 26-6A) for machining operations where the cutting tool must withstand heavy cutting forces.

- Use a lead or side-cutting edge angle of 15° or larger (Figure 26-6B). Producing a thinner chip along a wide cutting edge section reduces tool loading.

- Hone a radius or slightly chamfer the cutting edge of the superabrasive cutter. A very fine grit diamond stone is used to hone a stronger cutting tool point. A honed radius produces a high quality surface finish and provides for the wide distribution of heat.

- Use a chamfer angle of 15° x 0.008" (0.20mm) for roughing operations on hardened steels, and for roughing and finishing cuts on cast iron and hard facing alloys (Figure 26-7). No chamfer dimensions are required for machining superalloys. The recommended chamfer dimensions for finishing operations on hardened steels are: 20° x 0.004" (0.10mm).

Table 26-6 Recommended Cutting Speed and Feed Rates for Polycrystalline CBN Cutting Tools (Partial Table)

Workpiece Materials			Cutting Speed		Feed Rate	
Classification	Hardness Range	Example of Material	ft/min	m/min*	in./rev	mm/rev
Cast Irons (Soft)	180–240 BHN	Pearlitic Gray Iron	1500–3000	450–900	0.006–0.025	0.15–0.60*
Hardened Ferrous Metals	Greater than Rc45	Hardened Steels Hard, Chilled, Cast Iron	250–350	75–100	0.006–0.020	0.15–0.50*
Superalloys	Greater than Rc35	Alloys: Nickel and Cobalt Base (Examples: Stellite, Inconnel)	650–800	200–250	0.006–0.010	0.15–0.25*
Flame-Sprayed Metals	Greater than Rc45	Hard Facing (Examples: Nickel, Chromium)	200–350	60–100	0.006–0.012	0.15–0.30*
Cold-Sprayed Metals	Less than Rc45		350–500	100–150	0.006–0.013	0.15–0.30*

*Rounded-off metric speeds and feeds

PCBN CUTTING TOOL SPEEDS, FEEDS, AND DEPTH OF CUT

Cutting speeds, feeds, and depth of cut control material-removing costs and cutting-tool life. These directly affect *tool costs* and *tool-change* (maintenance and setup) costs.

Manufacturer's tables provide guidelines for maximum cutting speeds, feeds, and heaviest depths of cut for PCBN cutting tools, consistent with the ruggedness and capability of the machine tool. Tables 26–6 relates to recommended cutting speeds and feed rates for general machining operations when using PCBN cutting tools. Handbook tables and technical bulletins provide a more complete coverage.

FACTORS AFFECTING PCBN CUTTING TOOL LIFE

During machining with PCBN cutting tools, continuous, discontinuous and segmental chips are produced. The material being cut is sheared (deformed) just ahead of the cutting tool edge. On ductile metals, there is a continuous, plastic flow of chips. Discontinuous (segmented) chips are formed when the chips produced in machining hard, brittle metals rupture.

The flow or movement of metal chips along the interface of the cutting tool and workpiece affects the rate of wear. Wear usually takes place on the cutting tool nose, flank, and a short distance from the chip-tool interface (known as *crater wear*). Chipping, fracturing, thermal cracking, and abrasive wear (as for cast irons) are other forms of tool wear. Examples of cutting tool wear are illustrated in Figure 26–8A, B, and C.

How to Prepare for Machining with PCBN Cutting Tools

Machine Selection and Tooling Setup

STEP 1 Select a heavy-duty machine tool with adequate horsepower to fully utilize the potential of CBN cutting tools.

STEP 2 Select the appropriate form and size of PCBN insert and cutting toolholder.

STEP 3 Mount and secure the toolholder with minimum overhang.

Preparation of the Cutting Tool

STEP 1 Install the PCBN insert in a clean pocket in the toolholder.

STEP 2 Clamp the insert securely.

Caution: Use a chipbreaker to distribute the clamping forces. Avoid clamping directly on the PCBN layer.

STEP 3 Check the wear on the cutting tool insert at regularly planned intervals.

Note: Inserts are reground as soon as there is any dulling or inserts are replaced according to a prearranged schedule.

STEP 4 Set the toolholder with the cutting tool on center.

STEP 5 Set the side-cutting edge angle (usually at a lead angle up to 45°).

Note: The exiting edge of a cut with superabrasive PCBN must be chamfered. The depth of the chamfer equals the depth of cut.

Note: Use as large a nose radius as the workpiece permits.

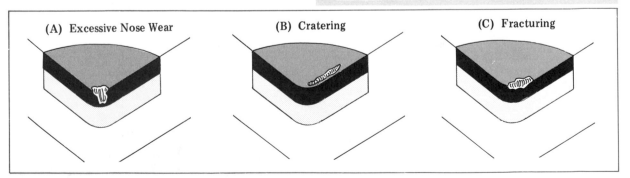

Figure 26–8 Cutting Problems that Affect Cutter Wear Life, Statistical Process Control (SPC), and Productivity

Setting Speed and Feed Rates

STEP 1 Check the manufacturer's recommendations or a handbook table. Select the speed range according to the material to be machined, nature of the operation, and grade of the PCBN cutting tool or insert.

STEP 2 Use a feed rate based on utilizing the full potential of the PCBN cutting tool and tool wear life.

STEP 3 Select a maximum depth of cut to produce heavy chips that rapidly carry away heat generated during machining.

Note: Factors such as machine rigidity, capacity of machine tool, and quality of surface finish are considered when selecting depth of cut and feed rate.

Selecting and Applying Coolants

STEP 1 Determine whether the machining process requires a coolant to remove heat from the cutting zone to retard cutter wear.

Note: Milling processes are performed *dry*.

Note: Use a coolant for very deep cutting operations to flow chips and heat away and for heavy cuts with tipped inserts or brazed-shank cutting tools.

STEP 2 Check the coolant system and the condition of the coolant. Bring the coolant up to specifications, if needed.

STEP 3 Direct the flow and quantity of cutting fluid to flood the area where the cutting tool and workpiece interface.

Machine Tool and Personal Safety

STEP 1 Adjust and secure all splash and chip guards.

STEP 2 Set all machine enclosures in required positions.

Caution: All personal and machine safety precautions, industrial codes, and manufacturer's specifications must be observed before start-up, during machining, and at close-down.

Safe Practices in Superabrasive Grinding and Other Machining

- Avoid clamping PCBN cutting tools directly on the superabrasive layer.

- Regrind (resharpen) or replace superabrasive inserts before the normal wear limit to prevent rapid deterioration of the cutting edges.

- Chamfer the exiting edge on the workpiece, especially for heavy-duty machining to prevent excessive burring at the end of the cut.

- Recheck the position, clearance area, and the security of all tooling; rates, speeds, and depth of cut; and all guards and safety trips, before start-up.

- Direct the coolant (when wet grinding) to direct and to flood the work-cutter interface area.

- Mount CBN grinding wheels to be within a maximum runout limit of 0.001″ (0.20mm); then true and dress the wheel. Note that electroplated-bond wheels do not require dressing or truing.

- Check wheel performance regularly for cutting action, metal and chip removal rates, and quality of surface finish.

- Extend the wear life of honing stones by using the appropriate grit, bond, abrasive concentration, and honing cutting oil.

- Use negative rake PCBN cutting tools and a lead or side-cutting edge angle greater than $15°$.

- Select the heaviest depth of cut and feed rate possible when using PCBN cutting tools. The rapid production of heavy chips causes up to 75% of the heat generated to be carried away from the cutting area; the balance, through the cutting tool.

- Use safety goggles and observe all precautions recommended by machine tool, superabrasive, and other suppliers, and industrial safety codes.

TERMS USED WITH SUPERABRASIVES: TECHNOLOGY AND PROCESSES

Cutting Tool—Chip Interface	An area of contact that is bounded by the cutting tool and the chip.
Cleavage	The splitting (fracturing) of an abrasive grain along one or more internal planes.
Concentration of Contained Abrasive	A quantitative measure of abrasive within a unit mass of a bonding material.
Conventional (Standard) Abrasives	Generally refers to silicon carbide and aluminum oxide abrasives.
Crater Wear	A deformed area on the top of the cutting tool near the cutting edge. Wear produced by the friction and abrasion of chips flowing over the cutting tool.
Friability	The tendency of an abrasive grain to fracture from forces produced by grinding.
Impregnated Bond	The bonding of superabrasive grains (that are uniformly distributed throughout the bond material) around the rim of a grinding wheel.
Microfracture	Cleavage of an abrasive grain along a principal plane.
Microcrystalline	Micron size crystals.
PCBN and PCD Superabrasives	Abbreviations for polycrystalline cubic boron nitride (PCBN) and polycrystalline diamond (PCD) wheels and other cutting tools.
Polycrystalline	A crystalline grain composed of randomly oriented small crystals that are identifiable by microscope.
Resin Bond	A hot-pressed phenolic resin that is used for impregnated bonding of superabrasive materials to the periphery of the grinding wheel.
Substrate (Superabrasives)	A tough, strong, cemented carbide base onto which a layer of polycrystalline superabrasive is bonded.
Superabrasives	Diamond and cubic boron nitride grains and grinding wheels and PCBN and PCD cutting tool inserts and blanks.
Vitrified Bond	The securing of superabrasive grains around a grinding wheel rim by sintering glassy materials.

SUMMARY

- Superabrasives are used for grinding, honing, lapping, polishing and superfinishing processes and for shaping, turning, milling, form cutting, sawing, slotting, threading, and other machining processes.
  - General categories of cutting tool materials include: high speed steels, coated and uncoated carbides, ceramics, cubic boron nitride (CBN), natural diamonds, and polycrystalline CBN and PCD.
- The cutting ability and tool wear life of superabrasives depend on four major physical properties: hardness, compressive strength, abrasion resistance, and thermal conductivity.
  - Manufactured diamond abrasive grains, supported in a proper sub strate, are three to four times harder than silicon carbide.
- Diamond grinding wheels are produced by using regular friable, nickel-coated friable, copper-coated, and blocky, nickel-coated diamond crystals and an appropriate bond.
  - CBN abrasive crystals are produced to meet the requirements of different bond systems, machining processes, material removal rates, quality of surface finish, and dimensional and geometric tolerances. CBN crystals are designated by *type;* such as: I, II, 500, 510, 550, 560, and 570, and as micron powders.
- The concentration of superabrasive by unit volume of bond material influences tool wear life, the free cutting ability of the grinding wheel, and the quality of surface finish.
  - CBN grinding wheels are used for dry or wet grinding. Water soluble and sulphurized straight grinding oils are commonly used for wet grinding. Generally, richer concentrations of soluble oils permit higher metal removal rates and increase wheel wear life.
- Superabrasive grains are bonded to a wheel core by resin, vitreous, metal, and electroplating bonding systems.
  - CBN micron powders are designed for lapping and polishing operations on hardened, tough materials under extremely high pressure and temperature in the work zone.
- Superabrasive honing stones have high material removal rates, slow tool wear, are free cutting at low machine pressure, and generate less heat.
  - PCBN cutting tools are commercially available as full-face inserts, tipped inserts, and brazed-shank inserts.
- Tool geometry for rake, side cutting edge angle, and cutting edge chamfer angle, influence product quality and machining characteristics of PCBN cutting tools.
  - The cutting (rupture) of a material and flow of chips on a cutting tool produces wear on the nose, flank, and a short distance from the work-tool interface area. Cratering, chipping, thermal cracking, and abrasion wear, are common forms of tool wear.
- Preparation of machining with PCBN cutting tools requires specific procedures for machine selection and tooling setup, cutting speeds and feed rates, selecting and applying coolants, and (importantly) following manufacturer's safety guidelines, industrial safety codes, and personal and machine tool safe practices.

UNIT 26 REVIEW AND SELF-TEST

1. List (a) four abrading (grinding) processes and (b) four machining processes that are performed by using either conventional or superabrasive cutting tools.

2. a. Refer to a Table of Properties of CBN and Conventional Abrasives.
   b. Give the Knoop Hardness value of cubic boron nitride (CBN) super-abrasives and conventional aluminum oxide abrasives.
   c. Make a simple statement comparing the hardness of these two abrasives.

3. a. Explain briefly what the property of thermal conductivity means in relation to CBN cutting tools.
   b. State the thermal conductivity value of a CBN cutting tool in relation to a silicon carbide cutting tool and compare the two materials.

4. a. Explain the meaning of the term *concentration of a contained abrasive.*
   b. Tell what effect increasing the concentration of the abrasive grains in an impregnated-bond superabrasive wheel has on the grinding process.

5. a. Describe briefly how resin-bond wheels are produced.
   b. Name two properties of vitrified-bond CBN abrasive wheels.

6. Identify three different workpiece materials that are best ground with super-abrasive grinding wheels.

7. State three advantages of using polycrystalline CBN honing stones in comparison with aluminum oxide honing stones.

8. Identify two materials for which CBN micron powders are ideally suited for polishing and lapping.

9. Describe the function of a tungsten carbide substrate for PCBN cutting tools.

10. Give two reasons why the point of a PCBN cutting tool is honed to a small radius.

11. Name four different types of PCBN cutter wear.

12. a. Secure a Table of Speeds and Feeds for Machining with PCBN Cutting Tools.
    b. Locate and give the range of cutting speeds in ft/min for machining (1) hard cast iron parts (Rc 50) and (b) stellite superalloy parts (Rc 30).
    c. Give the feed rates in mm/rev for machining (1) soft gray iron castings (200 BHN) and (2) nickel-base superalloy parts (Rc 35).

13. State two reasons why diamond grinding wheels are used for grinding cemented carbide tools.

14. Identify each feature and give each dimension of a PCBN cutting tool insert that is specified as follows.

B T N U – 6 4 4

15. Simulate setting-up procedures for machining with PCBN cutting tools. *Note:* Either demonstrate the procedures at the workplace or list the major areas in which clusters of related steps are taken by the machine operator.

16. List three personal safety practices that must be followed before any grinding or other machining operations are performed with PCBN and PCD cutting tools.

SECTION ONE

Precision Machine Tools and Tooling

Three of the most widely used machine tools for the laying out, machining, and measuring of production tooling include jig boring, jig grinding, and universal measuring machines. Older models of these high-precision machines are being retrofitted for NC and CNC modes of control, operation, and readout. Tool-making techniques are examined in this section for the building and measuring of tooling commonly used on conventional and NC machine tools.

UNIT 27

Jig Boring, Jig Grinding, and Universal Measuring Machines and Processes

OBJECTIVES

After satisfactorily completing this unit, you will be able to:

- Relate major design features, functions, and applications of jig borers, jig grinders, and universal measuring machines to precision machining and measurement processes.
- Identify machine accessories, including precision boring heads, locating microscopes, digital readout, rotary table, micro-sine table, index center, and measuring devices.
- Understand basic horizontal (planetary) and vertical grinding processes and wheel dressing attachments; including, radius, angle, cross-slide, spherical socket, and pantograph types.
- Determine functions and applications of universal measuring machines and accessories such as bidirectional gaging systems, TV microscope, and small angle divider.
- Simulate by production planning all steps to perform the following processes.
  - Accurately Positioning and Setting Up a Workpiece on a Jig Borer.
  - Using a Locating Microscope.
  - Setting Up and Using a Jig Borer.
  - Dressing a Jig Grinder Wheel.
  - Grinding Holes on a Jig Grinder.
  - Grinding a Tapered Hole.
  - Grinding Shoulder Areas.
- Follow recommended *Safe Practices* for personal, machine tool, and workpiece protection.
- Correctly use each new *Term*.

415

PRECISION JIG BORING MACHINE TOOLS, PROCESSES, AND ACCESSORIES

The jig borer is especially adapted to machine tool operations requiring extreme dimensional measurement accuracies. Linear and machining accuracies are expressed in terms of millionths of an inch (0.025μm). Repeatability in maintaining dimensional accuracy and complete positioning and machining controls make the jig borer an ideal machine tool for precision machining.

Jig borers are widely used in toolrooms for the construction of jigs and fixtures, punches and dies, special cutting tools, gages and other fine machine work. The basic processes include centering, drilling, reaming, through and step boring, counterboring, and contouring. Holes as small as 0.013″ (0.33mm) may be drilled without spot drilling. The spindles are rigidly constructed for power and strength in taking heavy hogging cuts. For example, the power, strength, and rigidity of a hole-hog tool in taking a 1/2″ (12.7mm) cut on a jig borer is shown in Figure 27–1.

Since a great variety and number of toolroom applications require constant tool changing, a special shank design permits easy removal and replacement of cutting tools and noncutting, locating, and positioning devices. The Moore taper-shank, fast-lead square thread tool illustrated in Figure 27–2 ensures accurate, quick-change tooling and permits the removal and replacement of a tool to repeat the hole size on any diameter. Most jig borers have an infinitely variable spindle speed. Speed is read directly on a tachometer.

JIG BORING TOOLS AND ACCESSORIES

Precision Boring Chuck. The most common accessory on the jig borer is the precision boring chuck. Hole diameters may be increased 0.001″ (0.02mm) per graduation. Vernier settings permit hole diameter increments of 0.0001″ (0.002mm) per graduation.

Microbore® Boring Bars. A Microbore® boring bar series is available for jig boring work. The Microbore® boring bar, illustrated in Figure 27–2, is used for roughing and finishing opera-

Courtesy of MOORE SPECIAL TOOL COMPANY INC.

Figure 27–1 Hole-Hog Tool Taking a 1/2″ (12.7mm) Cut on a Jig Borer

tions on a jig borer where close dimensional limits and accuracy are to be maintained. The micrometer dial permits size changes to be read easily. These boring bars are available in overlapping sizes from 3/4″ to 4 5/16″ (19mm to 109.5mm) diameter. Complete bars are furnished in high-speed steel and carbide.

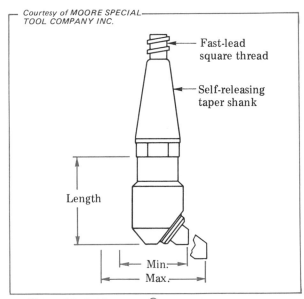

Courtesy of MOORE SPECIAL TOOL COMPANY INC.

Fast-lead square thread

Self-releasing taper shank

Length

Min.

Max.

Figure 27–2 Microbore® Boring Bar for Roughing and Precision Finishing Operations on a Jig Borer

Chucks and Collets. Key and keyless chucks are used. Chuck shanks may be straight fitted for insertion in collets or tapered for easy mounting in the quill. Collets are designed for use with spotting drills, conventional drills, reamer drills, and end reamers. The hole tolerance on standard-quality collets is +0.0005″ (0.01mm) and -0.0000″. The hole concentricity of the shank to the collet hole is 0.001″ (0.02mm). The precision quality is finer. The hole tolerance is +0.00005″ (0.001mm); the hole concentricity, 0.0002″ (0.005mm). Collet hole sizes range from 1/8″ to 49/64″ diameter. Collet sets are also available in metric sizes in standard and precision quality.

Tapping Heads. One widely used model of tapping head permits the tapping of holes up to 5/16″ (6.4mm) in mild steel and 1/4″ (6.4mm) in tool steel. The tapping head permits tooling without changing the position of the workpiece. This instant-reversing, speed-reducing tapping attachment is held in a standard collet or drill chuck.

Locating Microscopes. Jig borer setups and other workpiece setting needs require the use of a locating microscope. The microscope is readily adapted to locating edges, irregular shapes, and holes that are too small to be located with an indicator.

The microscope (Figure 27–3), with suitable attachments, is interchangeable for application on the jig borer, jig grinder, and universal measuring machine.

Digital Readout and Printer. The digital readout and printer accessory verifies measurement accuracy for the X and Y axes to assure precise hole location. Readout modes are available for inch-standard screw, metric-standard screw, or switchable inch/metric screws. Readout accuracies to 0.00005″ (0.0012mm) are provided. The X and Y channel readouts are ± through seven digits.

The digital readout accessory includes positive backlash take-up, manual zero reset, and entering of slide travel distance for reference to readout. Visual signals indicate problems of digital readout overflow and loss of power, when all numbers return to zero and there is no counting until the unit is reset.

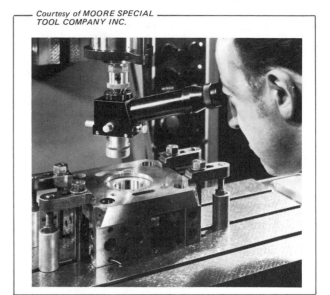

Figure 27–3 Application of a Locating Microscope on a Jig Borer

TABLE MOUNTING ACCESSORIES

Center and Micro-Sine Table. A tailstock and adjustable center permit work on vertically mounted rotary tables to be held between centers. Boring, grinding, inspection, and other operations may be performed. The tailstock is adjustable laterally and vertically.

A *micro-sine table* is a jig borer accessory used to position and hold workpieces for accurately machining or measuring angles. A set of adjusting rods permits angular adjustments for fine settings from 0° through 95°.

Precision Index Center. This table mounting accessory facilitates jig boring, milling, grinding, and other work processes requiring the workpiece to be rotated accurately on centers.

Measuring and Locating Accessories. The standard indicator set consists of an indicator, indicator holder, edge finder, and line finder. Indicators are of the dial indicator type with readings to within 50 millionths of an inch (0.00005″). Metric dial indicators are also available.

Figure 27-4 Tilting Rotary Table for Jig Boring Work

Rotary Tables. Standard, precision, and ultraprecise rotary tables permit indexing within tolerances of ±12 seconds, ±6 seconds, and ±2 seconds of arc, respectively. Rotary tables are available for hand-operated mechanical indexing. Rotary tables for NC and CNC machine tools are driven by stepping motors that eliminate mechanical indexing.

A *tilting rotary table* (Figure 27-4) is available with a tilt axis dial and rotary axis dial capable of direct readings to 1 second of arc. A disengageable worm permits quick alignment during setup and engagement for precise angle positioning.

How to Accurately Position and Set Up a Workpiece on a Jig Borer

Aligning a Work Edge by Using an Indicator

Clean the base of the workpiece and jig borer table. Check for burrs and nicks.

STEP 2 Position the workpiece on parallels and as near the center of the table as practical.

STEP 3 Select and mount an indicator holder in a collet in the spindle. Add the indicator to the holder so that the leg almost touches the edge to be indicated.

STEP 4 Bring the indicator leg into contact with the machined edge to produce a reading on the dial face.

STEP 5 Turn the machine spindle slowly in one and then the opposite direction for only a fraction of a turn. Stop the turning when the highest reading is registered.

Note: This step is taken to establish that the axis of the dial indicator is at 90° to the work edge.

STEP 6 Turn the dial indicator scale to the zero position.

STEP 7 Move the table slowly and carefully longitudinally. Note any variation in the dial indicator reading and the direction (±) of movement. The setup for aligning an edge parallel to the longitudinal (X) axis by using a dial indicator is shown in Figure 27-5.

STEP 8 Use a soft-face hammer to tap the workpiece gently away from the dial indicator.

Note: Any abrupt force or jamming against the leg of the indicator may cause damage to the instrument.

Figure 27-5 Aligning an Edge Parallel to Longitudinal (X) Axis by Using a Dial Indicator

STEP 9   Tighten the setup when a zero reading is indicated.

STEP 10   Recheck the alignment to be sure the workpiece has not shifted during tightening.

Aligning a Work Edge by Using Parallel Setup Blocks

STEP 1   Set up the straight edge on the front or back face of the table.

STEP 2   Make the usual checks for burrs, nicks, and cleanliness of surface on each parallel or setup block, table, and workpiece.

STEP 3   Place two setup blocks between a machined edge of the workpiece (either on the front or back side, depending on the nature of the operation to be performed) and the location of the straight edge.

STEP 4   Select appropriate clamps, screws, and heel rests. Secure the workpiece to the table.

STEP 5   Recheck the setup to see that each setup block or the workpiece has not shifted. Then, remove the setup blocks.

Aligning a Work Edge by Using a Parallel

STEP 1   Proceed to set up the workpiece on parallels and to mount a dial holder and indicator in the spindle.

STEP 2   Adjust the dial indicator with the point as close as possible to the center of the machine spindle.

STEP 3   Bring the indicator into contact with the edge of the workpiece until there is a reading.

STEP 4   Test to see that the indicator is at a right angle to the edge of the workpiece (Figure 27–6). Set the dial reading at zero.

STEP 5   Raise the dial indicator and spindle so that the point clears the top edge of the workpiece (Figure 27–7).

STEP 6   Rotate the spindle 180°. Secure a parallel against the edge of the workpiece.

STEP 7   Turn the spindle 180°. Lower the spindle carefully. Note any reading change from the zero setting. Adjust the table longitudinally for any variation.

> Note: A zero reading in both indicator positions (180° to each other) indicates the axis of the spindle and the edge of the workpiece are aligned.

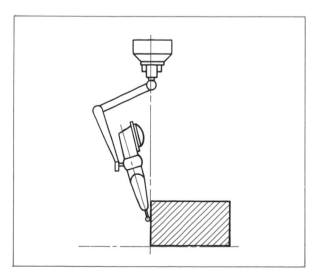

Figure 27-6 Dial Indicator (Locator) Set at Right Angle to Edge of Workpiece

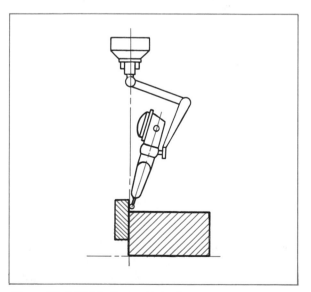

Figure 27-7 Dial Indicator Raised to Clear Workpiece and Rotated 180° (Setting Checked against Face of Parallel)

DESIGN FEATURES OF A LOCATING MICROSCOPE

A locating microscope is used when small or partial holes, slots, or irregular contours serve as reference points. A locating microscope with magnification power of 20X or 50X locates edges, contours, and holes that are too small to be indicated with a dial indicator.

The reticle on the microscope has a number of concentric circles and two pairs of cross lines. These vary between 0.0025″ on the 20X magnification to 0.001″ on the 50X magnification. On the 20X microscope, six concentric circles range in diameter from 0.005″ to 0.030″ in increments of 0.005″. Another seventeen circles continue from 0.030″ to 0.200″ in increments of 0.01″. Finally, eight additional concentric circles are spaced 0.020″ apart from 0.200″ to 0.360″.

How to Use a Locating Microscope

STEP 1   Mount the locating microscope with an adapter in the machine spindle. Determine the magnification (20X or 50X).

STEP 2   Mount, position, and lightly clamp the workpiece to the table.

STEP 3   Position the hole, edge, or contour within the range of the microscope.

STEP 4   Determine the position of the workpiece in relation to the concentric circles in the reticle.

Note: The cross lines are used to zero in on the intersection of scribed lines on the workpiece when identifying the center point.

STEP 5   Move the table longitudinally or the cross feed travel handwheel to position the workpiece concentrically with the concentric circles within the microscope.

How to Set Up and Use a Jig Borer

Positioning the Workpiece and Drilling

STEP 1   Place the work on parallels. Align the work with respect to X and Y coordinates.

STEP 2   Align the axis of the machine spindle with the starting reference point.

STEP 3   Set the longitudinal and cross feed micrometer collars at zero.

STEP 4   Determine the machining sequence steps and tooling changes in advance.

STEP 5   Position the spindle at the X and Y axes for the first hole. Center drill or use a spotting drill.

STEP 6   Position the spindle and center drill at the location of each hole. Center or spot drill each hole location.

STEP 7   Return the spindle to the starting point. Make a tool change.

STEP 8   Position and drill all holes of a particular size. Change drill sizes as may be required.

STEP 9   Check the work alignment and the setup to be sure the workpiece does not shift if any heavy rough machine drilling takes place.

Boring Concentric, Parallel Holes

STEP 1   Select a Microbore® boring head or a boring chuck and a straight or offset boring tool. Replace the drill chuck with the boring chuck and cutting tool or boring bar setup.

STEP 2   Set the cutting speed (spindle RPM) according to the job requirements for rough boring.

STEP 3   Take a series of roughing cuts. Rough bore all holes to within 0.003″ to 0.005″ (0.08mm to 0.1mm) of finish size.

STEP 4   Replace the rough boring tool with a finish boring tool for the final cut. Increase the spindle speed. Decrease the tool feed.

STEP 5   Use a solid plug, leaf taper gage, or other measuring instrument to check the bore size. The surface finish must meet the specified requirements.

PRECISION JIG GRINDING MACHINE TOOLS AND PROCESSES

The jig grinder is used to grind straight and tapered holes and to grind contour forms. Such forms combine straight, angle, round, radii, and tangent surfaces. Since a jig grinder is an abrasive machining machine tool, the process may be used on soft or hardened metals.

Jig grinders are widely used in toolrooms for finishing punches, dies, jigs, fixtures, special cutting tools, and gages to exact size and high quality surface finish. Jig grinding eliminates many earlier hand fitting processes.

GRINDING IN A HORIZONTAL PLANE

Jig grinders provide versatility in planetary jig grinding in a horizontal plane. They are used for out-feed, wipe, chop, plunge, taper, and shoulder grinding.

In *out-feed grinding* (Figure 27–8), the grinding wheel spindle moves in a planetary path at a slow rate of rotation. The axis of the wheel spindle moves at a high grinding wheel speed within the planetary path. The diameter of a hole is enlarged by continually *out-feeding* the wheel while grinding.

In *wipe grinding*, the workpiece is fed past the wheel without any oscillating motion. By contrast, in *chop grinding* the grinding wheel has an oscillating movement. Metal is ground as the fast-revolving wheel oscillates and the work is fed past it. Chop grinding is used in contour grinding for fast stock removal (Figure 27–9).

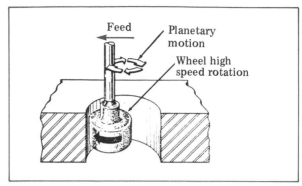

Figure 27–8 Out-Feed Grinding

Plunge grinding (Figure 27–10) is done with the bottom edge of the grinding wheel and is used for rapid stock removal. The wheel spindle travels in a planetary path with the wheel traveling at a high rate of speed. The wheel is set radially at the required diameter. Feeding is axially into the workpiece.

Taper grinding requires the wheel axis to be inclined at the required taper angle. Straight-sided wheels are usually used in jig grinding tapers.

Shoulder grinding requires grinding with a concaved-end grinding wheel. Increments of down-feed are controlled by a positive stop or precision depth stop on the machine.

SLOT GRINDING IN A VERTICAL PLANE

Many tooling requirements call for the grinding of angles, slots, and corners that cannot be

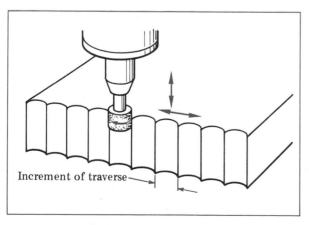

Figure 27–9 Chop Grinding

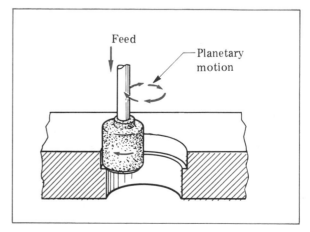

Figure 27–10 Plunge Grinding with Bottom Edge of Wheel

generated by vertical-spindle grinding wheel action. A *slot grinding attachment* (Figure 27–11) is used in a slot grinding operation to produce corners, radii, slots, and concave sections.

The wheel motion is vertical as contrasted with planetary and horizontal motion used to grind straight, cylindrical, and chordal sections of regular shape.

The grinding wheel face is dressed at a double angle. The wheel and attachment are set at the proper angle. A taper-setting feature permits grinding draft on both flanks of the angle at the same time. The inside tapered contour surfaces are ground by vertical oscillating motion of the high-speed revolving grinding wheel.

DESIGN FEATURES OF JIG GRINDERS

The major features on jig grinders, such as the 11″ × 18″ (280mm × 450mm) precision jig grinder, are illustrated in Figure 27–12.

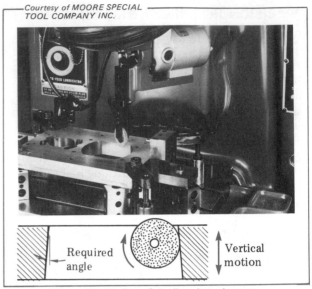

Figure 27–11 Slot Grinding Attachment Positioned for Slot Grinding Operation

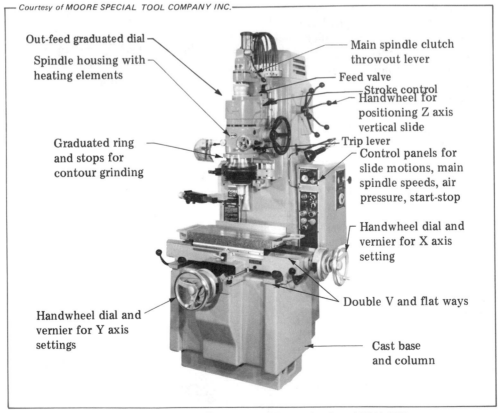

Figure 27–12 Major Features and Controls of an 11″ × 18″ (280mm × 450mm) Precision Jig Grinder

JIG GRINDING HEADS

Grinding heads are available for speeds ranging from 9,000 RPM for grinding operations requiring power and rigidity for large, deep holes. The 40,000 RPM heads are designed for high thrust capacity for bottom, face, and shoulder grinding. The 175,000 RPM air-driven heads are applied to grinding ultrasmall holes. These heads are self-cooling. The exhaust air tends to keep the work cool. Dimensional errors caused by thermal expansion are thus minimized.

DRESSING ATTACHMENTS FOR JIG GRINDING WHEELS

The radius angle, cross slide, spherical socket, angle, and pantograph dressing attachments are each designed for dressing jig grinding wheels to particular shapes. Each dressing attachment, except for the angle-type dresser, is designed for use with a basic universal wheel dresser unit. Precise micrometer adjustments of 0.001″ to 0.0002″ (0.025mm to 0.005mm) and finer vernier adjustments of 0.000020″ (0.5μm) are made. The repeatability of positioning is within 0.0002″ (0.005mm).

WHEEL AND DIAMOND-CHARGED MANDREL SPEEDS

As a general rule, grinding wheels operate efficiently at speeds of 6,000 sfm (1,828 m/min). The spindle speed RPM is found by dividing the sfm by the circumference of the wheel diameter. Therefore, the spindle speed range of jig grinders is from 9,000 to 175,000 RPM. Diamond-charged mandrels, which are used for small-hole grinding, are generally operated at 1,500 sfm (457 m/min).

MATERIAL ALLOWANCE FOR GRINDING

General grinding considerations apply to jig grinding. Also, since hardened materials and complex shapes are ground on the jig grinder, adequate material allowance must be made to compensate for distortion resulting from heat-treating processes. The stock allowance for hole sizes smaller than 1/2″ (12.7mm) is from 0.004″ to 0.008″ (0.1mm to 0.2mm). Hole sizes larger than 1/2″ are left from 0.008″ to 0.015″ (0.2mm to 0.4mm) undersize.

How to Dress a Jig Grinder Wheel

Out-Feed Grinding Processes

STEP 1 Dress the top and bottom face of the wheel by hand with an abrasive stick dresser.

STEP 2 Dress the wheel diameter with a sharp diamond held in a nib holder.

STEP 3 Relieve the cutting face so that about one-fourth of the width remains for cutting.

Plunge Grinding Processes

STEP 1 Repeat steps 1 and 2 as for out-feed grinding.

STEP 2 Use an abrasive stick to dress a concave form in the bottom face. Leave a narrow, flat rim for faster and higher-quality grinding.

How to Grind Holes on a Jig Grinder

Setting Up and Locating the Workpiece

STEP 1 Mount the workpiece on parallels and position it in the center of the table area.

STEP 2 Place strap clamps on the workpiece over the area of the work-supporting parallels. Apply a slight force on the clamps so that the workpiece is temporarily held in place.

Note: Many jobs require simply seating the workpiece in a precision vise.

STEP 3 Use the straight edge, edge finder, microscope, or indicator method of accurately locating the workpiece.

Note: Heat-treated workpieces are often positioned by indicating two or more holes. The average location of a group of holes is selected.

General Grinding Sequence

STEP 1 Rough grind all through holes first.

STEP 2 Dress the grinding wheel for finish grinding. Finish grind all holes that can be machined with the same grinding head to avoid changing grinding heads.

STEP 3 Grind all holes that are closely related in importance as a continuous series.

STEP 4 Grind stepped or shoulder surfaces only once to avoid making more than one depth setting.

How to Grind a Tapered Hole

Using the Taper Setting Plate

STEP 1 Loosen the spindle taper adjustment screw and tighten the other adjusting screw. Continue until the required taper angle is reached on the taper setting plate.

Using a Dial Indicator

STEP 1 Determine the required taper. If given in degrees, convert this value into customary inch or metric linear measurements.

STEP 2 Position a precision angle plate or a master square on the table.

STEP 3 Mount an indicator on the machine spindle.

STEP 4 Bring the indicator point into contact with the square (vertical) surface of the angle plate. Set the reading at zero.

STEP 5 Set the spindle down-feed dial at zero. Move the spindle 1".

STEP 6 Note any variation and the direction in which there is any movement of reading from the original indicator dial reading.

STEP 7 Loosen and tighten the opposite adjusting screws until the difference in indicator dial readings from the starting point to the 1" height equals the offset calculated to produce the required angular setting.

How to Grind Shoulder Areas

STEP 1 Select the largest size grinding wheel that meets the job requirements.

STEP 2 Dress the bottom face with an abrasive stick to produce a slightly concave face. True and dress the cylindrical face of the wheel.

Caution: Position the machine safety shield before truing, dressing, or grinding.

Caution: Protect the spindle housing, table, and front slides with molded machine bibs. Use machine aprons to further protect machined surfaces.

Note: If possible, the outer bottom edge is dressed to a slight radius that blends the side and bottom faces. This dressing strengthens the wheel edge and tends to prevent scoring.

STEP 3 Position the wheel to start the cut on the face of the workpiece.

STEP 4 Set the depth stop at the required shoulder depth.

STEP 5 Rough grind the outside diameter of the shoulder area and the shoulder itself at the same time.

STEP 6 Dress the wheel for finish grinding.

STEP 7 Finish grind the outside diameter and the shoulder to depth.

STEP 8 Check the dimensional measurements and surface finish.

STEP 9 Disassemble the setup. Return all work-holding, tool-holding, and machine accessories to proper storage areas. Clean each item and check for burrs and nicks. Leave the machine in an operable condition.

Courtesy of MOORE SPECIAL TOOL COMPANY INC.

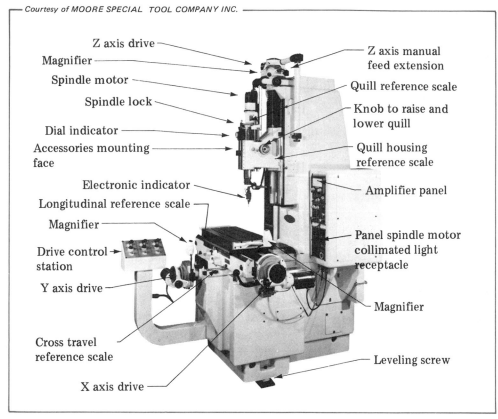

Figure 27-13 Major Components of a Universal Measuring Machine with Motorized Lead Screw Drive

UNIVERSAL MEASURING MACHINES AND ACCESSORIES

The fine tolerances between punches and dies; the precise fitting of components in different mechanisms; the precision requirements of fixtures, cutting tools, gages, and gaging systems; and ever higher precision requirements between mating parts of mechanical, pneumatic, and other movements mean that finer measurements are required. Such measurements are in millionths of an inch (0.025μm) and angular dimensioning, within seconds of arc.

The universal measuring machine is a widely used measuring device. The universal measuring machine illustrated in Figure 27-13 is equipped with a motorized lead screw drive. This universal measuring machine may also be equipped for numerical control. Additional accessories, similar to jig borer and jig grinder accessories, are avail-

able for preselect positioning and readout and printer information.

The extreme accuracies to which these machines measure demand that all components of a machine are held to equally precise tolerances at 68°F (20°C). The total accumulative positioning accuracy of the X and Y axes is 35 millionths of an inch (0.9μm). The greatest amount of positioning error in any 1″ (30mm) is 0.000015″ (0.4μm). The straightness of travel longitudinally is 25 millionths of an inch (0.6μm). The cross travel accuracy is 15 millionths of an inch. The spindle accuracy in terms of trueness of rotation is 0.000005″ (0.15μm).

Part drawings, using coordinate systems of measurement, are commonly used in jig boring, jig grinding, and machine measurement. NC and CNC universal measuring machines are programmed for two- or three-axis measurements.

In addition to accuracy of movement (geometric accuracy), as measured in either a vertical or horizontal plane, squareness of travel is built into the universal measuring machine. There is extreme accuracy in the parallelism of the table to the X and Y axes through the full travel. Squareness of the quill vertical travel and trueness of spindle affect the degree of measurement accuracy. Backlash compensation features are incorporated in the machine design.

BIDIRECTIONAL GAGING SYSTEM

The bidirectional type of gaging system permits dimensions to be measured without having to rotate the gaging probe. Also, the probe diameter is not subtracted from an observed measurement. The bidirectional gaging system consists of a series of gage head inputs with probe tips, centering controls, and measuring scale. The gaging system is switchable between inch and metric measurements.

TV MICROSCOPES

The *TV microscope* is one of the newer advances in micromeasurement technology. The TV microscope employs closed-circuit television for microscopic observation and measurement. The system provides images that have exceptional fidelity and accuracy. Details may be magnified up to 2500X. The TV microscope is particularly adapted for inspection and measurement of actual components of microminiature parts and circuits.

ANGULAR MEASUREMENT INSTRUMENTS

While conventional angle dividing heads and precision rotary tables are used on universal measuring machines, a *small angle divider* permits greater versatility. For measurement and layout purposes, the small angle divider has the capacity to quickly and accurately divide a circle into 12,960,000 parts. Measurements of ±10 seconds of arc are read directly on a vernier dial.

Other precision angular measurement instruments include the index center, rotary table and tailstock, and the micro-sine table. The precision *spin table* is power operated. The table turns at infinitely variable speeds from 5 to 100 RPM. The spin table is suited for inspection of roundness to tolerances within 0.000005" (0.15μm).

Safe Practices for Operating Jig Borers, Jig Grinders, and Measuring Machines

---

• Position the leg of a dial indicator carefully against the edge of a workpiece on a jig borer or a jig grinder. Use care in moving the table to obtain an indicator reading.

• Place parallels and setup blocks so that through holes may be drilled, reamed, bored, or tapped without cutting into these positioning tools.

• Limit the amount of force required to strap workpieces securely on parallels on a jig borer, jig grinder, or accessory table. Excessive or uneven force may cause distortion or cracking if the part is hardened.

• Use the shortest possible shank or diamond mandrel length to reduce wheel overhang.

• Dress the side and bottom wheel faces to a small corner radius. This technique provides a stronger grinding wheel edge and tends to prevent the scoring of ground shoulder surfaces.

• Cycle the NC positioning movements and machining processes for programming accuracy before actual part production.

• Position the jig grinder safety shield for visibility of the grinding process and for eye protection.

• Maintain a constant room and workpiece temperature of 68°F (20°C) for ultraprecise measurements.

• Use machine bibs and aprons to protect housings, tables, slides, graduated collars, and other machine elements from abrasive particles and dust.

TERMS USED WITH JIG BORERS, JIG GRINDERS, AND UNIVERSAL MEASURING MACHINES

Jig borer	A machine tool adapted to precision drilling, boring, reaming, shoulder forming, indexing, and precision measurement and inspection.
Jig grinder	A precision grinder for generating straight, circular, chordal, and contour surfaces by planetary horizontal and vertical axial movement and high-speed grinding wheel action.
Microbore® boring bar	Trade name of a series of boring bars used for precision operations on jig borers. An adjustable head for rough boring and precision finishing operations.
Leaf taper gages	Flat, thin metal gage pieces. Accurately ground gage pieces for measuring diametral sizes, concentricity, parallelism, and bell-mouth of bored and/or ground holes.
Wipe grinding, plunge grinding, shoulder grinding	Abrasive machining processes requiring a combination of planetary motion, the high-speed revolving of a grinding wheel in the planetary path, and vertical movement of the spindle.
Out-feed grinding	Enlarging a hole by continually feeding the grinding wheel outward.
Chop grinding	Grinding with an oscillating spindle as the workpiece is fed past a high-speed revolving grinding wheel.
Slot grinding	Abrasive machining jig grinding operation using a vertical reciprocating motion and slot grinding attachment. Grinding internal and external vertical slots, grooves, radii, and angles, as in the vertical faces of a die.
Taper setting plate	An angle-graduated plate for small angle setting of the spindle.
Pantograph dresser	Grinding wheel dresser that forms the grinding face by guiding a diamond nib in a fixed path according to a template. A jig grinder attachment using a linkage system to reproduce a required form from a greatly magnified template.
Universal measuring machine	A highly precise measuring instrument capable of making measurements in millionths of an inch (0.025μm).
Locating microscope	A spindle-mounted locating accessory. An optical measuring device with which features of a workpiece are positioned in reference to the spindle. Cross lines and concentric circles are used to locate a specific part feature.
TV microscope (universal measuring machine)	Micromeasurement device for magnifying part details up to 2500X in closed-circuit TV. An advanced system for visual inspection of microminiature parts, circuits, and other conventional-size parts.

SUMMARY

- Jig borers are also used for measurement and layout work and all machining processes associated with jig and fixture, punch and die, special tooling, instrument making, and gage work.

    - Common jig boring tools and accessories include precision boring chucks, Microbore® boring bars, extenders, adapters, chucks, and collets.

- Hole centers are spotted with center drills, standard drills ground for NC spotting, and spotter drills.

    - Digital readouts verify measurement accuracy along X and Y axes. Readout equipment covers backlash take-up, manual zero reset, and digital readout overflow.

- Rotary tables provide for angular indexing within tolerances of ±12 seconds, ±6 seconds, and ±2 seconds, depending on the precision quality. The tilting rotary table has the capacity to set an angle to 1 second of arc.

    - The micro-sine table uses gaging rods for accurate work angle setups from 0° to 95°.

- The precision index center permits the rotation of a workpiece on centers for boring, milling, grinding, and other processes.

    - The indicator, line finder, and microscope are widely used on jig borers and jig grinders for aligning work edges.

- Jig grinders provide the versatility and extreme accuracy required to grind the cutting and forming tools, gages, instruments, and precision parts produced on the jig borer.

    - The speeds of jig grinding heads range from 9,000 RPM for heavy grinding operations to 175,000 RPM for grinding ultrasmall holes.

- Wheel dressing attachments for jig grinders include radius angle, cross slide, spherical socket, angle, and pantograph attachments.

    - Extremely small holes are ground with diamond-charged mandrels.

- Jig grinders are used for out-feed, wipe, chop, plunge, taper, and shoulder grinding.

    - Angles, slots, corners, and other intricate forms may be slot ground in a vertical plane.

- The jig grinder spindle may be set to grind a tapered hole by either the taper setting plate or the dial indicator method.

    - Universal measuring machines are used for ultraprecise measurements on flat, plane surfaces; cylinders; irregular contours; and complex single and compound angle settings.

UNIT 27 REVIEW AND SELF-TEST

1. List four classifications of toolroom work that are performed on a jig borer.

2. Identify two important design features of drilling tools that are required for jig boring work.

3. a. Identify the function served by a micro-sine table.
   b. Describe briefly how a micro-sine table is set.

4. a. List the steps for setting up a locating microscope on a jig borer to center hole A in the figure shown below. The hole has been drilled and reamed to size.
   b. Tell how hole B is manually positioned for centering, drilling, and boring.

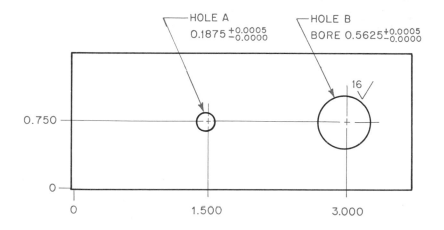

5. State two conditions or factors that differentiate slot grinding from plunge grinding.

6. Identify two methods of manually controlling spindle depth movement on a jig borer or jig grinder.

7. List the general sequence of grinding steps that are followed on a jig grinder.

8. Describe two applications of fixture, punch and die, or gage parts that may be measured on a universal measuring machine.

9. Tell what function a spin table serves when used on a universal measuring machine.

10. State two personal safety precautions to observe when using either a jig borer or a jig grinder.

Design Principles Applied to Production Tooling

This section relates design technology to construction features of tooling that is widely used in mass production. Three categories of such tooling include: jigs and fixtures; punches and dies; and basic workholding devices and tool setups for NC/CNC machine tools and machining centers.

UNIT 28

Jigs and Fixtures, Dies, and Cutting Tools

OBJECTIVES

After satisfactorily completing this unit you will be able to:

- Apply positioning and clamping design techniques to regular, circular, and irregular surface positions, using standardized locating devices.
- Recognize principles of clamping and use strap, quick-acting swing, guide-acting cam, toggle-action, latch, and wedge clamps.
- Describe design features of: leaf, box, open-plate, indexing, and vise jigs.
- Discuss tool design principles applied to: vise, milling, universal and progressive, and face-plate fixtures, and magnetic and vacuum chucks.
- Understand construction features and applications of such metal die-cutting and forming processes as: blanking, piercing, lancing, cutting-off and notching, and shaving and trimming.
- Refer to manufacturer's data sheets and select appropriate tool and die steels to meet particular tooling requirements.
- Identify power presses for stamping and forming parts, including: inclinable, single-action top-drive, hydraulic, and special-purpose punch presses.
- Translate the functions of: inverted, progressive, compound, and combination dies; punch and die components, and die-making principles of design.
- Tell about factors affecting metal flow, and design features applied to sheet metal bending, curling, embossing, and drawing dies.
- Select workholding devices and basic cutting tools for NC/CNC setups and machining, including tool presetting of end- and radial-cutting tools.
- Apply personal and machine tool *Safe Practices* and correctly use each new production tooling *Term*.

POSITIONING AND CLAMPING (WORKHOLDING) METHODS

The term *positioning* is used to refer to the dimensional relationship between a workpiece and a cutting tool. Consideration is given in positioning to characteristics of the workpiece. The shape (straight, circular, or contour), surface condition (plane, curved, or rough), size, and construction features are considered for both positioning and clamping the workpiece.

Locating devices in jigs are designed with regard to a basic reference plane. The machine table provides a reference plane surface. The table prevents cutting forces from moving the workpiece. However, longitudinal and cross feed forces may cause the workpiece to shift. In general practice, three *locators* are required in jig and fixture work to establish the same location of the workpiece. Figure 28–1 shows a typical work-positioning setup in which the workpiece is positioned on the machine table against locator pins in relation to the plane of the base, the longitudinal movement of the machine table, and the cross feed movement.

LOCATING FROM CIRCULAR SURFACES

Round surfaces require location from an axis of a circular section. Round workpieces are usually located and clamped by *concentric location.* Piece parts (workpieces) that are drilled or that are to be located according to holes in the workpiece may be located by *conical location.* Machine centers, nesting beveled holders, and locat-ing pins (cones) provide common locating devices. V-blocks provide another common locating device.

LOCATING FROM IRREGULAR SURFACES

Weldments, castings, and forgings often have irregular surfaces. In order to prevent deformation (stress forces that alter the shape) when clamping forces are applied, it is necessary to shim or jack the part. *Adjustable rest pins* are used to compensate for the condition of the workpiece surface. Poppets and other adjustable locators provide for surface and size variations. Once positioned, adjustable locators are usually secured by some form of lock nut. *Sight location* is used, particularly for first-operation positioning. Sight features (scribed lines, punch marks, or sight holes) are incorporated in the tooling design for rough location of the part.

TYPES OF LOCATING DEVICES AND METHODS

Locating devices have been standardized to the extent that they are commercially available in a wide variety of shapes and sizes. The devices that follow are *stock items.*

PIN, BUTTON, AND PLUG LOCATORS

Pins are used in jig, fixture, and die work. Pins are generally used for horizontal location. Larger sizes of pins are often called *plugs.* Shorter pins, generally used for vertical location, are

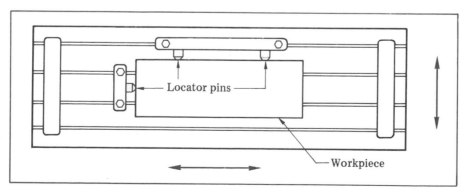

— Locator pins —

— Workpiece

Figure 28–1 Workpiece Positioned on Machine Table against Locator Pins as Reference Points

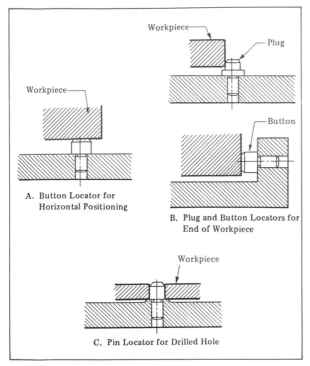

Figure 28-2 Applications of Button, Plug, and Pin Locators

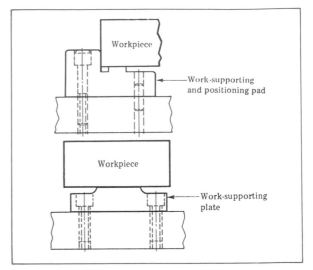

Figure 28-3 Applications of Rest Plates and Pads

known as *buttons*. Three common applications of button, plug, and pin locators are shown in Figure 28-2A, 28-2B, and 28-2C.

When two holes are used for locating, it is common practice to use two pins of different lengths or to use a *diamond locating pin*. The shape of this pin is a variation of the round pin. Two sides are relieved to provide for any center-to-center dimensional variations.

REST PLATES AND PADS

Rest pads and plates are used for work support and positioning. Two common applications of rest plates and pads are illustrated in Figure 28-3. Rest pads and plates are large bearing surfaces on or against which workpiece surfaces may be positioned. A pad may be grooved to permit easy chip removal and proper seating.

LOCATING NESTS

Locating nests are areas that receive a piece part (workpiece) and locate it without the need of any supplementary device. A nest accommodates size variations, permits easy removal of the piece part, and provides adequate space for chip control.

PRINCIPLES OF CLAMPING AND COMMON DESIGN FEATURES

Clamps fulfill four requirements:

- To hold the workpiece securely so that it will withstand the cutting forces;
- To provide a quick-acting device for efficient loading and unloading;
- To clamp without distortion or damage to the workpiece;
- To provide positive clamping, which resists vibration and chatter.

Therefore, certain conditions must be observed in clamping. Clamping forces must be directed toward nesting and locating surfaces. Clamps are positioned so that cutting forces are directed away from the clamping setup. Clamping forces must also be directed against areas of the workpiece or a pad that can safely handle the required holding force. The clamp design permits the loading of the workpiece in only one (the correct) way.

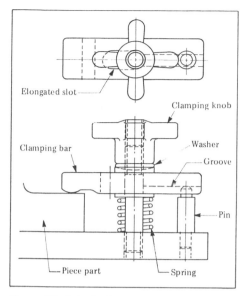

Figure 28–4 Main Features of a Strap Clamp

STRAP CLAMPS

Strap clamps work on a lever principle. While there are many designs of strap clamps, depending on the application, the main features are shown in Figure 28–4. The elongated slot and the groove permit the strap clamp to be slid away from the piece part for easy work removal or loading.

The strap clamp is generally tightened by a hand knob, hexagon nut, or knurled handle. More complex tooling requires a cam-locking device, pneumatic tightening, or other nonhand actuating clamping method.

QUICK-ACTING SWING AND CAM CLAMPS

Swing clamps pivot so that the clamping end may be swung over the workpiece. Once this position is reached, the lever rests at a fixed height to apply the correct clamping force. *Eccentric and spiral cams* are other forms of quick-locking devices. The end of the cam is designed as a spiral or an eccentric. The cam forces the piece part to nest against pins or other work-locating device. Once the piece part is in position, the cam action provides the holding force necessary to machine the part. Figure 28–5 shows a quick-acting swing and cam clamp, which combines the features of the standard strap clamp and the quick action of the cam clamp.

TOGGLE ACTION CLAMPS

Toggle action clamps are designed to produce a maximum, constant force to hold workpieces and fixturing devices securely and, at the same time, permit layout, forming, and machining processes to be carried on without any movement (shifting) of parts.

Toggle action clamps are readily available in a wide variety of sizes and designs. The range of holding forces is from less than 100 pounds to as great as a few tons of force as required on large, heavy power applications. Depending on application, toggle clamps are generally cam actuated or of compound leverage design. Toggle clamps are operated by hand or require pneumatic or hydraulic pressure.

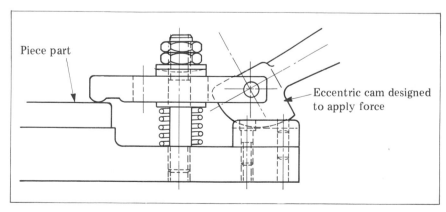

Figure 28-5 Quick-Acting Swing and Cam Clamp

There are three basic actions for toggle clamps.

- *Squeeze action.* Clamping force on a single, adjustable spindle (jaw) or double jaws is produced by squeezing the tool handles together.
- *Hold-down action.* Force exerted on the toggle clamp handle is multiplied and transmitted by cam action or levers to a leg (jaw) that is positioned on the workpiece. A downward holding force is produced.
- *Straight-line action.* This type of toggle clamp is adapted for applying a *push* or a *pull force.* The plungers of this clamp come in different lengths to accommodate the length of movement required to move into position and tightly secure a workpiece.

There are adaptations of each type of toggle clamp. For instance, on hand-operated models, the actuating (force) handle is designed for either vertical or horizontal position operation. Other clamps are operable around 360° (Figure 28–6). Power clamps are widely used on production lines. Clamping forces are regulated by adjusting input pressure to the clamp mechanism to control single or multiple clamping devices.

LATCH AND WEDGE CLAMPS

Latch clamps have the advantage of speed in clamping a piece part in position. One of the simplest forms is a thumbscrew latch. Figure 28–7 illustrates an example of a latch clamp ap-

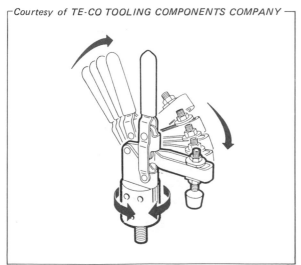

Figure 28–6 Rotating-Type Hold-Down Clamp

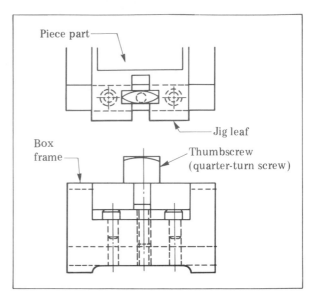

Figure 28–7 Example of a Latch Clamp Application

plication. A quarter turn of the thumbscrew locks a jig leaf in position.

Wedge clamps depend on a taper angle and levers. The wedge angle provides the holding force. The lever ensures that the wedge is locked securely to maintain a constant holding force against the piece part.

DRILL JIG DESIGN

Holes are incorporated in the design of most manufactured products. Holes may be cast, punched, flame cut, or sawed. Holes may be drilled, reamed, bored, counterbored, or threaded. Holes may be laid out and drilled on drilling, milling, turning, and other machine tools. Holes may be programmed and produced automatically on NC machines.

Drill jigs are still widely used in manufacturing to obtain precise hole location and accuracy. Accuracy relates to dimensional tolerances, concentricity, and parallelism. Drill jigs are used principally for drilling, reaming, and tapping. The function of a drill jig is to limit and control the path of the cutting tool in relation to fixed reference surfaces. While there are great numbers of drill jigs, the most common standard types include leaf, box, indexing, and universal designs. The vise is the most common type of work-holding and positioning device.

LEAF JIGS

One of the simplest *leaf jigs* has a hinged cover (leaf) that is swung to load or unload the jig. Once the workpiece is nested, the leaf is closed and locked in position. Holes may be drilled in one or more surfaces. Drill bushings may be replaced with reamer bushings. The bushings may be located in the hinged plate or base. Figure 28–8 shows a hinged-plate leaf jig for drilling and reaming. Fixed drill bushings are secured in the hinged plate. Reamer bushings are directly opposite the drill bushings but are located in the base. The jig and piece part are turned over once the holes have been drilled in preparation for the reaming process.

BOX JIGS

Box jigs, as the name suggests, are box shaped. Bushings may be included on one, two, or more sides. Some box jigs are designed with jig feet on opposite sides of the part. Once a hole is machined, the jig and piece part are turned to the next machining position. When the second hole is machined, the jig and piece part may again be rotated to expose the third jig face to the tooling in the machine spindle.

A *channel jig* is an adaptation of a box jig. The jig is designed in a three-sided channel form.

OPEN-PLATE JIGS

Open-plate jigs may or may not have legs. The main construction feature is a plate. The plate is designed with liner bushings that permit the use of replaceable bushings for drilling, reaming, or tapping. Plate jigs are designed for carrying on a single machining process at one loading.

INDEXING JIGS

Indexing jigs permit angle setting of a piece part for positioning successive holes in a circular pattern. The indexing may be done by using a standard indexing device. The piece part may be held in a standard chuck mounted on a graduated indexing device. The part is indexed for each location. A drill bushing, secured in an overarm on the knee of a drill fixture for the indexing mechanism, may be adjusted for height and position.

UNIVERSAL FIXTURES

Universal fixtures are commercially available for adaptation to particular hole-forming processes. For example, the fixture illustrated in Figure 28–9 is designed to be a drill jig and features a double post, rack and pinion, and simple clamping and release mechanism. The fixture is open

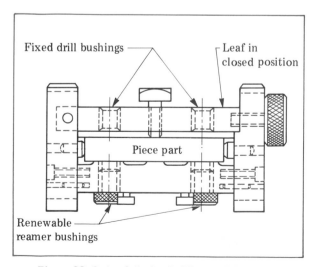

Figure 28–8 Leaf Jig for Drilling and Reaming

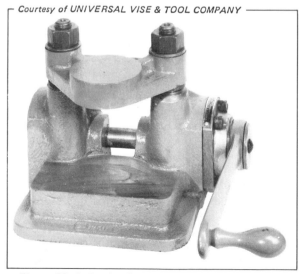

Courtesy of UNIVERSAL VISE & TOOL COMPANY

Figure 28–9 Double-Post, Rack and Pinion Fixture

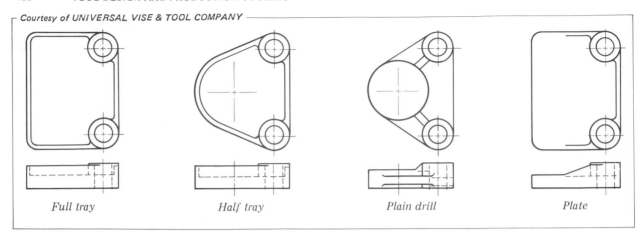

Full tray   *Half tray*   *Plain drill*   *Plate*

Figure 28–10 Four Basic Types of Fixture Heads

through the guide post bosses. This opening permits parts to project through the fixture and to extend to the right and left of the base. This particular fixture is commercially available with a vertical movement range from 1/2″ (12.7mm) on the 3 1/2″ (90mm) opening height model to 1″ (25.4mm) on the 9″ (230mm) height model.

Vertical movement is precisely aligned—that is, the head and base are accurately positioned. Wear on bushings and cutting tools is eliminated. The downward movement of the lever clamps the piece part. The upward stroke releases the work-holding force and clears the piece part and fixture.

The *shut height* provides for a wide range of workpiece sizes. The adapter position of the head may be changed by turning the lever until the rack is disengaged from the pinion. Once the new adapter height position is established, the rack and pinion are reengaged. Figure 29–10 shows four basic types of fixture heads.

The manual hand crank on some drill jigs and fixtures is replaced with a *rotary actuator*. The fixture shown in Figure 29–11 is pneumatically operated (automated) by a rotary actuator. The construction and design features of the actuator are illustrated by the line drawing. The actuator is used to convert fluid (air) energy that moves a piston rod in a cylinder (linear motion) into rotary motion. The rotary motion substitutes for the manual motion normally required

to open and close the drill jig. The rotary actuator is adaptable to advanced automation techniques and NC applications.

VISE JIGS

Short run, low-production drilling is often performed using a cam-actuated vise. Special jig plates are secured to the solid vise jaw. The nesting feature fits between the two jaws.

DRILL JIG DESIGN CONSIDERATIONS

Design considerations are influenced by the required machining specifications. Rigidity, simplicity, clamping forces, chip control, and jig feet and bushings are five prime concerns of the designer, toolmaker, and machine operator.

RIGIDITY

Rigidity relates to the ability of the jig to withstand workholding and cutting forces. The jig must also be designed to support the piece part so that it does not bend during machining.

SIMPLICITY AND CLAMPING FORCES

The jig is designed to nest the piece part so that it may be easily loaded and, when machined, removed. Clamping forces are directed toward the nesting feature. The locating points must permit loading in one way only.

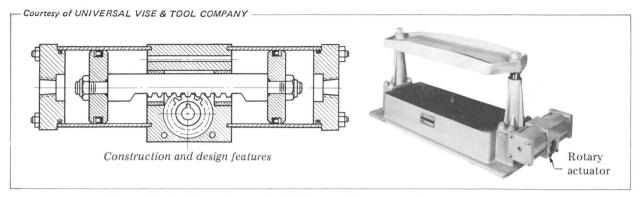

Courtesy of UNIVERSAL VISE & TOOL COMPANY

Construction and design features

Rotary actuator

Figure 28-11 Fixture Pneumatically Operated (Automated) by a Rotary Actuator

CHIP CONTROL

Chip control requires, where practical, that a segmental type of chip be produced in preference to long, stringy continuous chips. Adequate space is provided between the piece part and the drill bushing to allow chips to flow freely between the work and the bushing plate. Jigs are provided with corner relief to help prevent chip buildup in corners.

DRILL JIG FEET (LEGS)

A drill jig seats better on a plane reference surface when the base is cut away to form legs (feet). There are usually four feet. They are ground to form the plane reference surface on the jig. The bases on the different sides on box jigs are also designed with legs.

PRESS-FIT AND RENEWABLE BUSHINGS

In drilling, reaming, and tapping operations in jig work, bushings serve to position the cutting tool and to guide it. A clearance between 0.0005″ (0.0125mm) and 0.001″ (0.025mm) clearance is provided between the cutting tool and the bushing diameter. Larger clearance tends to cause drill margins to chip and to produce machining inaccuracies.

The basic types of press-fit and renewable bushings are shown in Figure 28-12. *Press-fit wearing bushings* (Figure 28-12A) are designed for permanent press-fit applications. The *headless type* is for flush mounting with the jig plate. This type is used for light axial loads. The permanent press-fit *head type* resists the effects of heavy axial loads. The drill plate may be coun-

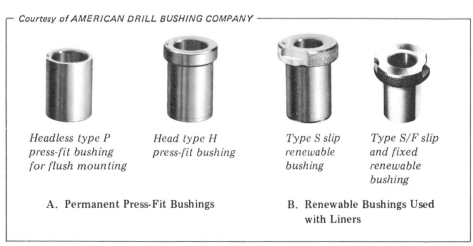

Courtesy of AMERICAN DRILL BUSHING COMPANY

Headless type P press-fit bushing for flush mounting

Head type H press-fit bushing

Type S slip renewable bushing

Type S/F slip and fixed renewable bushing

A. Permanent Press-Fit Bushings

B. Renewable Bushings Used with Liners

Figure 28-12 Basic Types of Press-Fit and Renewable Bushings

terbored to permit the head to be flush with the jig plate.

Renewable wearing bushings (Figure 28–12B) are used with *liners* on long production runs and when it is necessary to change bushing size to perform more than one operation in a hole. The outside diameter of the *slip type* is ground to a slip fit with the liner. The bushing is changed without removing the lock screw. The *fixed type* of slip renewable bushing is used for long production, single process (drilling or reaming) runs. The worn bushing is quickly replaced by removing the lock screw. The lock screw recess secures the new bushing in place.

Bushings are hardened to provide a wear-resisting surface. Bushings are ground externally and internally and are lapped to within a concentricity of 0.0003" (0.007mm). ANSI bushing classifications and size standards are used. There are many modifications of bushings—for example, thin-walled bushings for close hole drilling and double-guide bushings for aligning the starting and body pilots of a reamer that requires such guiding.

ADDITIONAL JIG CONSTRUCTION FEATURES

The components of a jig are usually assembled mechanically. The parts are machined before assembly. Worn parts may be replaced. Jigs constructed in this manner are designed with *dowels* for alignment and *fillister and socket-head cap screws* to secure the components. Dowel pin reamers, which are generally ground slightly undersize, are used to ream dowel pin holes. Dowels are press fit into matching parts.

Drill jigs are sometimes designed with welded components. Such jigs are stress-relieved before the components are machined to size.

NC MACHINING TECHNOLOGY IN JIG WORK

Some of the functions performed by drill jigs since the beginning of interchangeable manu-facturing are now being done by precision drilling, boring, and other NC machine tools. Precision point-to-point and continuous-path positioning capability makes it possible to precisely locate hole positions. On NC drilling and boring machines, once the processes are programmed and a tape is produced, just a simple positioning fixture may be required to reference all slide movements and machining processes that are controlled by the tape. While jig borers and precision boring and NC machine tools eliminate the need for some drill jigs in production work, drill jigs are still widely used.

FIXTURE DESIGN

A *fixture* is a workholding device strapped securely to a machine tool. The fixture holds a piece part in a fixed-position relationship with one or more cutting tools during machining operations. Welding and assembly fixtures are used where parts are positioned in relation to each other in order to be fabricated.

Fixtures are generally classified as follows:

• According to the machine tool and type of operation performed—for example, face milling, straddle milling, and slotting fixtures used with a milling machine;

• According to the manner in which the workpiece is clamped—for example, hand clamping for pneumatic (power) clamping fixtures. Automatic fixtures permit loading and unloading according to fixed cycles of machining processes;

• According to the techniques of locating the part—for example, V-block fixtures and center fixtures;

• According to the method of feeding the workpiece to the cutter—for example, indexing fixtures for permitting parts to be rotated (indexed) to the next position during the machining cycle; rotary drum fixtures for securing parts around a drum and permitting separate operations to be performed as the drum is rotated.

VISE FIXTURES

Standard machine vises are widely used as fixtures. The jaws may be fitted with inserts designed to accommodate regular or irregular-shaped parts. Workpieces may be held in special jaws for machining surfaces in a horizontal plane. Surfaces may also be machined in second and third planes by swiveling a vise on its base or by positioning a vise and workpiece at a vertical angle when a compound angle vise is used.

MILLING FIXTURES

Milling fixtures are rigidly held to the work table. A part is secured in a fixture as it is moved past one or more cutting tools, as for gang or straddle milling. Most milling fixtures include a base plate for strapping to the table; nesting areas for supporting and holding the workpiece; locating points; and gaging surfaces. The base plate is usually slotted and keyed to ensure alignment with the table T-slots.

Milling fixtures are designed to permit rigid clamping while at the same time not obstructing cutter movement. *Feeler (set block) surfaces* are designed as part of the fixture base. Set block surfaces are machined lower than the cutter depth so that a feeler gage may be used to set the cutter. Caution must be used in setting a cutter with a feeler gage. The cutter must be checked in advance for concentricity between the bore and the ground peripheral teeth. Cutter runout produces an inaccurate depth setting. Also, the machine spindle must be stopped in setting the cutter to depth.

UNIVERSAL AND PROGRESSIVE FIXTURES

Families of parts are held in *universal fixtures*. Workpieces in *progressive fixtures* are located in different positions and are moved progressively between machining stations until all processes are completed. The progressive machining stations on a rotary fixture are shown in Figure 28–13.

MAGNETIC AND VACUUM CHECKS

Magnetic chucks are a practical clamping device for ferrous metal parts. Newer low-voltage magnetic chucks provide increased heat-free holding power and minimized distortion of the workpiece. Magnetic chuck fixtures are designed with fixed workpiece stops against which cutting forces are directed.

Courtesy of CINCINNATI MILACRON INC.

Figure 28–13 Progressive Machining Stations on a Rotary Fixture

Vacuum chucks, like magnetic chucks, have many advantages over mechanical, hydraulic, and other clamping devices. A vacuum chuck provides a nondistorting and nondamaging force on thin or fragile materials. One prime advantage is that, since less clearance space is required for vacuum chucking because clamping parts are not used, smaller-diameter cutters may be used.

Applications of two principal vacuum chucks are illustrated in Figures 28–14A and B. The rectangular-shaped, top-plate vacuum chuck (Figure 28–14A) is used primarily for conventional and numerical control milling machines, surface grinders, jig borers, and similar machine tools. The rotary vacuum chuck (Figure 28–14B) is adapted to faceplate work. This chuck is made for lathes and other turning machines, rotary abrasive grinding, and similar machining processes.

Vacuum chucks operate with a series of *valve ports* (openings). Each port is controlled with a valve screw to open or seal a port. Sectioning or controlling the vacuum area is thus permitted. The knurled nut on the vacuum fixture is turned to vary the holding power. After machining, the part is released by pushing the release valve. Vacuum chucks are especially adapted to hold ferrous metals and a wide variety of nonferrous and non-magnetic materials.

TURNING MACHINE FIXTURES

Turning machine fixtures are specially designed for holding castings, forgings, and many irregular-shaped parts that cannot be held by conventional chucks. While general principles of fixture making apply, a number of other design and machine operator considerations are necessary:

- Since fast-revolving workpieces produce centrifugal forces, clamps and other work-holding devices must operate under excessive force conditions without loosening;
- Static and/or dynamic fixture balancing must be done before the workpiece is rotated under power. Counterweights may be required to prevent vibration;
- Projections and sharp corners must be eliminated because a rotating fixture may be dangerous;
- Cutting forces require that the workpiece be gripped on the largest diameter of the workpiece (thin workpiece sections may need to be supported against cutting tool forces).

Courtesy of DUNHAM TOOL COMPANY, INC.

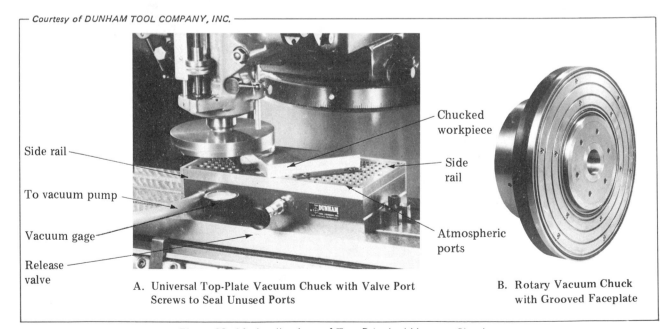

Side rail

To vacuum pump

Vacuum gage

Release valve

Chucked workpiece

Side rail

Atmospheric ports

A. Universal Top-Plate Vacuum Chuck with Valve Port Screws to Seal Unused Ports

B. Rotary Vacuum Chuck with Grooved Faceplate

Figure 28-14 Applications of Two Principal Vacuum Chucks

The use of chucks with standard jaws and horizontal rocking jaws was discussed earlier in relation to work-holding devices for turret lathe work. Other jaws are designed to rock in a vertical direction. The vertical jaws (on a three-jaw chuck) grip a cylindrical surface in six equally spaced sections. Thus, the gripping force is equally distributed. Wrap-around jaws are used to minimize distortion by gripping over a greater area. These jaws are bored to within a thousandth of an inch of the turned diameter of the workpiece in order to evenly distribute the gripping force.

FACEPLATE FIXTURES

Simple *faceplate fixtures* employ locating pins and require direct strapping using the faceplate slots and standard clamps. Other fixtures use a right angle bracket on which the workpiece is secured. Angle-type fixtures require balancing with a counterpoise to compensate for weight imbalance and centrifugal forces.

METAL DIE-CUTTING AND FORMING PROCESSES

The punch press is another machine tool that continues to make significant contributions to the manufacture of interchangeable parts. Parts are cut to internal and external sizes and are formed into simple and complex shapes by die-cutting processes such as blanking, piercing, lancing, bending, forming, cutting off, shaving, trimming, and drawing.

National security, national economy, and standards of everyday living depend on the output of parts, components, and devices that are manufactured by using punches and dies. Ferrous and nonferrous parts are economically mass-produced by dies that are identified by the processes performed.

Single or combined processes may be required to produce or assemble a simple or a complex part. The processes are referred to as *die cutting* and *die forming*. The cutting and forming tools are identified as *punches* and *dies*. The stamping and forming machines on which punches and dies are used are simply called *presses*, *punch presses*, or *power presses*. Categories of such presses and major components are discussed later.

BLANKING PROCESSES

Blanking refers to the process of cutting out (usually from a sheet, strip, or roll) a part to a specified shape and size. A punch of the required size and shape forces against the stock and, with the aid of a mating die section, cuts a blank. The blank is moved by successive cuttings through the die section cavity. The relieved sides of the die permit the blanks to fall free and be removed. On the return stroke, the excess material (strip stock) surrounding the punch is removed as the punch moves through a stripper plate. Subsequent pieces are blanked by positioning the stock against a stop pin and repeating the cutting and return stroke cycle. Figure 28–15 illustrates design features of a plain blanking punch and die set.

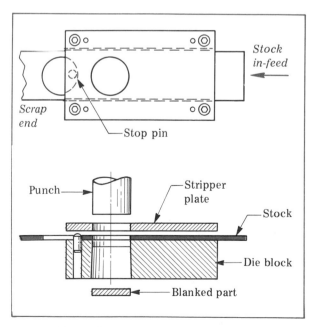

Figure 28–15 Design Features of a Plain Blanking Punch and Die Set

PIERCING PROCESSES

Piercing usually refers to the stamping of one or more holes in a strip of material. The holes are design features of the part that may be blanked simultaneously or at a different time.

LANCING PROCESSES

During *lancing*, a workpiece is cut on two or three sides and formed along one or more sides. A blanking process may follow lancing.

CUTTING-OFF AND NOTCHING PROCESSES

Cutting off refers to separating a blanked workpiece from the scrap strip. Cutting off is generally the final step in progressive die work. Cutting-off processes are straight-line cuts. *Notching* is used to remove material from one or both edges of the strip. Notching serves to remove excess material so that a metal formed part flows as it is being drawn or formed to shape.

SHAVING AND TRIMMING PROCESSES

Shaving removes a very thin (small) area of metal from a previously cut edge. Shaving cuts the part to close dimensional tolerances and thus produces a smooth finished edge. *Trimming* is used to remove surplus, irregularly formed material from around the edges of a shaped or drawn part. A smooth accurate surface is produced. Both shaving and trimming are second operations.

POWER PRESSES FOR STAMPING AND FORMING PARTS

Punch presses range from simple bench presses for stamping small, fine parts that are a fractional part of an inch or centimeter to automated systems for large-volume, integrated production units. Such presses are designed with quick-change bolsters; floor-mounted, self-contained materials handling devices; and automated controls for mass-producing complicated panels and forms. While the sizes, tooling, and controls differ, there are a number of common features. The following descriptions of a few basic types of presses provide an overview of the range, capabilities, and characteristics of these machines.

INCLINABLE PRESSES

Inclinable presses are designed for metal cutting and forming operations that include blanking, forming, bending, drawing, assembling, and combination progressive die processes. The inclinable press illustrated in Figure 28–16 is available in standard sizes from 22 tons to 250 tons. The smaller sizes are adapted to production requirements for small and medium-sized stamped parts. High-speed gap frame presses, which are adaptations of the inclinable press, are operated at higher speeds up to 1,200 strokes per minute. Continuous automated feeding is employed.

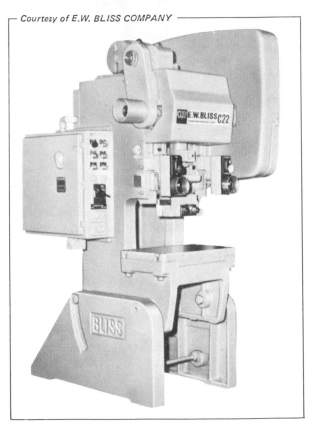

Courtesy of E.W. BLISS COMPANY

Figure 28–16 Inclinable Press for Blanking, Forming, and Assembly Operations

Inclinable presses may be fitted with optional feed devices such as single or double roll, dial, and transfer devices. A combination air-friction clutch and brake are provided for fast safety stops. All models are available with variable-speed drives. The presses up to 110 tons are of cast Meehanite construction. Larger-capacity presses are of welded steel frame construction.

SINGLE-ACTION TOP-DRIVE PRESSES

Single-action top-drive presses are adapted to press operations that require dies with greatly varying heights. The construction of the operating surfaces between the two vertical outer frame members permits medium-sized parts to be fed through the press from front to back. Figure 28–17 illustrates a straight-side, single-action, eccentric gear design press that is widely used for blanking, piercing, stamping, forming, and similar operations on medium-sized parts. This press is available in capacities from 200 tons to 2,000 tons for use in aircraft, automotive, appliance, hardware, furniture making, and farm machinery parts production. The press, as illustrated, has a 300-ton capacity, 120″ × 84″ (3048mm × 2134mm) die area, and *rolling bolsters* that permit rapid, accurate die changes. The rolling bolster (with die attached) is moved into the press and automatically positioned and clamped in place. The bolsters are self-powered, interlocked, and controlled at the master panel.

HYDRAULIC PRESSES

Hydraulic presses serve functions similar to the mechanical types. Hydraulic presses are available from small, general-purpose machines to extrusion presses of 35,000 tons capacity. The machines are designed to utilize die slides, feed mechanisms, sliding or rolling bolsters, and many other accessories. Protection is provided against overloading either the dies or the press.

A universal electric control permits speed change and reversal of ram position. A stop button on the cycle control permits instant stopping at any point in the cycle. Die setup is aided by inching with push buttons. The press stops on release of the button. The hydraulic components are protected against wear by a positive filtering and cooling system that provides clean fluid at all times.

Courtesy of E.W. BLISS COMPANY

First rolling bolster

Master control panel

Second rolling bolster

Figure 28–17 Single-Action, Eccentric Gear Design Press for Blanking, Piercing, Stamping, Forming, and Similar Operations

SPECIAL-PURPOSE PRESSES

In addition to standard presses, metal products manufacturing requires *special-purpose presses*. A few examples include: *cold extrusion presses*, for extrusion and piercing operations; *automatic foil presses*, designed for long feed lengths of wide but very thin material; *mechanical forging presses*, built to absorb heavy work load forces imposed by forging operations; and *welding presses*, to speed production of body parts and appliance panels that are to be permanently assembled by welding.

MAJOR COMPONENTS OF THE OPEN-BACK INCLINABLE PRESS

There are five basic components of an inclinable press. These components include the *frame and bed*, *slide (ram)*, *clutch*, *crankshaft*, and *variable-speed drive*.

FRAME AND BED

The frame and bed on the inclinable press is mounted on vertical uprights. The frame may be adjusted and inclined at an angle. The frame is machined to receive a rectangular bed plate called a *bolster* or *bolster plate*. The die plate of a die set is mounted and secured to the bolster plate of the bed. Ways are machined at a vertical right angle to the bolster plate in order to accommodate the slide (ram) and to permit accurate vertical travel. The frame is generally cast of Meehanite and is ruggedly constructed to continuously withstand cutting and forming forces.

VERTICAL SLIDE (RAM)

Vertical movement on a press is produced by the slide (ram) component. The slide is designed to accommodate a crankshaft or eccentric drive at the upper end. The bottom of the slide receives the punch section of the die set. The slide is adjustable for positioning dies of varying heights.

CRANKSHAFT

The crankshaft produces the vertical movement of the ram. The crankshaft is dynamically balanced and usually nongeared to permit smooth high-speed operation.

CLUTCH MECHANISM

The clutch is generally a combination of a friction clutch and a brake. The clutch is crankshaft mounted and requires only a fraction of an inch travel between full clutch engagement and full brake. Some heavier models have an *unsticker feature*. A press that is stuck on bottom or just past bottom dead center may be freed by turning the press over under power, using this specially designed clutch. A reversing feature is used when the press stops before dead center.

VARIABLE-SPEED DRIVE

The variable-speed drive permits the operator to fine tune the press speed to the particular process and other cutting and/or forming requirements. Some models are equipped with a mechanical or an eddy-current type of drive. While electrical drives permit speed adjustments from zero to maximum, as the speed is reduced, the horsepower decreases. Mechanical drives produce almost a constant drive but are limited to a particular ratio, such as 2:1 or 3:1.

Standard speed ranges are provided from 45 strokes per minute on presses with 8″ to 10″ (203mm to 254mm) lengths of stroke to 750 strokes on smaller presses with 1/2″ (12.7mm) length of stroke. High-speed presses of from 22 tons to 60 tons capacity may be operated at 1,200 strokes per minute. Press speeds are read on some models on a tachometer. Stepless speed adjustments may be made on the eddy-current type of clutch and the variable pitch pulley system from a machine control unit console while the press is running.

OPTIONAL EQUIPMENT

Roll feeds are available in either single roll feeds or double roll feeds. *Single roll feeds* push or pull the stock through the dies. This design is used where the scrap is cut up by the die. *Double roll feeds* provide for positive control of the strip at both entry and exit ends of dies. Single and double roll feed drives are either rack-and-pinion or lever type. The lever type is used for shorter feed lengths and higher press speed ranges.

Other feed accessories include *special roll finishes* (chrome plated, milled, sand blasted, or coated). Shear-type, press-actuated *scrap cutters* are available. *Slide feeds* and other special arrangements for front-to-back feeding, spray-type automatic *stock oilers*, *power run-in rolls* to start a new strip into the die, *strip-straightening devices*, and motor-driven and unpowered *coil winding equipment* are examples of additional feed accessories.

STROKE, SHUT HEIGHT, AND DIE SPACE

The toolmaker deals with stroke, shut height, and die space in constructing a die and in setting it up on the press for initial tryout. The *stroke* represents the reciprocating motion of the ram. Stroke is the distance between the position of the ram at the top and the bottom ends of a complete cycle. The stroke is the dimensional distance in inches or millimeters.

The *shut height* is measured with the ram at the bottom of the stroke and the stroke adjustment up. The shut height of a die must be less than the shut height of the press. The *die space* denotes the area available on the press for mounting dies.

DIE CLEARANCE AND CUTTING ACTION

The cutting action of punches and dies is similar to the cutting action of single- and multiple-point cutters used on other machine tools. The punch forces the workpiece against the die face. At the point where the *elastic limit* of the material is exceeded, the metal flows due to *plastic deformation*. At the start of the flow, depending on the clearance between the punch and die and other characteristics of the material, a minute radius is turned and a fine burr is produced. As the punched *slug* moves into and passes the land of the die, a burnishing action takes place on the sides. The burr side of the blanked part (slug) is on the upper side toward the punch. Conversely, the burred side of the stock is on the bottom surface next to the die face.

DIE CLEARANCE

Die clearance relates to the planned difference in size between the cutting edges of the punch and die. Clearance is most generally expressed in relation to each side. Clearance is determined by the thickness of stock, working properties of the material, and the nature of the blanked edges.

Too great a clearance produces a larger roll-over edge and unsafe burred edges. Insufficient cutting clearance produces excessive cutting forces, reduces tool wear life, and may cause the die to fracture.

The size of the die opening establishes the dimensional size of the blank. In the case of a sheared hole, the size of the pierced hole is the size of the punch. The clearance is placed on the punch when the slug is to be blanked to the required size. The clearance is placed on the die when the pierced hole is to be the required punched part.

ANGULAR WORK CLEARANCE

Angular work clearance is represented by the number of degrees per side below the narrow cutting land on the die. The angular work clearance permits a stamped part that fits tightly on the die land to be moved through the die. The tight fit is due to springback produced by grain structures being stressed below the elastic limit. Recommended angular clearances for band machining a punch and die from the same piece were given in an earlier unit on band machining.

CUTTING FORCES AND SHAPES OF CUTTING FACES

Cutting force for punch press operations is identified as the total amount of force required to completely blank a workpiece. While the designer is concerned with cutting forces within the capacity of the press, the toolmaker considers changes that may be needed to reduce cutting forces of a punch and die.

Cutting forces may be reduced by altering the heights of punches when several are included in the design. Another common technique is to grind the punch at a slight shear angle to the horizontal plane. Shear reduces the cutting forces. Punches and dies may also be ground with a double-angle shear. For blanking operations, the double-angle shear is on the die plate; for piercing, on the punch. Shear distorts the work material.

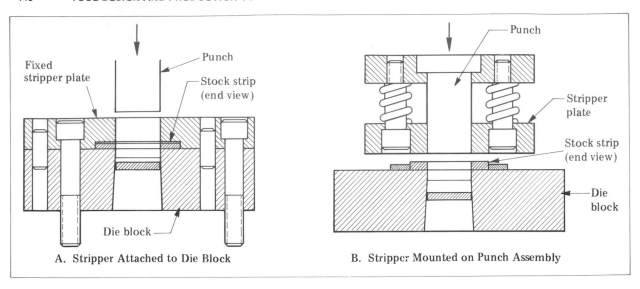

Figure 28-18 Basic Types of Stripper Plates

STRIPPER PLATES AND EXCESS STOCK REMOVAL

As a punch cuts through the metal, the slug remains in the die while the strip clings to the punch. The excess stock is removed from the punch by using a *stripper plate*.

The two basic types of stripper plates are shown in Figure 28-18. The first type of stripper plate is fixed and is attached to the die block (Figure 28-18A). The bottom of the stripper plate is designed to provide a track for the stock. After a part is punched, on the return stroke, the stock and punch move a short distance toward the stripper plate. The stripper plate holds the stock as the punch continues its upward movement. Once freed, the strip is then moved to its next position as the stamping cycle continues.

The second type of stripper plate is mounted on the punch assembly and fitted around the punch with springs (Figure 28-18B). Once the blank is punched and the punch leaves the die block, the stripper plate forces the stock off the punch. The thickness of the stripper plate and the size of springs to use depend on the nature and thickness of the stock, the part size, the complexity of the part itself, and the surface finish of the punch.

DIE SETS AND PUNCH AND DIE MOUNTINGS

Standardized *die sets* are widely used to maintain precise alignment between the punch and die. A standardized two-post die set is shown in Figure 28-19. A die set consists of a base that is flanged for mounting on the bolster plate of the press. The base is called the *die shoe*. *Guide pins* are pressed in the base and serve to keep the punch and die aligned. The guide pins ride in *guide bushings* in the top plate of the die set. The top plate is known as the *punch holder*. The punch and die sections are mounted between the die shoe and punch holder in the *die area*. The shut height of the die set is the vertical distance between the bottom surface of the die shoe and the top face of the punch holder when the die is in a closed position.

Die sets are commercially available with or without a shank on the punch. The guide pins and guide bushings may be centered, mounted at the back end of the die set, or positioned at diagonal corners. Die sets are made with two or four guide pins and bushings. When the shank-type holder is used, the punch is secured to the ram by two tightening nuts in front of the ram.

Standard die sets are available in commercial and precision grades of accuracy. Grade usually relates to the quality of fit between the guide pins and the bushings. The tolerances for the precision grade are finer and provide for extremely accurate alignment between multiple punches and corresponding holes in a die block.

The die shoe and punch holder in standard die sets are made of a high-quality cast iron or alloy steel. The cast iron die sets are used where the processes produce a limited amount of shock. Excessive shock would crack the cast iron parts. Steel die sets are capable of withstanding greater shock loads. Large and special die sets, which are subjected to sever shock loads, are constructed of rolled steel. Manufacturers' catalogs are available to give die set specifications and ordering information.

BALL BEARING BUSHING DIE SETS

Ball bearing bushing die sets are designed to meet the following requirements.

• Smoothness and accuracy of alignment to reduce die wear and increase the productive wear life of a punch and die.

• Accurate control of pre-load and ball cage engagement to ensure maximum load capacity for all power presses, including the reclinable types.

• Tough, free-rolling guide post surfaces.

Figure 28–20 illustrates the design features of a typical ball bearing bushing die set. In addition to the punch plate (holder) and the bottom (die shoe) plate, the assembly includes: *guide posts* that are secured in the punch holder. *Straight sleeves* or *demountable bushings* and *ball cages* are mounted on the die plate.

The demountable bushings facilitate assembly and disassembly. Straight sleeve bushings assure a uniform fit. All parts of the punch and die assembly are interchangeable.

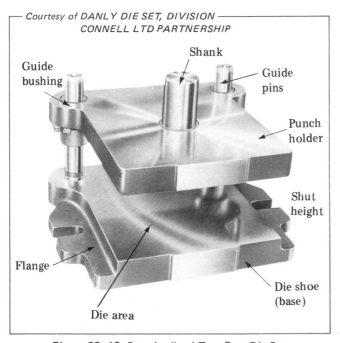

Courtesy of DANLY DIE SET, DIVISION CONNELL LTD PARTNERSHIP

Figure 28–19 Standardized Two-Post Die Set

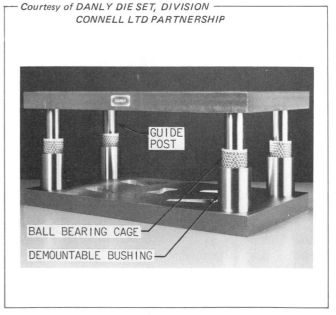

Courtesy of DANLY DIE SET, DIVISION CONNELL LTD PARTNERSHIP

Figure 28–20 Assembly of Ball Bearing Bushing Die Set

CONVENTIONAL, COMPOUND, AND COMBINATION DIES

INVERTED DIES

The dies discussed thus far were designed with the die block secured to the die shoe and the punch fastened to the punch holder. The punch and die positions are reversed on *inverted dies*. In the inverted die set shown in Figure 28–21, the punch is doweled and fastened to the die shoe. Similarly, the die section is doweled and held securely to the punch holder. Note in this design that on the return stroke a stripper pad with an ejector plug in the die forces the blank to be ejected. The ejector is moved in the die cavity by a knockout rod.

Inverted dies are used to keep the cutting edges clear of chips to minimize grinding. Also, each blank is removed as it is cut instead of being forced through the die.

PROGRESSIVE DIES

Progressive dies are used when two or more press operations are performed for each stroke.

The strip is moved from one station to another. At each station, the part is formed progressively. At the final stage, a completed part is produced. After the strip has once been cycled through each station, each ram stroke produces a finished part.

A simple two-station progressive die is illustrated in Figure 28–22. The strip is fed from the right side. In sequencing operations, piercing is placed first. Pierced holes permit accurate locating for subsequent operations. Bending and forming operations are planned for the last stations. In Figure 28–22, the first operation is performed by a piercing punch. On the next stroke, the workpiece is advanced to the next station where the pilot on the punch centers in the pierced hole. The rectangular blanking punch cuts the part to shape. At the first station, the strip is positioned by the primary die stop. For the second stroke, the strip is advanced to the die-stop pin. At each successive stroke thereafter, the strip is positioned against the die-stop pin. The scrap strip emerges from the left side of the die. The stripper plate is mounted on the die block. A two-post rear position die set is used for the punch and die mounting.

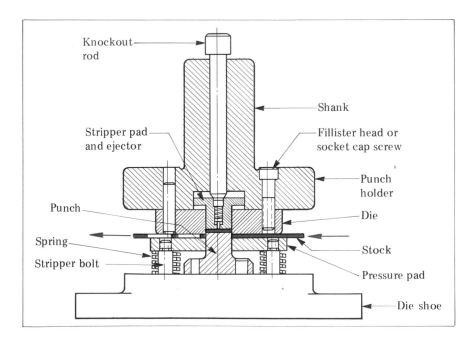

Figure 28–21 Example of an Inverted Die Set

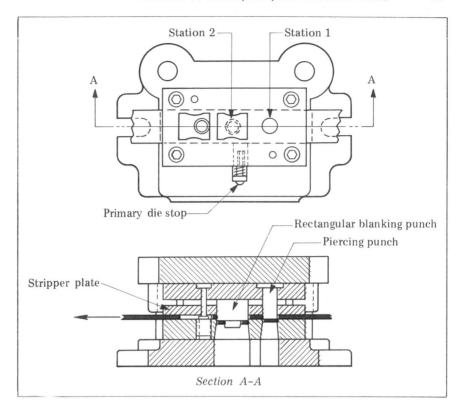

Station 2

Station 1

A ——— A

Primary die stop

Rectangular blanking punch

Piercing punch

Stripper plate

Section A–A

Figure 28–22 Example of a Simple Two-Station Progressive Die

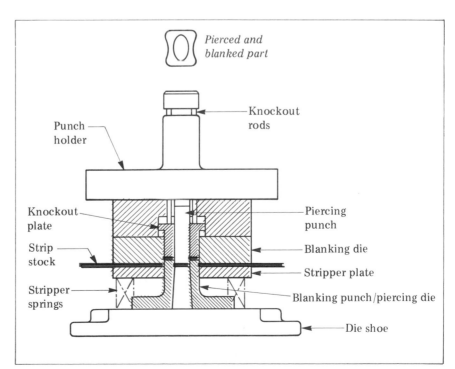

Pierced and blanked part

Knockout rods

Punch holder

Knockout plate

Piercing punch

Strip stock

Blanking die

Stripper plate

Stripper springs

Blanking punch/piercing die

Die shoe

Figure 28–23 Example of a Compound Die for Piercing and Blanking a Part at One Station

Additional stations involving a greater number of piercing, forming, and blanking operations are common. The final product may be of regular or irregular shape. On some progressive dies, the part is often blanked before the final operation. The blank is then forced back into the strip and carried along to subsequent stations for bending or forming.

Progressive dies take the place of a number of separate dies. A whole sequence of operations may be performed simultaneously on the same strip. One main disadvantage with parts that have small supporting areas is that the final part may become dished as it is forced through the die block.

COMPOUND DIES

Compound dies perform two or more operations at one station during one press stroke. Figure 28–23 shows the setup for piercing and blanking a part at one station of a compound die. Punching and blanking elements are mounted directly opposite each other on the punch plate and die shoe of the die set. One of the simplest compound dies is a cutting die for a washer. The hole must be concentric with the outside diameter. The blanking punch for the outside diameter is also the piercing die for the concentric hole. This part of the cutting tool is mounted on the die shoe. The blanking die for the outside diameter and the piercing punch are mounted on the punch plate.

The stripper plate on a combination blanking punch and piercing die is assembled on the die shoe. A knockout plate is included as part of the punch plate assembly. The knockout plate forces each blanked washer out of the blanking die.

Compound dies are used for stamping processes where parts (with pierced holes) are to be held to close dimensional tolerances and must be flat. Compound dies require less die space so that smaller presses may be used.

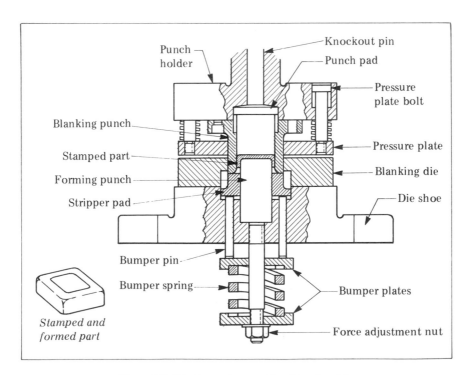

Figure 28–24 Example of a Combination Die

COMBINATION DIES

Combination dies are used when a cutting and noncutting operation are combined. For example, the combination die shown in Figure 28–24 may be used to produce a stamped and formed part. The outside rectangular slug is blanked first. As the die sections continue to move together on the downward stroke of the ram, the cup is formed by the forming punch in the die shoe moving the blanked part into the hollowed blanking punch in the punch section.

On the upward stroke, the stripper plug in the blanking punch and forming die (mounted on the punch plate) forces the formed cup out of the die. The blanking die and forming punch are mounted on the die shoe. The blanking punch for the outside rectangular form is designed internally to become the forming die section. This blanking punch/forming die combination is mounted on the punch holder section of the die set.

SHEET METAL BENDING, FORMING, AND DRAWING DIES

Bending refers to the shaping of metal from one plane to another and completely across the part. For example, a part made of flat steel plate is bent at a right angle across its width.

Forming deals with the flow of material along a curved path, usually a closed path. The formed part takes the shape of the punch or die.

Drawing is the process of producing shell forms. These forms may be cylindrical, square, or rectangular in shape. The sides may be square or at an angle.

BENDING DIES

Where practical, metal is formed by *bending dies* parallel to the grain direction produced in rolling the stock. The metal is bent without fracturing. In bending, the workpiece is usually held against the die block by a pressure pad. The edge over which the bend is to be formed on the die is rounded to the required internal radius. The forming (leading) edge of the punch is rounded. As the punch moves downward, the metal is caused to flow in the area between the die and punch. This action is sometimes called *wiping* and the die, a *wiping die.*

Large parts, such as panels, are often formed by *press-brake bending dies* and *forming dies.* The long punch and die sections, which conform to the shape to be formed, are mounted on a type of machine tool called a *press brake.* Examples of sheet metal press-brake bends and punch and die forms are illustrated in Figure 28–25.

In most metal bending operations, elastic stresses develop at the bend areas. As a result, the part tends to spring back to produce a smaller bend. This problem is overcome by *overbending* —that is, the part is bent through a greater angle so that it springs back to the required angle. Elastic stress varies according to the composition of the metal, thickness, radius of the bend, part size, and the design of the bend.

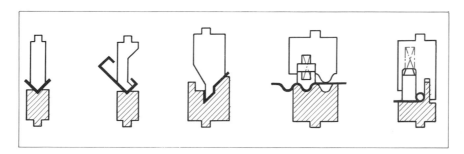

Figure 28–25 Examples of Sheet Metal Press-Brake Bends and Forms Using Brake Bending and Forming Dies

Pressure Pads. *Pressure pads* are used in die work to control the position of a part being formed so that sharp, intricate details may be produced. Pressure pads are applied above or under the strip and/or workpiece, depending on the bend and die design. Force is applied against the pressure pad by springs, hydraulics, or fluids (air). While springs are commonly used, the force against the pressure pad is not as constant as the other methods. As the springs are compressed, the force increases so that it becomes difficult to control the stretching and tearing of metal.

CURLING DIES

Many parts require a raw blanked edge to be rolled for safety, for appearance, or for strengthening the edge. Curled edges are formed by a group of dies known as *curling dies*. These dies are used on soft, ductile metals that may be rolled. Tempered metals tend to form irregular curves. Usually a forming lubricant is applied to help the metal flow over the highly polished edge curling surfaces of the punch and die. A common design of a curling die is illustrated in Figure 28–26.

EMBOSSING DIES

Embossing dies (also called coining dies) produce surfaces in which a detail is impressed on one side and raised on the other side of a stamped part. Embossing dies are produced on tracer milling machines; three-dimensional contouring machines; and embossing, die-sinking, and other detailing machines. The punch and die are machined with the same details but with allowance for the thickness of the metal. Pressure pads and/or ejector pins are incorporated in the die design to control metal flow during embossing and for ease in removing the stamped part. Embossing may be done in a single-process die, or the operation may be included at one of the stations of a progressive die.

DRAWING DIES

Drawing dies form a round cup or shell from a round blank. The blank is centered over a

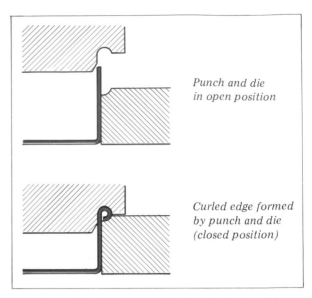

Punch and die in open position

Curled edge formed by punch and die (closed position)

Figure 28–26 Common Design of a Curling Die

drawing die of the required outside diameter. The punch, which conforms to the inside shape and dimension, is vented so that no vacuum is created to hold the drawn part to the punch on the return stroke. As the formed cup tends to spring back when it clears the bottom of the drawing die, the cup is held in the die recessed hole area and disengages on the return stroke. A drawing die for forming a shell is illustrated in Figure 28–27. In this case, the pressure pad over the workpiece controls the flow of metal.

Drawing dies are classified as shallow when the drawn shell is less than half the diameter of the part. Deeper draws require the metal to be

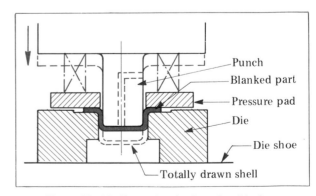

Punch
Blanked part
Pressure pad
Die
Die shoe
Totally drawn shell

Figure 28–27 Example of a Drawing Die for Forming a Shell

confined and force to be applied on the blank. These conditions cause the metal to flow without wrinkling or distortion. Pressure-loaded pads, which extend over the nesting area, hold the metal against the die face as the metal is drawn into the die. Exceedingly deep draws require a blanked part to be drawn in a number of stages.

Some products require the shell to be reduced in thickness as it is drawn to shape (lengthened). Such reduction processes call for forcing the shell through a *tight die*. Drawn parts often call for heat treating during drawing stages so that the metal remains ductile.

FACTORS AFFECTING METAL FLOW

PUNCH RADIUS

The punch radius is an important factor in terms of pressure required to draw and the possibility of thinning or tearing the metal at the bottom radius of the shell. A small radius requires greater forming forces. Similarly, the larger the radius on the draw ring (die), the freer the metal flows plastically. Too large a radius may cause wrinkles to form. Too small a radius tends to produce uneven thinning of the shell wall. The amount of force exerted by the pressure pad is controlled as the workpiece is drawn to shape. The frictional force must be adequate to permit drawing without wrinkling.

LUBRICANTS

Frictional force is overcome in some applications by using a lubricant. A number of specially prepared lubricants provide chemical protection films. Examples include chlorinated oils, sulfurized oils, soap-fat compounds, and soluble mineral oils. These lubricants are recommended for both stamping and deep drawing press operations. Other lubricants such as chalk and graphite act as physical separators for the workpiece and die sections.

Lubricants increase the wear life of dies, improve the quality of both the process and product, and increase production efficiency. Some

lubricants also provide rust protection. Most lubricants leave a residue and parts require vapor or liquid degreasing. Some lubricants produce stains on nonferrous metals. Lubricants are applied to the preformed blank by roller coating, swabbing, spraying, and brushing.

PART MATERIAL

Consideration must be given in drawing operations to the ductility and yield strength of the material. *Ductility* refers to the ability of the metal to change shape without fracturing. *Yield strength* relates to the property of the metal to the point at which permanent change in shape takes place. These two metal properties must be considered by the designer. Consideration must also be given when selecting the correct material to the following factors:

- Shape of the shell (cylindrical, rectangular, conical);
- Material thickness;
- Amount of work material to be reduced for each draw;
- Grain size and other properties of the material;
- Surface finish;
- Other stamping and forming processes (such as flanging requirements).

Metals handbooks provide manufacturers' recommendations on the classes of commercially available low-carbon hot and cold rolled sheet or strip steels and other nonferrous metals. Mechanical properties such as yield strength, reduction percent in area, and hardness values are also given in trade tables.

TOOL AND DIE STEELS

The tool designer, diemaker, and parts manufacturer are concerned with tool and die steels that possess specific physical properties for punches and dies, cutting tools, and gages. The end products of tools and dies are interchangeable parts that are stamped and formed within fine dimensional and geometric tolerances and with limited maintenance of the production tools. In other words, tool wear life is maximized.

Tool and die steels may be grouped into six general types, as follows.

- Water hardening carbon tool steels.

- Oil hardening.

- Air hardening.

- Shock resisting.

- Hot work.

- High speed (tungsten, cobalt-tungsten, and molybdenum-tungsten).

The properties of each different type of tool steel (that affect the selection of a specific tool steel) include a combination of: non-warping, safety in hardening, strength and toughness, depth of hardening capability, and wear resistance. Other factors that are considered relate to design features of punches and dies to meet particular manufacturing needs. Figure 28–28 shows a number of applications of tool and die steels for a multiple-operation, multiple-stage progressive punch and die.

Chart 28–1 identifies five different types of tool steels according to AISI classification, gives some of the characteristics of each type, and, then, provides examples of different applications in tool and die work and precision toolmaking.

Chart 28–1 Examples of AISI Types of Tool Steels, Characteristics, and Applications in Tool and Die Work and Toolmaking

Type	Classification	Characteristics	Applications
AISI S–7	Air hardening tool steel	• Good resistance to softening at moderately high temperatures • Suitable for hot-work/cold-work applications	• Blanking, forming, and swaging dies • General-purpose tools • Punches and pneumatic tools
AISI L 6	Oil hardening/tool steel	• Hard, tough, versatile tool steel that resists chipping and breaking problems	• Punches • Blanking, swaging, and embossing dies
AISI D 2	High-carbon, high-chromium tool steel	• Extremely high wear resistance properties • Deep hardening and mild corrosion resistance properties • Free from size change after heat treatment	• Blanking, forming, extrusion, and drawing dies • Forming and edging rolls
AISI A 2	Air-hardening tool steel	• Hardening throughout heavy sections • Safety in hardening • Extreme accuracy of size in hardening	• Large forming dies • Thread roller dies • Turning, forming, and coining dies • Precision tools and gages
AISI A 6	Combination air- and oil-hardening steels	• Combines deep hardening of air-hardening steels with low temperature heat treatment of oil-hardening steels • Dimensional stability	• Very large sections: blanking/forming dies • Turning and notching dies • Precision tools • Bending tools

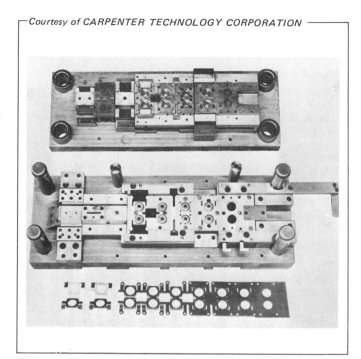

Figure 28–28 Progressive Piercing, Blanking, Forming, and Shearing Punch and Die

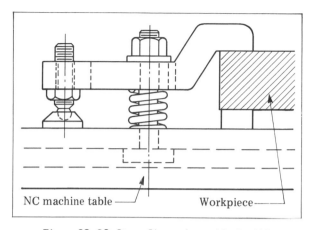

Figure 28–29 Strap Clamp Assembly for NC Work-Holding Setup

WORKHOLDING DEVICES FOR NC AND CNC MACHINE TOOLS

The principles of design for strap clamps, work supports, and locators for conventional machine tools apply to tooling for NC machine tools. One main difference is that these work-holding devices are brought together as an assembly. The assembled parts can be relocated to accommodate different sizes of workpieces. The strap clamp assembly illustrated in Figure 28–29 shows a NC workholding setup using conventional parts. The unit is designed for use directly on the table. Normally, such parts are designed for mounting on a fixture base plate. On NC machine tools, the table may replace the base plate.

SPECIAL NC ADJUSTABLE VISES

Standard vises used on drilling and milling machines may be used to hold workpieces for NC machining. Some vises are fitted with an adjustable part locator, which is attached to the vise. The adjusting feature accommodates variations in workpiece size.

Vises designed especially for NC machine tools consist of three major parts, as shown in Figure 28–30. The solid jaw and a side aligning plate serve as stationary work locators. The movable jaw is the clamping device. All three parts are keyed to slots and are secured to the machine table.

GRID-PATTERN BASE PLATES

Base plates are built to be table mounted. Base plates contain a grid pattern of holes. One

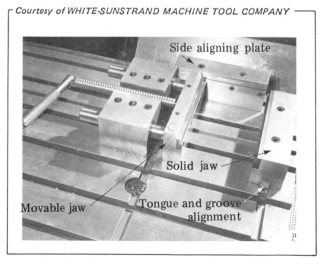

Figure 28–30 Adjustable Vise Designed for NC Machine Tools

set of holes receives hardened and ground dowel pins. The dowel pins serve as locators. The dowel pin holes are jig bored to an accuracy of 0.0002″ (0.05mm). Alternate holes are tapped to receive hold-down screws or studs. This type of base plate provides numerous locations for the locating pins. The designer or programmer usually identifies locator positions for workpiece setup. Slide movements are positioned from a starting reference point. Auxiliary fixtures and workholding devices may be mounted on the grid-pattern base plate, or *grid plate*.

Some NC machine tool builders provide a *grid system* as standard workholding tooling. The grid system consists of the base plate with a fixed pattern of locating and screw holes. Special angle blocks, or *knees*, of various sizes are included. Knees are designed with the same dowel hole and screw hole pattern as the base plate. Adjustable jacks are provided. They, too, may be screwed into the base plate.

The base plate hole series is marked along X and Y axes. As the part is programmed, the dowel

locations are identified on a grid drawing. Information is also given for locations of stops, clamps, and the workpiece. Positioning the machining processes are then programmed in conformance with the location on the workpiece. One advantage of this system is that a record is kept of the setup and it can be precisely duplicated any number of times after disassembly.

The tool setup sheet for the NC machine tool or machining center craftsperson provides information for assembling the tooling components. If a tool management program is followed, information about pregaged and measured cutting tool lengths, diameters, and other features are programmed into the tape. The assembled tool components are arranged in turret stations or in an automatic storage drum. NC machining centers are designed with one or more automatic storage chains. A 48-tool chain storage matrix is pictured in Figure 28–31. Tool storage permits random and shortest-path tool selection. Random select allows tool usage in any operation sequence and as frequently as required.

Turret tooling on turning centers in a number of manufacturing systems are set up so that every tool tip around the turret lies in the same plane. Figure 28–32 shows the tip alignment of cutting tools on the vertical turret for inside operations and the single-point, outside-diameter cutting tools on the crown turret. Each tool clears the workpiece with minimal retraction. The inside-diameter cutting tools are preset by means of a tool setting gage.

BASIC CUTTING TOOLS FOR NUMERICAL CONTROL MACHINING

TWIST DRILLS

NC drilling operations are performed without the aid of jigs and drill bushings. Therefore, the accuracy of hole location, concentricity, size, and quality depends on the drill point. The sharpness of the drill, rigidity, machining conditions, uneven cutting lip angles, a drill running out-of-true, and other problems of drill tool geometry all affect accuracy.

Courtesy of CINCINNATI MILACRON INC.

Figure 28–31 48-Tool Chain Storage Matrix for Random, Shortest-Path Selection and Automatic Switching between Chains

Figure 28–32 Tool Tip Alignment for NC Turret Lathe Application

NC drilling requires a drill point that is self-centering. The spiral-point drill provides a self-centering drill point. The point design is produced by a drill sharpener in which a generating system grinds the point with an accurate spiral.

Spotting drills and center drills are often used to precede a drilling process for accurately starting standard twist drills. The use of spotting drills and center drills requires a second drilling operation.

A design factor that affects drilling accuracy is shank length and torsional rigidity. Drills used for NC machining should be as short as practical. A collet-type holder is available to grip a twist drill. Twist drills are also engineered with shorter lengths for added strength and greater drilling accuracy.

Holes that are larger than 1″ (25.4mm) diameter are often drilled with spade drills. Spade

drills are capable of drilling accurate holes from the solid. These drills are of stubby design which adds to their rigidity and ability to withstand drilling forces. The small dead center on the spade drill minimizes end thrust. Chips are broken as they are formed by including chip breakers in the tool point geometry.

MILLING CUTTERS

Cutting speeds, feed rate, and depths of cut are generally higher on NC milling machines and machining centers. Therefore, the milling cutters must be more rigidly constructed and must maintain their cutting edges. The current trend is toward insert-type milling cutters of tungsten carbide. Solid carbide end mills are commercially available for NC machining. Cutters are designed with axial adjustment of the inserts. Finer-pitch insert-type cutters permit faster feeds without increasing the chip load.

HOLE THREADING TOOLS (TAPS)

NC tapping operations require a tap design that prevents chip accumulation in the flutes and has the greatest possible cross-sectional body area and strength. Gun or spiral-point taps are used on through holes. The spiral point produces a shearing, cutting action. The chips are projected ahead of the tap. When blind holes are tapped with gun or spiral-point taps, care must be taken to provide space below the threads where the chips may accumulate.

Ductile and soft metals are tapped with spiral-fluted taps. The helix of the spiral flute tends to lift chips out of the threaded hole. Bottoming taps are used for blind holes. A torque-limiting adjustment is important when tapping blind holes on NC machines.

In NC tapping operations, the feed rate is manually selected or programmed. If the feed rate of the NC machine does not correspond accurately with the lead of the tap, the feed rate is set at slightly less. The tap floats axially to compensate for any minor lead error. At the end of the tapping process, as the spindle reaches the preset depth, the tap driver spindle reverses and allows the tap to feed out.

CONVENTIONAL AND NC TOOL CHANGE AND SETUP METHODS

QUICK-CHANGE TOOLING SYSTEMS

Quick-change tooling systems for rotating cutting tools permit throwaway inserts to be indexed to a new cutting edge. Holders are designed so that the new cutting edge is in the same relative position with the workpiece and the original reference point.

Quick-change toolholders consist basically of a master toolholder, which becomes an integral part of the spindle design. Adapters are then designed to adapt various cutting and holding tools to the master toolholder. One design consists primarily of a taper spindle nose and quick-locking/releasing screw for the mating spindle and adapter. A second design incorporates an eccentric cam and a grooved area in the adapter. Figure 28–33 shows how a tool is aligned and locked in the adapter by turning the eccentric cam with an Allen or other wrench. The tool is ejected when the eccentric is turned in the opposite direction.

Power drawbars provide another mechanical method of securing cutting tools in adapters.

The quick-change tooling system employed on Moore jig borers consists of a taper-shank adapter with a fast-lead square thread end. The adapter is seated in the spindle by a quick turn. The process is reversed to remove the adapter. Such an adapter is used for lighter cutting operations.

NC TOOL POSITIONING

Conventional machine tools require the manual positioning of tools. NC machine tools permit cycling the tools once they are preset for height and, where applicable, the diametral dimensions are established. The NC machining center shown in Figure 28–34 has a tool positioning capability to automatically drill, ream, bore, counterbore, tap, and mill. Each operation is selected and controlled by tape. Hole depths are automatically controlled. A single tool may be

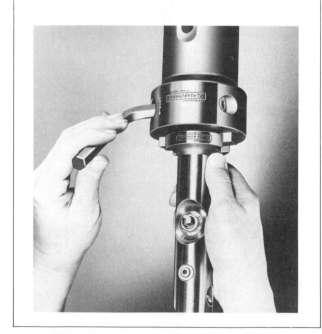

Figure 28–33 Locking Tool in Adapter with an Eccentric Cam Movement

Figure 28–34 NC Machining Center with Tool Positioning Capability to Drill, Ream, Bore, Counterbore, Tap, and Mill

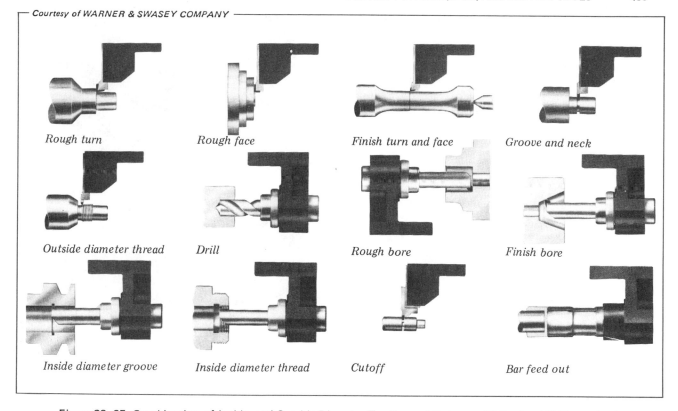

Rough turn   *Rough face*   *Finish turn and face*   *Groove and neck*

Outside diameter thread   *Drill*   *Rough bore*   *Finish bore*

Inside diameter groove   *Inside diameter thread*   *Cutoff*   *Bar feed out*

Figure 28–35 Combination of Inside- and Outside-Diameter Tooling and Cuts on a 12-Station NC Turret Lathe

used to cut to several depths during the cycle without manual adjustment.

The turret drilling machine may be used as an example of automatic tool cycling. The cycle covers rapid traverse to position the spindle, power feed control, rapid spindle retraction, automatic indexing, adjustable feed rates, and automatic tap reversal. The combination of inside- and/or outside-diameter tooling and cuts that can be performed on a 12-station NC turret lathe is shown in Figure 28–35. The NC turret tooling at each of the twelve stations is shown in Figure 28–36. On this machine tool, the turret is placed alongside of rather than aligned axially with the spindle axis. The turret moves left and right (at a right angle to the spindle/workpiece) and axially (parallel to the workpiece centerline). Tapers and contours are cut by numerically controlling both axes.

Figure 28–36 12-Station NC Turret Tooling

Each station on the turret may be tooled for either internal or external cutting operations. The turret may be indexed in either direction. The two basic toolholders that are required for this 12-station NC turret lathe are illustrated in Figure 28–37. One toolholder is for inside-diameter tools. The other toolholder is for outside-diameter tools. Solid-shank cutter holders with insert-type cutters may be slid into the holder and clamped. Roughing cutters, whose edge locations are established to within ±0.003″ (0.07mm), require no further adjustments. Finish insert cutters are brought to exact size by dialing the required amount of tool offset on the machine control unit console.

As discussed earlier in the unit, the turret tooling setup places the edges of all inside-diameter end working tools in a single plane. Outside-diameter cutters automatically fall into known locations.

MANUALLY CONTROLLED TOOL CHANGES

Once the machine is automatically cycled, all machine functions are tape controlled. However, for trial cutting or for changing inserts, the machine functions may be manually controlled. Pushbuttons, dials, and switches, mounted on the operator's panel on the control console, permit manual control.

TOOL PRESETTING

END CUTTING TOOLS

Presetting means that the end of a cutting tool is set longitudinally in relation to a master reference surface. The diameter of the tool establishes the radial relationship. Tool management systems permit precise, rapid measurement of tool length and feed information for programming as related to the end cutting surface and the diameter. Digital readouts on the measuring machines provide instant dimensional measurement. A commercial universal presetting height gage may also be used to set various tool lengths. A micrometer head attached to an arm that may be positioned at 1″ intervals on the height gage

Courtesy of WARNER & SWASEY COMPANY

For outside-diameter cutters

For inside-diameter tools

Figure 28–37 Two Basic Toolholders for 12-Station NC Turret Lathe

permits adjustments to be made within ±0.001″ (0.02mm) along the 1″ (25.4mm) travel of the micrometer spindle.

RADIAL CUTTING TOOLS

Boring bars require presetting to required bore diameters prior to use. Precision presetting spindles provide a method of accurately holding the boring bar and adapter so that the cutter teeth edges may be indicated. The preset spindle fixture and inserted cutter are mounted on a surface plate. Gage blocks representing the radius dimension may be set at the required radius.

Precision presetting machines, using a tool-setting microscope and optical reading scales, are used for measurements in increments of 0.0001″ (0.002mm) in either horizontal or vertical direction. The optical comparator is another tool setting/measuring machine. The comparator requires a stage that is suitable for holding the tool adapter.

Safe Practices for Applications of Production Tooling

- Inspect every punch and die part against possible hardening cracks.
- Check the tolerance between a punch and die. No undue forces should be applied to produce the required blanked or formed part.
- Make sure the die design and press conditions permit a stamped and/or formed part to be ejected properly.
- Check the shut height of the die set to ensure there is adequate clearance to safely carry on the power press operations.
- Make certain all safety guards and protective devices on the press are operable and in place before power is applied.
- Balance work-holding fixtures dynamically when the fixtures are used on turning machines. Centrifugal forces must be reduced to minimum for safe operation at machining speed.
- Check to see there are no projections and/or sharp corners on rotating fixtures.
- Secure machining guards before a fixture (and spindle) is rotated by power.
- Use adequate clamping force to prevent a rotating fixture from moving during a machining process.
- Check the valves under the workpiece and adjust the pressure to ensure that the holding forces on a vacuum fixture are adequate to hold the workpiece against cutting forces.

- Check the valve chuck areas. Wipe the chuck face to free chips and foreign matter from around the valve openings.
- Use end and side plates on magnetic fixtures to absorb the cutting forces.
- Cover the faces of base plates and other work-holding/positioning parts of a grid system as protection against chips.
- Recheck the tip alignment of cutting tools for turret lathe applications to be sure all cutting edges are in the same plane.
- Use only self-centering drill types when drill points on NC machine tools are to be used to precisely locate and form holes.
- Select taps for NC threading with maximum tooth strength and flutes that force chips ahead of the tap and out of the workpiece.
- Operate a NC machine tool by manual control to cycle all operations before a production run. Check for operating clearances, positioning, surface finish, and dimensional accuracy.
- Check the accuracy of all preset end- and radial-cutting tools.
- Follow standard hand tool, layout, and machine safety precautions.
- See that all hand and eye protection devices are in place and operating before starting sheet metal blanking and forming operations.
- Handle metal sheets, strips, and rolls; setups; and stamped and scrap metals carefully. The sharp edges can cause serious cuts.

TERMS USED WITH PRODUCTION TOOLING

Positioning The dimensional relationship between the cutting tool and a specified reference point on the workpiece. The process of moving a table and mounted workpiece to be aligned at a particular reference point.

Locator (jigs and fixtures) Pins or other members against which a piece part rests in order to perform another process at a specific distance from the reference surface. Work-positioning members against which successive parts are brought to establish reference surfaces for second operations.

Rest plates	Parts used in jig and fixture design as bearing surfaces for positioning a piece part.
Leaf, box, and open-plate jigs	Three basic types of drill jigs used primarily for hole forming processes such as drilling, reaming, and tapping.
Shut height (jigs and fixtures)	The clamping range of a fixture for holding a piece part. The shut height is adjustable within the range of the fixture.
Blanking, piercing, and cutting-off (die work)	Three basic sheet metal production stamping processes.
Inclinable presses	A family of power production machines that utilize punches and dies to cut, form, and emboss parts by automatic cold metal stamping processes.
Roll feed (power press)	A power feed attachment that controls the automatic feeding of roll stock into a die.
Shut height (press work)	The opening at the bottom of the ram stroke when the machine stroke adjustment is in the up position.
Elastic limit	The point within the metal at which forming or cutting tools cause the metal to flow mechanically.
Stripper (stripper plate)	A plate extending a short distance from the stock. A plate that holds the stock on the return ram stroke so that the punch or die may be freed and the stock positioned to produce the next part.
Progressive and compound dies	Classifications of punch and die tooling in which two or more press operations are performed each stroke.
Pressure pad	A component of a die that forces against a blanked part to control movement (metal flow) during a forming or drawing process.
Embossing (press work)	The process of impressing a design into one side of a stamping while simultaneously raising a duplicate (but altered by the metal thickness) reverse reading design on the opposite side.
Shallow and deep draw	Punch and die tooling for forming a shell by drawing stock through a die between the die cavity and the outside surface of the punch.
Grid system	NC positioning and holding system. Precision-machined base plate and accessories that use a particular pattern of holes for locating and holding workpieces in horizontal, vertical, and angular planes.
Tip alignment	A system of presetting machine tooling in which the point of each turret tool is set in the same plane.
Spade drill (NC tool point geometry)	Drill of stubby design that provides chip breakers in the cutting edges and lips.

SUMMARY

- Three locators are generally used in jig and fixture work to establish the piece part in the same location each time.

  - At least two pins are used in combination to fix the location of jig or die elements in the fixture or die set.

- Locating nests are areas formed to hold a piece part in a specific location for clamping and machining.

  - Quick-acting swing, lever, and cam clamps are based on lever principles for applying force to securely hold a part in position in a jig or fixture.

- Drill jigs limit the path of a cutting tool to cut concentrically and parallel and to produce dimensionally accurate holes.

  - Open-plate jigs are primarily for single-hole drilling, reaming, or tapping processes.

- Box jigs permit nesting a piece part within the jig and performing multiple operations with the jig positioned on any one of its sides.

  - Leaf jigs use a hinged plate to open the jig, secure the workpiece, and position and guide a cutting tool.

- Indexing jigs provide for setting a piece part a specified circular distance for each successive operation.

  - The shut height of a universal fixture may be adjusted to accommodate the height of the jig or fixture components and the piece part.

- Segmental chips are preferred in jig operations to prevent interference in loading, unloading, and performing the machining operation.

  - Drilling, reaming, tapping, and other hole-forming operations in jig work depend on the use of press-fit and renewable bushings.

- Dowels are used for accurately aligning mating parts. Components are usually secured with fillister and socket-head cap screws.

  - Magnetic and vacuum chucks serve as positioning and holding fixtures.

- Blanking, piercing, cutting off, shaving, and trimming dies perform metal cutting operations.

  - Piercing refers to the stamping of holes in a strip or plate.

- Shaving is a second stamping operation. When parts are shaved, a small amount of material is cut to produce a part to close dimensional tolerances.

  - Power press ram faces and bolster plates are designed for mounting commercial die sets and for adjusting the machine shut height.

- Die clearance refers to the dimensional difference between the size of the punch and the die.

  - Stripper plates are either fitted around a punch or hold the stock near the die face as the punch is withdrawn on the up ram stroke.

- Commercially available cast iron and alloy steel die sets include: a die shoe, guide pins, guide bushings, and the punch holder.

    - A progressive die replaces a number of separate dies in performing a series of operations simultaneously on the same strip or sheet.

- A compound die performs two or more stamping and/or forming operations at one station during one press stroke.

    - Bending, forming, drawing, and embossing dies are used for mass production of parts that require other than a flat, plane surface.

- Raw, blanked edges may be curled for appearance, safety, and strengthening.

    - Embossing (coining) dies produce a depressed design that is raised and reversed on the second side of a stamped part.

- Pressure pads control the force on a stamped part so that it may be formed to a required shape without wrinkling or distortion.

    - Shallow and deep draw dies form a shell or cup between the inside surface of the die and the outside surface of the punch.

- Metal flow is affected by the radius of the punch, characteristics and thickness of the metal, the nature of the forming process, the finish of the die surfaces, speed, and the lubricant used.

    - NC vises are especially designed for table alignment and positioning of the solid jaw, as a right angle reference (locator) surface, and as a movable jaw.

- Grid patterns on base plates and knees permit precise work positioning and mounting along X, Y, and Z axes.

    - NC drills are designed to be self-centering and to drill concentrically, parallel, and true diameters.

- NC spindle adapters permit the use of many shank designs on tools. A quick-locking/unlocking device is provided in the design of a tool adapter.

    - NC tool management systems may be programmed to compensate for tool length and tool diameter variations.

- Precision presetting spindles provide a fixturing device for setting boring bars to required boring diameters.

    - Presetting machines provide greater flexibility and more precise tool setting measurements than fixture devices. Setting accuracies within 0.0001" (0.002mm) are made using tool-setting microscopes.

- Standard hand, machine, tool, and personal safety precautions and ANSI safety codes must be followed when making trial runs on sheet metal blanking and forming production tooling, jigs and fixtures, and cutting tools.

    - Extra precautions are to be taken to properly handle sheets, rolls, strips, and scrap for power press operations.

- ANSI safety codes must be followed for all machine tools, particularly power presses.

UNIT 28 REVIEW AND SELF-TEST

1. a. Identify two forms of locators that are commonly used in jig and fixture work.
    b. State the function served by each locator.

2. State two design features that differentiate a box jig from an indexing jig.

3. a. List two different classifications of jigs or fixtures.
    b. Give an example in each classification.

4. a Name circled parts ① through ⑨ of the combination shell blanking and drawing die shown below.
    b. State the functions of parts ②, ④, ⑥, ⑦, and ⑩.

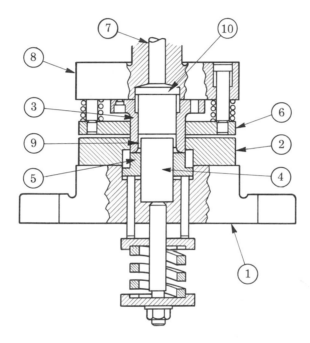

5. Differentiate between (a) blanking and forming dies, (b) progressive and compound dies, and (c) forming and embossing dies.

6. Explain briefly the effect of punch radius on metal flow for drawing processes.

7. Describe briefly one method that is used for tool point setting for a NC turret lathe.

8. Cite two milling cutter design features that are especially important to NC milling setups and operations.

9. Identify three features of a modern NC production machine tool.

10. Describe one tool presetting method for end-cutting and radial-cutting tools.

11. State two personal safety precautions that the toolmaker takes in testing a punch and die on a power press.

PART 9 Numerical Control (NC), CNC, CAM, and Flexible Manufacturing Systems (FMS)

SECTION ONE

Numerical Control Machine Tools

This section includes two units. Unit 29 deals with components of numerical control systems (NC), functions, and applications, and provides examples of NC programming for drilling and milling processes. Building on this foundation, unit 30 covers machine control units (MCU), NC data processing technology and applications, CNC systems, computer-assisted programming (CAP), and CAD/CAM.

UNIT 29

NC Systems, Principles, and Programming

OBJECTIVES

After satisfactorily completing this unit, you will be able to:

- Discuss applications of NC to standard machine tools; constant and intermittent cycling of automated processes, and open- and closed-loop systems.
- Describe positioning the spindle by incremental and absolute measurement and a floating zero point; linear, circular, parabolic, helical, and cubical interpolation factors in programming; NC tapes, codes, and preparation; and electromechanical, photoelectric, and pneumatic tape readers.
- Set up tab sequential format programs, including console presets and sequencing statements.
- Translate coded functions and applications of NC word address format programs to continuous-path machining and fixed-block format programs.
- Determine applications of mirror images.
- Develop step-by-step work plans to perform the following processes.
    - Programming a Single-Axis, Single-Machining Process.
    - Programming for Two Axes and Tool Changes.
    - Programming for Three Axes.
    - Programming for Milling Processes Involving Speed/Feed Changes.
- Read NC drawings with incremental and absolute dimensioning. Interpret X, Y, and Z axes and the reference point, rotational axes, and quadrants.
- Follow each *Safe Practice* relating to personal safety and machine tool and workpiece protection.
- Correctly use each new *Term*.

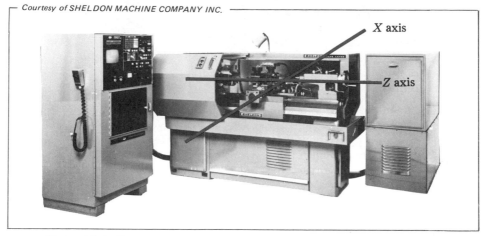

Figure 29-1 Two Basic Axes of an NC Lathe Equipped with a CNC System

NUMERICAL CONTROL FUNCTIONS

Numerical control is a complete system of taped or computerized instructions. The basic functions of the system include the following:

- Controlling movements to position cutting and forming tools in relation to a fixed reference point,
- Controlling movements of cutting tools for setups and machining,
- Establishing sequences of operations and time intervals,
- Setting feeds and speeds,
- Monitoring accuracy and cutting tool performance or other machining functions,
- Providing readouts of machining accuracy,
- Changing the nature and sequencing of processes,
- Actuating shutdown or recycling.

The term *numerical control* designates a numbering system of letters and numbers that regulate an entire machining process and the sequence of operations to machine a workpiece. Movements of the workpiece and tools are programmed on tapes or the storage banks of computers. *Programming* means that, *on command* (when required), impulses of energy move by electrical means from an electronic controller to interface at the machine where electronic/electrical energy is converted into mechanical energy.

The interfacing of energy from electrical/electronic to mechanical or pneumatic energy actuates motors, switches, clutches, and brakes. The actuating devices control, guide, and time every movement of the processing tools and the machine elements. All machine tools are capable of being controlled by tape or computer. Drilling, turning, shaping, abrasive machining, band machining, and production equipment are examples of NC applications.

ADVANTAGES OF NUMERICAL CONTROL MACHINES

Control functions formerly performed on machine tools by the machine operator are now translated into functions of the numerical control system. The longitudinal distance and direction movement of a machine table (X axis), the traverse cross feed movement (Y axis), and the vertical or angular movement of a spindle (Z axis) may be numerically controlled.

The two basic axes (X, cross slide; Z, longitudinal) of a NC lathe are shown in Figure 29-1. Figure 29-2 illustrates the three basic axes (X, table; Y, saddle; Z, spindle) of a computerized numerical control (CNC) vertical milling machine.

Speed and feed rates for particular cutting tools, cycling for each process, and controls of cutting fluids may also be numerically controlled. In each of these examples, the machine control functions are performed by synchronized motors that respond to pulse commands.

Numerical control machines have many advantages. Five significant advantages include (1) productivity, (2) repeatability, (3) flexibility, (4) reduced tooling, and (5) increased machining capability.

PRODUCTIVITY

Once a numerical control machine has been programmed and the total cycle of processes has been checked, machining continues without the possibility of operator error and at a fixed rate of production. This rate is usually a higher production rate than a machine operator can maintain.

A factor of productivity is the ratio between cutting and noncutting time. Some noncutting steps that can be handled more efficiently by programming than by manual operation include: positioning for different operations; changing spindle speeds and tool feeds; machine start-stop; releasing, retracting, and returning the cutting tool to the starting point; and indexing.

REPEATABILITY

A numerical control machine tool has the capability to reproduce parts with extremely small variations of accuracy. The variations within close dimensional measurement and surface texture requirements result from tool wear. Since cutting tools are constantly monitored and replaced when worn by accurately sharpened tools, the accuracy is repeated throughout a production run. The tape instructions are unvarying in controlling all processes. Therefore, with precision repeatability and the reliability of the system, inspection costs may be reduced.

Once programmed, point-to-point positioning accuracies are obtainable for each linear axis to within ±0.0005″ (±0.013mm). The repeatability accuracy is within ±0.0003″ (±0.008mm). Moreover, point-to-point errors are not cumulative.

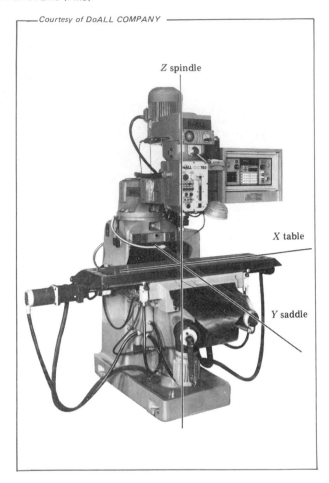

Figure 29–2 Three Basic Axes of a Computerized Numerically Controlled (CNC) Vertical Milling Machine

FLEXIBILITY (VERSATILITY)

Numerical control machines may be programmed to produce a single workpiece, a short run, or a production run. Machining functions are changed by using a new tape. The tapes provide permanent programming and are stored between runs. With computer numerical control (CNC), the programs are stored in computer memory. CNC machines can be operated directly without the use of a tape. CNC machines are discussed in the next unit.

With tape or computerized programming, it is possible to interrupt a production run, set up to produce an altogether different workpiece, and then return to the machining of the original part. The changeover is accomplished in a minimum of setup time. In many cases, jigs and fixtures are eliminated.

REDUCED TOOLING

Shop layouts are required for conventional machining when one part or a few parts are to be machined without jigs or fixtures. On NC machines, jigs and fixtures are often replaced because the accuracy programmed into a tape or computer permits precise machining. Programming and tape preparation and duplication require less time than the construction of jigs, fixtures, and positioning gages. Further, a tape has a longer wear life.

NC machines have a built-in check system. At the end of a block of commands, the control returns the spindle (cutting tool) to a zero starting reference point. On most systems, the operator is alerted by a warning system if the control fails to return to zero.

INCREASED MACHINING CAPABILITY

Tool changers are widely used on NC machine tools. Tool changers serve the function of holding the tooling required for machining a particular part. A *tool storage drum* may be turned until the required tool is in position so that it may be pivoted 90° downward. A *tool changing arm* grasps the pivoted tool and places it in the spindle. Simultaneously, the tool in the spindle is released to the tool changing arm, is returned to the pivot arm, and replaced in the drum.

The tool drum may be located vertically or it may be top mounted. Figure 29–3 shows a tool changing application with a top-mounted tool drum and horizontal spindle. A drill is being inserted in the spindle and the previously used shell mill is in position to be returned to the tool drum.

The cutting tools are usually held in specially designed taper-shank toolholders. The taper shank fits the spindle taper (Figure 29–3). The locking mechanism serves to draw the toolholder and tool concentrically into the spindle taper

Figure 29–3 Tool Changing with Top-Mounted Tool Drum and Horizontal Spindle

and hold it securely. On command, after the machining process, the mechanism releases the holder and tool to the tool changing arm.

DISADVANTAGES OF NUMERICAL CONTROL MACHINES

Numerical control machines do have some disadvantages. Unless the nature of the machining processes fully utilizes the capability of NC tooling and machines, the initial cost is higher than for conventional machines. Consideration must be given to necessary complementary equipment such as the tape writer, electronic control console, and other accessories. NC and CNC equipment requires considerably more floor space. Also, NC machines must be built more rigidly than conventional machines in order to withstand the accelerations dictated by servocontrols. Small jobbing shops, not able to justify a programmer, may need to farm out the programming of the system.

Maintenance for the NC system requires knowledge and skills that cut across mechanical, electronic/electrical, and pneumatic systems. Machine downtime on NC equipment can be more expensive than the maintenance of regular machine tools and production machines.

Considerable analysis must be made to establish that a train of parts may be moved through an automated series of processes economically for profit. It must also be established that the market is ready to receive the products of NC or CNC machine tools.

APPLICATION OF NUMERICAL CONTROL TO STANDARD MACHINE TOOLS

Applications of numerical control vary from a single-spindle drill press requiring positioning along X and Y axes to complex machining on a multipurpose machining center. The drill press applications may include machine capability to follow tape instructions for cycling the quill and spindle. Speeds and depth on the drill press may be hand controlled.

Some numerical control systems are designed to be added to existing lathes, drilling machines, abrasive machining machines, milling machines, and other machine tools. This adaptation is called *retrofit*.

In order to combine a wide variety of machining processes that otherwise would require movement of workpieces among several machines, the whole movement toward numerical control has brought on the concept of the *machining center*. Total machining is performed on a single composite machine tool—that is, a machining center. With the machining center, the several spindle speeds and feed rates; turret indexing and workpiece indexing; positioning of machine components along three, four, and five axes; and continuous-path machine cuts for profile cutting in three dimensions are functions that may be controlled from tape instructions.

FOUNDATIONAL PROGRAMMING INFORMATION

The starting point for NC programming is the part drawing. The part drawing supplies information about design features and dimensional and other machining specifications. The numerical control programmer is concerned with control functions and machine positioning information. This information is coupled with a knowledge of machine technology. The programmer needs to know the capability of the NC machine tool to be used, work-holding devices, tooling, speeds, cutting feeds, coolant flow, and the processes and setups needed to manufacture the part.

The NC program requires information to position a spindle and work table. The tape instructions then specify the desired machining processes. For example, the specific tape letter code used in one system identifies *machining tasks* as *M functions*. This letter code is followed by a numeric code such as 01, 02, or 06. The following are a few common M functions and their meaning:

M01	index
M02	end of program (calls for tape rewind)
M06	tool change
M07	coolant #2 on
M08	coolant #1 on
M09	coolant off
M10	clamp
M11	unclamp
M50	spindle up
M51	spindle down
M57	override milling rate

SPECIAL MACHINE DESIGN CONSIDERATIONS

Under conventional conditions of machine tool operation, the craftsperson is responsible for compensating for tool wear. On numerical control machines, since final movements are not subject to human controls, the manufacturer must include many features to maintain and prolong the accuracy of the machine.

One of the common problems in table, saddle, and knee movements on a milling machine or in saddle, cross slide, and compound rest movements on lathes and other machine tools that depend on a screw and nut movement is backlash. When numerical control is added to standard machine tools, electromechanical backlash compensation is built into the machine control unit. Where numerical control is an integral part of the design, backlash is overcome by a *recirculating ball screw.*

The limitations of using thread mechanisms for movement of machine components is taken care of on other machine tools by using hydraulic positioning systems instead of thread mechanisms. The introduction of hydraulic, pneumatic, and electronic controls to automated machining was accelerated during the early 1940s. However, some of the concepts of automatically controlling movements of machine components date back to the first quarter of the eighteenth century. At that time, punched cards were used in England to control weaving and knitting machines. Today's added machine design features relate to computerizing numerical control systems and to the establishment of multipurpose machining centers.

TYPES OF AUTOMATED PROCESSES

All machining requires either *constant* or *intermittent movement (travel).* Intermittent means the workpiece travels at different lengths of time for a number of processes. When the entire processing is operated by mechanisms, the machine is automated and the workpiece is automatically processed.

CONSTANT CYCLING

Grinding, milling, and turning are operations that are adapted to constant cycling travel. These operations may require straight-line or circular movement. Normally, the workpiece is held in a fixture while one or more operations are performed. In constant cycling processing, the workpieces are removed from moving fixtures. The workpieces move in a designated sequence for machining.

INTERMITTENT CYCLING (TRAVEL AND WORK STATION INDEXING)

Drilling, boring, reaming, and counterboring operations in a single workpiece provide an example of intermittent cycling. Uneven time periods are required for each operation, and, within an operation, holes of different sizes may be machined.

Some parts are straight-line indexed. The workpiece moves from one station to the next according to function, sequence, time, and machining requirements. Production machines are arranged according to function and sequence of processing. A number of machines are self-contained, making it possible for a machined part to be processed to completion. In other production, a series of machines may be grouped and the workpieces fed by automatic feeding devices.

NUMERICAL CONTROL SYSTEMS

Automation requires every type of movement of the cutting tools and the workpiece to be plotted and programmed into a controlling unit. Automation also requires the mechanization of every movement according to time factors in response to impulses of energy. The predetermined manner in which machining processes, machine components, and tooling respond to and are controlled by impulses of energy is established through the use of tapes and computers.

CLOSED LOOP NC SYSTEM

In general, there are two numerical control systems: *closed loop* and *open loop.* The main components of a closed loop NC system are illustrated in Figure 29–4. A signal from the numerical control unit is fed through the machine control unit (MCU) to provide a specific instruction (motion command) to the servo drive unit. The lead screw is actuated by a servomotor. If the signal to the MCU directs the servo drive to feed a machine table 6″, the table is moved this distance. A *sensor* on the servomotor (drive motor) feeds back a signal

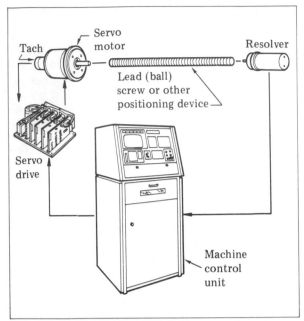

Figure 29-4 Major Components of a Closed Loop NC System

(through the encoder) to the MCU to indicate the table has moved the instructed distance of 6″. In the closed loop system, the machine control unit is provided with a check on the accuracy of the machine movement.

More circuits or loops may be added to control more motors. The motors drive the tools or move tables on any axis the machine can handle. The movements may also be controlled simultaneously to permit the cutting tool and/or table to be moved in any direction on any axis. Control information is used instantaneously to make any cut in any contour within the capability of the machine.

OPEN LOOP NC SYSTEM

The open loop NC system is used on many installations of numerical control *(retrofitting)* to existing machine tools. In the open loop system, *stepping motors* are used to control the movements of the machine components. There is no feedback system.

The MCU supplies the electric current impulse to the stepping motor. The number of pulses of the MCU is determined by the number

a table a specific distance within a time limit. Each current pulse causes the motor rotor to turn a fraction of a revolution. Some stepping motors advance a machine table 0.001″ (0.025mm) each pulse. For example, if the table is to advance 1.000″ (25.4mm), the MCU directs 1,000 pulses to the stepping motor.

ADDITIONAL CIRCUITS AND LOOPS

The simplest NC system includes a *control unit*, an *information feeder* (tape or computer), an *actuating/control unit* to supply power to each movement, and a *feedback device (transducer)* that tells how much movement is taking place.

NUMERICAL CONTROL DRAWINGS AND INTERPRETATION

Numerical control systems rely on drawings to furnish full engineering/design information. The craftsperson produces a part according to a drawing to meet exacting dimensional requirements, surface finish standards, and other specifications. Common guidelines have been established for preparing NC drawings for a programmer to produce the required part program. The NC machine tool technician must also interpret the drawing to establish that all operations are being performed according to requirements. Dimensional and other machining processes on a drawing may also be coded in numerals and letters that are foundational to programming and numerical control.

NC RECTANGULAR COORDINATE SYSTEM

Numerical control part drawings are based on what is called the *rectangular (cartesian) coordinate system*. Coordinate dimensions are used on drawings to represent the part. The coordinates are also readily adapted to define the position of machine slides that are designed to move in mutually perpendicular directions. The system of rectangular (cartesian) coordinates makes it possible to describe any position of a point in terms of distance from an original reference point. The distance may be along two or three mutually perpendicular axes.

X, Y, AND Z AXES AND ZERO REFERENCE POINT

Within the rectangular coordinate system, there are two basic axes: X and Y. These axes are perpendicular, lie in the same plane, and are known as *coordinate axes*. A third *spindle axis*, Z, is perpendicular to the X-Y plane. A three-dimensional part (mass) can be described accurately according to relationships with the X, Y, and Z axes. The three axes intersect at a point. The point is identified as the *origin* or *reference point*. The numerical value assigned to this origin point is *zero*. Figure 29–5 illustrates the basic X and Y axes, spindle axis Z, and zero reference point.

The X, Y, and Z notations are applied to machine tools as identification of the basic machine axes. The Z axis is reserved for the machine spindle axis regardless of whether the spindle is positioned horizontally or vertically.

The two basic axes of a lathe are identified as X axis for cross slide movements and Y axis for longitudinal carriage movements. The spindle axis is the Z axis. On a vertical-spindle milling machine, the X axis relates to longitudinal table movements; Y, to saddle (transverse) movements; and Z, to spindle movements.

ROTATIONAL AXES

Numerically controlled motion around the basic X, Y, and Z axes may be related to *rotational axes*. A rotary table or indexing mechanism may be operated from tape or computer instructions around a basic axis. Information must be provided for (1) direction of rotation and (2) the basic axis around which rotation takes place. Lowercase letters a, b, and c identify the rotational axes (Figure 29–5).

Two- and three-axis numerical control machine tools are used in this unit to explain NC principles, systems, and applications. Common applications of two-axis numerical control machine tools include drilling machines (standard

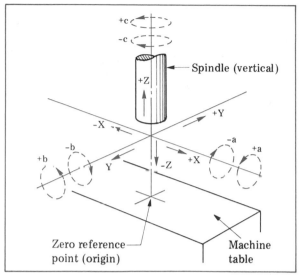

Figure 29–5 Basic X and Y Axes, Spindle Axis Z, Zero Reference Point, and Rotational Axes a, b, and c

spindle and turret heads), engine lathes, jig borers, and horizontal boring mills. Examples of three-axis numerical control machine tools include turret lathes, vertical turret lathes, planer mills, and horizontal-spindle milling machines. Machining centers are designed and operated as three-, four-, and five-axis numerical control production machines.

QUADRANTS AND NC SYSTEM POINT VALUES

The X and Y coordinate axes, which are at 90° to each other and are in the same plane, form four quadrants. The quadrants are numbered QI, QII, QIII, and QIV. The origin point is zero. Any point (position or dimension) within a quadrant has a plus or minus value (Figure 29–5) depending on the direction a measurement or distance is taken from the point of origin. Since the X and Y axes are in the same plane, there are no measurement points above or below the plane.

Quadrant I	+X and +Y
Quadrant II	−X and +Y
Quadrant III	−X and −Y
Quadrant IV	+X and −Y

Quadrant point values are important because they are used as directions or specify locations for particular operations or setups. These directions are programmed on tape and fed into the MCU. The commands control the movements of the machine tool spindle or work table. The movement in one direction is positive; in another quadrant it may be negative. Similarly, spindle movement in the Z axis has positive and negative directions. When the spindle (cutting tool) moves toward the machine table, the movement is defined in terms of a negative (–Z) direction. Conversely, any movement of the spindle away from the machine table is in a positive (+Z) direction.

BINARY SYSTEM INPUT TO NUMERICAL CONTROL

Numerical control is a precise electronic control system. It depends on a two-word numerical vocabulary. The numerals communicate information to "stop" and "go" control pulses. Numerical data are entered on a tape by *binary numbers* and *binary notation*. The following brief description of the binary numbering system, binary notation, and binary code decimal system provides background information for the programmer and NC machine tool technician.

TWO-WORD BINARY NUMBERING SYSTEM VOCABULARY

The binary numbering system is based on the powers of the Arabic number 2. In the binary system, 2 is written 2^1, 4 is written 2^2, and so on. In mathematical terms, the *law of exponents* requires that any number raised to the zero power equals 1. Thus, $2^0 = 1$.

The ten basic numbers in the standard Arabic numbering system use a binary code. Numbers 1 through 10 fit into two-word binary number combinations. Any conventional number may be written as the sum of particular binary numbers. The numbers 1 through 10 are expressed as binary numbers as follows:

Arabic Numbers	Binary Numbers
1	2^0
2	2^1
3	$2^1 + 2^0$
4	2^2
5	$2^2 + 2^0$
6	$2^2 + 2^1$
7	$2^2 + 2^1 + 2^0$
8	2^3
9	$2^3 + 2^0$
10	$2^3 + 2^2$

The binary numbers are punched according to a particular pattern into one of the channels of a tape. The punched binary numbers transmit control information to the tape reader. The pulses that are produced control the machine tool motors or devices for positioning and machining processes.

BINARY NOTATION

The binary numbers are arranged in an organized pattern to express the original numerical values. Two characters, 1 and 0, are usually used to express a binary number. Each character is called a *bit*. The binary character 1 shows that the binary number *is present* and is to be counted. The character 0 indicates the binary number *is not present* and is not to be counted. Binary numbers are placed in an order of increasing power from right to left. The binary numbers start with 2^0. The sum of the binary notation in a row is its numerical value. For example, the binary notation

$$\begin{array}{cccccc} 2^4 & 2^3 & 2^2 & 2^1 & 2^0 \\ 0 & 1 & 1 & 0 & 1 \end{array}$$

equals 13:

$$0 + 2^3 + 2^2 + 0 + 2^0 = 0 + 8 + 4 + 0 + 1 = 13$$

The two characters, or bits, of a binary notation may be represented electronically by a switch. The switch may be open or closed. The bits may have a plus (+) or minus (–) charge of a ferrite core or magnetic film in a computer.

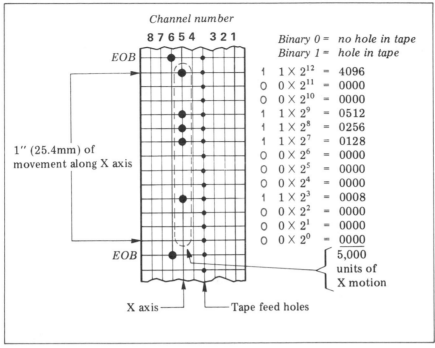

Figure 29-6 Portion of a Tape for Positioning a Spindle One Inch (25.4mm) along X Axis by Using a Straight Binary Number

A portion of a tape for positioning a spindle by using a straight binary number is illustrated in Figure 29-6. In this figure, channel 5 is used to program the input information needed to advance a workpiece along the X table axis for 1″ (25.4mm). In this case, each unit of X motion is 0.0002″. Thus, 5,000 units of X motion are required to produce the 1″ of slide motion. A hole in the tape in channel 5 equals a binary 1. The lack of a hole in the channel equals a binary 0. The binary notation for the 5,000 units of movement is expressed as:

1 0 0 1 1 1 0 0 0 1 0 0 0

Thirteen lines are used in channel 5 on the tape. Binary numerals are identified by a punched hole or no hole. The binary notation for the 5,000 units of motion is punched between the beginning and end-of-block holes in channel 6 of the tape.

BINARY CODE DECIMAL (BCD) SYSTEM

Tape formats for numerical control provide standardized information. The EIA (Electronics Industrial Association) and the ASCII (American Society for Computer Information Interchange) tape formats use the *binary code decimal (BCD) system* of digit coding. Each channel on the tape is assigned a value, as follows:

Channel	Assigned Numerical Value
1	1
2	2
3	4
4	8
6	0

These five channels may be used to designate any number between 0 and 9.

Numerical quantities are expressed in binary notation running along the length of the tape. Each number is expressed as a number of digits, usually six. While a decimal point is not shown, it is understood to be between the second and third digit.

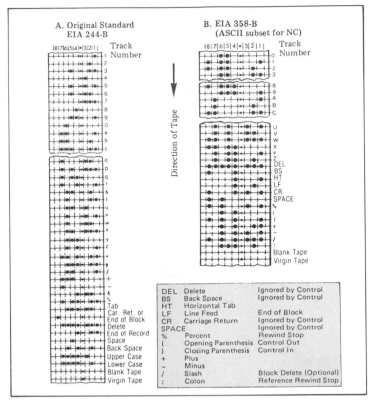

Figure 29-7 EIA Original 244-B Standard and ASCII Subset
(EIA 358-B Standard) NC Punch Tape Codes

EIA tapes are encoded with numbers (0 through 9), letters (A through Z), signs, and other symbols. The codes are arranged in horizontal rows. Two examples of EIA tape formats are illustrated in Figure 29-7.

Tape (A) identifies the original EIA Standard (EIA 244-B) punch tape codes. The *odd number* of punches at every level on the tape provide for a parity check. EIA Standard 358-B has an *even number* parity and represents the ASCII subset for numerical control. Both EIA Standard codes EIA 244-B and EIA 358-B are in current use in NC. The feed track on both tapes is between the tracks numbered (3) and (4).

NUMERICAL CONTROL WORD LANGUAGE

A *NC word* is a set of letter and numeric characters arranged in a prescribed way. The two basic types of words in a NC word language are (1) *dimension* and (2) *nondimension*.

DIMENSION WORDS

Each dimension word begins with the axis address code (letter). Each dimension word has three parts: the axis address code (letter), the appropriate (+ or -) directional sign, and the distance of movement. If there is no + directional sign in the word, the control interprets the word as positive.

NONDIMENSION WORDS AND FUNCTIONS

Nondimension (direction) words fall into categories such as: (1) sequence, (2) preparatory function, (3) feed rate, (4) spindle speed, (5) tool (turret) selection, and (6) miscellaneous function. In each case, an *address character* (designated letter) is followed by a given number of digits.

The *sequence number* is designated by the address character N and three numeric digits.

The word indicates the start of a specific sequence of operations. It is the first word for the programming sequence within the block.

The *preparatory function* is designated by the character G and two numeric digits. This word immediately follows the sequence number word. The G word prepares the numerical control unit for a specific mode of operation. Examples of G words include:

G00	point-to-point positioning
G01	linear interpolation
G02	circular interpolation, Arc CW (arc clockwise)
G03	circular interpolation, Arc CCW (arc counterclockwise)
G04	dwell
G13–G16	axis selection
G33	thread cutting, constant lead
G40	cutter compensation cancel
G80–G89	fixed cycles 1 through 9

Dimension words follow the preparatory function word. For multiple-axis systems, the dimension words follow in order: X, Y, Z; U, V, W; P, Q, R; I, J, K; A, B, C, D, E.

The *feed function* is designated by the letter F and a maximum of eight numeric digits. This word follows the last dimension word. The F is programmed on tape as a coded feed rate number. The EIA feed rate system consists of a series of two-digit code numbers, each representing the linear motion feed rate in inches per minute (ipm).

The programmer follows the same guidelines for establishing feed rates as are applied to conventional machining. Material to be cut, the machining process, and characteristics of the machine tool are considerations. The control units of NC machines are designed with automatic, smooth acceleration or deceleration to a new higher or lower programmed feed rate.

The *spindle speed* is designated by the letter S and three numeric digits. This word follows the last dimension word or feed rate word. The spindle speed is expressed in a coded three-digit number.

The *tool (turret) function* is designated by the letter T and a maximum of five numeric digits. This word immediately follows the spindle speed word. The digits selected must be

compatible with the particular numerical control system being used.

The *miscellaneous function* is designated by the letter M and two numeric digits. This word follows the tool function word and immediately precedes the end-of-block (EOB) character.

OTHER SELECTED NC FUNCTIONS

Arc clockwise (Arc CW) specifies the path of curvature generated by coordinating the movement of a cutting tool in a clockwise direction along two axes.

Arc counterclockwise (Arc CCW) specifies the path of curvature generated by a cutting tool whose movements along two axes are coordinated in a counterclockwise direction in the plane of motion.

Automatic acceleration (G08) means accelerating the feed rate from the starting feed rate within a block. A starting feed rate, in any block in which the G08 code is used, of 10% of the programmed feed rate may be accelerated, according to a time constant, to 100%.

Plane selection (G17, G18, or G19) is generally used for X–Y, X–Z, or Y–Z plane selection for circular interpolation and cutter compensation functions.

Program stop is an M00 word that stops the spindle, coolant flow, and feed after completion of all commands in the block. The remainder of the program may continue after the machine operator pushes a button.

Spindle clockwise is an M03 command that starts the machine spindle rotation to advance a right-hand screw into the workpiece. An M04 command starts the spindle to retract a right-hand screw from the workpiece.

Spindle off is an M05 command that stops the spindle as efficiently as possible and turns off the coolant.

FACTORS TO CONSIDER IN NC PROGRAMMING

LINEAR INTERPOLATION

Linear interpolation relates to the control of a travel rate in two directions. The travel rate is proportional to the distance traveled. The axis drive motors must be capable of operating at different rates of speed. Thus, in linear interpolation, the stepping motors on the axes drives permit a cutting tool to move along an angular, circular, or arc path.

In programming a contour, the type of interpolation directly affects the calculations involved in traversing an arc or circle. When an arc is generated in a series of straight lines, the greater the number of lines computed to traverse a given arc, the finer the arc.

CIRCULAR INTERPOLATION

Circular interpolation is defined as the ability of a control unit to generate a circular arc of maximum 99.99° span in one block. A circle or an arc is generated as a continuous curve rather than as a series of straight lines. In circular interpolation, the start and the end of an arc are programmed in only one block of tape. One type of circular interpolation is the EIA standardized method. Some of the newer CNC machines are designed to generate a 360° arc in one block.

HELICAL INTERPOLATION

Helical interpolation relates to two axes circular interpolation coupled with the linear movement of a third axis. The simultaneous movement of a cutting tool along the three axes produces a helical spiral form. Helical interpolation is used for large diameter internal thread milling and other NC/CNC form machining applications where a helical spiral form is to be generated.

PARABOLIC INTERPOLATION

Parabolic interpolation was used for programming tool path movement for producing sculptured molds and dies, free-form designs, and special curves, requiring a complete parabola or part of a parabola. Such forms depend on two end points and a mid-point (three curved line positions).

Parabolic interpolation has the advantage over linear interpolation in producing curved sections by using a considerably smaller number (as many as 50 to 1) of part program points. Today, cubic interpolation has largely superceded parabolic interpolation.

CUBIC INTERPOLATION

Cubic interpolation is used in programming to describe and to blend a continuous segment of a curve with another segment without producing any surface interruptions in the generated curve. In this method of generating complex cutter paths (for example, form dies), a comparatively small number of input data points are required. Software routines for cubic interpolation are available for the part programmer.

One purpose served by each different interpolation program relates to the elimination of as many coordinate points in the part program as possible. Helical, parabolic, and cubic interpolation forms provide for a programmer to input the description of a specific form. The control unit then generates the series of points that define the path along which a cutting tool is directed.

DIMENSION COMMAND PULSE WEIGHT

All dimension words must be divisible by the command pulse weight of the numerical control system. The *pulse weight of the system* is the smallest increment of a machine slide movement caused by one single command pulse. For example, if the control unit has a command pulse weight of 0.0002″ (0.005mm), each electronically generated command pulse causes a movement of a machine slide of the same magnitude.

ACCELERATION AND DECELERATION FOR CUTS

The NC programmer must recognize that upon approaching the end of a cut, such as an inside corner, the cutting tool may have to be slowed down (decelerated) to prevent *overshoot*. Overshoot causes the cutter to cut deeper than required and to leave an undesirable indentation in the workpiece. The deceleration is block programmed with a reduced feed in order to machine precision inside corners.

POSITIONING THE SPINDLE
INCREMENTAL MEASUREMENT

Incremental measurement means that the spindle measures the distance to its next location from its last position. Incremental measurement

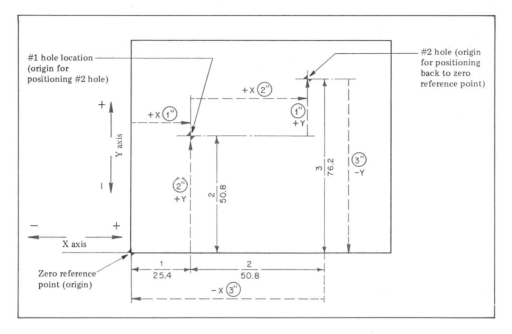

Figure 29-8 Spindle Positioning Using Incremental Measurements

(positioning) utilizes positive and negative directions. Spindle positioning using incremental measurements is illustrated in Figure 29-8. For example, the dimensions show the first drilling location (centerline) to be 1″ (25.4mm) from the point of origin along the X axis. The hole is 2″ (50.8mm) along the Y axis. The second hole is centered an additional +2″ (50.8mm) along the X axis (same direction of movement). This hole is also 1″ (25.4mm) farther along the Y axis in the same direction of movement.

In this example, then, to position a cutting tool by the incremental method, the spindle moves +X 1″ (25.4mm) and +Y 2″ (50.8mm) from the zero point of origin for the first operation. To reach the second operation, the spindle moves +X for another 2″ (50.8mm) and +Y for an additional 1″ (25.4mm). That is, to position the spindle for the second operation, location 1 (the first operation) is used as the new point of origin from which movements are measured.

After completing the second operation in this example, the spindle is returned to the original zero point by a –Y movement of 3″ (76.2mm) and a –X movement of 3″ (76.2mm). Note that the second location is the new origin from which

the measurements are made back to point zero.

It should be reemphasized that in numerical control language the spindle position is considered to be moved. Actually, X and Y movements and positions result from moving the table.

ABSOLUTE OR COORDINATE MEASUREMENT

The spindle of a NC machine tool may also be positioned by *absolute or coordinate measurement*. One system is identified in relation to a fixed zero—that is, all measurements are taken from the same reference point. The advantage of using a fixed zero is that the spindle operates only in quadrant I. All movements for positioning locations have positive values. All coordinate location points are specified in relation to distance from the coordinate axes.

Spindle positioning using absolute or coordinate measurements is illustrated in Figure 29-9. The zero reference point is located at the bottom left corner of the workpiece. Location 1 is identified at point 1X, 2Y from point zero. Location 2 is at point 3X, 3Y from the same point zero.

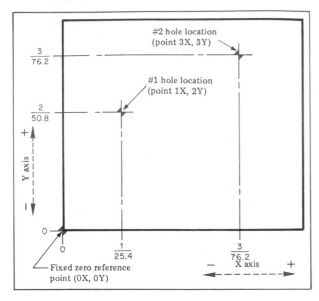

Figure 29-9 Spindle Positioning Using Absolute
or Coordinate Measurements

instruction is for the spindle to return to the point of origin, which is coordinate location 0X, 0Y.

FLOATING ZERO POINT

NC machine tools may also be programmed to permit a *floating zero* to be used as absolute zero. The floating zero may be established as any point that will make programming easier. Spindle positioning using a floating zero point is illustrated. Figure 29–10A is an ordinate drawing of a workpiece that requires the drilling of four holes symmetrically located. Figure 29–10B shows that the absolute zero may be floated to the intersection of X and Y axis center lines. The four holes are positioned by coordinate locations in each of the four quadrants. Thus the X and Y values change.

The spindle is instructed in NC to position at coordinate location 1X, 2Y. At the completion of this operation, the spindle is instructed to position at coordinate location 3X, 3Y. The third

Absolute positioning has the same advantages over incremental positioning with respect to machining accuracy. Errors that accumulate from incremental positioning are not a problem in absolute positioning.

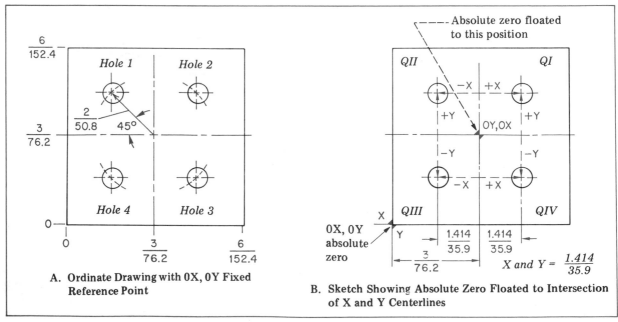

**A. Ordinate Drawing with 0X, 0Y Fixed
Reference Point**

**B. Sketch Showing Absolute Zero Floated to Intersection
of X and Y Centerlines**

Figure 29-10 Spindle Positioning Using a Floating Zero Point

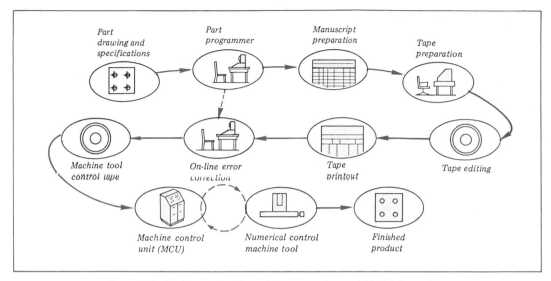

Figure 29–11 Events in Manual Programming and Machining a Part

NUMERICAL CONTROL TAPE PROGRAMMING, PREPARATION, AND CORRECTION

MANUAL PROGRAMMING

The first event that must take place to machine a part by using a numerical control machine tool is the preparation of the program. The programmer reads the part print. With expert knowledge and skills, the programmer translates every machine and cutting movement and setup to machine the part according to the specified dimensions and requirements. All cutting tool and machine tool functions are listed in a logical sequence to produce a numerical control program *manuscript*. The program is written either (1) manually or (2) with computer assistance.

The events in manual programming and machining a part are shown in Figure 29–11. The events for computer-assisted programming are described in the next unit.

Manual programming requires that all cutter, machine function, and numerical coordinate data be given. The cutter positions must be calculated and specified on the manuscript.

The manuscript is then processed by a *tape preparation unit*. The program for the part is encoded in English letters and Arabic numerals

on a tape. The preparation unit also prints a copy or *printout*, of the program. The printout is corrected or changes are made in the tape. Once the tape is accurate, it is fed to the machine control unit (MCU). The MCU feeds control information to the actuating and movement systems, devices, and mechanisms incorporated in the machine tool.

TAPE MATERIAL AND FEATURES

A NC tape is a common method of giving control instructions to a NC machine. NC tape materials include durable paper, paper-plastic, aluminum, and plastic laminates. The nonpaper tapes are able to withstand greater usage with less wear and are not subject to soil from materials used in the shop. Regardless of material, tape sizes are standardized and are manufactured to close tolerances.

The dimensions and design features of a standard NC tape appear in Figure 29–12. The 0.046″ (1.17mm) diameter feed holes are punched off center to permit easy alignment of the tape in the tape reader. Note that eight positions are available for code holes. The positions, or *channels*, are numbered 8, 7, 6, 5, 4, 3, 2, 1. The feed holes between channels 4 and 3 assure the proper positioning of the tape for both punching and reading in the tape reader. The feed holes are in line with the code holes.

TAPE BLOCKS AND REWIND STOP CODE

The information coded on the tape provides input data to the MCU. The MCU directs the machine tool through its various functions. The input coded information on the tape is sectioned in units referred to as *blocks*. Each block represents a complete entity: a machining operation, machine function, or a combination. Each block is separated from a succeeding block by an *end-of-block (EOB) code*. The EOB code is punched on channel 8 (Figure 29–12).

Each block consists of *words* (where a word address type of format is used). The words are the characters typed on the keyboard of the tape punch machine. The character is usually a letter. This letter identifies the numerical work data that follows. For example, the M02 code is usually the first instruction. When the MCU reader reads an M02 end-of-program function, the tape is rewound in preparation for a new run.

PREPARATION OF THE TAPE

A *tape punch machine* resembles a typewriter. A letter and number keyboard is used to punch a particular pattern of holes on a NC tape. Several extra symbols and control keys for the tape punch are included. When a key is pressed, a unique pattern of holes is produced in the tape for each tape punch key symbol. A printed record is typed simultaneously on paper in the tape punch machine. The *tape feed key* is punched to cause blank tape to feed through the tape punch machine as feed holes are punched.

A *tape reading head (reader)* is an integral part of the tape punch machine. The function of the reading head is to operate the tape punch machine typewriter from the punched tape. This operation provides a printed record of the tape or produces duplicate tapes. The punched tape actuates the typewriter to prepare a record of information.

CORRECTION OF ERRORS AND CHANGES

Any typing errors detected by the tape punch operator are corrected by pressing the *delete key*

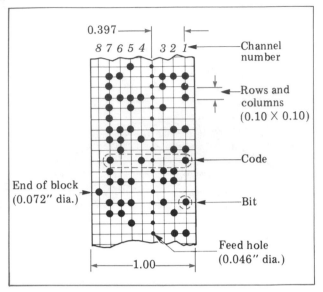

Figure 29–12 Dimensions and Design Features of a Standard NC Tape

and punching out all seven rows on the tape. The corrected information is then typed, followed by the balance of information. A new or a spliced tape will be produced if the error is not picked up until a tape is completed. The incorrect tape is inserted in the reader and advanced to duplicate a new tape to the point of error. The incorrect tape is stopped as the last correct entry is reached. The correct information is typed on the keyboard while the original tape is advanced through the reader by hand. The duplicating of the balance of the correct tape is then continued. When a punched tape requires splicing, care must be taken to align the feed holes and the tape information.

TAPE READERS

A *tape reader* has two major components: a *tape reading head* and a *tape transport system*. The transport system includes mechanisms for moving the tape and tension arms to maintain uniform tautness of the tape as it passes across the reading head. The tape reader is generally designed as part of the machine control unit. The MCU may be designed as an integral part of the machine tool or it may be wired to the machine.

There are three different means of reading the tape: (1) *electromechanical*, (2) *photoelectric*, and (3) *pneumatic*.

ELECTROMECHANICAL READER

The output of an electromechanical reader is controlled by mechanical fingers. These fingers are operated through the punched holes in the tape.

PHOTOELECTRIC READER

A photoelectric reader utilizes the photoelectric principle of a light beam activating sensitive photoelectric cells. A concentrated light beam passing through the punched holes permits fast reading of the information on the tape.

Photoelectric readers are applied on numerical controls that are designed for continuous-path machines. Since continuous-path machining generally requires a series of cuts to form an angle, radius, or other irregular shape, photoelectric readers are used for such directions.

A machine control unit may be equipped with *memory storage*. The tape reader in this MCU reads and stores one instruction in advance and thus conserves the waiting time that otherwise is required while the instructions are read from the tape. Memory storage is also known as *buffer storage*. As one cut is being taken, the MCU stores the next set of instructions the tape reader is reading. At the completion of one cut, the instructions for the next cut are instantly available.

PNEUMATIC READER

A pneumatic reader depends on a flow of a fluid (in this case, air) to flow through the punched holes in a tape. The flow activates electromechanical switches. A pneumatic reader requires precise alignment of the tape reader air jets over the tape columns. This type of reader is slower than either the electromechanical or photoelectric readers.

TAB SEQUENTIAL BASIC FORMAT PROGRAMS

Three basic formats for programming follow. These formats include: tab sequential, word address, and fixed block programs.

Tab sequential is a basic NC format used for point-to-point NC applications. It may also be adapted to continuous-path contour programming. The code produced on the tape is *tabbed* by the tape punch typewriter to give separate information about axis positioning, machining, and other functions. The MCU is able to differentiate in the electrical sections between X, Y, and Z axis positioning, tooling functions, and machining processes.

In tab sequential, the information follows a particular sequence. The first information relates to the positioning or operation step. A tab code follows to separate this information from X axis information.

The next tab code separates X positioning from Y positioning. The third tab separates Y positioning information from machining M functions. Additional tab codes are used to separate Z positioning and/or rotational axis positioning (if required).

Tab sequential programming is used in the following four examples. These examples apply to drilling and milling.

SINGLE-AXIS, SINGLE MACHINING PROCESS

Example 1: Point-to-Point Program. Drilling processes are a good example of point-to-point machining. The zero reference point for numerical control machining is usually planned to be off the workpiece. The spindle is usually positioned by manual control over the zero point.

An example of a single-axis, single machining process for drilling a part is shown in Figure 29–13. The part, represented by the line drawing, requires the drilling of four holes along the X axis. For convenience, the zero reference point is located 1″ from the left work surface (Figure 29–13A). The drilling of the four holes involves

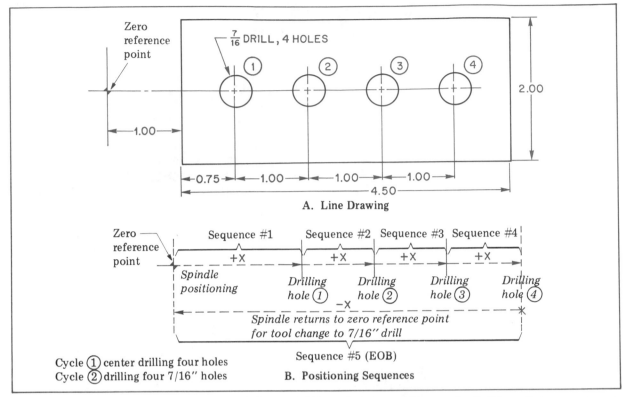

Figure 29–13 Example 1: Tab Sequential Numerical Control Program for Drilling a Part

five program sequences. Four positioning sequences (Figure 29–13B) are required for drilling. Sequence #5 returns the spindle to the starting position for the next workpiece. The sequences involve +X positioning for the holes and –X positioning to return the drill (spindle) to the starting position after the tape run. The total of the +X positioning movements must equal the –X return movement in order to move the spindle back to the point of origin.

A drawing often includes notes, which must be considered in programming. An example of such notes is included in Figure 29–13 for cycle 1 and cycle 2.

A simple numerical control tape program form is completed. The tab sequential numerical control tape program for producing the drilled plate is illustrated in Figure 29–14.

How to Program a Single-Axis, Single-Machining Process

Console Presets

1. The tool mode switch is set to the automatic position. The cutting tool (in this example, the center drill and the 7/16″ diameter drill in turn) is to machine and retract automatically at each position.
2. The tool positioning rate (feed rate) is set at the high (Hi) rate of speed.
3. Lead screw backlash compensation is provided. (In this example, a #2 compensation provides the degree of precise positioning required.)

Beginning of Program, Rewind Stop Code, and Sequencing Statements

The NC program (Figure 29–14) begins with the end-of-block (EOB) code. This code is followed

Company: C. G. Whitehurst Machine Tool Works

Dept. **169** | Part **DRILLED PLATE** | Part # **16** | Oper. # **1-A**

Prepared by Date **CTO-91**

Ck'd by Date **TPO-91**

Tape # **12206**

Sheet **1** of **1**

Remarks

TOOL MODE SWITCH- AUTO
FEED RATE- HI
BACKLASH- TAKEUP #2

Tools

CENTER DRILL
7/16 DIAMETER DRILL

Seq. #	Tab or EOB	+ or –	(X) Increment	Tab or EOB	+ or –	(Y) Increment	Tab or EOB	(M) Function	EOB	Instructions
									EOB	
0	RWS								EOB	CHANGE TOOL; LOAD; START
1	TAB		1750						EOB	
2	TAB		1000						EOB	
3	TAB		1000						EOB	
4	TAB		1000						EOB	
5	TAB	–	4750	TAB			TAB	02	EOB	

Figure 29–14 Example 1: Tab Sequential Numerical Control Tape Program for Drilling Process

by the rewind stop code. Before there are any program statements, the sequence statement #0 identifies the rewind stop code instruction to the tape reader. The tape is stopped at this point after rewinding and the program is recycled.

1. Sequence statement #1 gives the spindle (table) movement along the X axis from the reference point to the center of the first hole to be center drilled. The 1.75" movement is indicated as 1750. The first hole is center drilled automatically.
2. Sequence #2 gives the spindle movement (1000) along the X axis from the first hole to the second hole. The second hole is center drilled automatically.
3. Sequence #3 provides positioning information or the spindle movement (1000) from the second hole to the third hole. The third hole is center drilled automatically.
4. Sequence #4 prescribes positioning information (1000) to move the spindle from the center of the third center-drilled hole to the fourth hole. The fourth hole is center drilled automatically.
5. Sequence #5 gives the –X increment (–4750). This positioning information moves the spindle back along the X axis from the fourth center-drilled hole to the reference (origin) point.

Further, the M02 function instructs the tape reader to rewind the tape. The tape does not call for any drilling at the point of origin.
6. The center drill is replaced with the 7/16" (10.9mm) drill.
7. Sequences #1 through #5 are repeated. The four center-drilled holes are drilled to the required 7/16" (10.9mm) diameter. At the end of this cycle, the 7/16" (10.9mm) drill is replaced with the center drill.
8. The M02 function signifies an end-of-program instruction.

X AND Y AXES AND TOOL CHANGES

Example 2: Point-to-Point Program. An example of point-to-point positioning along the X and Y axes, with tool changes for drilling four holes of two different sizes, is illustrated in Figure 29–15. The part drawing shows positioning sequences for a tab sequential program.

How to Program for Two Axes and Tool Changes

Beginning of Program, Rewind Stop Code, and Sequencing Statements

1. After the console presets are made, the instructions start with the loading of the part. Sequence position 0 on the NC tape program RWS for recycling the program.

2. Sequence numbers 1 and 2 give the X axis and/or Y axis increments for positioning and drilling the 9.5mm (0.38″) diameter holes.

3. Sequence number 3 of the program provides information about positioning the spindle at the original zero point. The M06 code M function stops the control and lights the tool change lamp. The instructions column on the tape program indicates that the machine tool operator makes a tool change to the 5.6mm (0.22″) stub drill. –X and –Y increments return the spindle to the starting position.

4. Sequence numbers 4 and 5 position the spindle to drill the two 5.6mm (0.22″) diameter holes.

THIRD AXIS FUNCTION

Example 3: Tab Sequential Program. The third axis function normally controls a spindle component, a Z axis movement of a quill or knee, or an accessory such as a rotary table. Figures 29–16 and 29–17 provide an example of a drilling application using a rotary table for the

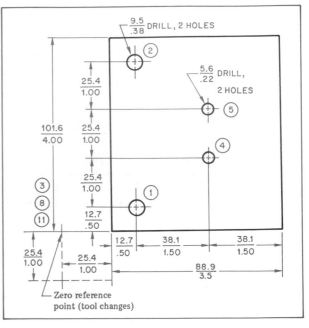

Figure 29–15 Example 2: Part Drawing Showing Positioning Sequences for Point-to-Point Positioning along X and Y Axes, with Drilling Tool Changes

third axis. An angle positioning plate is to be programmed for drilling (Figure 29–16). Seven sequences are involved in drilling the four holes (Figure 29–17). Note from the numerical control tape program (Figure 29–18) that there are seven sequence numbers in the block.

Console presets cover setting the tool switch at automatic, feed rate at high, and backlash switch at #2. The control programming for the three axes in this example starts at a reference point that falls on the horizontal centerline 1.00″ from the left machined edge.

How to Program for Three Axes

1. Sequence #1 moves the spindle (cutting tool) from the reference point along the X axis (centerline) to the center of the first hole. The first hole is drilled automatically.

2. Sequence #2 contains a 54 code in the M function column. The M54 code informs

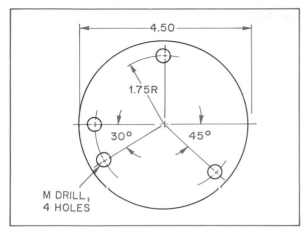

Figure 29–16 Example 3: Part to be Programmed for Drilling Application Using Rotary Table

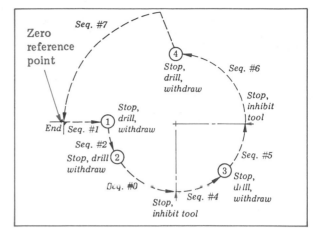

Figure 29–17 Example 3: Schematic Sequence Statements for Drilling Applications Using Rotary Table

MCU that the numerical information is for the third (Z) axis. Movement of the rotary axis (3000) positions the spindle at the 30° location. The second hole is drilled automatically.

3. Sequence #3 moves the rotary table 60°. The 6000 represents 0.01° of arc for each motor step along the Z increment. Note code 546 in sequence #3. It is a combination of the code 54 (third axis) and code 56 (tool inhibit). The "5" in both codes is entered only once to produce the code 546.

4. Sequence #4 continues the movement of the rotary table 45° (4500) to move the

spindle and drill to the 105° position. The third hole is drilled automatically.

5. Sequence #5 includes a tool inhibit code because the fourth hole position is 135° counterclockwise from the third hole. The rotary motion requires two sequences. Sequence #5 advances the tool 45°.

6. Sequence #6 continues the rotary motion another 90° (9000) to the 135° position. The fourth hole is drilled automatically.

7. Sequence #7 performs the 0254 function of rewinding the tape and returning the spindle (drill) to the reference point. The X increment is in the fourth quadrant.

Company: C. G. Whitehurst Machine Tool Works

| Dept. | 169 | Part | ANGLE POSITIONING PLATE | Part # 208 NM | Oper. #4 |

Prepared by Date CTO-91

Ck'd by Date TPO-91

Remarks: TOOL MADE SWITCH - AUTO FEED RATE - HI BACKLASH #2

Tools: M DRILL

Tape # 13301

Sheet 1 OF 1

Seq. #	Tab or EOB	+ or –	(X) Increment	Tab or EOB	+ or –	(Y) Increment	Tab or EOB	(M) Function	EOB	Instructions
									EOB	
0	RWS								EOB	CHANGE TOOL; LOAD; START
1	TAB		1500						EOB	
2	TAB			TAB		3000	TAB	54	EOB	
3	TAB			TAB		6000	TAB	546	EOB	
4	TAB			TAB		4500	TAB	54	EOB	
5	TAB			TAB		4500	TAB	546	EOB	
6	TAB			TAB		9000	TAB	54	EOB	
7	TAB	–	1500	TAB		9000	TAB	0254	EOB	

Figure 29–18 Example 3: Tab Sequential Numerical Control Tape Program for Drilling Using Rotary Table

In the system described, one complete revolution about the rotary axis (Z) totals 36,000. The X axis increments add to zero.

MILLING PROCESSES INVOLVING SPEED/FEED CHANGES

Example 4: Tab Sequential Program. An example of using tab sequential programming in a milling process is provided in Figures 29–19 and 20. The part drawing (Figure 29–19) shows the position of the zero reference point. Three sequences (Figure 29–20) are involved in the procedure for milling the elongated slot. Three new features—M55 (override milling feed rate), M52 (quill down), and M53 (quill up)—are included in the tab sequential program (Figure 29–21).

NC machine tools are designed to permit the change from rapid traverse in a slow quill (and cutting tool) down-feed and finally to a still different table feed rate.

The milling is completed in sequence #2 and the end mill clears the workpiece. Sequence #3 returns the spindle (end mill) to the starting reference point by using rapid traverse.

How to Program for Milling Processes Involving Speed/Feed Changes

1. Sequence #1 calls for positioning at high feed rate along the X and Y axes to the point where the milling operation begins.
2. Sequence #2 actuates the spindle so that the end mill feeds through the workpiece. The Y axis movement to mill the elongated slot is also identified.
3. Sequence #3 provides information for removing the end mill (upward movement of the quill) from the workpiece. Further directions are given for the rapid traverse (or high speed return) table movement to reposition the spindle at the reference starting point. Instructions are indicated to rewind the tape.

WORD ADDRESS FORMAT PROGRAMS

Word address is another NC format. The word address format requires a single letter code (A to Z) to deliver an address. Thus, control information is differentiated by the single letter address. By contrast, tab sequential format provides control information by a specific sequence that is separated by tab codes.

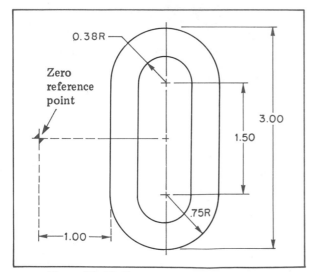

Figure 29–19 Example 4: Part Drawing of Tab Sequential Program for Milling Process

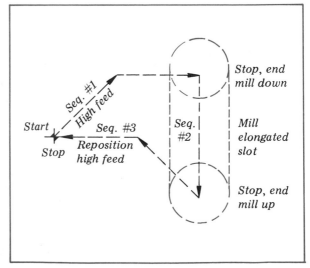

Figure 29–20 Example 4: Schematic of Tab Sequence Statements for Milling Process

Company: C. G. Whitehurst Machine Tool Works										
Dept. 169			Part ELONGATED COVER				Part # EX4		Oper. #6	
Prepared by Date CTO-91			Remarks				Tools			
Ck'd by Date JEF-91			TOOL MADE SWITCH - OFF FEED RATE- 6 ipm BACKLASH #1				0.750" END MILL			
Tape #12206										
Sheet 1 OF 2										
Seq. #	Tab or EOB	+ or -	(X) Increment	Tab or EOB	+ or -	(Y) Increment	Tab or EOB	(M) Function	EOB	Instructions
									EOB	
0	RWS								EOB	
1	TAB		1750	TAB		750	TAB	55	EOB	
2	TAB			TAB	—	1500	TAB	52	EOB	
3	TAB	—	1750	TAB		750	TAB	02535	EOB	

Figure 29–21 Example 4: Tab Sequential Numerical Control Tape Program for Milling

SAMPLE CODED FUNCTIONS

In word address format, specific instructions are given by adding a numeric code to the letter address word. Both the letter and the numeric code are used in word address. For example, a feed rate of 8 ipm is identified in a word address program as F8.

Word address format programs use the letter G for preparatory functions; M, for miscellaneous functions; N, for program sequence numbers; and F, for feed rates. Common letter words of X, Y, and Z for axis distances and + or – directions are included. Examples of coded functions in word address format are as follows:

- A G81 preparatory function calls for the cycling of a milling machine spindle (or the quill of a drilling machine) to perform a milling operation;
- A rapid-traverse feed rate may be programmed into the F address by adding the coded feed rate;
- At the end of a program, the M02 instruction cancels the first sequence so that the tool is not actuated;
- In using the third axis, information is programmed into the Z address;
- Information to indicate circular interpolation for a contouring program is provided through I and J data.

APPLICATION TO CONTINUOUS-PATH MACHINING

Word address may also be used for continuous-path machining. The G01 preparatory function information may be used to call up an interpolation cycle. Interpolation cycles on some NC machines require sending pulses to stepping motors. The stepping motor on the X or Y axis "steps" according to a pattern of tape punches passing the tape reader. Radii and angles may be approximated by the cuts in incremental steps. The cuts depend on the rate and relationship of movement of the motors. The motors may be stepped separately or together.

FIXED BLOCK FORMAT PROGRAMS

Fixed block is a third common type of NC format. As the term implies, each tape block in a fixed block format is the same length. All spaces in a fixed block format must be filled with a symbol. The features of a 20-digit fixed block program tape are illustrated in Figure 29–22.

- Tape rows 1, 2, and 3 relate to the sequence number (from 000 to 999);

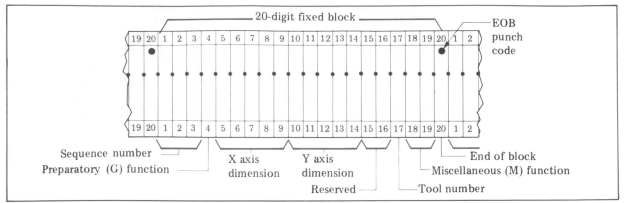

Figure 29-22 Features of a 20-Digit Fixed Block Program Tape

• Tape row 4 is a one-digit preparatory G function the same as the G function in word address (programming by the fixed block method follows each manufacturer's programming handbook of code digits for specific preparatory functions);

• Tape rows 5, 6, 7, 8, and 9 relate to an X axis dimension (decimal point values begin after the first digit (tape row 5) without using a decimal point);

• Tape rows 10, 11, 12, 13, and 14 provide similar information for a Y axis dimension;

• Tape rows 15 and 16 are reserved (the two rows are filled with zeros);

• Tape row 17 is a one-digit tool number. The coded information operates an automatic tool changer or provides a signal indicator for the machine operator to change the tool;

• Tape rows 18 and 19 provide spaces for a two-digit numeric portion of the miscellaneous M function;

• Tape row 20 is the EOB (end-of-block) code for carriage return.

MIRROR IMAGE PROGRAMMING FEATURE

Symmetrical and right-hand and left-hand parts may be efficiently programmed from a single tape by using a *mirror image (axis inversion) feature.* A programmed mirror image produces a **reverse** duplicate of a part, as illustrated in Figure 29-23. The machine control unit (MCU) is programmed to direct the X and/or Y axis mirror image to electrically reverse the direction of travel. All positive direction (+) movements around a program zero become negative (−). All negative direction movements become positive.

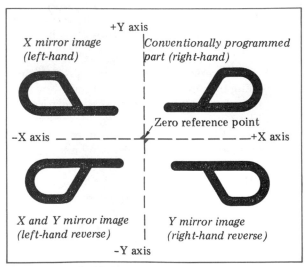

Figure 29-23 Mirror Image to Produce Reverse Duplicate of Part

Safe Practices Relating to Numerical Control Machine Tools

- Reduce the overhang of each cutting tool or the extension of the spindle. A shortened position provides greater rigidity for machining.
- Study tool position heights, especially when multilevel surfaces are to be machined. There must be adequate clearance between cutter, workpiece, and work-holding device to permit safe slide movement between all processes, including loading and unloading.
- Set cutting speeds and feed rates for each tool process within limits recommended for each process. Although processes are tape programmed, the operator still needs to observe the cutting conditions.
- Read the tape printout. Check for axis positioning accuracy (spindle location) and correctness of each machining process.
- Make a dry run for safety checks, spindle positioning, and each machining process.
- Check in advance on how the MCU and machine tool may be stopped immediately in an emergency.
- Use safety goggles or a protective eye shield and follow personal safety precautions.
- Remove burrs carefully. Use standard safety practices for cutters, work-holding devices, and machine tool processes.

TERMS USED WITH NUMERICAL CONTROL SYSTEMS, PRINCIPLES, AND MACHINE TOOLS

Absolute data	Data for measurements that are made from a fixed starting point or zero reference point.
Address	A number, name, or label that provides a method for the NC programmer, machine operator, or the NC program to identify information or to refer to a position location.
Binary code	A code in which each allowable position permits a choice between two alternatives.
Block	Words grouped together as a unit to provide complete information for a cutting operation. One or more rows of punched holes separated from other words by an end-of-block character.
Block address format	An address code or system of identifying words that specifies the format as well as the meaning of the words in a block.
Closed loop system	A NC system that provides output feedback for comparison to input command.
Continuous-path system (contour control)	The continuous and independent control of two or more instantaneous tool motions.
Control signal	The application of energy to actuate the device that makes corrective changes.
Fixed sequential format	A plan for identifying a word by location in the tape block. Presentation of all words in a particular order in a block.
Incremental data	Data from which each movement is referenced from the prior movement.

Interpolation — The path and rate of cutting tool travel or machine tool slide movement that is defined by establishing intermediate points between programmed end points.

Miscellaneous function — On/off functions of a machine tool as related to workholding and machining processes.

Point-to-point positioning — A positioning control system in which the controlled motion requires no path control in moving from one end point to the next.

Preparatory function — A command for changing the mode of operation of the control.

Tab sequential format — A method of identifying a word by the number of characters in the block preceding the word. The first character in each word is a tab character.

Tool function — A separate tool change command. The automatic or manual selection of a tool by a command.

Zero offset — The ability of a numerical machine tool control to shift the zero point on an axis over a specified range.

SUMMARY

■ Numerical controls deal with tool positioning along two or more axes, tool transfers, machining setups, and sequencing operations; controlling speeds, feeds, and coolant; diagnoses of processes; and other machining controls.

  ■ Once a tape is programmed accurately, is dry run, and tested by a complete cycling, successive machining processes are cycled automatically. System repeatability and reliability are ensured.

■ Every movement and process is programmed in NC to control impulses of energy. These impulses activate drive mechanisms and systems.

  ■ A closed loop NC system feeds a signal back to the MCU. The signal provides an accuracy check of the machine movement.

■ An open loop system provides controls of movements without feedback.

  ■ Part drawings for NC use the rectangular (cartesian) coordinate system.

■ Each point and feature may be identified along two or three mutually perpendicular planes (axes) in relation to a zero reference point.

  ■ Rotational axes describe circular motion around each normal axis. Rotational motion may be provided for two, three, four, or more axes.

- Preset tools and the use of tool changers increase the range of machining capability of NC machine tools.
- The binary system consists of a two-word vocabulary. Binary numbers and binary notations control the energy pulses to produce every programmed slide, tooling, and machine motion required to produce a part.
  - Nondimension words relate to sequence number (N), preparatory function (G), feed rate (F), spindle speed (S), tool selection (T), and miscellaneous function (M).
- Incremental positioning utilizes + and – directions to position a tool from the location of one process to the next process.
  - Absolute or coordinate measurements are made from the same reference point.
- A floating zero point may be used as an absolute zero. The floating zero may be located at any point that makes programming easier.
  - Manual programming involves the writing of the program, processing of the manuscript by tape preparation, correcting printout information on the tape, and in-feeding to the MCU.
- The MCU actuates movement systems, devices, tools, and mechanisms to numerically control the machine tool.
  - The tape consists of channels in which holes are punched to control specific machine functions. A complete entity of coded information is contained in a block.
- Tape readers are classified according to the form of energy used (electromechanical, photoelectric, or pneumatic).
  - Linear interpolation permits stepping motors to control spindle positioning in a straight-line movement.
- Circular interpolation permits generating a circular arc (99.99° maximum) in one block. Helical interpolation permits simultaneous three-axes movement to produce a helical spiral form. Cubic interpolation permits blending continuous curve segments without producing any machined surface interruptions.
  - Tab sequential programs are adapted to machining in two, three, or more axes, including tool, speed, feed rate, and other changes.
- The specific sequence of control information in a tab sequential format is separated by tab codes.
  - In a word address format, all activities are programmed by a code letter and a numerical word. For example, a G81 preparatory function controls the cycling of a machine spindle and M02 is a miscellaneous function code for cancelling a sequence so that a tool is not actuated.
- Fixed block format programs utilize a constant number of spaces that must be filled in a block.
  - A mirror image (axis inversion) feature permits machining a reverse duplicate by using a single tape. The MCU may be programmed to reverse the direction of travel to produce a symmetrical or right- or left-hand part.
- Added safety practices must be observed for emergency stopping of the MCU and NC machine tool. Working clearances are especially important for freedom to position the spindle and each cutting tool without interference.

UNIT 29 REVIEW AND SELF-TEST

1. State four functions, performed by a craftsperson on a conventional machine tool, that may be manually programmed for numerical control.

2. Describe briefly two advantages (excluding repeatability) of NC machine tools over conventional machine tools.

3. Cite three tooling-up and machining functions that may be performed on NC machine tools.

4. Make a simple sketch showing the point values of X and Y coordinates in quadrants I through IV.

5. Differentiate between a fixed zero reference point and a floating zero point.

6. List the main steps required to manually program a NC machine tool to produce a specific part.

7. a. Prepare a simple sketch to show the programming sequence for drilling the three holes in the drill plate shown in the part drawing below. Locate the zero reference point along the center line, 1.00″ from the left end of the workpiece.
   b. List the console presets.
   c. Prepare the statements for producing the tape according to the tab sequential format.

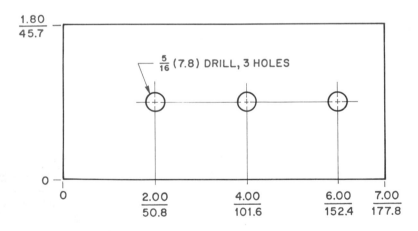

8. a. State the functions performed by the letters G, F, and M in word address format programs.
   b. List examples of code numbers used with each letter.
   c. Give an application of each word format.

9. Describe mirror image.

10. List two personal safety features that apply particularly to the operation of NC machine tools.

NC and Computer-Assisted Programming (CAP) for CNC Machine Tools

This section extends the basic principles of numerical control to on-line machine tool applications, tool gaging and management control, and tool-holding devices. Then, basic descriptions are given of computer-assisted programming (CAP) and computer numerical control (CNC). Step-by-step procedures are included for planning and preparing a computer-assisted program manuscript. The section concludes with descriptions of major components in a sophisticated CNC machining center. In the future, these centers may be adapted to complete computer-aided manufacturing (CAM) systems.

UNIT 30

Basic NC/CNC Machine Tools and Machining Processes

OBJECTIVES

After satisfactorily completing this unit, you will be able to:

- Describe tape-controlled two-axis drilling machines, contour programming, and four-axis drilling machines.
- Translate manual data to point-to-point two-axis NC engine lathes.
- Explain continuous-path systems and NC applications to bar and chucking machines and machining centers.
- Apply point-to-point, continuous path, and combination types of NC systems to two-, three- and four-axis milling machines.
- Describe applications of NC to boring machines, jig borers, machining centers, and inspection machines and on-line and off-line data processing, part programming, and postprocessors.
- Use common languages and functions of computers in computer-assisted programming, including APT programming.
- Apply CNC design features to computer-aided manufacturing centers; cutting tool size computations, electronic tool length gaging with digital readout, and tool management.
- Prepare an APT program manuscript.
- Interpret part drawings for starting point, end-of-cut points, end point, and circles on programming layouts for the program manuscript.
- Follow recommended *Safe Practices* for personal safety and machine tool, cutting tool, and workpiece protection.
- Correctly use each related *Term*.

NC MACHINES AND CONTROLS

Five of the most common categories of machining processes include (1) drilling, (2) turning, (3) milling, (4) grinding, and (5) boring. The machine tools by which these processes are performed may be accurately controlled by numerical information. This information is usually fed to a machine control unit (MCU).

The MCU accepts the dimensional, tool control, and other information and processes it through electrical signals. The electrical signals govern the direction, distance, and rate of movements of the prime movers of the NC machine tool. Other auxiliary functions such as coolant flow, spindle direction, and backlash compensation are included in the NC program to produce a part according to design specifications.

Each machine tool is designed with a positioning control system. Drilling machines require movement of a table from one point location to another—that is, *point-to-point positioning.* In milling processes, such as in contour milling, the motion and path of a tool usually needs to be controlled at all times. A *continuous-path control system* is required in many applications for simultaneous motion in two or more axes. A *combination control system* is designed for point-to-point as well as continuous-path machining processes.

COMMON CONTROL FEATURES

Regardless of the type of system, there are certain basic control features, as follows:

- Electronic components generate heat and therefore require temperature controls from $50°F$ to $120°F$ ($10°C$ to $48°C$), humidity control up to 95%, and the use of fans and air conditioning;

- The control panel area must include additional space for the machine-tool builder to mount additional or parallel machine function control switches and other accessories;

- The control system incorporates safeguards to ensure an accurate read-in of the tape;

- The NC system is able to control linear and rotary movements in any combination and simultaneously for all axes of machine motion. The smallest increment of motion, depending on the machine tool and processes, ranges from 0.000020" or 0.0005mm to 0.0001" or 0.0025mm, and the smallest increment of rotary motion ranges from 1.396 seconds of arc, or 0.000001 of a revolution, to 3.6 seconds of arc);

- The overall accuracy of the control system with an increment of motion of 0.0001", operating at a constant temperature, should be within ±0.0002" (0.005mm). Finer accuracies are required for machine tools with smaller increments of motion;

- The feedback system permits the control system to compare the tape command with the actual NC machine positioning or other functions;

- Backlash compensation is provided regardless of direction of motion from the previous position;

- Complete documentation is provided for programming, operating, and maintaining the NC system.

NC DRILLING MACHINES

The three broad classes of drilling machines are (1) *simple* (bench or pedestal), (2) *general*, and (3) *complex* (multi-operation). All machines have a spindle to hold and drive a cutting tool; a stationary or movable table; and a mechanism for feeding a drill, reamer, or other cutting tool into the workpiece.

Simple NC Drilling Machine (Two Axes). The X and Y axes movements of a tape-controlled drilling machine are shown in Figure 30–1. The axes represent the movements of the table. In the case of a simple drilling machine, the positioning of the spindle axis (Z) is controlled mechanically or manually. The X and Y movements may be independent or simultaneous.

The control system provides a *complete signal* so that the X axis process may be cycled mechanically. The NC system may include a sequence number display to guide the operator

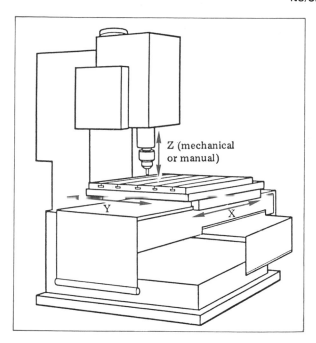

Figure 30–1 Tape-Controlled Drilling Machine (Two Axes)

on the operations being performed. The number following the N code on the visual display may be compared with the program printout.

NC machines may be provided with a control panel display, which indicates the current dimensions on the tape, the machine slide position, or both. The tape program on some NC machines may be interrupted to insert additional machine operations, to move the tool away from the point of machining for inspection, and to carry on other troubleshooting and checking procedures. Manual data input controls are used for moving the tool and repositioning it at the exact machining point.

General NC Drilling Machine. The classification of a general drilling machine differs from a simple drilling machine in that the design permits the drilling machine to be used for some basic milling processes. Thus, continuous-path machining capability is added to the point-to-point system. Contour milling requires tape-controlled feed rates. The MCU reads the coded number on the tape and outputs it to the machine tool.

The NC system for a general drilling machine may include continuous-path NC units for linear or circular interpolation. Arcs up to 90° may be generated within a single quadrant from tape commands in a single block. The following control features may also be included:

- Spindle speed selection, which stores the coded set of digits and then relays the information to the speed changing mechanism on call;
- Tool select feature, which provides for the automatic selection, changing, and storing of tools for multiple-sequence operations;
- Mirror image, which, through axis symmetry switches, permits the sign reversal of dimensional tape information and thus permits a single tape to be used to produce mirror image parts;
- Auxiliary function display, which, through a function selector switch, permits the operator to select an auxiliary function such as the feed number, preparatory function number, feed rate, spindle speed code, and tool number for display.

Complex NC Drilling Machine. A complex drilling machine includes the control of depth axis (Z) and, if required, a rotary fourth axis. The complex class of NC drilling machines uses combination positioning and contour control systems. A drilling machine with turret and tape control for three axes is illustrated in Figure 30–2.

The third axis is tape controlled so that the spindle is programmed for final depth. In addition, there is a *feed engage point* at which a rapid-traverse feed is changed to a slower programmed feed.

The control system automatically programs preparatory G functions. Thus, the Z axis may be programmed for drilling, boring, reaming, tapping, and other processes.

Since most cutting tools vary in length and may be of a different length than programmed, *tool length compensation* for general machining on some NC machines requires manual setting. The difference in dimensions is set up on switches for each specific tool number. A compensation dial is automatically activated when the tool is selected.

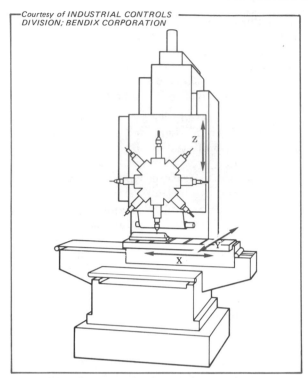

Figure 30–2 Tape-Controlled Drilling Machine with Turret (Three Axes)

NC TURNING MACHINES (POINT-TO-POINT POSITIONING)

The engine lathe and the turret lathe (bar and chucking machine) are two basic types of numerical control turning machines. NC principles and machine control units are adaptable to vertical turret and other types of lathes.

NC Engine Lathe (Two Axes). A two-axis NC positioning system may be used for general turning processes that do not require taper turning, thread cutting, or contour forming. Where contours are required, the NC system must have continuous-path capability. The general NC features, including backlash compensation as described earlier, apply to turning machines. Four additional features increase the control or machine operation capability:

- *Sequence number display*, which provides the NC machine operator with a visual display that may be checked with the program printout. An operation coded with N and a three-digit number is displayed at the time the given sequence of operations is being performed;
- *Dimension display*, which visually indicates either the dimensions on tape, or the actual slide position, or both;
- *Manual data input*, which permits the interruption of a tape program for insertion of additional machine operations, movement of the cutting tool out of position for checking and inspection of the workpiece, and troubleshooting and checking procedures;
- *Tape-controlled spindle speed selector*, which makes it possible to program for multiple-speed operations. The control unit stores the code and digits and relays the commands as required to the speed changing device on the lathe.

Continuous-Path System NC Engine Lathe. The continuous-path system permits the machining of tapers, threads, radii, and other contour cutting processes. Contour cutting requires the feed rates to be tape controlled. The control unit reads the specified rate from the coded number on the tape and processes it. Other features that are included with a continuous-path system are the following:

- *Linear or circular interpolation*, which provides for straight-line or smooth-curve movement of a cutting tool in machining a contour
- *Internal and external thread cutting*, which is available on numerical contouring control to eliminate manual processes and cumulative errors from thread to thread;
- *Tape-controlled spindle speed selector and auxiliary function display*, a combination of features serving similar functions as described for applications on drilling machines and for drilling processes.

NC Bar and Chucking Machine and Machining Centers. Production turning machines may be tape controlled for two, three, and four axes, depending on the complexity of the machining processes. Some of the common combinations include the following tape controls:

- Simultaneous control of carriage and cross slide along two axes. The turret position selection may also be tape controlled;
- Control of carriage, cross slide, and turret ram (three axes), with turret position selection;
- Simultaneous control of any two of the three axes, with indexing of stations on the turret ram or square tool turret;
- Simultaneous tape control of four axes.

Some typical NC features include a sequence number display, dimension display, manual data input, tool select, spindle speed control, and auxiliary function display. In addition, a *tool offset* feature permits the cutting tool to be moved a preset distance in the automatic tape mode. Tool offset compensates for changes in tool length due to resharpening a cutter or to a position change of the tool in a different holder. The NC machine operator dials in required off-sets so that any one tool may be programmed to pick up any given offset.

NC MILLING MACHINES
(TWO, THREE, AND FOUR AXES)

When classified according to position of the cutter driving spindle, milling machines are of two types: horizontal and vertical. These types may be programmed for tape control of at least two axes of simultaneous motion. Control of three axes (Figure 30–3) and four axes is general.

A NC system with continuous-path capabilities is practical for angle and other contour milling jobs that are common in jobbing shops. A continuous-path NC system with three axes of control has controls that relate to M functions of table movement, continuous-path controls, and linear and circular interpolation for generating arcs. Additional features, such as sequence number display, dimension display, manual data input, tool selection, and spindle speed control, are available to improve control capability. An auxiliary function display may be included in conjunction with a function selector switch. The operator is able to select any auxiliary function and display it. Auxiliary functions relate to sequence number, preparatory function number, feed rate or spindle speed code, and tool number.

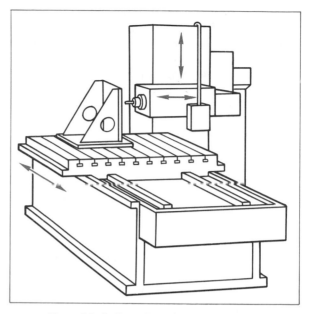

Figure 30–3 Tape-Controlled Horizontal Milling Machine (Three Axes)

If a considerable amount of drilling, boring, reaming, and tapping is to be done, as in the case of universal vertical turret milling machines, a combination of point-to-point and continuous-path systems is desirable. Multiple-operation milling machines are designed with a combination positioning and continuous-path system. This design provides simultaneous control of all axes, accurate positioning of the spindle and cutting tool, built-in cycles for the depth axis, and continuous path for curved or contour machining. All of the previously described basic control requirements and features of the continuous-path control system apply to the combination system. Predetermined fixed cycles of the spindle axis (Z) are required. Drilling, boring, reaming, tapping, and other processes are programmed on tape under the G preparatory function code. The appropriate digits call out a specified cyclic action.

NC BORING MACHINES

There are two general classifications of boring machines. The first class includes vertical turret lathes and boring mills. The cutting tool, when once positioned for a cut, remains stationary.

The workpiece revolves while the cutting tool feeds in to take a cut. In the second classification, the action is reversed. The workpiece is moved into position while the cutting tool revolves and is fed in the direction into the workpiece. Horizontal boring machines and the more widely used jig borer are included in the second class.

Three axes of control are common for horizontal boring machines. The three axes are the longitudinal (X), cross feed or transverse (Y), and the depth axis (Z). Other boring mills have four axes of control. The controls, systems, displays, features, and functions for milling machines are applicable to boring machines. The preparatory G function may be programmed on tape for predetermined fixed cycles of the spindle (Z) axis. The specific cyclic action is called out as required for drilling, boring, reaming, and other operations.

Courtesy of MOORE SPECIAL TOOL COMPANY INC.

Figure 30–4 NC Jig Borer (Equipped with a Rotary Table) and Machine Control Unit

NC Jig Borer. The jig borer is regarded by industry as an extremely precise machine tool. The accuracy of positioning for longitudinal and cross travel is within 0.000030″ in any inch or 0.0008mm (0.8μm) in any 30mm. The compound slide is square (along its full travel) to within 0.000060″ or 0.0015mm (1.5μm). The alignment of the spindle travel in 5″ (127mm) is 0.000090″ or 0.0023mm (2.3μm).

The NC jig borer shown in Figure 30–4 has an 11″ × 18″ (280mm × 450mm) table travel. The jig borer is equipped with point-to-point positioning on X and Y axes. The rotary table is controlled by a preset indexer. The table is actuated by the NC system's auxiliary functions. The rotary table may serve as a fully numerically controlled axis. Variable point-to-point, varying angle positioning, and controlled feed rates are provided. Contour positioning is an optional feature. The manufacturer's specifications identify the following features as common to this particular NC jig borer:

- Complete solid-state integrated circuit system;
- Photoelectric tape reader with capacity of 125 characters per second;
- Temperature-controlled operating range from 50°F to 120°F (10°C to 48°C);
- Auto, single, and manual modes;
- Data input of 1″ (25.4mm), 8-track perforated tape (EIA standards RS–227);
- Word address, variable block format (EIA RS–244 and RS–274B);
- Automatic backlash compensation;
- Miscellaneous functions M00 through M99;
- Reference, set, and grid zeroing;
- Absolute programming;
- Test mode and test circuits;
- Incremental feed and programmed feed rates;
- 100% manual feed rate override in 10% steps;
- Milling control of continuous, one-axis-at-a-time capability in a straight line parallel to the X and Y axis;
- Spindle cycle control;
- Multiple depth selection.

The jig borer has program data resolution accuracy of 0.00001″ (0.001mm). Movements are provided by DC servomotors. Parity checking for accuracy is provided. Position may be read from machine scales and dials. The measuring methods involve lead screw and rotary resolvers. The machine is equipped for mirror image (X–Y plane).

NC MACHINING CENTERS

NC machining centers have tape control for three, four, and five axes. It is possible to move the column, spindle head, table traverse, and table rotation and tilt simultaneously under tape control. The NC system may be point-to-point, continuous-path, or a combination system.

The basic control features include simultaneous movements along all axes, programmable feed rates, spindle speeds, and fixed cycle. Typical additional features that increase the control capability or machine operation include: sequence number display, auxiliary function display, manual data input, tool gaging and selection, and tool length compensation. NC machining centers with circular interpolation are capable of generating arcs up to 90° in one quadrant from single-block tape commands. An addition is available on some circular interpolation features for generating 360° of arc from single-block tape commands. Like the controls on milling and other NC machine tools, the spindle depth axis and the feed engage point are programmed. Specific cyclic preparatory functions are programmed to control the axis for particular operations and to call out the action as required.

NC DATA PROCESSING

NC data processing requires mathematical computations by the programmer and checking of tool travel, feed rates, work-to-cutter clearances, speeds, and so on. Special machine tool functions (tooling requirements, coolant flow, predetermined packaged operation sequences), appropriate miscellaneous or preparatory functions, and codes must be established. Thus, a considerable amount of data processing is completed in advance of the actual machining.

OFF-LINE DATA PROCESSING

Off-line data processing is completed prior to using a control unit in combination with a machine tool to produce the part described through programming. Off-line data processing, when once completed, may be used repeatedly in tooling up for any number of manufacturing processes in which the control data is required.

ON-LINE DATA PROCESSING

On-line data processing relates to translating information from a control tape into input signals that feed directly to a machine tool. The data processing takes place in a time interval that is so minute as to almost coincide with the machine tool's execution of a process. On-line data processing must be repeated each time the machine tool is set up for a new part to be manufactured.

The input to on-line data processing is often a part of programming language describing desired machining. The instructions are transcribed to punched cards or a control tape or are entered into a computer manually. The cards may be transcribed onto magnetic tape on a large-scale computer.

A previously written computer program executes a sequence of instructions for processing the input of part programming. Computer processing results in information that may either be put on a control tape or be fed directly to a numerical control unit at the machine center. The computer program gives the part programmer a printout of the control tape contents and, in many cases, indicates errors.

As the machine control unit reads the control tape, the information is processed on-line. A series of input signals are produced. The signals go directly to the machine control unit to call out control path movement and to execute appropriate miscellaneous and preparatory functions. The combination of the control unit and machine tool has the capability and reliability to manufacture the part repeatedly.

PART PROGRAMMING

After studying the design features and specifications of a part to be manufactured, a part programmer decides on the part programming language and the processes, tooling, and setups required to machine the part. If there are geometric calculations involved in calculating the tool path and tool offset movement, these calculations may be included in the program for solution by the computer. The part programmer's statements, which are produced on cards, punched tape, or a magnetic tape, are processed by a previously written computer program.

PROCESSOR

A *processor* is an NC computer program that performs computations that are *workpiece oriented* or tool offsets. The processor program, as part of the software of numerical control, places the cutting tools on the workpiece. The processor program ignores all control unit information and items that are machine tool oriented.

Additional information required to manufacture a part relates to controls such as spindle speeds, feed rates, spindle direction, coolant requirements, tool selection, and other items that are *machine tool oriented*. The processor assumes that further processing of additional information from another computer program will need to be interlocked to produce the appropriate control tape for the ultimate manufacture of the part.

POSTPROCESSOR

A *postprocessor* is a computer program that accepts a processor program of information about the tool located on a part, machining setups, and machine processes. A postprocessor program makes additional computations to ensure compatibility of information among the MCU, NC machine tool, and the part specifications, including tolerances and machine limitations. The product of postprocessing is a control tape or information that may be supplied to produce a control tape.

The greatest advantage of postprocessing is one of economics. Changes to or adaptations of a computer program from one MCU or NC machine to another are readily made. Another advantage is that the postprocessor information may be postprocessed for multiple applications on additional MCU and NC machine tool applications.

There are five major elements in each postprocessor. These elements are: (1) input, (2) motion, (3) auxiliary, (4) output, and (5) control.

Input Element. This element reads the cutter location. It checks other output information of the processor computer program and makes a diagnostic printout of nonprocessable information. Accepted input information is transferred for subsequent postprocessing.

Motion Element. This element of the postprocessor relates to characteristics of the machine tool and geometric functions. Machine dynamics deal with items such as feed rate, cornering velocities, and tolerances.

The geometry portion translates all coordinate information into the two-, three-, four-, or five-axis coordinates, depending on the processes and the machine tool. Further, the geometry portion ensures that when a combination of linear and rotary motions is executed, the path is geometrically accurate and meets the tolerance requirements of the job. Geometry also deals with tool and workpiece clearance and safe operating spaces.

Auxiliary Element. This element is associated with miscellaneous and preparatory functions. When the part programmer needs auxiliary information, the postprocessor searches the computer memory to locate the appropriate control tape code for the desired function. The auxiliary element passes the coded information along to the output element to be incorporated at a specified time.

Certain programmer statements are interpreted to establish conditions within the postprocessor. Examples of these statements include spindle speed and directions, coolant status, tolerances, and feed rates.

Output Element. Information from the motion element is accepted and converted in the output element into codes. The codes output directly to a control tape or they may be produced in a form and format to be readily converted to the input control tape for the MCU.

The output element also accepts miscellaneous and preparatory information from the auxiliary function element. This information is coded and is available for output at a specified time. A printed listing of the tape information, comments, and diagnostic data is generated by the output routine.

Control Element. This element controls the timing for the processing of the information and the appearance of any diagnostic data in proper sequence.

COMPUTER-ASSISTED NUMERICAL CONTROL PROGRAMMING

Thus far, numerical control has been considered in terms of manual programming. All cutter movements and machine setups and machine functions were listed in sequence in the manuscript. All cutter positions were calculated by the programmer and specified on the manuscript. A tape was prepared, a printout was edited, and a control tape was prepared. All of this information was translated in the MCU to command pulses to the NC machine tool. At the end of the program, the machine was reloaded. The processing cycle between the MCU and the NC machine tool was repeated.

Numerical control may also be considered in terms of *computer-assisted programming* (CAP).

This method of writing a program is used for continuous-path and other operations requiring complex and time-consuming calculations. CAP eliminates the need to make numerical coordinate data calculations. CAP calculations are made rapidly and are error-free. A number of different processor languages are used for CAP, depending on the complexity of the computations and the combinations of machining processes.

COMPUTER-ASSISTED PROGRAMMING (CAP) LANGUAGE

CAP requires the use of symbols and modified English words, each of which has a precise meaning. The language is simplified to reduce the number of entries in the manuscript and to simplify writing the program. However, different equipment manufacturers use different program-oriented languages. Two of the widely used and powerful languages are APT, for *Automatically Programmed Tools*, and Compact II®. Less powerful languages are used to control basic machine tools such as a lathe, drilling machine, grinding machine, or milling machine or to produce a particular part on a single machine. The events and additional input in computer-assisted programming for producing a part are illustrated in Figure 30–5.

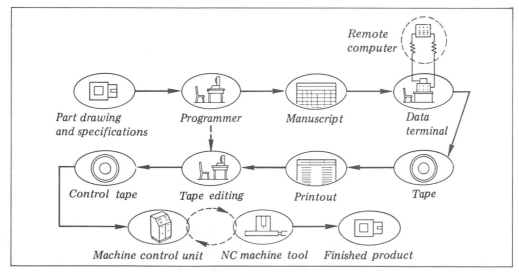

Figure 30–5 Events in Computer-Assisted Programming Using Additional Remote Computer Input

PROGRAM EDITING

Point-to-point programs may be edited to correct mistakes or to change the part program. When a tape is produced and a printout is provided, the printed letters and numerals, which are coded on the tape by a pattern of punched holes, are easily edited. The numerical data on the printout reports exact cutter positions and machining operations.

Continuous-path programs may be edited directly when the numerical positions of the cutter describe exact cutter positions for machining. When compensation must be made for cutter offsets, nose radius effects, and difficult-to-follow cutter positions, it is necessary to use other methods of tape editing. One method plots the path of the cutting tool on paper. Another editing method displays the cutter path on a cathode ray tube (CRT). When a sample part run is performed, the machine tool is run through the programmed cycle. No part is used in this *dry run*. Errors or changes are noted and the tape is corrected accordingly.

Corrections and changes may be made in the part program while it is stored in computer memory. A corrected tape is then made from the corrected stored part program. The corrected tape becomes the permanently stored tape and is used, when required, to reenter the program into computer memory. A tape may also be edited after the operation is started for changes in cutter speed, feed rate, cutter position, or small errors.

A *parity check* permits automatically checking a tape for malfunction errors of the tape punch. Depending on the tape format used, each horizontal row on the tape must have either an odd or an even number of holes. The parity check on the EIA RS-244-B tape format has an odd number of holes; the EIA RS-358-B format, an even number of holes. If a row does not have the required odd or even holes, the control system stops to indicate an error.

FUNCTIONS OF THE MACHINE CONTROL UNIT (MCU)

The instructions within the part program are converted in the machine control unit (MCU) into a form of energy that controls the machine tool. Two basic types of machine control units are available: (1) *hard-wired* integrated circuits and (2) *soft-wired* computer numerical control (CNC) units.

HARD-WIRED UNITS

Hard-wired units use *digital logic packages* mounted on plug-in printed circuit boards (PCB) in a fixed and permanent arrangement. PCB connectors receive the circuit boards and are also wired together to connect the electronic components in a permanent, fixed manner. One hard-wired unit controls one type of machine tool. Input signals are derived from the tape reader as the tape is run through each time a part is to be machined. The input signals, in turn, activate the machine control functions.

SOFT-WIRED UNITS

Each CNC unit is able to control more than one type of machine tool. Soft-wired CNC units have a control, or *executive*, program adapted to control a particular type of machine tool. The executive program is entered by the builder in the computer memory. The executive program is a computer-based control system capable of executing the commands of the program for a part. The program is occasionally modified by the computer or machine tool builder.

MCU DESIGN FEATURES AND FUNCTIONS

An edited punched tape is run through a tape reader in order to enter the program into the computer and to store it in computer memory. One or more part programs may be stored within the computer storage capacity. One feature of a computer is *random access memory* (RAM). This feature permits any stored program to be called up when needed.

When a CNC unit has a *cathode ray tube* (CRT), the part programmer or machine tool operator may visually view positioning and operational information. The CRT is capable of displaying axes positions for all machine slide movements and other machine functions. Messages to the operator may be programmed for display on the CRT screen at a specified time. Other functions displayed on a CRT screen include:

- Part numbers of all programs stored in computer memory;
- Used and available capacity for further storage;
- Compensations for cutter radius, tool lengths, tool offsets, and tool fixture offsets;
- Diagnostic information when the CNC units are able to isolate and identify malfunctions in the numerical control system;
- Part program information when editing.

FUNCTIONS OF A COMPUTER (CPU) IN COMPUTER-ASSISTED PROGRAMMING

The computer used for computer-assisted programming has the computational ability to generate numerical part program data in a useful form by the MCU of a machine. Some computers are located a distance from the MCU. Other smaller computers (often called *minicomputers*) are an integral unit of the MCU.

CNC computers store part programs and process them to generate output signals for the control of a machine tool. Computers use a two-digit binary notation. The two binary digits correspond to two conditions of operation of the electronic components: on or off, charged or discharged, positive or negative charge, and conducting or nonconducting.

The *central processing unit* (CPU) includes all the circuitry controlling the processing and execution of instructions entered into the computer. The circuits relate to basic memory for storage and retrieval and *logic*. The four basic arithmetical processes of addition, subtraction, multiplication, and division are accomplished by adding positive and/or negative numbers. Simple computations are made in billionths of a second.

Input information may be entered into a computer in several forms: punched tape, mag-

netic tape, diskette, or signals from other computers. Similarly, *output information* may be received in these same forms or as printout sheets or electrical signals that control an NC machine tool operation. Buffer storage is provided on some machines so that the stored advance information is available at the point where it is needed.

ROLE OF THE PROGRAMMER IN COMPUTER-ASSISTED PART PROGRAMMING

The Automatically Programmed Tools (APT) system was one of the first computer processor languages. APT, Compact II®, and many other different processor languages are used depending on the machine tool or the complexity of producing parts requiring several machines to be incorporated in a complex machining center.

PROGRAM PLANNING

The part programmer first views the part to be machined in terms of axes, type of cutting tool, the NC machine tool, and the path of the cutting tool. For purposes of providing an example of computer-assisted programming, the part drawing in Figure 30–6 illustrates a workpiece to be turned on a NC lathe. In this instance, the spindle axis (longitudinal feed) is Z and the transverse (in and out feed) axis is X. Plus and minus coordinates are used. The part drawing uses incremental coordinates for dimensioning.

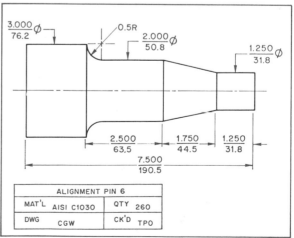

Figure 30–6 Part Drawing of Workpieces to Be Turned on a NC Lathe

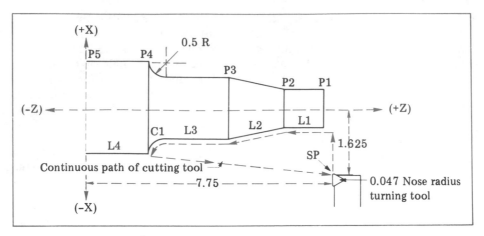

Figure 30–7 Part Programming Layout with Turned Surfaces, End Points, and Starting Points

After viewing the part to be machined, the part programmer prepares a programming layout, as shown in Figure 30–7. The cutting tool has a nose radius of 0.047″ (1.2mm). This layout identifies each process of straight turning (L1), angular turning (L2), straight turning (L3), and radius turning (C1), as well as the reference surface (L4). The starting and ending points (P1, P2, P3, and P4) for each cut are indicated. The reference point (P5) and the starting (SP) point of the cutting tool are given. This reference point is 0.250″ from the end of the workpiece and 1.625″ from the center line.

MANUSCRIPT PREPARATION

After completing a sketch of the programming layout, the part programmer writes the manuscript by preparing a series of statements. An APT program uses four types of statements:

- *Motion* statements to describe cutting tool position,
- *Geometry* statements to describe the features of the part,
- *Postprocessor* statements to identify machine tool and control system data,
- *Auxiliary* statements to provide additional information that is not given in any other statement.

The statements must follow grammatical rules for construction of statements and words, punctuation, and spelling. When the APT system is used, the spelling of an APT word must be exactly as it is spelled in the APT *system dictionary*. This dictionary specifies the only form the computer understands. The computer will indicate an inaccurately spelled word by an undefined symbol.

The following allowable words to be used for programming the turned part shown in Figure 30–7 are selected from and are written according to an APT dictionary. The spelling of each word to be used in programming the part is accompanied with a brief identification of the process, function, or location. An application is then given to show the usage of the word. The program planner first lists postprocessor and auxiliary statements in the manuscript:

PARTNØ   *Part number.* For information. The statement must be the first statement or appear immediately after END.

MACHIN   *Machine identification.* Example: MACHIN/HARDINGE 6. The computer is to make a control tape for a Hardinge turning machine number 6, equipped with a compatible input system.

INTØL   *Inside tolerance.* Example: INTØL/ .001. The computer is to stay within 0.001″ on the inside of curves in making straight-line approximations of curves.

ØUTØL   *Outside tolerance.* Example: ØUTØL /.001. The computer is to stay within 0.001" on the outside of curves in making straight-line approximations of curves.

CUTTER   *Cutter.* Example: CUTTER/.94. The nose radius of the cutting tool is designated by the diameter of a theoretical circle. The 0.047" (3/64") radius of the cutting tool as shown on the drawing in Figure 30–7 is designated as CUTTER/.094.

CØØLNT   *Coolant.* Example: CØØLNT/ØN. Turn the coolant on. It will remain on until CØØLNT/ØFF or STOP command.

CLPRNT   *Print out.* Coordinate dimensions of all end points and straight-line moves are called to be printed out.

FEDRAT   *Feed rate.* Example: FEDRAT/4, IPM. Feed rate in all directions, including Z, will be 4" per minute.

SPINDL   *Spindle.* Example: SPINDL/400, RPM. Start spindle at 400 RPM. Spindle stays on until SPINDL/ØFF, STØP, or END.

Postprocessor and auxiliary statements are followed by geometry and then motion statements:

PØINT   *Point.* Example: P2 = PØINT/6.25, 2.20, 1.50. P2 is the point with coordinates X = 6.25, Y = 2.20, and Z = 1.50.

LINE   *Line.* Example: L1 = LINE/P1, P2. Line 1 is the line through points P1 and P2. Example: L3 = LINE/P3, RIGHT, TANTØ, C1. Line 3 is a line, tangent to circle 1 (C1), right of P3, through points P3 and P4.

The program and machining cycle are completed by using closing statements:

CØØLNT   *Coolant.* Example: CØØLNT/ØFF. Turn the coolant off.

FINI   *Part program is completed.* FINI must be the only word in the statement.

How to Prepare an APT Program Manuscript

Planning the Program

Note: The alignment pin shown in Figure 30–6 is used to illustrate the preliminary steps and the kind of input required to prepare an APT program manuscript. The part is to be turned on a NC lathe.

STEP 1   Study the part drawing in terms of design features, machining processes, tooling, and machine characteristics.

STEP 2   Prepare a sketch to identify the path of the cutting tool in machining the workpiece.

STEP 3   Add the programming layout to the sketch (as shown in Figure 30–7).

STEP 4   Determine the sequence of each APT word to be used in the program.

STEP 5   Check an APT language dictionary for the correct spelling of each word.

Prepare the Program Manuscript

STEP 1   Prepare the computer-assisted program in APT processor language. Start with postprocessor and auxiliary statements—for example,

```
PARTNØ    ALIGNMENT PIN NØ6
MACHIN/HARDINGE 6
INTØL/.001
ØUTØL/.001
CUTTER/.094
CØØLNT/ØN
CLPRNT
FEDRAT/4, IPM
SPINDL/400, RPM
```

STEP 2   Add the geometry statements in APT language—for example,

```
SP = PØINT/7.75, -1.625
P1 = PØINT/7.5, -.613
P2 = PØINT/6.25, -.613
P3 = PØINT/4.5, -1.0
P4 = PØINT/2.0, -1.5
P5 = PØINT/0, -1.5
```

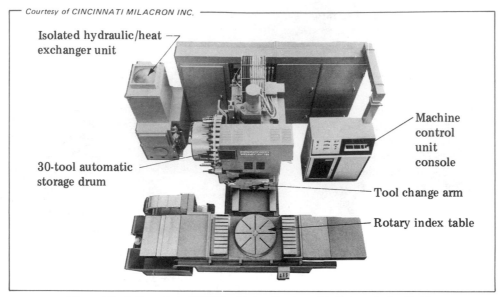

Courtesy of CINCINNATI MILACRON INC.

Isolated hydraulic/heat exchanger unit

Machine control unit console

30-tool automatic storage drum

Tool change arm

Rotary index table

Figure 30–8 Complete CNC Machining Center with Rotary Index Table, 30-Tool Automatic Storage, and Tool Change Arm

```
        L1  =  LINE/P1, P2
        L2  =  LINE/P2, P3
        C1  =  CIRCLE/2.5, -1.5, .5
        L3  =  LINE/P3, RIGHT, TANTØ, C1
        L4  =  LINE/P4, P5
```
STEP 3 Follow the geometry statements with the motion statements—for example,
```
        FRØM/SP
        GØ/TØ, L1
        GØLFT/L1, TØ, L2
        GØLFT/L2, PAST, L3
        GØRGT/L3, TANTØ, C1
        GØFWD/C1, PAST, L4
        GØTØ/SP
```
STEP 4 Conclude the program with two closing statements—for example,
```
        CØØLNT/ØFF
        FINI
```

BASIC COMPONENTS OF A CNC SYSTEM

The maximum productivity of a machining center depends on the versatility of the computer numerical control system. For example, the Acramatic® CNC system is designed with solid-state circuitry and a minicomputer. This combination of hardware and software systems provides flexibility of application to a complete CNC machining center (Figure 30–8). The CNC unit may be adapted at a later date to a complete computer-aided manufacturing center (CAM). The unit is designed to allow for the addition of advanced features, data input/output devices, monitors, and sensors; tool path modification calculations; and tool management information. The CNC system generally has four basic components: (1) tape reader, (2) cathode ray tube (CRT) display, (3) keyboard, and (4) manual and NC panel.

DESIGN FEATURES OF A CNC SYSTEM

FIXED MACHINING CYCLES

Defined EIA machining cycles—for example, drill, ream, bore, and tap, including dwell—are used to initiate a complete automatic sequence of machining processes within one programmed block of information. Other systems require a separate block for each function in the cycle.

A package of EIA machining cycles saves programming, tape punching, and reader time. A part program may also be stored with less memory and tape length.

BUFFER STORAGE

The machine tool operates from one block of information executed in the control. At the same time, up to 400 characters of data are read and held in advance in buffer storage.

LINEAR AND CIRCULAR INTERPOLATION

Any two axes (X–Y, X–Z, Y–Z) may be moved simultaneously. All three axes (X, Y, Z) may be moved in a straight-line path at conventional or rapid feed rates. High-precision angle cuts may be machined. A curve may be machined by cutting a series of short, straight-line segments along an arc.

The CNC system has the capability to machine arcs in a two-axis plane in any circular quadrant. Program input of start point, end point, and centerline of circular arc is used. Instantaneous switching is provided from circular and linear interpolation modes.

AXIS INVERSION

A keyboard key is used to invert the sign of the X and/or Y dimensional data. The axis inversion feature permits using a single tape to machine symmetrical or right- or left-hand parts.

PROGRAMMED SLIDE DIRECTIONS AND MOVEMENT MODE

Directions of linear slide movement along X, Y, and Z axes (planes) may be programmed. The direction is controlled by preceding programmed directions by a + or – sign. Programs may be written in the absolute or incremental dimensional system or in a combination of these systems. Programmed instructions automatically switch to the required code.

The Acramatic® CNC system accepts punched tape programmed to either EIA RS-244-B or (ASCII) EIA RS-358-B standard. Inch-standard or metric programs for each controlled axis are initiated by a control panel push button. Minimum input units of 0.001″ or 0.001mm (as selected) are automatically displayed on the CRT.

TAPE-CONTROLLED PROGRAMMING

Tape-controlled programming features include:

- Tape-controlled feed rates, programmed in ipm or mm/min;
- Tape-controlled spindle speeds, programmed directly in RPM;
- Rotary table index positions, programmed directly in degrees (noninterpolated rotary axis movement is simultaneous with any X-Y-Z plane movement);
- Preselected program pockets or tool numbers, tape controlled for random tool selection;
- Machine slides, programmed for axis dwell that may be varied from 0.01 to 99.99 seconds in 0.01 second increments. The X, Y, Z slides may be kept motionless in a cut to permit a higher-quality surface finish and greater accuracy for certain machining operations;
- Program interruption, permitting a manual tool change.

OTHER STANDARD CNC FEATURES

A few other CNC system features include:

- Slash (/) code, which is an operator selection feature;
- Block delete, which determines whether or not blocks of programmed data, prefixed by the code, will be ignored;
- Two levels of manual control;
- Spindle keyboard override which permits on-site adjustments for material hardness or cutting action through keyboard modification of the programmed spindle speed;
- Gage height feature, which establishes the limit, may be programmed as a six-digit R word;
- Data search, which permits running the tape in forward or reverse direction;
- Manual power axis feed, which permits independent control of each linear axis. Three feed ranges may be manually selected from 0.1 ipm to 1.0 ipm, 1.0 to 10.0 ipm, and 10.0 to 100.0 ipm;

- Incremental jog, which permits independent control of each linear axis;
- Target point align, which provides automatic positioning of X, Y, and Z axes;
- Position set, which controls floating zero capacity;
- Inhibit feature, which permits the operator to interrupt machine activities;
- Push button emergency stop, which initiates stoppage of power to servodrives and feed commands;
- Dry run feature, which allows the program to be cycled through a noncutting tryout at higher than programmed feed rates in order to cycle the numerical control machine tool through all movements and machining processes and check the program before a part is produced.

EXPANDING PRODUCTIVITY AND FLEXIBILITY OF A CNC SYSTEM

The flexibility and productivity of a NC machine tool, machining center, or computer-assisted machining system may be expanded through additional compensation features for cutter diameter, tool length storage, and workpiece irregularities. A cutter diameter compensation (CDC) feature permits the program to be altered mechanically for variations in cutter diameter. Compensation may be made for undersized or oversized cutter diameters. The keyboard permits compensation in increments of 0.0001″ up to a maximum of ±1.0000″. The CDC feature is operable in linear and circular interpolation modes, but only in the X–Y plane.

The tool length storage/compensation feature permits accurate tool depth adjustment. This feature eliminates the preset tooling stage. The feature also permits storing a six-digit length of tool dimension in core memory for each available tool pocket. Tool length compensation may be made for variables in increments of 0.0001″ up to a maximum of 99.9999″.

Variations in setup or irregularities of workpieces may be compensated for by an assignable tool length trim feature. This feature is available for groups of 16 program selectable tool trims. The trim range is ±0.0001″ to ±1.0000″.

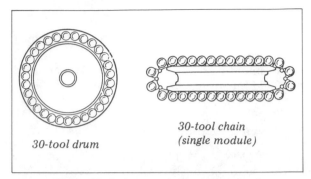

Figure 30–9 Automatic Tool Drum and Tool Chain Single Modules

PRESETTING NUMERICAL CONTROL CUTTING TOOLS

The cutting tools for machining a specific part on a NC machine tool are usually preset for length. The complete set of preadjusted tools are then stored for future use. When needed, the cutting tool set is placed in the NC changer, a tool drum, or a tool chain. Figure 30–9 shows an automatic 30-tool drum and a single-module 30-tool chain. Often two or more modules are combined. For example, two 30-tool modules are used to accommodate 60 tools on a machining center; three modules for 90 tools, etc.

TOOL LENGTH GAGE

The length a cutting tool extends from its toolholder may be measured by a *tool length gage*. One example of a NC *dial indicator gage* for tool length and diameter has a vertical column that includes a series of accurately spaced collars. The collars are usually 1″ (25.4mm) apart. A micrometer head with 1″ (25.4mm) travel is used to cover all measurements in increments of 0.0001″ (0.002mm) within the range of the gage.

Each cutting tool is set in its holder. The distance the cutting tool projects beyond the holder is adjusted to meet job requirements.

ELECTRONIC TOOL GAGE WITH DIGITAL READOUT

An *electronic tool gage* provides highly discriminating tool length measurements and adjustments. Figure 30–10 shows an electronic tool gage for length and diameter with digital readout.

Figure 30–10 Electronic Tool Gage for Length and Diameter with Digital Readout

An advantage over dial indicator types is the ability of the electronic gage to provide instantaneous readout in either customary inch or metric units of measure.

The digital tool length gage may be used for the off-line gaging of tool lengths and diameters. A tool is gaged for length by inserting it into the gage spindle socket. The feeler tip of the tool gage is then brought into contact with the tool point or edge. The enter button is pressed for storage in memory of the CNC system. Figure 30–11 shows an application of the electronic tool gage in setting an insert-tooth milling cutter for diameter.

When used in conjunction with hard-wired numerical control systems, the electronic tool gage permits tool dimensions to be transferred into the control by means of compensation switches, a keyboard, or punched tape entry of tool data. All tool data, including storage location, tool identification number, tool length, and cutter diameter compensation, are preselected off-line.

Figure 30–11 Application of Electronic Tool Gage in Setting an Insert-Tooth Milling Cutter for Diameter

Safe Practices for Computer Numerical Control Machine Tools

- Establish whether the MCU console has an emergency work table and/or spindle-positioning stop control. Use the stop control as needed or whenever there is possibility of a personal accident or damage to the cutting tools, machine, table, or other accessories.
- Check the clearance of the work table, vises, clamps, or spindle when manually positioning the work table or spindle. This clearance check is required before positioning data is entered in the MCU console.
- Check the cutter and workpiece clearance before lowering the quill (and cutting tool) on a drill press or vertical mill.

- Reduce the overhang of a quill to ensure maximum rigidity.
- Check the quill movement to be sure a cutter may be raised to clear the workpiece when the quill-up control is actuated by the MCU.
- Verify the tape input by a dry run on the machine tool.

- Observe each operation as it is performed on the NC machine tool, including checks for accuracy.
- Maintain a duplicate, permanent tape record in a separate storage location.

TERMS USED WITH COMPUTER-ASSISTED NUMERICAL CONTROL MACHINE TOOLS

APT	A computer system for Automatically Programmed Tools. A multi-axis contouring program. A collection of computer programs in which a shortened English language vocabulary enables the part programmer to direct a cutter and its path.
Automatic programming	Use of a digital computer to transform what a part programmer is able to program into a computer programming activity.
Block count readout	NC cabinet display showing the number of blocks that have been read from the tape, counting each block as it is read.
Digital input data	Pulses, digits, or other coding elements that supply information to a machine control.
Encoding	Translating to a coded form with limited loss of information.
End-point control	Continuous automatic analysis of the final product. Changing the process as required for quality control.
Off-line data processing	Computations of a geometric and mathematical nature, tool offsets, feed codes, paths, and others; miscellaneous and preparatory function tasks. Generation of a control tape.
On-line data processing	Translation of the information on a control tape into input signals on a NC machine tool. On-line data processing is repeated each time the control tape is used to produce a part.
Processor	A computer program that performs computations that are part oriented, computes tool offsets, and places the cutting tool on the part. A portion of the computer that controls input and output devices. A unit that operates on received, stored, and transmitted data.
Postprocessor	A computer program that translates the results of a basic computer program into a control tape.
Readout	A numerical display showing the actual position of a machine slide or tool.
Readout (command)	A display showing the absolute position. An absolute position derived from a position command. Readout information taken directly from the dimension storage command or as a summation of command departures.

SUMMARY

- A machine control unit (MCU) may be hard wired to permanently connect electronic components to control one type of machine tool. Soft-wired CNC units include a control (executive) program entered into computer memory.

    - A control processing unit (CPU) executes output instructions from a computer, as required.

- APT is a multi-axis contouring program. It uses an abridged English word vocabulary to direct the path of a cutter, to completely program all machining processes, and to make dimensional accuracy checks.

    - A parity check permits a review of tape information. Errors in tape preparation and other possible malfunctions are identified.

- CNC systems are designed with combination hardware and software flexibility. The CNC capability is extended by adding advanced data input/output devices, monitors, sensors, tool path modification, tool management, and other functions.

    - Advanced CNC systems may be adapted to a complete computer-aided manufacturing (CAM) center.

- CNC standard design features provide for multiple-axis contouring, fixed machining cycles, linear and circular interpolation, buffer storage, axis inversion, programming slide direction, and movement mode.

    - A few other CNC system features include: block delete, slash (/) code, spindle keyboard override, data search, incremental jog, zero position set, dry run, and emergency stoppage.

- CNC system productivity may be extended by the addition of cutter diameter compensation, tool length compensation, and tool length trim features.

    - Features of NC engine lathes are applicable to general bar, chucker, and turret lathe machining. The features include but are not limited to: dimension display, sequence number display, manual data input, tool select, tool offset, spindle speed and feed controls, and auxiliary function display.

- A combination NC system permits simultaneous control of all axes, positioning of spindle, built-in cycles for depth axis, and continuous-path control.

    - NC jig borers may be positioned within 0.000030″ for longitudinal and traverse travel. The machine may be equipped with a rotary table to serve as a fully numerically controlled axis.

- NC machining centers combine the processes performed on a number of machines into one center. Such a center may have computer control of three, four, or five axes and simultaneous movement of the spindle head, table traverse, table rotation, and table tilt.

UNIT 30 REVIEW AND SELF-TEST

1. Identify three basic control features that apply to all NC machine tools.

2. Indicate three functions that may be displayed on a control panel.

3. Cite one advantage of using a circular interpolation feature to generate curved surfaces in comparison to a linear interpolation feature.

4. State the reason for parity check.

5. Identify two major features for computers used in computer-assisted programming.

6. The contour on follower cam No. 12 as shown on the part drawing below is to be end milled on a NC vertical milling machine. A 1" diameter end mill is to be used.

    a. Make a program planning sketch. Indicate the sequence of events.

    b. Write the program in the APT system. Group the statements according to functions.

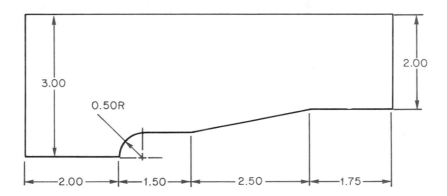

7. State three CNC machine functions that may be tape controlled.

8. Explain the function of a cutter diameter compensation (CDC) feature on a CNC machine tool.

9. State two safety precautions that should be taken with tape information.

New Materials and Hardness Testing: Technology and Processes

Particle steel and powdered metals metallurgy processes are used in the manufacture of cutting tool materials with better machinability, grindability, wear life, toughness, hardness, and other special properties. This unit includes an overview of the basic processes of producing particle metallurgy steels and powder metallurgy materials, their properties, and some general applications.

Other technical information is provided about the properties, advantages, and applications of titanium nitride (TiN) coatings for cutting tools, plastic cavity molds and tools, and other parts applications. A description is provided about physical vapor deposition (PVD) principles and coating process.

While hardness testing is important to the measurement of hardness properties of materials, the data gathered are significant to all of manufacturing. Hardness testing provides the database for establishing cutting speeds and feeds, surface finish standards, machinability of materials, and other technical information. This information is essential to tool design, machine tooling setups, and manufacturing. This unit builds upon earlier units that related to metallurgy and heat treatment.

UNIT 31

Particle and Powder Metallurgy, Coating Materials and Hardness Testing

OBJECTIVES

After satisfactorily completing this unit, you will be able to:

- Discuss different applications of particle and powdered metallurgy steels and titanium nitride (TiN) coatings.
- Identify the materials and describe the manufacturing processes for producing PM steels, powdered metal products, and TiN coatings.
- List the advantages and disadvantages of using PM steels, powdered metals, and TiN coatings.
- Make comparisons of cost effectiveness when using conventional steels, PM steels, powdered metallurgy steels, and as a result of coating cutting and forming tools and parts with TiN coatings.
- Interpret designations of hardness values and convert such values.
- Describe the functions and general applications of Rockwell, Brinell, Vickers, scleroscope, microhardness, and Knoop instruments and systems of hardness testing.
- Refer to handbook tables and convert hardness numbers among the major hardness testing systems.
- Select heat treating temperatures for particular grades of steels and determine melting points of metals and alloys in general use.
- Identify applications of diamond and hardened steel ball indentors.
- Perform or simulate each of the following processes.
  - Testing for Hardness by Using a Rockwell Hardness Tester.
  - Testing for Hardness Using a Brinell Hardness Tester.
  - Testing for Hardness Using a Scleroscope.
- Follow *Safe Practices* relating to personal safety in using PM, powdered metals, and TiN cutting tools and the protection of hardness testing instruments.
- Correctly use related *Terms*.

A. MANUFACTURE OF PARTICLE METALLURGY (PM) STEELS

The *particle metallurgy* (PM) process requires fluid molten alloy steel to be passed through a high-pressure atomizer. At this stage, the fluid metal is broken down into small metal droplets. The droplets solidify rapidly to form fine steel particles.

Upon cooling, the steel powder is screened for particle size. Particles of uniform size are then loaded into a steel canister. All gasses are evacuated from the canister before it is sealed. The container is then heated and pressed to compact the uniform metal particles.

The compacted mass is then worked into standard sizes and shapes by conventional forging, rolling, and other processing methods. Particle metallurgy steels are available in standard flat, square, round, sheet, wire, and coil shapes.

CHARACTERISTICS OF PM STEELS

The particle metallurgy process produces steel particles that are uniform throughout a cross section of the final product. This uniformity is in contrast with conventional steels wherein there is alloy segregation. Segregation is produced by the slow cooling of ingots.

The photomicrographs in Figure 31–1 show acid-etched cross sections of a conventional alloy composition steel bar at (A) and a particle metallurgy steel bar at (B). Alloy segregation may be identified in Figure 31–1A by the darker areas. These represent a denser alloy composition. The uniformity of the particles in a particle metallurgy steel is shown in the cross section in Figure 31–1B.

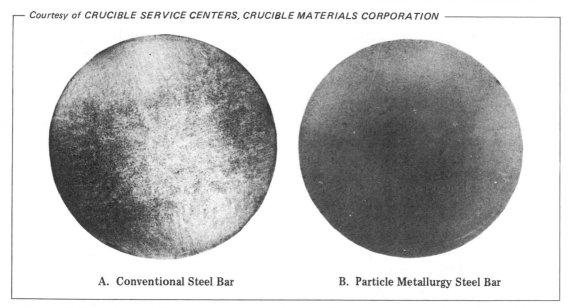

A. Conventional Steel Bar   B. Particle Metallurgy Steel Bar

Figure 31–1 Acid-etched Photomicrograph Cross Sections Showing Variations in Alloy Compositions

EFFECT OF CARBIDES ON WEAR RESISTANCE

The wear resistance of a hardened tool steel is determined by the amount and the hardness of the carbides present in the microstructure of the steel. Carbides in conventional steels are relatively large and, generally, form in clusters (Figure 31–2A). Carbides in particle metallurgy steels are evenly dispersed throughout a cross section and have a fine structure. The fine carbide structure and dispersion (shown at Figure 31–2B) increases the machinability (grindability) and toughness in a hardened state of a PM steel.

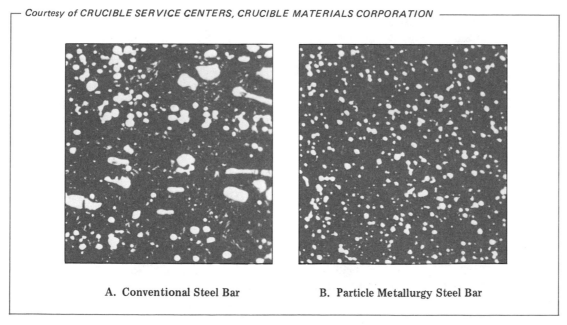

A. Conventional Steel Bar   B. Particle Metallurgy Steel Bar

Figure 31–2 Photomicrographs Showing Contrasting Carbide Structures and Dispersion

Carbides are formed metallurgically when carbon molecules combine with tungsten, molybdenum, vanadium, iron, or chromium. Of these, vanadium is the hardest element. As the vanadium content in a conventional ingot-cast tool steel increases beyond three percent, the steel becomes brittle and its grindability decreases.

By contrast, some grades of particle metallurgy steels (with almost ten percent vanadium by weight) exhibit greater toughness, wear resistance, and grindability.

PROPERTIES/CHARACTERISTICS OF PM STEELS

The large size and uneven clusters of tungsten or vanadium carbides in ingot-cast steels (Figure 31–2A) makes these steels difficult to grind due to the fact that large carbide particles tend to dig into aluminum oxide and silicon carbide grinding wheels. Cubic boron nitride (CBN) wheels are generally used for such applications.

The fine, evenly distributed carbides in PM steels (Figure 31–2B) permit the small grains to be flowed away in the grinding process without damage to the grinding wheel. Importantly, good grindability properties of PM steels do not affect wear resistance.

The machinability of PM steels is also enhanced by the amount of sulphur that is added to the alloy. The addition of sulphur (within limits) to steels improves machinability but sacrifices toughness. Like the fine, uniform dispersion of vanadium and other alloys in PM steels, the fine, uniform dispersion of sulphides in PM steels improves machinability without sacrificing toughness. Similarly, PM steels show a greater consistency and uniformity in heat treating hardness properties with less geometric distortion.

ADVANTAGES OF PARTICLE METALLURGY STEELS

The following advantages are claimed for PM steels as compared with conventional tool steels.

- Better machinability in the unhardened state.

- Improved grindability in the hardened state.

- Reduced part unit manufacturing costs. While the initial cost of PM steels is higher, this cost is offset by greater productivity, longer tool wear life, and decreased machine down time.
- Greater uniformity and consistency in heat treating.
- Less geometric distortion in heat treating.
- Increased toughness and resistance to breakage produced by shock loads or tool deflection.
- Consistency of composition and metal structure.

B. POWDER METALLURGY: PROCESSES AND PRODUCTS

Powder metallurgy relates to the cold pressing of steel powders and additives and sintering. Powder metallurgy processes may also be used to produce nonmetallic products.

One of the best examples of powder metallurgy applications is the manufacture of carbide inserts, cutters, and preformed machine parts. The three basic powder materials that are required to produce cemented carbides are: tungsten, carbon, and cobalt. When pure tungsten powder is mixed with carbon powder and heated, tungsten carbide is produced. Powdered cobalt serves as a binder to cement the mixed tungsten and carbon powders, yielding cemented carbide. However, before a final product is produced there are other processes to be performed such as: selection, compressing, forming, and sintering.

POWDER METALLURGY: MANUFACTURING CARBIDE TOOLS

The properties and requirements for cutting tools depend on the selection of a number of powders that are *blended* to produce a specific composition. The powders may be air blended, liquid blended to control oxidation, or blended under controlled atmospheric conditions.

Completely blended powders are transported from hoppers to forming dies (Figure 31–3). The

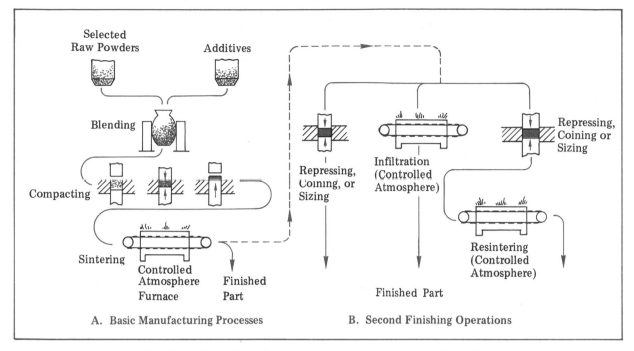

Figure 31-3 Manufacturing Process in Producing Powdered Metal Parts

powder mixture is subjected to tremendous mechanical/hydraulic forces (100 tons per square inch) during *compacting*. Blanks are formed by using forming dies on power presses, by extrusion, or by other hydrostatic press operations.

The carbide blanks are fragile and *green* at the compacted stage and may be broken easily. A solid mass carbide blank is produced by *sintering* at temperatures below the melting point of the metals in the alloy. Sintering is the process of using heat to sinter (cement) the carbide particles together. Sintering takes place in a non-oxidizing controlled atmosphere furnace to prevent surface scaling and discoloration. Some sintering is done without pressure. Other sintered products require *hot pressing*.

OTHER FINISHED POWDERED METAL PRODUCTS

The specifications of finished powder metal products determine additional processes that are to be performed. Examples include: accuracy (dimensional, geometric tolerances, and surface texture); titanium coating of a blank or insert, regrinding the cutting edges with CBN grinding wheels, and the brazing of inserts to carbide shanks.

Power metallurgy processes are also used in manufacturing to produce parts that either require no additional machining or a limited number of conventional machining processes. Parts of regular or intricate shapes may be formed by compacting blended metal powders in a die and then sintering to produce a finished formed part.

The porosity of powdered metal parts may be controlled. The porosity may be varied from a highly porous part like an impregnated bearing to a highly dense structure as required for cemented carbide and other cutting tools.

SECOND OPERATIONS

The term *second operations* means that in order to meet exacting specifications a powdered metal part must undergo further processing. Sizing, impregnating, infiltrating, and surface coating are examples of second operations (Figure 31-3B).

Sizing, coining, and **repressing processes** are required when a part is to be held to close dimensional and geometric tolerances (sizing); to a higher quality of surface finish (coining); or to increase density and to give added strength (repressing). Parts are repressed in a die similar to the original forming die.

Impregnation refers to the process of filling voids among the powder metal grains within the formed part. Oils and graphite are examples of nonmetallic impregnating materials used with powdered metal parts that require constant lubrication.

Infiltration is the process of filling the pores of a powdered metal part with a metal or alloy. Parts are infiltrated for purposes of increasing ductility, strength, impact resistance, and hardness. Infiltration is carried on under controlled atmospheric conditions. The melted lower temperature infiltrating material flows among the pores and fills the voids of the powdered metal part.

Machining, heat treating, metal joining, plating, and surface coating are some of the other second processes that may be required to produce the finished powdered metal product.

ADVANTAGES OF POWDERED METAL PRODUCTS

• Economy of manufacturing; preformed parts require minimum second operations.

• Versatility in being able to accurately form complex part features.

• Controlled porosity that permits impregnating movable parts.

• Ease of producing tapered and other hole shapes and openings, recesses, and internal details.

• Quality of product material is closely controlled.

• Elimination of second processes. Parts may be held to dimensional tolerances of ±0.001″ (0.02mm) with high quality surface finish.

DISADVANTAGES OF POWDERED METALLURGY

• Not all part design features may be produced: for example, internal threads and other surface indentations like grooves.

• The abrasive characteristics of certain powder materials produce severe wear on punches and forming dies.

• Press capacities. The compression ratio and extreme pressure of up to 200,000 psi (1380 MPa) limit the size of the parts.

• Equipment and furnace costs require high production runs to be cost effective.

• Difficulty in producing design features such as sharp corners and widely varying part thicknesses.

• Part shapes are limited to the plastic flow of powders around corners.

C. CUTTING TOOL COATINGS: TECHNOLOGY AND PROCESSES

Coatings are used on metal cutting and forming tools, plastic molds, wear parts, and other surfaces to protect them from abrasive wear and heat-producing friction; to permit greater productivity; to produce high-quality surface finishes and dimensionally accurate parts; and to reduce cutting tool forces by providing easy chip flow.

Titanium nitride (TiN) is a popular coating. Titanium nitride produces a superhard (over 80 Rc value), harder than carbide, coating for cutting and forming tools. Typical form tool applications include: punches and draw dies, cold forming, stamping, and fine blanking dies.

Extremely hard TiN coatings increase the surface hardness of cutting tools. Also, since titanium nitride has a considerably lower coefficient of friction than uncoated steels, the added *lubricity* characteristic permits easier material flow and stripping of machined parts from molds.

Titanium nitride coated cutting tools and forming tools have longer tool wear life. This means lower maintenance and manufacturing costs and higher productivity.

CHARACTERISTICS OF TITANIUM NITRIDE (TiN) COATINGS

High hardness. The Rc hardness scale value of TiN of 80 Rc and higher indicates its extreme hardness over uncoated high speed steel cutting tools that range from Rc 62 to Rc 65 in hardness.

Lubricity. Unlike conventional steels, a TiN surface coating significantly reduces the tendency of materials to adhere to cutting edges, within clearance areas, and along cutting tool flutes. As chips are deflected away from a cutting tool less heat is generated by friction.

Higher operating speeds. The improved cutting action of TiN coated tools permits cutting operations to be performed at increased speeds and feeds that begin from 20% to 30% higher than for uncoated tools.

Increased tool wear life. TiN coated cutting tools maintain machining capabilities (to produce parts within acceptable limits of surface texture and dimensional and geometric tolerances) that are three to eight times longer than uncoated cutting tools.

Improved surface finish. Surface textures are often improved by using titanium nitride coated cutting tools. Due to the low coefficient of friction of TiN, chips flow away more easily from the cutting edges. The cutting tool edges remain sharper as contrasted with cutting edges of an uncoated tool which become ragged as chips tend to adhere. The easier flow of chips away from the workpiece and cutting tool edges improves surface finish.

The effect of a TiN coated cutter on the quality of surface finish is illustrated in Figure 31–4. During machining, the cutting edge of an uncoated tool builds up and continuously changes its geometry and structure. The built-up cutter edge is forced into the work surface together with fine chip particles that break off. The condition is shown at (A).

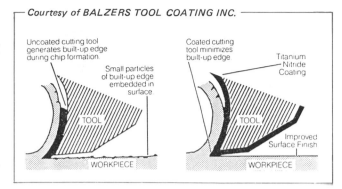

Courtesy of BALZERS TOOL COATING INC.

Figure 31–4 Effect of a Titanium Nitride Coating on the Improved Quality of Surface Finish

In the case of a TiN coated cutter (Figure 31–4B), the lubricity characteristics retard a chip buildup on the cutter edge. A higher quality surface finish is produced as the coated cutter edge remains unchanged throughout the machining process.

Reduced tool deflection. TiN coated cutting tools have less tool deflection since easier cutting results in reduced horsepower requirements. With less cutting force there is less tool deflection in contrast with similar tooling and machining conditions with uncoated tools. The same general principles for tool setups with the least possible overhang and using the largest possible diameter holder are followed with coated cutting tools.

Elimination of second process operations. With the greater ability of TiN coated cutting tools to maintain tolerance and surface finish accuracy, it is possible to eliminate many second operations.

Reduced cutting tool down time. This is the end result of all the advantages just cited. Also, less frequent sharpenings means fewer tool changes, greater tool wear life, and increased productivity.

Cost effectiveness. The additional cost of coating a cutting tool ranges between 25% to 50% of the original cost. However, with the increase in productivity over the wear life of the cutting tool, the unit cost in part manufacturing is substantially below machining costs when uncoated cutting tools are used. Generally, the wear life of a properly resharpened coated cut-

Table 31–1 Effects of Titanium Nitride (TiN) Coating of Taps on Tapping Speeds and Production

	User	Material	Tap Type	Conditions	Before	After
A	Computer Manufacturer	Plastic	No. 4-48 four-flute straight-flute taper tap	132 sfm	uncoated— 1350 holes	TiN coated— 7200 holes
B	Machine-Tool Builder	Cast Iron	1/4"–20 two-flute spiral-point plug tap	46 sfm, 0.625" deep hole, 69 percent thread	uncoated— 40 holes (inconsistent gaging)	TiN coated— 525 holes (consistent gaging)
C	Metal Products Manufacturer	Monel	No. 6-32 two-flute spiral-point plug tap	(see before and after)	chrome plated— 8 to 10 holes @ 19 sfm	TiN coated— 35 to 40 holes @ 25 sfm
D	Aerospace Manufacturer	4340 alloy steel	1/4"–20 two-flute spiral-point plug tap	1/2" deep hole	Nitrided and steam oxide coated— 3 to 5 holes	TiN coated—21 holes (17 holes after resharpening)
E	Computer Manufacturer	Aluminum	No. 10-32 forming tap	7/8" deep hole, water soluble coolant	uncoated— 127 holes @ 75 sfm	TiN coated— 205 holes @ 75 sfm 283 holes @ 136 sfm

*Data Courtesy of VERMONT TAP & DIE DIVISION, VERMONT AMERICAN CORPORATION

ting tool is from two to three times longer than an uncoated tool.

Table 31–1 provides information on cost effectiveness. Productivity data is given by five different manufacturers comparing output in machine tapping using uncoated and TiN coated taps. Note the variety of machining conditions and the fact that regardless of the metal or nonmetallic parts that are tapped, there is a significant increase in productivity in each instance.

CUTTING TOOL MAINTENANCE

Regardless of whether a cutting tool is coated or uncoated, the same principles of maintenance prevail. Cutters must be resharpened before heavy craters, or peripheral, or flank wear occurs, or the cutting edges become worn, chipped, or gouged. These conditions must be corrected before a TiN coating is applied. While the coating is ground off a cutting tool face or land, the coating remains along the cutting edge. This resharpening condition is illustrated in Figure 31–5 using a form-relieved milling cutter example at (A) and a profile ground end mill at (B).

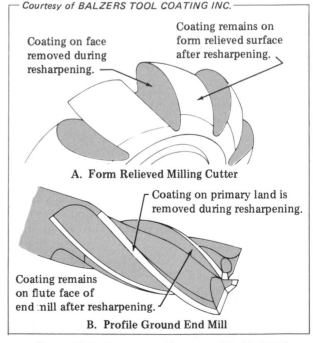

Courtesy of BALZERS TOOL COATING INC.

Coating on face removed during resharpening.

Coating remains on form relieved surface after resharpening.

A. Form Relieved Milling Cutter

Coating on primary land is removed during resharpening.

Coating remains on flute face of end mill after resharpening.

B. Profile Ground End Mill

Figure 31–5 Examples of Titanium Nitride (TiN) Coatings on Cutting Tool Edges after Resharpening

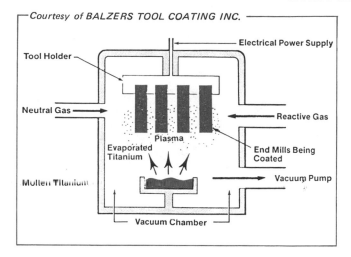

Figure 31-6 Schematic of Physical Vapor Deposition (PVD) Process for Titanium Nitride (TiN) Coating

PHYSICAL VAPOR DEPOSITION (PVD) COATING SYSTEM

SURFACE PREPARATION

There is excellent coating adhesion of titanium nitride when the surfaces to be coated are finely ground, have a bright finish, are free of burns, and have no surface treatment (like nitride or black oxide). Any foreign hardening salts, oils, or highly polished die molds or extruding tool surfaces may produce inferior coating adhesion.

APPLICATION OF TITANIUM NITRIDE (TiN) COATINGS

One common process of coating is called *Physical Vapor Deposition* (PVD). In this process, physically and chemically clean tools are placed in a reaction chamber. The tools become the cathode of a high voltage circuit. The chamber is then evacuated and charged with argon gas. As positive argon ions are propelled by a high voltage field to blast each tool, the tools become atomically clean. Figure 31-6 provides a schematic view of a physical vapor deposition chamber.

Within the reaction chamber, an electron beam gun heats the titanium until it evaporates. Nitrogen is then introduced into the chamber. Titanium ions accelerate electrically toward the tools. The nitrogen gas combines with the titanium bombardment within a 900°F temperature range to form a titanium nitride coating on the tools. Generally, a 0.0001″ (0.0025mm) TiN coating is applied. This coating is so securely bonded to the cutting tool that the bond is not affected by torsion, deflection, or other forces and stresses.

Since the PCD process occurs at temperatures below the tempering ranges of high speed steels and most tool steels, there is limited risk of tool deformation, metallurgical changes, or the softening of hardened tools. An advantage of applying TiN using the PCD process on solid carbide tools and inserts is that the process actually reduces the embrittlement of the cutting edges.

D. HARDNESS TESTING

A hardened steel is tested to define its capacity to resist wear and deformation. The hardness of steel, alloys, and other materials may also be used to establish other properties and performance.

Hardness testing is performed using measuring instruments of two basic types. The first set of instruments measures the depth of a penetrator for a given load. A few of the instruments that operate under this principle include: *Rockwell, Brinnel, Vickers, Knoop,* and *microhardness testers.* The second set of instruments measures the height of rebound of a given weight hammer of a special shape when dropped from a fixed height. *Scleroscope testers* represent this type of instrument. Each of the two types of instruments is covered in this unit.

Most instrument manufacturers indicate hardness by number scales. The number is related to the size and shape of the penetrator, load, height, and indentation or rebound. Reference tables provide numbers within each major system. These numbers permit conversion of hardness values from one system to another. For example, a Rockwell C (R_C) hardness of 50 is equivalent to 513 on the Vickers scale, 67 on a scleroscope, and BHN 481 on a Brinell scale when a 10mm (0.400″) diameter carbide ball and a 3,000 kg (6,600 lb) load are used. A 75.9 reading is recorded on the Rockwell A (R_A) scale using a diamond penetrator 50 kg (132 lb) load.

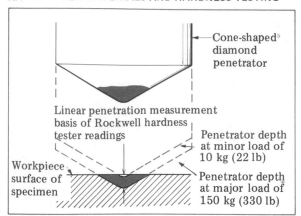

Figure 31–7 Linear Penetration Measured by a Rockwell Hardness Tester

Tables for Rockwell C (R$_C$) hardness numbers 68 through 20 are based on extensive tests with carbon and alloy steels primarily in a heat-treated condition. The hardness numbers may be reliably applied to tool steels and alloy steels in annealed, normalized, and tempered condition. Additional Rockwell B and other hardness scale numbers are provided in handbooks for unhardened or soft temper steels, gray and malleable cast irons, and nonferrous alloys.

ROCKWELL HARDNESS TESTING

Tests are made with the Rockwell hardness tester by applying two loads on the part to be tested. The first load is the *minor load;* the second, the *major load.* The Rockwell tester measures the linear depth of penetration, as shown in Figure 31–7. The difference in depth of penetration produced by applying a minor load and then a major load is translated into a hardness number. This number may be read directly on the indicator dial of a portable tester, a manual or motorized bench model, or on an digital display unit (Figure 31–8).

SIZES AND TYPES OF PENETRATORS

A shallow penetration indicates a high degree of hardness and a high hardness number. A deep penetration indicates a softer degree of hardness and a low hardness number. In general, the harder the material, the greater its ability to resist deformation.

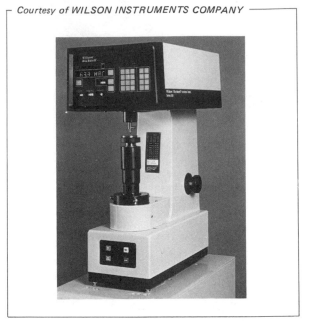

Figure 31–8 Motorized Rockwell Hardness Tester with Automatic Elevating Screw, Function/Numeric Keypad Controls, and Digital Display SPC Values

Different penetrator points are used on Rockwell testers, depending on the hardness of the material. Diamond- and ball-point penetrators are used on Rockwell hardness testers. Hard materials such as hardened steels; white, hard cast irons; and nitrided steels require a cone-shaped diamond. The trademark of this form of penetrator is registered as Brale®. A hardened steel ball-shaped point is used for unhardened steels, cast irons, and nonferrous metals.

C and A 120° Diamond Penetrator Points. The C Brale penetrator is used with a major load of 150 kilograms of force (kgf) or 330 pounds of force (lb/f) for hard materials. The *Rockwell C scale* is used with the C Brale penetrator.

Cemented carbides, shallow casehardened steels, and thin steels require the use of the A Brale penetrator and a 60 kgf (132 lb/f). The hardness is read on the *Rockwell A scale.*

Ball Penetrator Points. The second basic type of penetrator has a ball-shaped point and is made of hardened steel. The two diameters are 1/16″ and 1/8″ (1.5mm and 3.0mm). A *Rockwell B scale* is used for hardness readings of

Table 31-2 Rockwell Hardness Scales, Penetrators, Major Loads, and Reading Dials for Typical Hardness Test Applications

Scale	Type of Penetrator	Major Load (kgf)	Dial Color (Numbers)	Typical Hardness Test Applications
			Standard Scales	
B	1/16" (1.5mm) ball	100	Red	Soft steels, malleable iron, copper and aluminum alloys
C	Brale (diamond)	150	Black	Steel, deep casehardened steel, hard cast iron, pearlitic malleable iron, other materials harder than B-100
			Special Scales	
A	Brale (diamond)	60	Black	Thin steel, shallow casehardened steel, cemented carbides
D	Brale (diamond)	100	Black	Thin steel, medium casehardened steel, pearlitic malleable iron
E	1/8" (3.0mm) ball	100	Red	Cast iron, aluminum and magnesium alloys, bearing metals
F	1/16" (1.5mm) ball	60	Red	Thin soft sheet metals, annealed copper alloys
G	1/16" (1.5mm) ball	150	Red	Malleable iron, phosophor bronze, beryllium copper, other metals within an upper hardness limit of G-92
H	1/8" (3.0mm) ball	60	Red	Aluminum, lead, zinc
K	1/8" (3.0mm) ball	150	Red	Harder grades of aluminum, lead, and zinc

L, M, P, R, S and V scales with 1/4" (6.35mm) or 1/2" (12.70mm) diameter balls and major loads of 60, 100, or 150 kgf are used on very soft and thin materials.

			Superficial Hardness Scales	
15 N	Brale (diamond)	15	(N) Green	Casehardened and nitrided parts, thin metal sheets, highly finished surfaces, metal parts requiring lighter loads of 15, 30, or 45 kgf
30 N		30		
45 N		45		
15 T	1/16" (1.5mm) ball	15	(T) Green	Soft steels, cast iron, nonferrous metals capable of withstanding maximum loads of 15, 30, or 45 kgf
30 T		30		
45 T		45		

unhardened steels, cast irons, and nonferrous metals. The major load with the 1/16" (1.5mm) diameter ball is 100 kg (220 lb). The minor load is 10 kg (22 lb). The red dial hardness numbers are used for the R_B scale.

HARDNESS SCALES

Table 31-2 furnishes information about standard, special, and superficial Rockwell hardness scales, hardened steel ball-point and conical-shaped diamond-point penetrators, applications, major kg (lb/f) loads, scale symbols, and dial col-

or. The superficial scales are used on parts where only a light, shallow penetration is permitted.

Casehardened parts, nitrided steels, thin metal sheets, and highly finished surfaces are measured for hardness by using a lighter load. Rockwell superficial hardness scales are used to measure such loads. The *Rockwell N scale* requires the use of a special N Brale penetrator; the *Rockwell T scale*, a 1/16" (1.5mm) diameter hardened steel ball penetrator. The 15, 30, and 45 prefixes indicate the major kilogram load in each case. The minor load is 3 kg (6.6 lb).

FEATURES OF ROCKWELL HARDNESS TESTER

The main features of a Rockwell hardness tester with standard anvil forms are identified in Figure 31-9. A small-diameter plane anvil is shown on the instrument. "V," small- and large-diameter, and roller designs are furnished with this instrument. Cylindrical parts are supported on the V-type centering anvil. Tubing is usually mounted on a mandrel to prevent damage to the walls. Long, overhanging parts are supported by an adjustable jack rest so that the surface of the workpiece is in a horizontal plane with the anvil.

The practice under actual testing conditions is to take readings at three different places along the workpiece. The hardness number is the average of the three readings.

How to Test for Metal Hardness by Using a Rockwell Hardness Tester

STEP 1 Select and mount the appropriate penetrator and anvil in the hardness tester.

STEP 2 Turn the dial on the indicator to position the indicator hand for starting a test.

STEP 3 Place the workpiece or specimen on the anvil. Use a jack rest to support overhanging and long parts.

STEP 4 Raise the anvil or lower the penetrator until it just touches the workpiece.

STEP 5 Apply a 10 kg (22 lb) minor load. This load is shown on the indicator dial. Set the hardness reading (indicator) dial at zero.

STEP 6 Apply the appropriate major load.

STEP 7 Reduce the kgf of the major load to the setting of the minor load.

STEP 8 Read the hardness on the appropriate color scale.

STEP 9 Release the minor load. Move the workpiece or specimen to a second then a third location. Repeat steps 2 through 8 to obtain the second and third hardness readings

STEP 10 Average the three hardness readings.

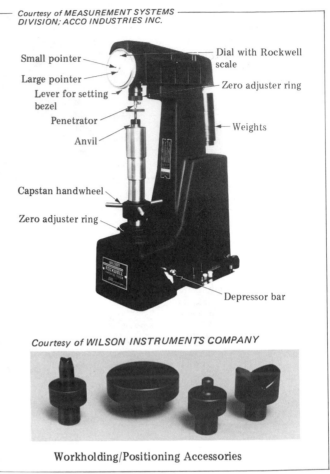

Courtesy of MEASUREMENT SYSTEMS DIVISION; ACCO INDUSTRIES INC.

Courtesy of WILSON INSTRUMENTS COMPANY

Workholding/Positioning Accessories

Figure 31-9 Main Features of a Rockwell Hardness Tester

BRINELL HARDNESS TESTING

The Brinell hardness tester produces an impression by pressing a hardened steel ball under a known applied force into the part to be tested. A microscope is then used to establish the diameter of the impression. The ball is 0.394″ (10mm) diameter. It may be made of hardened steel or carbide, or it may be identified as a Hultgren ball. The main features of a Brinell hardness tester with a measuring microscope are shown in Figure 31-10.

A Brinell hardness number (Bhn) is identified on a Brinell table according to the diameter of the impression and the specified applied load. The Brinell hardness value equals the applied load in kilograms divided by the square millimeter area (impression). The load for hardened steel

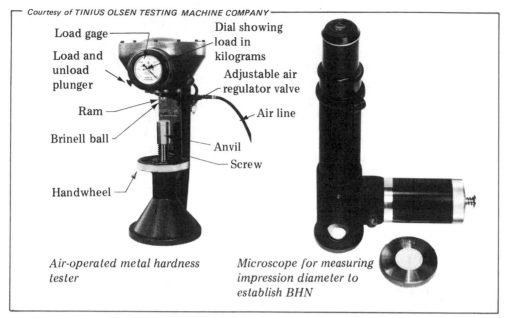

Courtesy of TINIUS OLSEN TESTING MACHINE COMPANY

Load gage

Load and unload plunger

Ram

Brinell ball

Handwheel

Dial showing load in kilograms

Adjustable air regulator valve

Air line

Anvil

Screw

Air-operated metal hardness tester

Microscope for measuring impression diameter to establish BHN

Figure 31-10 Main Features of a Brinell Hardness Tester with a Measuring Microscope

parts is generally 3,000 kg (6,600 lb). A standard load of 500 kg (1,100 lb) is used for hardness testing nonferrous metals. Lower numbers on the scale are assigned for softer metals and deeper impressions.

RECOMMENDED APPLICATIONS OF BRINELL HARDNESS TESTING

The range of Brinell hardness testing is between Bhn 150 for low-carbon annealed steels to Bhn 740 for hardened high-carbon steels. A carbide ball is required for upper ranges of hardness to Bhn 630. A Hultgren ball accommodates hardness testing up to Bhn 500. The hardened steel ball-point penetrator is adapted to hardnesses to Bhn 450. The Brinell hardness tester works best on nonferrous metals, soft steels, and hardened steels through the medium-hard tool steel range.

How to Test for Metal Hardness by Using a Brinell Hardness Tester

STEP 1 Select the load. Adjust the air regulator until the required load is obtained.

STEP 2 Place the part to be tested on the anvil. Raise the setup to within 5/8″ (15mm) of the penetrator ball.

STEP 3 Pull out the plunger control to bring the penetrator ball to the work surface to apply the load (Figure 37-14).

STEP 4 Maintain a steady load for at least 15 seconds for iron or steel; 30 seconds, for nonferrous metals.

Note: Generally, a standard 500 kg (1,100 lb) load is applied for nonferrous metals in the Bhn 26-100 hardness range. The standard load for steel and iron is 3,000 kg (6,600 lb).

STEP 5 Release the load and remove the part.

Note: The penetrator and ball retract in readiness for the next test.

STEP 6 Measure the diameter of the impression with a Brinell or hardness testing microscope.

STEP 7 Use a table to establish the Brinell hardness number (Bhn). Otherwise, calculate the Bhn by formula according to the diameter of the impression and the load.

VICKERS HARDNESS TESTING

The Vickers hardness test requires a square-based diamond pyramid whose sides are at an angle of 136°. A load of from 5 kg to 120 kg (11 lb to 264 lb) is applied generally for 30 seconds. The diagonal length of the square impression is measured.

The Vickers hardness number equals the applied load divided by the area of the pyramid-shaped impression. The Vickers test is very accurate. It is adapted to large sections as well as thin sheets.

SCLEROSCOPE HARDNESS TESTING

The scleroscope instrument measures hardness in terms of the elasticity of the workpiece. There are two types of scleroscopes. One type has a direct-reading *scale.* The other type has a direct *dial* recording. Figure 31–11 shows the main features of a direct-reading scleroscope. Scleroscopes are used for hardness testing of ferrous

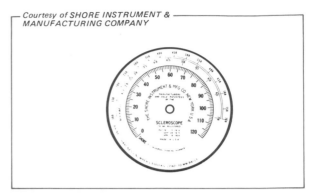

Courtesy of SHORE INSTRUMENT & MANUFACTURING COMPANY

Figure 31–12 Direct-Reading Scleroscope Dial with Equivalent Rockwell C and Brinell Hardness Numbers

and nonferrous metals. Unlike the Brinell and Rockwell hardness testers, no crater is produced.

A diamond hammer, dropped from a fixed height, rebounds. The rebound distance varies in proportion to the hardness of the metal being tested. The harder the metal, the higher the rebound distance. This movement is either read on a scale or a dial.

The scale is calibrated from the average rebound height of a tool steel of maximum hardness that is divided into 100 parts. The range of rebounds is from 95 to 105. The dial of a direct-reading scleroscope is shown in Figure 31–12. Readings are carried beyond 100 to 120. This range covers super-hard metals. Note that the conversion Rockwell C and Brinell hardness values are given on the dial face; a 3,000 kg (6,600 lb) load and a 10mm diameter ball are used.

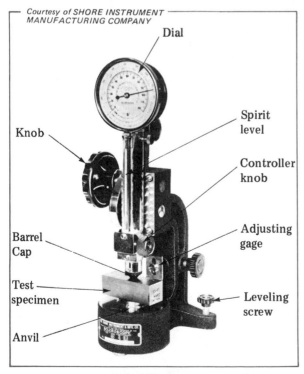

Courtesy of SHORE INSTRUMENT MANUFACTURING COMPANY

Dial

Spirit level

Controller knob

Knob

Adjusting gage

Barrel Cap

Test specimen

Leveling screw

Anvil

Figure 31–11 Main Features of a Direct-Reading Scleroscope

How to Test for Metal Hardness by Using a Scleroscope

STEP 1 — Level the scleroscope

STEP 2 — Turn the knob to bring the barrel cap in firm contact with the part being tested. This contact is maintained throughout the test.

STEP 3 — Draw the hammer to the up position.

STEP 4 Release the hammer with the control knob. Note the scale or dial reading to which the hammer rebounds.

STEP 5 Change the position of the workpiece on the anvil. Repeat steps 2 through 4 to obtain a second reading.

STEP 6 Average the rebound dial readings for several tests at different positions along the part. The average number represents the scleroscope hardness reading.

Note: A regular check of the accuracy of the instrument should be made against the reference test bar supplied by the instrument maker.

MICROHARDNESS TESTING

The Microhardness tester permits hardness testing of minerals, abrasives, extremely hard metals, very small or thin precision parts, and thinly hardened surfaces. Some of these materials and parts cannot be tested by other methods. Microhardness testers are also used to determine the hardness of grains in the microstructure of the material.

Hardness tests are made with a diamond penetrator. The penetrator is pressed into the specimen with extremely light loads of from 25 to 3,600 grams (0.7 oz. to 7.9 lb). Minute impressions are produced under appropriate low loads.

The microscope feature of the tester permits measuring the fine impression. The depth of penetration is determined by the applied load and the hardness of the material. A number value is assigned to each degree of hardness.

KNOOP HARDNESS TESTING

The Knoop hardness test is adapted to the same hard materials and thin parts as the Microhardness tester. The load range in grams is also from 25 to 3,600 grams (0.7 oz. to 7.9 lb). The plane surfaces must be free from scratches.

A *Knoop indentor* is used. The indentor is a diamond in an elongated pyramid form. The indentor acts under a fixed load that is applied for a definite period of time to produce an indentation that has long and short diagonals. The Knoop indentor is used in a fully automatic, electronically controlled Tukon tester.

Knoop hardness numbers are used. The numbers are equal to the load in kilograms divided by the area of the indentation in square millimeters. Tables are available that give the indentation number corresponding to the long diagonal and for a given load.

Safe Practices Related to Particle and Powdered Metallurgy Steels, TiN Coatings and Hardness Testing

- Avoid sharp corners, radii, fillets, and bevels in the design of powdered metal dies, punches, and other formed products.
- Determine the flowability of pre-alloyed metal powders that ensure the manufacture of powdered metal parts that are within the limits of the process and the forming presses.
- Prepare a cutting tool for application of a TiN coating by first grinding away periphery, corner, or flank wear; chipped or cratered edges; or gouged surfaces.
- Check the condition of a cutting tool that is to be coated with titanium nitride (TiN) to be sure there is no surface discoloration or burn and all foreign matter that may prevent proper adhesion is removed.
- Select the size and style of the penetrator and load which are appropriate to the material and condition of the part to be tested.
- Examine the part for any possible fractures caused by heat treatment.
- Check for hardness values at at least three different places on the workpiece. The *average* represents the hardness number.
- Use an adjustable jack for overhanging parts and a tube support for long, thin, pieces. The surface area must rest solidly on the hardness tester anvil to obtain an accurate reading and to prevent damage to the penetrator.

TERMS APPLIED TO PARTICLE AND POWDER METALLURGY, TiN COATINGS, AND HARDNESS TESTING

Particle metallurgy (PM) steels	Steels produced by breaking fluid, molten steel into small droplets that form fine particles, compacting uniformly sized particles that are proportioned according to the required composition, and formed into required shapes and sizes.
Dispersion of carbides (PM steels)	A pattern of distribution of carbide particles within a cross section of steel. In the case of PM steels, dispersion relates to the uniform dispersion (distribution) of carbides.
Embrittlement (tooling applications)	The result of including three percent (or more) of vanadium in conventional ingot-cast steel manufacturing. A brittle condition that also produces poor grindability.
Grinding ratio	A ratio formed by dividing the volume (mass) of metal removed by grinding divided by the volume of wear on a grinding wheel.
Sulphides (steels)	Sulphur that it added to a steel alloy for purposes of improving machinability.
Powder metallurgy (cemented carbides)	A manufacturing process requiring the mixing and heating of pure tungsten and carbon powders with powdered cobalt to produce cemented carbide.
Sintering (cemented carbides)	Heating *green* compact pressed forms in a nonoxidizing, controlled atmosphere furnace at temperatures below the melting points of the alloy metal. Application of heat to sinter (cement) carbide particles together.
Impregnation	A process of filling voids within a porous metal part with a lubricating oil, graphite, or other material.
Infiltration	Strengthening and increasing the ductility, impact resistance, and hardness of powdered metals by flowing a metal or alloy throughout the porous structure (where planned) of powder metallurgy steels.
Titanium nitride (TiN) coating	An extremely hard (Rc 80+) fine (0.0001″) thickness of titanium nitride tightly bonded to the surfaces of cutting and forming tools and parts. A surface coating to improve the wear life, cutting characteristics, and quality of surface texture produced by a coated tool.
Lubricity (TiN coating)	A low coefficient of friction property of TiN coatings that enhances the free flow of chips away from a tool cutter edge and a workpiece.
Reduced tool deflection	One of the end results of lubricity is the reduction of cutting forces during machining throughout the cutting tool. With less cutting force there is less internal tool stress and unnecessary tool movement.

Cratering, gouging; peripheral, flank, and corner wear	Wear conditions on a cutting tool that must be corrected before applying a TiN coating.
Physical vapor deposition (PVD)	A process by which atomically clean tools are bombarded in a vacuum deposition chamber by nitrogen gas combined with titanium (at temperatures within a 900°F range) to produce a TiN surface coating.
Major and minor loads	Standardized initial and subsequent loads applied on a hardness tester to a diamond or steel ball-shaped penetrator. Two reference points from which the linear depth of penetration is established to measure hardness.
Hardness testing scales	Numerical values related to depth or area penetration of a hardness tester penetrator for a series of fixed loads, according to material. Bhn, Rockwell, Vickers, Knoop, scleroscope, microhardness and other hardness testing measurement systems. Numerical values that may be converted from one system to the equivalent in another system.
Scleroscope	A hardness testing instrument where a hammer of known weight is dropped from a fixed height and rebounds. An instrument that records the (hardness) rebound of a hammer according to the elasticity of the metal.
Microhardness tester	A precision hardness testing instrument that uses a diamond penetrator under very light loads to obtain a hardness value for extremely hard minerals, metals, and grain microstructures.

SUMMARY

- Particle metallurgy (PM) steels have small size carbide grains that are evenly distributed through the mass.
    - PM steels permit greater uniformity and consistency of heat treatment, have exceptionally good wear resistance, toughness, and machinability in hardened or unhardened states, and have limited distortion.
- The powdered metallurgy process requires the blending of preselected powdered materials, including a binder; the compacting of cutting tool blanks or other formed parts; and the sintering in atmospheric-controlled furnaces.
    - Second finishing operations of powdered metal parts are performed to add special properties (infiltration); to manufacture parts to finer geometric and dimensional accuracy (repressing, coining, sizing); or to provide a different surface coating.
- Several advantages of the powdered metallurgy process include: the ability to accurately produce intricate shapes within close tolerances, to control porosity and other physical properties of the material, and cost effectiveness in producing parts without the need for second machining operations.
    - Disadvantages of powdered metallurgy processes include: abrasiveness of some powdered materials, difficulty in replicating sharp corners and other design features, and limitations of materials to plastic flow.

- Metal coatings are used to coat cutting tools, forming plastic dies, and other parts that require an extremely hard, abrasive resistant, low co-efficient of friction surface coating.
  - TiN coatings have high hardness properties (Rc80+), high lubricity to flow chips from cutting edges, and improved machinability characteristics.
- TiN coatings increase tool wear life, the quality of surface finish, productivity and cost effectiveness, and decrease machine down time.
  - A preventative maintenance tool program requires that craters, tooth and flank wear, and other tooth imperfections be corrected before recoating.
- Multiple scales are used on the Rockwell hardness tester, depending on the hardness range of ferrous and nonferrous metals and other characteristics.
  - The Brinell hardness tester uses a microscope to establish the diameter of an impression. A hardened steel ball, carbide ball, or Hultgren ball are used for different metals and hardness measurements.
- Vickers hardness testing is extremely accurate and is adapted to parts that require light loads from 5kg to 120kg (11 lb to 264 lb).
  - The scleroscope tester measures hardness by the elasticity of the hardened part.
- Microhardness testers measure the hardness of grains in minerals, abrasives, and extremely hard steels. Exceptionally fine loads of 25 to 3,600 grams (0.7 oz. to 7.9 lb) are applied. A microscope feature is required to measure the fine surface indentation.
  - The Knoop hardness tester measures the long and short diagonals produced by an elongated pyramid-form diamond using an extremely fine load.

UNIT 31 REVIEW AND SELF-TEST

A. PARTICLE AND POWDERED METALLURGY STEELS AND TiN TOOL COATINGS

1. Describe briefly how powdered metallurgy cemented carbide cutting tool blanks are manufactured.
2. Identify two unique advantages to the manufacturing of intricately-shaped parts by powdered metallurgy.
3. List three advantages of using particle metallurgy (PM) tool steels for cutting tools and punches and dies over conventional tool steels.
4. Explain briefly what the effect of increasing the amount of sulphur has on the properties of particle metallurgy (PM) steels.
5. Tell what effect the low coefficient of friction of titanium nitride has on the machining of high quality surface finishes.
6. State two conditions that must be corrected in preparing a cutting tool for coating with titanium nitride (TiN).

B. HARDNESS TESTING

1. a. Explain the difference between a Rockwell C scale and a Rockwell superficial hardness scale.
   b. Give an application of each scale.
   c. Indicate the type of penetrator for the R_C and superficial scales.

2. a. Harden and temper the SAE 1041 Aligning Block shown in the illustration. The part is to be drawn to a Bhn 243 hardness. Consult handbook tables to establish the (1) quenching medium, (2) hardening and tempering temperatures, and (3) soaking time.
   b. Test the block after the surfaces are finish ground to check the Bhn 243 hardness. The hardness may be tested on any hardness tester available. Values other than Bhn are to be checked against the required Bhn 243 hardness.
   c. List three personal safety and/or precaution measures to take in carrying out the heat-treating operations.

Note: Another part may be used or the processes simulated by preparing a step-by-step production plan for the aligning block or the new part.

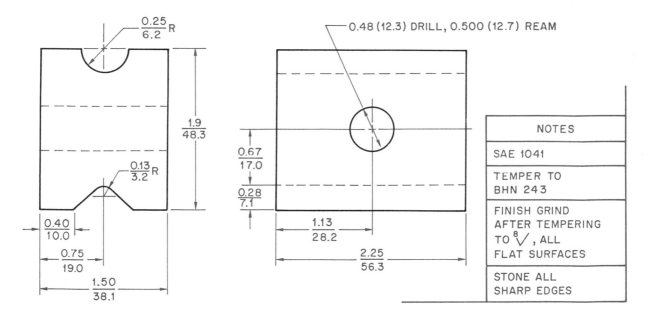

3. a. Make a second hardness test of the Aligning Block shown in the illustration. (A Rockwell hardness test may be substituted, if this process was not used previously.)
   b. Compare the Bhn scale value in test item 1 with the R_C scale reading.
   c. Check the accuracy of the hardness readings against a handbook table.
   d. Establish the comparable scleroscope and Vickers hardness numbers from the table.

4. List three precautions to take to avoid common problems such as soft spots, uneven hardening, and fracturing.

PART 11 New Manufacturing Technology and Processes

Nontraditional Machining

Laser Beam and Electrical Discharge Machining

OBJECTIVES

A. CO_2 LASER BEAM MACHINING (LBM)

After satisfactorily completing this unit, you will be able to:

- Describe the functions of hard surfacing, how such surfaces are produced by laser beam technology, and advantages of this method.
- Identify purposes served by surface alloying and explain the laser beam surface alloying process.
- Review basic principles of heat treatment; explain the laser beam surface hardening process, and state major advantages of such hardening.
- Describe CO_2 laser beam cutting and machining processes and cite advantages.
- Relate functions of laser cutting centers to multi-purpose and multi-axes machining applications and the potential of the centers to be interfaced within computer integrated manufacturing (CIM) systems.

B. NC AND WIRE-CUT CNC ELECTRICAL DISCHARGE MACHINING (EDM)

- Explain the principles of EDM machining and unique characteristics.
- Identify cutting tool materials and dielectric specifications and applications.
- Discuss overcut, electrical principles, and other conditions that are basic to EDM machining.
- Generate a list of advantages and disadvantages of EDM machining in comparison with conventional machining processes.
- Apply basic *Terms* used in laser beam machining (LBM) and EDM.
- Follow *Safe Practices* and health regulations according to Federal, State, and Local laws, equipment manufacturer's recommendations, and industry codes.

534

A. CO_2 LASER BEAM MACHINING (LBM)

LASER BEAM HARD SURFACING AND SURFACE ALLOYING

Hard surfacing is a welding process in which hard materials are alloyed and bonded to a softer metal substrate for purposes of improving surface hardness and wear resistance. Typical hard facing alloying materials include nickel-, chromium-, cobalt-, and manganese-base metals, tungsten carbides, and other alloys.

Hard surface alloying (also known as *cladding*) permits the use of a comparatively inexpensive softer metal (substrate) body of a part to which a hard metal alloy surface is to be strongly bonded. Hard surface facings improve corrosion and wear resistance.

ADVANTAGES OF LASER BEAM HARD SURFACING

Hard surfacing is usually applied by such conventional welding methods as tungsten inert gas (TIG), plasma arc, and flame spray. Many of these methods and applications are being displaced by laser beam hard surfacing technology and processes. Some of the major advantages of using laser processes to produce hard facings follow.

- Greater surface bonding *(dilution)* of hard facing alloys to the substrate (core of the part).
- Finer (less porous) hard surface grain structure with limited surface cracking.
- Cost effectiveness in reducing the amount of alloying material to be ground away from the outside surface of a hard surfaced part.
- Heat control of the process results in less heat input into a workpiece. As a result, there is limited warpage and distortion and reduced possibility of the part being softened.
- A higher degree of surface hardness reduces part wear.
- A finer, uniform surface texture provides greater wear resistance.

LASER BEAM HARD SURFACING PROCESS

Two of the requirements for hard surfacing of parts by the carbon dioxide (CO_2) laser energy

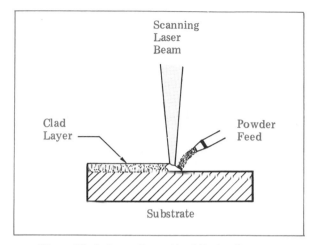

Figure 32-1 Laser Beam Hard Facing Process

cladding process include: (1) a concentrated, intense high temperature, and (2) control of the flow of the alloying elements (and the depth of dilution). *Dilution* relates to the blending of the hard surfacing alloy powder or chips with the substrate.

A hard surface is produced by controlling the movement of a high-heat intensity laser beam across the surface of a part that is to be coated. The controlled heat source is moved into and across an interaction zone that contains preplaced hard surface material chips. These conditions are shown in Figure 32-1.

The laser beam causes the cladding alloy and the substrate layer in the interaction zone to flow and to blend. The thickness of the hard surfacing deposit and the quality of the surface texture may be accurately controlled to within ±0.005'' to ±0.010'' (0.127 to 0.254mm).

As stated earlier, a laser-clad hard surface may be precision ground to produce a high quality surface texture and a part that is dimensionally accurate and falls within geometric tolerancing requirements with a minimum amount of stock removal.

SURFACE ALLOYING BY LASER BEAM

Surface alloying relates to the chemical changing of a substrate material. The change is produced by localizing the melting of the substrate metal and introducing alloying elements for dilution into the molten mass.

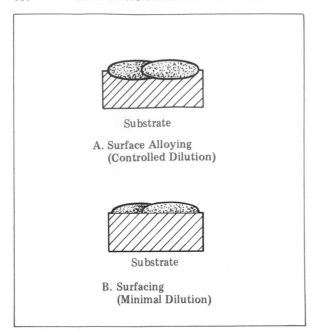

A. Surface Alloying
(Controlled Dilution)

B. Surfacing
(Minimal Dilution)

Figure 32–2 Comparison of Degree of Dilution for
Laser Hard Surfacing (Cladding) and
Laser Surface Alloying

Two major differences between laser hard facing and laser surface alloying relate to (1) changing the surface composition and (2) the depth or degree of dilution. Figure 32–2 provides a simple comparison of the degree of blending required for surface cladding and the greater depth of surface alloying.

The list of advantages of laser beam hard surfacing over conventional welding processes apply to laser beam alloying. Laser equipment may be interfaced with CAM, CIM, and CNC hardware and software in automated manufacturing.

LASER BEAM SURFACE HARDENING

To review, heat treating produces a transformation within the atomic cell structure of steel. The transformation occurs when steel is heated and cooled within certain critical points within a specified range of heating and quenching temperatures over a very short period of time. Maximum hardness and fatigue resistance are produced when a heated part is cooled to room temperature in a matter of seconds in order to change an austenitic grain structure at hardening temperature by cooling a part below $1000°F$ ($538°C$).

The austenite grain structure at hardening temperature is transformed to a martensite grain structure by rapidly cooling a workpiece below $1000°F$ ($538°C$). Martensite, as a saturated solid solution of carbon, produces maximum hardness in a heat treated part.

The limitations of carbon dioxide (CO_2) laser beams in the hardening of ferrous metals is noted in Table 32–1.

LASER BEAM SURFACE HARDENING PROCESS

A ferrous material surface is hardened by focusing a laser beam over an area that is to be hardened, heating a specific area, and rapidly cooling to room temperature. The shape and size of the laser beam is controlled by an *optical integrator*.

Figure 32–3 shows an annular (doughnut-formed) laser beam that is focused on a spherical-(segment) shaped mirror. The reflected laser rays are then transformed into a square area spot.

Table 32–1 Limitations of Carbon Dioxide (CO_2) Laser Beams in
Surface Hardening Steels and Steel Alloys

Good Hardenability	Marginal Hardenability	Not Hardenable
• Cast Iron • Medium Carbon (1045) Steel • High Carbon (1080) Steel • Tool Steel (52100) • Low Alloy (4140) Steel	• Ferritic Nodular Cast Iron • Annealed Carbon Steel • Spherodized Carbon Steel • SEA 1020 Steel	• Wrought Iron • Low Carbon (1010) Steel • Austenitic Stainless Steel • Nonferrous Metals and Alloys

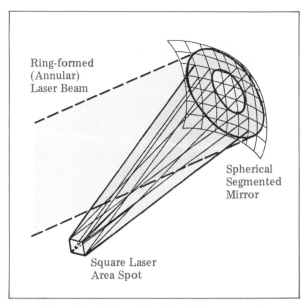

Figure 32–3 Use of An Optical Integrated to Form an Area Laser Spot

Hardening occurs as the square area laser beam moves across the workpiece surface. Hardening depth is obtained by controlling the laser power and spot transformation speed in accordance with the hardenability of the workpiece material. Other complex laser patterns may be formed by changing the initial laser output beam.

ADVANTAGES OF CO$_2$ LASER BEAM SURFACE HARDENING OVER CONVENTIONAL METHODS

- Ease of control; stable source of energy; and processing speed.
- Stability of the process.
- Accuracy of the laser beam to produce uniform hardening depths.
- Less workpiece distortion and warping.
- Flexibility to form localized surface hardness patterns.
- Adaptability of the process to the heat treatment of flat, round, and irregularly-shaped workpieces.
- Operating speed in moving from one workpiece location to another.

- Computer control of processes for laser positioning, laser power, and surface hardening speeds.
- Potential self-quenching heat produced at the surface of the workpiece is conducted away from the surface or into the cooler interior.
- Quenching equipment and media are eliminated.
- The large power range of laser beams permits surface hardening to depths that vary from 0.125″ to 0.156″ (3 to 4mm).
- Large laser surface spots may be produced to permit higher coverage rates, requiring a lesser number of overlap zones.
- Laser beam surface hardening is carried on without using a controlled, oxygen-free, atmosphere.

LASER BEAM CUTTING AND MACHINING PROCESSES

The principle of cutting with a high power carbon dioxide (CO$_2$) laser is demonstrated in Figure 32–4. A high-energy focused laser beam is

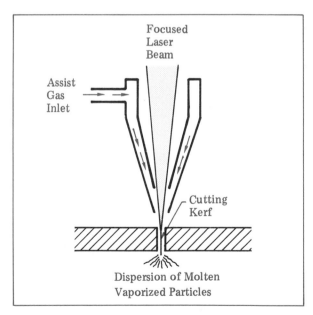

Figure 32–4 Principle of Cutting With a Carbon Dioxide Laser Beam

directed to a starting point to cause the temperature of the material at the interaction point to rise above the vaporization temperature. As additional energy is provided, the initial cavity penetrates the workpiece. The vaporized molten particles are blown (ejected) from the workpiece by a high-velocity inert gas like argon or helium. Oxygen gas is often used in the gas stream to increase the cutting rate. The inert gases also serve to speed the removal of undesirable oxides that may form on the cut surfaces.

The carbon dioxide (CO_2) laser beam provides a highly intensive beam of light energy to produce high temperatures that may easily be directed and controlled. The kerf produced is narrow and geometrically straight and has a minimally affected heat zone. A cut is produced (manually or CNC) starting at the laser induced cavity position and moving either the workpiece or the beam and inert assist gas jet nozzle along a controlled path.

ADVANTAGES OF CO_2 LASER BEAM CUTTING

The following are advantages of CO_2 laser beam cutting as contrasted with other gas torch or machine cutting processes.

- Mechanical distortion of the workpiece is negligible as the laser process does not produce mechanical forces during cutting.

- There are no chemical reactions to produce surface contaminants as other gas cutting processes do.

- Optical systems permit focusing to make cuts in otherwise limited access areas of a workpiece.

- The versatility of the CO_2 laser beam permits an extremely wide range of cutting from stainless steels and other hard-to-machine alloyed steels to nonferrous metals to soft textiles.

- The quality and accuracy of many cuts is adequate so that second cut grinding or other machining operations may not be required.

- Simplified, light weight workholding fixtures and clamping devices may be used due to the fact that there are minimal cutting forces.

- CO_2 laser beam cutting processes require no hard tooling setups or changeovers.

LASER CUTTING CENTERS

A laser cutting center provides alternative manufacturing methods. A laser beam cutting center combines such features as high machining speeds, totally enclosed laser beam guidance, water-cooled laser beam optics, an integrally designed exhaust system, and work-load devices.

Each operation and operating condition is monitored with graphic (screen) feedback to the operator. A CNC control unit permits quick processing of data for storage or to generate a new machining program while the laser cutting center is operational. Processes may be interfaced with other machining centers or group technology cells for conventional machining processes.

Courtesy of TRUMPF INDUSTRIAL LASERS INC.

Figure 32–5 Multi-Axes, Wrist Focusing, Universal Laser Beam Head

MULTIPLE CUTTING PROCESSES

The laser head (Figure 32–5) is the cutting tool. On some laser cutting machines, the workpiece moves past the laser cutting head. In the case of directional cutting centers, the laser head moves while the workpiece remains stationary.

The multiple axes movements of a directional laser head permits programming of the laser beam path. Programming makes it possible to machine contoured (sculptured) surfaces, and small holes, slots, and narrow webs, and perform other machining processes. Rapid positioning for successive cuts and quick acceleration and other design features are included in laser beam cutting systems.

B. NC AND CNC WIRE-CUT ELECTRICAL DISCHARGE MACHINING (EDM)

NC ELECTRICAL DISCHARGE MACHINING (NC-EDM)

The single-axis control of an electrical discharge machine provides for an EDM spark to jump vertically along the Z axis from the electrode to a workpiece. In other words, the electrode movement and electrical discharge pulses are in the Z-axis. More sophisticated EDM machines provide for cutting action along X and Y axes, permitting two axes machining.

Another axis of movement for C-axis cutting is made possible by an automatic rotary chuck accessory. Coupled with increasing CNC capability and more accurate movement controls of servomechanisms it is possible to do pattern machining, adjust the motion in each quadrant to independently accommodate different electrode shapes, and to switch machining planes from X–Y to Y–Z or Z–X, depending on the machining axis.

Figure 32–6 shows five different examples of EDM machining operations. Rotary chuck movement (as indicated by the circular symbol) permits internal contouring (Figure 32–6A). Sharp corners of a stepped surface are produced by simultaneously using three axes machining as illustrated at (B). Helical forms such as gears, worm shafts, and internal threads, may be EDM machined by combining rotating C-axis move-

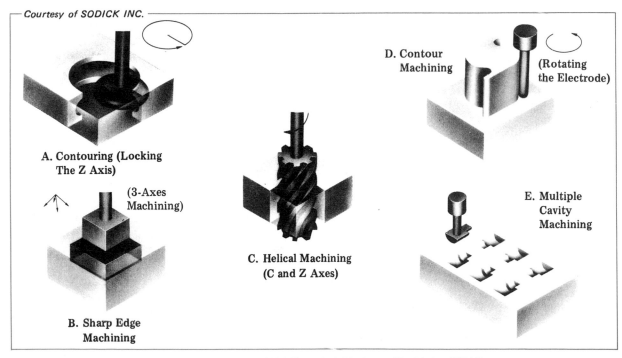

Courtesy of SODICK INC.

A. Contouring (Locking The Z Axis)

B. Sharp Edge Machining

C. Helical Machining (C and Z Axes)

D. Contour Machining (Rotating the Electrode)

E. Multiple Cavity Machining

(3-Axes Machining)

Figure 32–6 Examples of NC Electrical Discharge Machining (EDM)

ments and Z-axis depth movements as shown at (C). Contour machining may be performed by rotating the electrode (Figure 32–6D). Multiple cavity machining is displayed at (E).

ROTATION, ANGLE POSITIONING, AND TOOL CHANGING FUNCTIONS

Some EDM heads are designed as a compact rotating and axis unit. This means the head serves two electrode positioning functions: (1) rotating and (2) angle positioning. An automatic tool changer may be used with the rotating-axis head unit when a number of preformed (shaped) electrodes are to be used to automatically rough out and finish machine a complex shape (Figure 32–7).

NC OPERATED EDM SYSTEMS

General operating features of NC-EDM systems include: a CRT screen display of the part shape, editing the next program, program storage, the setting of machining parameters, and origin positioning. Multi-axes and through machining, precision boring, simultaneous three-axes machining, and multiple cavity machining are programmed according to the number of machining parameters.

Electrodes may be positioned by X, Y, and Z axis movements for outside and inside diameter centering, reference sphere positioning, and edge positioning.

Tape punch programs may be interlinked with an automatic programming unit to directly transfer machining programs. Programs may be safely stored on floppy disks to add search and editing capabilities.

MACHINING WITH AUTOMATIC PROGRAMS (MAP)

An NC program may be created by feeding input information and dimensions for a part together with the required quality of surface roughness (finish). The MAP unit is designed to select machining conditions, provide electrode size and estimated machining time, and create the machining program.

NC EDM MACHINE TOOL ACCESSORIES

Cutting quality, dimensional and surface finish accuracy, machining time, and cost-effectiveness depend on voltage regulation, dielectric cooling, removal of foreign particles and chips, and machining currents. The following are some of the accessories used to perform these functions.

Automatic voltage regulator. This unit corrects line voltage fluctuations for maximum machining performance.

Dielectric cooling unit. This unit functions to hold the dielectric temperature constant in order to maintain machining accuracy.

Vacuum unit. This unit removes dust and chips produced in EDM, maintains the dielectric in clean condition, and prevents gap widening due to secondary discharge.

Stone/ceramic plates. Accurate plates made of these materials are used on EDM machine tables to minimize thermal (heat) deformation, to ensure greater machining accuracy, and to improve surface texture.

NC power supply. An advanced numerically controlled power supply (combined with an NC system that has a full complement of functions to control the EDM) is essential to high machine tool productivity.

Courtesy of SODICK INC.

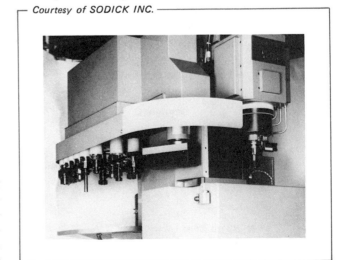

Figure 32–7 EDM Automatic Tool Changer (ATC)

EDM CUTTING TOOLS

A required property of EDM electrode tools is that of being a good electrical conductor. Common electrode materials include: graphite, copper, copper-graphite, tungsten, tungsten-copper, zinc alloys, and brass. Of these, graphite is the least expensive, is the easiest to form to shape, and has a good *wear ratio*. Wear ratio refers to the work produced (erosion) and tool wear (erosion).

Two disadvantages of graphite electrodes relate to brittleness (which makes it impossible to form sharp corner radii) and the porosity (which affects the quality of surface finish).

Stepped electrodes are often used on heavy-duty operations for roughing, second cut, and finish cuts. Some EDM machines have a *no-wear* cutting cycle. This cycle provides for using the same electrode for roughing and finishing cuts and producing high quality, dimensionally accurate surface finishes.

OVERCUT

Overcut relates to the difference between the size of an electrode tool and the width of the cavity produced. The general range of EDM overcut is from 0.0002″ to 0.007″ (0.005 to 0.1mm). The amount of the overcut depends on the following.

- The power settings (voltage and amperage).
- The quality (cleanliness) of the dielectric fluid.
- The dielectric fluid breakdown voltage.
- Machine spindle rigidity, resistance to rotation, and travel path.
- The extent to which suspended metal chips in the gap between the electrode and the work form a conductive path for the electrical discharge.

DIELECTRIC FLUIDS

Dielectric fluids serve in producing arcing frequency for EDM machining and to flow away chips and other foreign particles. Dielectric fluids must have excellent fluidity properties in order to flow along all surfaces of the electrode and through the arc gap to continuously remove chips during the erosion process. Properties of the dielectric fluid also affect the quality and accuracy of the surface, overcut, machining time, and cost.

Deionized water, low-viscosity petroleum oils and silicone oils, and ethylene glycol and water solutions are common EDM dielectric fluids.

EDM MACHINING RATES AND SURFACE FINISH

EDM machining rates depend on the electrode material, amperage, and arcing frequency. In general practice, graphite electrode cutting rates for steel range up to a 20:1 ratio in comparison with lower cutting ratios for other electrode materials. The average cutting rate of a graphite electrode on steel is around 0.05 in³ or 0.32 cubic centimeters per hour.

The rate of metal removal and the quality of surface finish depend on amperage and arcing frequency. The lower the amperage, the slower the metal removing rate. The higher the amperage, the higher the metal removal rate.

Low arcing frequencies produce a lower quality of surface finish as contrasted with a high quality surface finish at high arcing frequencies (at a constant amperage). Machinability tables provide comparative data on EDM surface finishes and machining rates and time with other conventional machining processes.

DISADVANTAGES OF EDM

A few of the disadvantages of EDM machining follow.

- The workpiece must possess electrical conductivity properties.
- Electrode wear during the cutting process may require a number of tool changes.
- The work surface in the machined area may be damaged metallurgically and otherwise to a depth of 0.0001″ to 0.0002″ (0.003 to 0.005mm).

WIRE-CUT ELECTRICAL DISCHARGE MACHINING (EDM)

Wire-cut EDM is used to produce shearing and cutting dies and punches and extrusion and plastic dies; enlarge the sizes of regular and special formed holes in parts; and machine irregular contours, gears and tapers; and machine other high precision, intricately machined parts. Ultra-precision machining is performed on wire-cut EDM machines under temperature-controlled dielectric conditions. Thermal deformation is minimized by holding the workpiece at a constant temperature during machining.

One of the most efficient wires used on wire-cut EDM is 0.007" (0.2mm) in diameter. Machining rates of up to 16 in^2 per hour (or 170mm^2/min) are possible with submersion wire-cut EDM machining processes.

WATER- AND OIL-SUBMERSION MACHINES

Some wire-cut EDM machines are designed for *water-* or *oil-submersion systems*. Water machining is used for high-speed roughing processes; oil machining, for finishing. Oil machining reduces the thickness of the softened layer on the workpiece and prevents loss of hardness. Oil-submersion machining is especially practical in micro-machining extremely accurate parts. Machining is done in temperature-controlled chambers.

BASIC COMPONENTS OF WIRE-CUT EDM MACHINES

The following are basic components of EDM machines.

- The *machine structure* consisting of the bed, column, and table. These provide immunity to thermal deformation, rigidity, stable machine accuracy, and high insulation performance.

- *Electrode wire system* with wire carrier mechanisms to control the wire traverse rate and wire tension, a wire storage unit, and an automatic wire-threader guide. A hole cutting accessory is available for making the initial hole in extremely hard-to-machine materials.

- The *NC* or *CNC unit* that controls such functions as: input increment, automatic positioning, program editing, backlash error compensation, automatic expanded contour machining, interpolation, and others.

- An *EDM power supply* for controlled spark energy.

- A *dielectric fluid* system and cooling unit for fluid storage, chip removal, temperature control, control of the flow rate, and ion exchanger.

WIRE-CUT EDM X–Y POSITIONING

In an X–Y axes positioning system, the electrode wire generally remains stationary and the workpiece travels along the X and Y axes. A variation is in the case where the electrode wire travels along the X axis; the workpiece, the Y axis.

Wire-cut EDM machines are designed with other variations. For example, the electrode wire may travel in X and Y directions or just along the X axis. Also, the workpiece may travel in only the Y direction. Direction, speed, and distance of travel are controlled by NC or CNC commands to X and Y axes servomotors.

WIRE-CUT EDM OVERCUT

Overcut for wire-cut EDM is the same as for NC EDM form and conventional machining. An overcut (space on both sides of a wire electrode) permits the cutting tool to be fed for machining purposes. The *kerf width* is equal to twice the spark length and the diameter of the wire.

WIRE-CUT METAL REMOVAL RATES

Two energy conditions control the rate of metal removal; (1) the amount of energy in each spark and (2) the time interval (closeness) between one spark and the next spark.

As the *ON time* of a spark is increased, the energy level of the spark increases, and the amount of metal removed increases. However, a coarser surface finish is produced.

WIRE-CUT EDM ELECTRODE WIRE

Copper, brass, molybdenum, and tungsten electrode wires are commonly used. Tungsten and molybdenum electrode wires are adapted to the machining of parts with small radius inside corners where a high tensile strength wire is required. The range of diameters of EDM electrode wires is from 0.002" to 0.016" (0.05 to 0.4mm).

In metal cutting, erosion occurs on both the electrode wire and the workpiece. Erosion occurs when the EDM sparking energy produces a spark between the electrode wire and the surface of the workpiece. The heat produced causes the removal of a greater amount of material from the workpiece in comparison with the electrode. The erosion that does take place on the electrode wire causes it to pit and to be reduced in diameter. Under these conditions, the wire is generally fed through once and is not reused.

WIRE-CUT EDM DIELECTRIC FLUID

To review, the dielectric fluid serves three functions.

- Flooding the sparking area with a conductor of electricity.

- Cooling the electrode wire and workpiece to maintain a stable machining temperature.

- Flowing metal particles away from the sparking area.

Dielectric surrounding the sparking area. Deionized water is used as the dielectric. In its normal state, deionized water is a nonconductor of electricity. However, when a high voltage is applied between the electrode wire and the workpiece, the dielectric that fills the area between the exposed surface of the workpiece and the electrode wire becomes *ionized* (when the ionization point is reached).

The water at this stage becomes a conductor of electricity. In the *OFF* position, as the voltage is removed and current flow ceases, the water returns to its original state as a dielectric. The sparks produced are at a microsecond rate.

Filtered dielectric fluid is pumped under pressure into fluid ports. The ports are located at the point above the workpiece where the electrode enters the workpiece and under the area where the electrode exits. The dielectric completely floods the electric wire and the workpiece where machining takes place.

THE EDM WIRE-CUT CHIP

A *chip* is produced when wire-cut machining takes place. The chip is formed as the spark energy heats and vaporizes the workpiece material. The dielectric serves to quickly remove heat and to form a small ball-shaped solid that is about 0.0005" (0.012mm) in diameter. This is the EDM chip.

The extra-fine chips are removed by a dielectric flush. In other words, the filtered dielectric fluid is moved through the sparking area to (1) flow the chips away from the machined surface and (2) to provide a fresh, clean supply of dielectric fluid for maximum cutting action and quality of surface texture.

DIELECTRIC CIRCULATING METHODS

There are four general methods by which dielectrics are circulated in EDM machining.

- **Forcing by pressure.** The dielectric is forced through a cored hole in the electrode to completely surround the machining area and to wash the particles away.

- **Circulating the dielectric.** The dielectric is forced up through the workpiece.

- **Creating a vacuum.** A vacuum is created in the gap to cause the dielectric to flow between the electrode and the workpiece. This method also reduces smoke and fumes in the work area.

- **Vibrating.** The vibration method requires a pumping and sucking action to remove chips from the spark gap. This method is used in blind cavity and small, deep hole EDM.

ADVANTAGES OF EDM MACHINING

In comparison with conventional machining processes, EDM has the following advantages.

- Regardless of hardness, any material that is electrically conductive may be machined by EDM.
- Machining processes are carried on without great cutting stresses and forces.
- Part deformation is exceedingly limited making EDM practical for machining fragile parts.
- A battery of EDM machines may be operrated by one person.
- Versatility of EDM to reproduce intricate forms.
- Finished machined parts that are free of burrs.
- Problems of hardening after machining are eliminated as the workpiece is machined directly from the hardened state.
- Like conventional machine tools, EDM is compatible with NC-CNC controls and may be interlocked into flexible integrated manufacturing (FIM) systems.

Two of the disadvantages of using wire-cut EDM machining practices relate to: (1) low metal removal rates and (2) limitations of electrode size. Electrodes under 0.003″ (0.07mm) diameter are not practical for machining purposes.

A. Safe Practices Applied to CO_2 Laser Beam Machining (LBM)

- Read, study, and follow all laser safety regulations prescribed by Federal, state, product manufacturers, and other applicable industry safety codes.
- Note the locations of *Danger Labels*. Read and follow safety directions relating to power level, emission indicator, shutter light, key-locked controls, high-temperature sensors, and other controls.

- Avoid exposure to direct or scattered radiation.
- Make sure that all plasma tube and power supply shields are secured before operating the equipment.
- Check the exhaust system for the removal of toxic fumes produced by the cutting processes.
- Check the interlocks on access panels to see that the work enclosure doors may not be opened during laser operation.
- Know the location of emergency shut-down controls and be sure the controls are easily accessible.
- Use safety glasses when operating CO_2 laser beam equipment. Plastic and glass safety lens are recommended. Other types of laser beam operations require different kinds of safety glasses.
- Make sure that adequate safety and health hazards training is provided by *authorized personnel before working at any laser beam station.*

B. Safe Practices Applied to NC and CNC Wire-Cut Electrical Discharge Machining (EDM)

- Turn the EDM machine *OFF* and check to see that the voltmeter *reads zero* before handling the electrode, electrode holder (NC-EDM), or machined part.
- Flood the tool and workpiece area to a depth of at least 1″ (25.4mm). This precaution is taken to avoid any possibility of producing a flame resulting from the igniting of any gases and arcing.
- Check to ensure that a proper class fire extinguisher is readily accessible and operable.
- Secure all protective shields and wear appropriate safety glasses.
- Protect the skin from contact and exposure to the dielectric fluid.

TERMS APPLIED TO CO_2 LASER BEAM MACHINING (LBM) AND ELECTRICAL DISCHARGE MACHINING (EDM)

A. CO_2 LASER BEAM MACHINING (LBM)

Hard surfacing (laser beam cladding) — The alloying and bonding of a hard facing metal to a softer metal substrate.

Dilution — A process of blending hard surface alloy powder or chips in molten state with melted steel from the surface of the substrate.

Laser beam hard surfacing requirements — Hard surfacing by controlling concentrated, intense, high-heat elements of a laser beam and the flow and depth of the hard alloying metal.

Interactive zone — An area of a workpiece within which laser beam operations are performed on the substrate surface, hard surface metal chips, or powder metal alloys. Also, an area in which other laser beam cutting, heat treating, or machining processes take place.

Surface hardening (CO_2 laser beam) — A laser beam heating process in which the temperature of the surface layer of a workpiece is brought up to hardening heat and then quickly quenched below the critical cooling temperature. Producing a hardened wear-resistant surface.

Laser power range — Limits within which the energy output capacity of a laser beam may be increased or decreased to meet heat depth penetration, area of flow, and varying workpiece material and product requirements.

Optical integrator — A system of mirrors for transforming an annular laser beam into a particular area (form and size) spot to meet specific job needs.

Initial laser cavity — A depression (cavity) at the interactive point where a high heat intensity laser beam penetrates the surface of a workpiece.

Kerf — A narrow and geometrically straight cut produced as a laser beam continues from the initial cavity through a workpiece.

Laser cutting center — Laser beam equipment with versatility in multiple cutting and machining processes along multiple axes under CNC control. Such machining centers may include monitoring sensors for graphic feedback and capacity for interfacing within a manufacturing cell.

B. ELECTRICAL DISCHARGE MACHINING (NC AND CNC WIRE-CUT, EDM)

Deionization	Returning a dielectric fluid to a nonconductive state.
Dielectric fluid (wire-cut EDM)	A fluid used in the work zone to fill the gap between the electrode wire and the workpiece. The fluid acts as an insulator until it is ionized to become an electrical conductor. A fluid used in EDM to cool the workpiece and flow chips away.
Dressing (EDM)	Process of cutting off the used end of an electrode wire that has been used to form a through hole or to sharpen the details of a three-dimension electrode for EDM machining.
Duty cycle	The sum of the *ON time* plus the *OFF time* of a specific cut, in microseconds. Expressed as a percentage, duty cycle *ON time* divided by the *OFF time*, multiplied by 100 (%).
Edge finder	An electronic/electrical device used to aid in accurately positioning the electrode in relation to a fixed reference point or surface.
Electrode (NC EDM machining)	An electrically-conductive material that is formed into a mirror image of a required finished form (with compensation for overcut).
Spark eroding	The process of metal removing in EDM machining.
Filtering (EDM)	The removal of EDM chips and other foreign materials from a dielectric fluid by using a filtering unit. Returning clean dielectric fluid to the EDM system work (fluid) tank.
Flushing	The rapid movement of dielectric fluid under pressure through the EDM gap for purposes of removing chips and other debris.
Gap voltage	Two different voltage readings over one complete cycle; (1) the *open* electrode-workpiece *gap voltage* (OGV) before the spark current begins to flow and (2) the *working gap voltage* (WGV) across the electrode-workpiece during the spark current flow.
Heat affected zone	The depth to which an EDM cut alters the metallurgical structure and characteristics of the workpiece material.
Ionized path	A channel of electrically-conductive dielectric molecules from the electrode to the workpiece. A pathway for the flow of spark current.
Total (diametral) overcut (wire-cut EDM)	The difference between the diameter of a wire-cut EDM electrode and the size of the cavity or hole.
Electrode wear	The volume of electrode wear divided by the machined volume, expressed as a percent.
Pulsator	A built-in EDM unit that causes the electrode to briefly pulsate (move up and down) to flush out a deep hole or cavity.
Stepped electrode	An NC-EDM electrode designed with increasingly larger diameters or surface areas. Electrodes that permit roughing out and finish cut machining of a through hole in one setup.

SUMMARY

A. CO₂ LASER BEAM MACHINING (LBM)

- Laser beam hard surfacing is used to surface clad a comparatively soft, inexpensive steel workpiece with a hard facing alloy.

  - A soft substrate metal workpiece is hard surfaced to improve its corrosion resistance, wear resistance, and other mechanical properties.

- The CO_2 laser beam hard surfacing process requires a concentrated, intense, high heat for the substrate surface and control of the flow rate and depth of the hard surfacing alloy.

  - The flowing and blending of an even layer of hard surfacing material may be precisely controlled to within ±0.005″ to 0.010″ (0.127 to 0.254mm) so that minimal stock removal is required to accurately meet dimensional and surface finish requirements.

- CO_2 laser beam surface alloying requires changing the chemical composition of the surface of a workpiece by heating and blending with an alloy steel.

  - Laser beam surface hardening is produced by localizing and concentrating a high intensity light beam to heat the surface of a workpiece to its critical hardening temperature and quickly dropping this temperature below the cooling range. The cell structure is transformed from an austenite to a hard martensite grain structure.

- Some of the advantages of using CO_2 laser beams for hard surfacing, surface alloying, surface hardening, and other cutting and machining processes include, but are not limited to: ease of control; stability, uniformity, and accuracy of laser processes; adaptability to manual and CNC programming; reduction of fixturing and workholding devices; absence of cutting forces, and reduction of workpiece distortion.

  - Laser beam cutting processes begin with the formation of an initial vaporized cavity followed by penetration through a workpiece.

- Inert gases are used to speed the removal of undesirable oxides from the cut surfaces. Oxygen gas in the stream accelerates the cutting process.

  - Laser cutting centers permit multi-axes and multiple cutting and machining operations. Sculptured surfaces, hole forming, slotting, and groove cutting are typical laser cutting center operations. The centers may be CNC interfaced within group technology manufacturing cells.

B. ELECTRICAL DISCHARGE MACHINING (NC AND WIRE-CUT EDM)

- Electrical discharge machining (EDM) is a thermal process for the removal of metal from a workpiece (in any state of hardness) with little metallurgical damage or distortion.

  - NC EDM machining requires a preformed electrode that will replicate its own form, produce a sculptured surface, form helical gears and forms, machine multiple cavity forms, and perform other machining processes using from two to six axes of movement or cutting action.

- A high voltage, when applied to a deionized water solution, changes the water to an ionized state where it serves as a conductor. Cutting action is produced by the microsecond changing from conductor to nonconductor states to generate a high-intensity electrical spark between the electrode and workpiece.

  - The electrode feed system of a wire-cut EDM machine is designed to feed the wire through the workpiece at a prescribed speed and rate of cut and under a constant tension.

- Overcut in EDM machining provides an allowable space between the surface of the workpiece where it is to be machined and the diameter or outer dimensions of a preformed electrode. The purpose of overcut is to provide clearance for the electrode and to permit flooding the area with dielectric fluid.

  - Increasing the spark energy increases the amount of material removed during each spark, decreases machining time, and produces a coarse surface texture. Decreasing the spark energy produces the opposite results.

- A wire-cut EDM dielectric fluid produces the required operating conditions between the electrode and the workpiece to machine a part, maintain a constant temperature, and to flow chips away from the work area.

  - Small, ball-shaped chips are produced in EDM machining as the dielectric quickly removes heat from the vapor formed by the spark energy.

- Dielectric circulation methods include: pressure application, rapid circulation of the dielectric, creating a gap, and vibrating.

  - Water-submersion wire-cut EDM processes are used for high-speed roughing cuts; oil submersion, finish machining.

UNIT 32 REVIEW AND SELF-TEST

A. CO₂ LASER BEAM HARD SURFACING, ALLOYING, HARDENING, AND CUTTING AND MACHINING PROCESSES

1. a. State the function that is served by hard surfacing.
   b. Describe briefly the CO_2 laser beam process of hard surfacing.
   c. List three advantages of using a laser beam to hard surface a workpiece.

2. a. Define surface alloying by the laser beam process.
   b. Describe breifly how to surface alloy a workpiece.

3. Tell how a part may be surface hardened using a laser beam.

4. Explain the function of (a) a laser cavity and (b) a laser kerf.

5. Identify the purposes of using (a) inert gases and (b) oxygen during a laser cutting process.

6. List three advantages of CO_2 laser beam cutting over conventional (welding) cutting processes.

7. Identify three design features of a laser cutting center that extends the use of laser beam processes in cutting and machining metal parts.

8. List four safety and/or health practices *that must be strictly followed* before any laser equipment is programmed or operated.

B. NC AND CNC WIRE-CUT ELECTRICAL DISCHARGE MACHINING (EDM)

1. Name three (a) NC EDM and (b) three wire-cut CNC EDM machining processes.

2. State the function a rotating head serves in NC EDM machining.

3. Identify three generally used NC EDM electrode materials.

4. List three factors upon which overcut (NC EDM) depends.

5. a. State two functions of the dielectric fluid in NC EDM machining.
   b. Name three dielectric fluids.

6. List three disadvantages of NC EDM machining.

7. Differentiate between the processes and quality of surface finish produced by water- and oil-submersion wire-cut CNC EDM processes.

8. Explain briefly the principle of erosion as applied to wire-cut CNC EDM machining.

9. Tell what purpose is served in EDM machining when the state of water is continuously changed between ionized and deionized states.

10. Describe how an EDM chip is formed.

11. Name three common methods for circulating dielectric fluids in EDM machining.

12. State four advantages of nontraditional EDM over conventional machining processes.

13. Identify two safety precautions to observe that are unique to EDM machining.

Interfaced Subsystems in Flexible Manufacturing Systems (FMS)

A computer-controlled, random-order, flexible manufacturing system (FMS) provides the capability to produce a specified range of parts or components. FMS also represents an advanced stage in the development of a completely automated factory. FMS requires a computer-integrated group of multiple machine tools and/or machining centers, work and storage stations, and an automated handling system. FMS involves subsystems of computer-aided design and drafting (CADD), computer-integrated manufacturing (CIM), variable-computer integrated manufacturing control (CIM/GEN), and others.

Subsystems in FMS are further interlocked with at least one robotic system. Robot functions relate to work handling and tool transfer, sensing, machining and tooling productivity, inspection, gaging, quality control, and others.

This concluding section provides the capstone to all preceding sections and units.

UNIT 33

CAD/CAM/CIM, Robotics, and Flexible Manufacturing Systems (FMS)

OBJECTIVES

After satisfactorily completing this unit, you will be able to:

- See the relationship of all the technology and processes covered in previous units with latest developments in random-order, flexible, fully-automated factory manufacturing systems (FMS), including robotics.
- Describe CAD/CADD, CIM/CAM, and CIM/GEN as components in FMS.

- Explain the functions of the following subsystems.

  - Automated materials handling subsystems.
  - Automated in-process and post-process gaging, inspection, and statistical process control (SPC).
  - Precision surface sensing probes and components.
  - Automated speed, feed, and/or adaptive control subsystems.
  - Group technology systems and cellular manufacturing.
  - Palletizing components and systems.
  - In-line, loop, ladder, and open-field layouts for FMS.

- Tell about construction features and applications of robots.

- Analyze robot components, versatility, and factors affecting the selection of robots in manufacturing.

- Discuss design features and the capabilities of CADD robots in quantity production.

- Use FMS subsystem component and robot *Terms* correctly.

- Recognize the importance of following *Safe Practices* for personal safety and machine tool, control, and other equipment according to Industrial Safety Codes and manufacturer's recommendations.

COMPUTER INTEGRATED MANUFACTURING SUBSYSTEMS

Computerized subsystems and robots, when fully integrated into conventional CNC and DNC machine tools and machining centers, represent an accelerating movement toward totally automated flexible manufacturing systems.

Computer-Aided Manufacturing (CAM) relates to plant-wide integrated systems which reach from market needs, to design concepts, to fabrication, and on to the marketing of the final product.

The computer has the capability to generate a manufacturing database. The database includes technical information about process planning: the generation of NC programs, materials requirements planning (MRP), routing, manufacturing, and other functions. With the transfer of database information at the design level, the designer may make necessary changes that affect manufacturing efficiency and productivity.

The terms CAD/CAM interlock part or all functions including part programming. *Computer integrated manufacturing* is displacing CAD/CAM. CIM relates to hardware, software, and inte-grated subsystems. CIM serves to make all manufacturing functions independent and interactive.

Computers are used to collect, file, analyze, process, and change geometric, engineering, and other data. It is this manufacturing database with the computer network that is the foundation for CAD/CAM and CIM.

Numerical control developments also contribute to CIM by being able to link machines, robots, and other forms of computer-controlled materials handling equipment.

CAD/CADD, CIM/CAM, AND CIM/GEN COMPONENTS

Descriptions of a few additional components commonly associated with computer-integrated manufacturing follow.

CAD/CADD. CAD refers to automated drafting. Drawings that normally would be prepared by a draftsperson manually are automatically programmed and are graphically displayed. When the CAD system has the added capacity to analyze and to design parts and components (using computer graphics to perform such engineering functions as calculations and design analyses), the system is referred to as computer-aided design and drafting (CADD). In CADD, coordinates are calculated and instructions are generated to direct NC operations.

CIM/CAM. This computer-graphics system creates the geometrics of a part, machining data, and other manufacturing input for multiple-axis machining on CAM controlled machine tools (or DNC to remote machine tools). Representative manufacturing processes performed on machine tool units which operate interactively include: milling, turning, punching, nontraditional machining, form cutting, and others.

CIM/GEN. These code letters relate to computer-integrated manufacturing (CIM) with a variable computer-integrated manufacturing control system (GEN). GEN operates on the selection of software routines to build and to control the system. Some modules include manufacturing specifications, inspection and quality control requirements, and inventory control. Other system modules relate to tooling and machine setups, planning material requirements, measurement standards setting, NC programming, production control, and shop floor control.

COMPUTER GRAPHICS APPLIED TO CIM

Computer graphics is central to computer-aided design. A computer graphics system includes the visual display of an image (workpiece) on a cathode ray tube (CRT). The image, accompanied by letters and numbers, may range from a simple one-view dimensioned drawing, to a tool cutter path, to a cutaway three-dimensional assembly drawing, to sketches, and to related documents.

Sophisticated drawings of irregularly-shaped parts that may be rotated for viewing different positions, and color, may be graphically displayed.

Designs and drawings are produced from the geometry database of the part; the same information used in manufacturing. Design changes and modified instructions are made instantaneously. The final designed product may then be memory stored. Drawings and documents (hard copy)

are readily generated by electronic plotters or line printers.

Computer graphics are also applied to complicated engineering software programs. Product information can be obtained before any prototype model is produced. Factors of stress, material strength, tolerances, and other design elements may also be established through computer graphics.

PROCESS PLANNING IN INTEGRATED MANUFACTURING

Simply defined, *process planning* relates to the processes and sequences needed to produce a product. Process plans provide part specifications, design information, required machine tools, and machining processes. The two basic computer-aided process planning (CAPP) systems are called: *variant process planning* and *generative process planning.*

Variant process planning relates to producing new plans based on changes of standard processes that are prestored in a database. While variant processing planning is unable to reflect the impact of fluctuating factors, it can automatically generate work orders, bills of materials, routing sheets, and other shop-floor documents.

Generative process planning develops a process plan based on part geometry and information about materials, equipment, tooling, fixturing/workholding devices, cost factors, and the like. Coupled with additional information in the database, it is possible through the generative process planning system to determine the most cost effective machine tool, tool, fixturing, etc. needed to produce particular parts.

EFFECTIVE COMPUTER-AIDED PROCESS PLANNING (CAPP) SYSTEMS

The following are advantages of an effective CAPP system.

• Improved productivity based on standardization and selection of the most efficient cutting tools, machine tools, and machining conditions.

• Process planning decisions are based on accumulated experiential data generated within a computer as contrasted with variations of expertise among individual program planners.

• Cost effectiveness in developing process planning materials.

• Efficiency in generating tool lists, process sequences, and other paperwork for a part.

• Immediate availability on the shop floor of process planning information from a central source.

MANAGEMENT OF MATERIALS: CIM

CIM makes it possible for a series of related functions such as: materials handling inventory control, cost, availability of materials, etc., to interact. This capability is needed for *just-in-time* manufacturing systems to serve effectively. Important cost-effective plans can be initiated from a CIM system where engineering standards, material purchases, inventory control, and other information is fed back to a control computer.

Computer capability is provided about material reorder information and printouts are computer prepared to requisition materials making them available without interruption in the production line.

A CIM system interlocks shop floor information on production with cost accounting reports and payments associated with manufacturing. CIM systems are used for remote controlled materials and parts handling, high-rise stacking towers, and track-oriented palletized units. Data about parts and materials locations, frequency and nature of usage, and the delivery and retrieval of parts are examples of computer technology is manufacutring.

INSPECTION AND STATISTICAL PROCESS (QUALITY) CONTROL: CIM

The computer can be programmed into automatic gaging systems for parts inspection and to gather data. The computer can accept or reject workpieces that fall within prescribed dimensional and geometric tolerancing limits. By interfacing a computer inspection (quality process control) system within a CNC machine tool it is possible for the computer to modify the CNC machine conditions and to take whatever corrective steps are needed to produce in-tolerance workpieces.

In addition to machine tools, the computer may also interface in quality process control (QPC) with functions performed with coordinate measuring machines.

ADVANTAGES OF CAD/CAM/CIM IN ENGINEERING AND MANUFACTURING

Computer-aided design and planning provide the following measurable benefits in *engineering graphics systems.*

• The design cycle and lead time required for producing a new product are shortened.

• The number of design protypes required for testing and assessment is reduced.

• Cost effectiveness in generating technical data, records, and reports.

• Design/engineering are based on group expertise and interaction.

CAD/CAM/CIM systems provide these benefits in the area of *manufacturing.*

• Better inventory scheduling, storage, and retrieval.

• Reduced inventory storage space required.

• Greater uniformity of product and production yield.

• More effective up-time in the utilization of machine tools and other accessories and equipment.

• Hazardous tasks may be automated for personal, machine tool, and product protection and safety.

GROUP TECHNOLOGY (GT) SYSTEMS

Group technology provides another system in the movement toward integrated, flexible manufacturing systems (FMS). Group technology requires engineering and machining data on a large population (number) of parts. The data are grouped into *families of workpieces* that have the same or similar characteristics, based primarily on manufacturing needs.

To structure a family-of-parts, a first selection is made to broadly group parts with similar form, size, and machining requirements. The groups are then subdivided a number of times until subgroups *(part family cells)* are formed of parts that may be produced on the same equipment using common tooling.

It should be noted that within each family of parts minor differences in sizes, numbers of operations, tooling, etc. may be accommodated. Special consideration is given in group technology to size.

PARTS CLASSIFICATIONS AND CODING

While computer software is available for parts classification and coding systems, adaptations are required to meet specific plant needs. Within each coding system, a multidigit code is used to identify particular features of parts in a part family. For example, a group technology (GT) code number is assigned for part geometry; others for classifying: material in a workpiece, manufacturing processes, surface finish, dimensional tolerances, and other features that describe a part.

Ten digits (0 through 9) are used in a simple GT code designation system. The first digit generally identifies the geometric features (configuration) of the part. The second and remaining digits identify each successive subgrouping. Values of zero (0) through nine (9) are used with each digit. More digits and numeric or alphanumeric (letters and numbers) are used for detailed descriptions of complex parts.

It is evident in commercially-available coded part designations that a high level of computer capability and interactive computer graphics is required.

Figure 33–1 provides an example of a part (A) that is coded and classified within a group technology system matrix for concentric shapes. The part description and features are identified by callouts in Figure 33–1B. Each vertical column identifies specific attributes of the product. Each horizontal row shows a variation of each attribute. The multidigit information for the sample part is interpreted as follows (Figure 33–2).

The computer reads back the part design for verification. A split CRT screen produces a code-generated part image.

The group technology data base is important in product design, process planning, production scheduling, and inventory management. Part families with common machine processing requirements are scheduled in group technology for specific tooling setups, NC machine tools, and for the sharing of jigs, workholding fixtures, dies, and other tooling that are keyed for particular part families. In group technology, design variations may be made in tooling to multiple machining processes that cut across more than one part family.

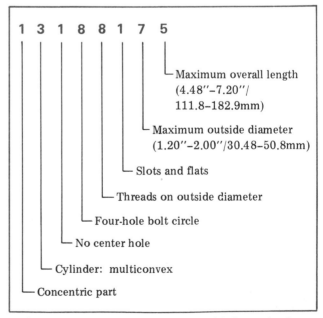

1 3 1 8 8 1 7 5

- Maximum overall length (4.48″–7.20″/ 111.8–182.9mm)
- Maximum outside diameter (1.20″–2.00″/30.48–50.8mm)
- Slots and flats
- Threads on outside diameter
- Four-hole bolt circle
- No center hole
- Cylinder: multiconvex
- Concentric part

Figure 33–2 Interpretation of Coded Information in a Group Technology (GT) System

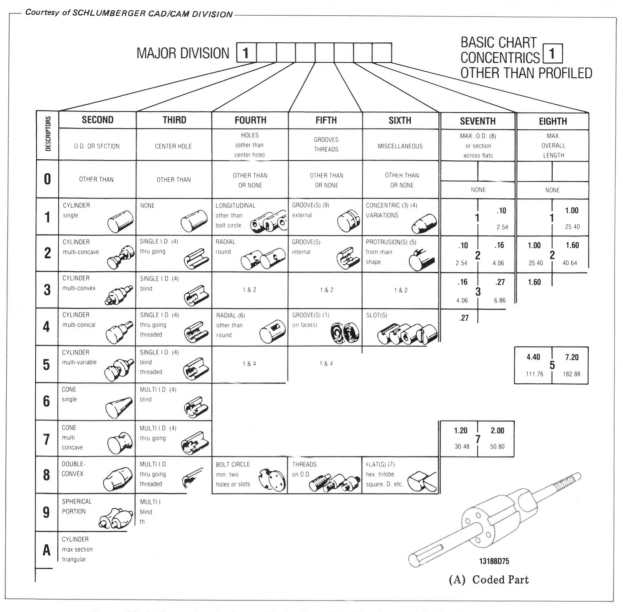

(A) Coded Part

Figure 33-1 Example of a Group Technology Classification and Coding System Matrix

CELLULAR MANUFACTURING

Group technology *cellular manufacturing* requires the clustering of a core of machines designated to process particular part families into a *cell* (particular grouping). The NC machines (complemented by conventional machines that may be added to a GT cell to perform operations that cannot be done on the core machine) are selected according to a major part classification.

There may be more than one of the same core machine in a cell, depending upon the nature of each machining process and the balancing of process work loads. In a simple straight process flow line in a group technology cell, the parts in a family move from one machine to another in a fixed sequence.

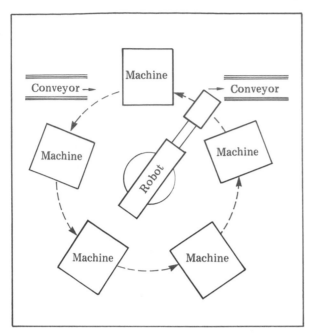

Figure 33-3 Roboticized Circular Flow Layout of a Group Technology Machining Cell

Figure 33–3 illustrates a *circular flow* line. This *roboticized automated cell* has the machines in group technology processing assembled around a robot. The work transfer function is performed by a computer-controlled robot that is programmed and tooled to accommodate the different parts within a family-of-parts.

BENEFITS OF GROUP TECHNOLOGY (GT) CELLULAR MANUFACTURING

The following benefits are derived from integrating group technology (GT) cellular manufacturing with CAD, NC design engineering and manufacturing database, and CNC machines.

- Cost effectiveness in generating, evaluating, and producing a final product design.

- Closer controls in specifying, purchasing, storing, and distributing part materials to meet just-in-time manufacturing needs.

- Availability of extensive design and manufacturing data to improve scheduling performance.

- Uniformity in interpreting and executing part processing plans.

- Cost reductions in materials purchasing, storage, handling, and record keeping.

- Better follow-through control of workpieces.

- Greater flexibility to level peak and valley production cycles.

- Reduction of materials inventories and work in process stockpiling.

- Lowered tool design and tooling construction costs.

PALLETIZING COMPONENTS AND SYSTEMS

Palletizing components and *systems* provide for the automated flow of parts to or from machining operations. Palletizing systems are designed to interface with horizontal and vertical machine tools and machining centers. Palletizing transports range from simple on-loading systems (Figure 33–4) to multiple station pallet pool systems.

The range of applications for different parts machining setups include the following.

- Palletizing one machine station.

- Palletizing two or more machine stations.

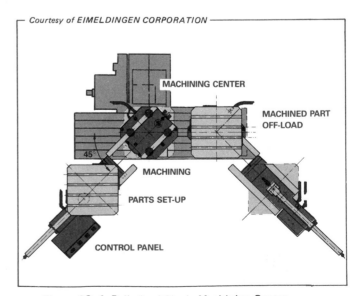

Courtesy of *EIMELDINGEN CORPORATION*

MACHINING CENTER

MACHINED PART OFF-LOAD

MACHINING

PARTS SET-UP

CONTROL PANEL

Figure 33-4 Palletized Single Machining Center

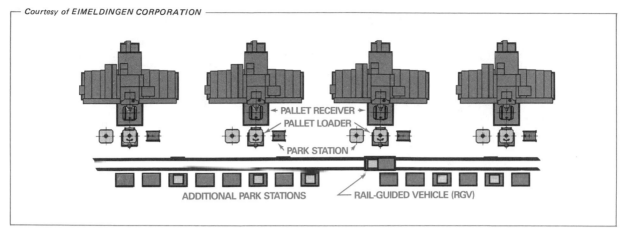

Figure 33-5 Fully-Automated Machining Cell of Four Machining Centers with a
Load/Unload Palletized System

- Integrating two or more similar or different machining stations as part of a fully-automated machining cell.
- Linking any number of machining cells in CIM or FMS systems.

Figure 33–5 illustrates a palletized, rail-guided vehicle system for loading and unloading four fully-automated machining centers.

MAJOR COMPONENTS OF A PALLET SYSTEM

The major components of a pallet system are shown in Figure 33–6 and are identified as:

- Round, square, and rectangular pallets.
- Pallet receivers and changers.
- Load/unload stations for straight or rotational alignment.
- Motorized pallet transfer mechanism.
- Rail guided vehicles.
- Park stations.
- Computer traveling pallet changer.
- Electrical push button control unit.

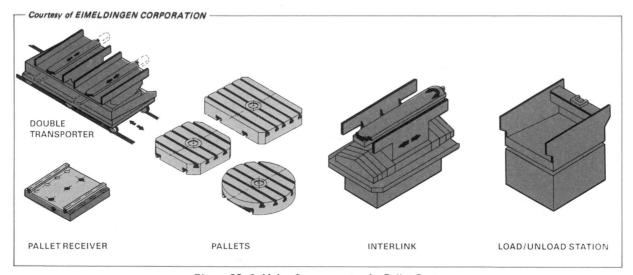

Figure 33-6 Major Components of a Pallet System

Courtesy of EIMELDINGEN CORPORATION

Figure 33–7 Inclinable Dividing (Rotary) Table

Accessories used for pallet systems for precise part positioning and machine auxiliary functions include precision rotary machine tool tables, combination horizontal and vertical rotary tables, inclinable dividing (rotary) tables (Figure 33–7), and coordinate inspection tables.

EQUIPMENT LAYOUTS FOR FLEXIBLE MANUFACTURING SYSTEMS

Each FMS plant layout is influenced by the diverse nature and quantity of parts to be manufactured, the number and positional relationship of machines, the palletized work transport system, robotic feeding devices, and machine control equipment.

Four typical shop layout plans *(configurations)* are presented in Figure 33–8 as: in-line, loop, ladder, and open-field.

IN-LINE FMS SYSTEM

Workpieces are processed in an *in-line FMS system* by being deposited and picked up at each machine stage, starting at one end and continuing throughout the line (Figure 33–8A).

LOOP FMS SYSTEM

Workpieces (routed on pallets that are driven by rollers on rails) are computer fed sequentially from machine to machine (Figure 33–8B). The machines are *formed into a loop.* The directional turntable at the end of the loop routes the pallet in the right direction.

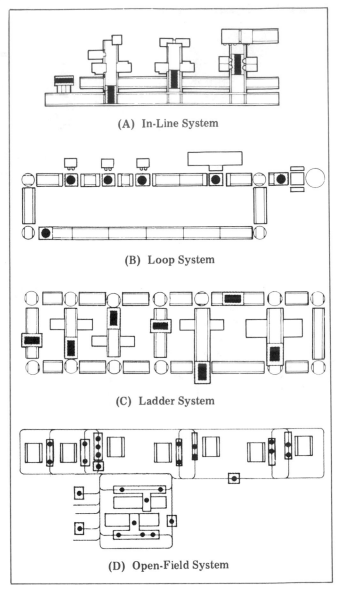

(A) In-Line System

(B) Loop System

(C) Ladder System

(D) Open-Field System

Figure 33–8 Four Basic Configurations for Flexible Manufacturing Systems

LADDER FMS SYSTEM

The machine tools in a *ladder configuration* (Figure 33–8C) are positioned at different locations on the cross rungs of a ladder pattern. Workpieces, that are mounted on pallets, are routed in any sequence to any of the machine tools by programmable computers on computer command.

OPEN-FIELD FMS SYSTEM

Generally, a random-order FMS which is directed by its own computer and interfaces with each machine control unit (MCU) and transport system, uses the *open-field* floor plan (Figure 33–8D). Within this plan, machines are located in any order according to type and use sequences. Materials that are positioned and secured on pallets on wire-guided carts are routed through the loop, stopping at designated machines.

ROBOT FUNCTIONS, CONTROLS, AND INFORMATION SYSTEMS

Robots are capable of performing handling or processor applications. *Handling applications* generally deal with the movement of materials and parts, components in loading and unloading, work activities associated with storage retrieval systems, conveyor systems, machine tools, and machining centers.

Processor applications are identified with the actual performance of programmed manufacturing tasks such as: welding, immersion of parts in heat treating quenching baths, assembly processes, and others. The robot is programmed in processor applications to manipulate equipment or tools in performing a process.

BASIC AXES OF MOTION

Robots are designed with linear, rotary, and an infinite variety of combinations of these movements. The six basic axes of movement are illustrated in Figure 33–9. Robots are built with hydraulic and/or electrical power units for each rotary actuator. The robot may be programmed for a predictable path movement, acceleration or decelleration, velocity, gripper holding force, load control, and other functions.

FACTORS AFFECTING THE SELECTION OF AN INDUSTRIAL ROBOT

Stand-apart robots serve multi-machine functions. Major specifications used in selecting an appropriate robot follow.

Degrees of Freedom. Each robot is identified by a number of *degrees of freedom*. This means,

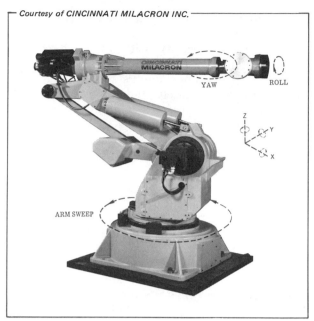

Figure 33–9 Six Basic Axes of Movement of an Industrial Robot

for example, that a general-purpose robot with *six degrees of freedom has six axes of motion.*

Operating Range and Reach. The maximum reach and operating range for a *three-dimensional outer envelope.* The reach of this envelope is measured from the tool center point. Manufacturers also specify an *inner envelope.* It is within this inner envelope that the robot can handle a maximum load, move it at maximum speed, and position a workpiece with the greatest accuracy.

Load Capacity. The combined weight of the grippers and workpiece are identified as the load capacity of a robot.

Speed of Programming. This factor relates to the capability of the robot to be flexibly programmed for operating speeds. The robot operation may be controlled within an infinitely-variable speed range in which operating speeds are automatically increased or decreased.

Loading and Unloading Design Features. Design considerations are based on safe-carrying capacity, maximum movements, clearance of the arm and workpiece, maximum reach, positioning repeatability, and compliance. *Compliance* relates to the security with which the robot hand accommodates irregularities in the workpiece. Equally important is the fact that the workpiece is accurately and safely guided and piloted into position for machining, assembling, or other work processes.

Robot Hand (End Effector). Generally, a robot hand has two sets of *grippers (fingers)*. One set is designed to accommodate an unmachined workpiece and to load it into a machine. After machining, the robot hand turns to permit the second set of fingers to grip and to remove the finished part.

In the case of flexible manufacturing systems where families-of-parts are produced, universal hands and fingers are designed to accommodate and to produce the full range of different parts in the family.

PROGRAMMING TEACH AND AUTOMATIC MODES

The *control system* diagrammed in Figure 33–10 identifies the functional relationship of the com-

puter within the control system. A robot is programmed through a control system that has two modes. One mode serves as the *teach mode;* the second, the *automatic mode*. In the teach mode, the robot is programmed for positional, function, and speed by data that is fed through the *teach pendant* (control) for a series (sequence) of *points*. After checking, the programming may be displayed on a CRT (cathode ray tube) where the operator may add or delete points.

Software within the computer directs the cathode ray tube display to verify whether the position, speed, and other functions of a robot are usable.

Once the taught functions are established, the automatic mode is programmed through the control system to replicate the taught sequences and functions at each point. The robot system is monitored during operation by computers to determine any malfunction. The computer also serves to perform the following tasks.

• Automatically put the robot in a shutdown mode unitl a problem is corrected.

• Make changes for the taught program that is contained in computer memory.

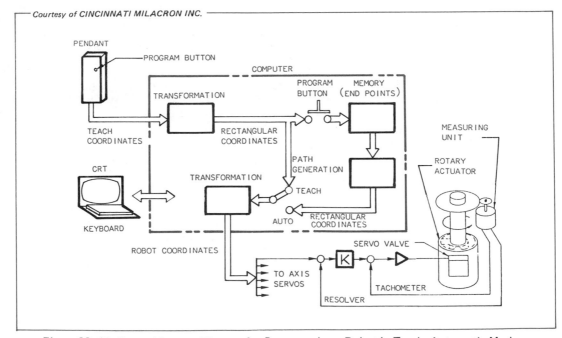

Courtesy of CINCINNATI MILACRON INC.

Figure 33–10 Control System Diagram for Programming a Robot in Teach–Automatic Modes

- Feed back information for the computer to check as to whether or not the changes are possible.

- Make the robot changes after it is once established that all changed movements and functions may be performed.

- Enter the sequences of operations and other changes into computer memory using a tape reader.

VERSATILITY OF ROBOTS IN PRODUCTION

The versatility of robots in production is illustrated in Figure 33–11. The *manipulator* (robot hand) in this application includes accessories used in combination with a laser unit.

CAUTION: "Safety equipment may have been removed or opened to clearly illustrate products and must be in place prior to operation."

Figure 33–11 Application of a Robot in Performing Arc Welding Processes

The robot hand serves as a fixture to position and to rotate the workpiece in relation to a fixed position of the laser unit. Cutting is performed by laser. In all such applications, *cautions* must be strictly observed. All protective guards and safety equipment must be in place prior to operation. All *Warning* notices on machine tools must be followed.

The robot illustrated has been programmed *online* by the operator who manually moves the manipulator through each successive step in the cycle. Each step is recorded in memory in the robot controller unit for automatic programming during production.

Robots may also be programmed *offline* and tested for accuracy and reliability. The program may then be loaded onto the robot controller and further integrated into a FMS.

CADD ROBOT MODEL

CADD has the following capabilities in relation to the design and development of robots.

- Storing a three-dimensional graphics model of a part in its *database*.

- Copying the model in the data base in order to design the work cell in which specified tasks are performed.

- Providing visual feedback against which design problems may be checked and corrected.

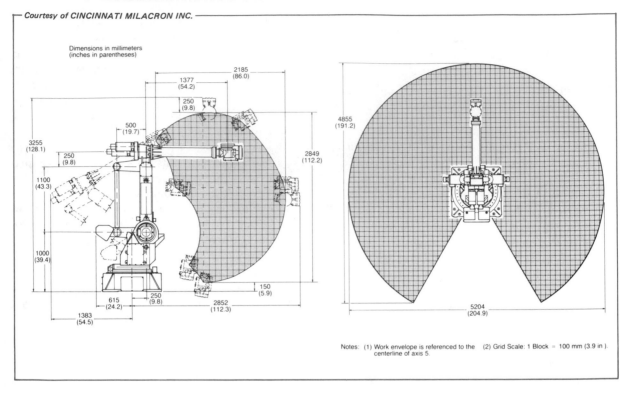

Figure 33–12 Parabolic-Shaped Envelope Describing the Range, Reach, and
Degrees of Freedom of a Robot, Including Required Floor Space

SPECIFICATIONS OF COMPUTER-CONTROLLED ROBOTS

Specifications of robots include information about the operating range, reach, and degrees of freedom. Figure 33–12 shows the effective parabolic-shaped volume (envelope) from the centerlines of the robot through to the tool center point (centerline of the gripper jaws).

APPLICATIONS OF COMPUTER-CONTROLLED ROBOTS

Industrial robots may be interfaced with other computers as components within manufacturing cells as subsystems in flexible integrated manufacturing. Some robots feed televised images to computers. Using computer logic, the images are converted to a digital code for comparison with digital representation of images stored in computer memory. This development makes it possible for a robot to discriminate in selecting the correct parts from a conveyor or other feed system, based on part features.

Attention is also directed to further applications of sensors for greater sensitivity of grippers in applying workholding forces and to speech recognition.

Figure 33–13 provides an application of a computer-controlled, six-axis robot in a *manufacturing cell*. Workpieces are brought to location on a transport car. From the car, the fingers of the robot hand are designed to transfer the material for a part to a preset location on a CNC lathe. Once machined, the finished part is removed and replaced.

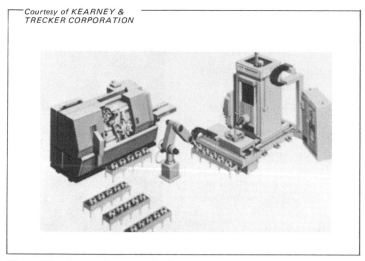

Figure 33–13 Application of a Computer-Controlled Six-Axis Robot

Another general application of a computer-controlled robot is illustrated in Figure 33–14. The robot is rear-loading one CNC turning machine while the second CNC machine is machining a part.

Each machined part is measured by a laser gage. The laser gage is interlocked with the computer control system. Information about out-of-tolerance machining is fed back to the machine control unit (MCU). Automatic tool offsets are then actuated to adjust for cutting tool wear and to bring the particular part dimension within tolerance.

Figure 33–14 Rear Loading/Unloading Two CNC Turning Machines with a Robot Leaving the Front of the Machines Open for Operator Monitoring

FUNCTIONS OF AUTOMATED SUBSYSTEMS IN FLEXIBLE MANUFACTURING SYSTEMS (FMS)

FMS requires a series of automated subsystems which serve the following functions.

- Complete control of the manufacturing system by the host computer.
- Handling and transporting materials.
- Positioning, adjusting, and changing workpieces.
- Selecting, changing, and handling tooling.
- Continuous cycling of each individual machine.
- In-process and post-process gaging.
- Controlling speeds and feeds.
- Precision multidirectional surface sensing.
- Washing and cleaning workpieces.
- Disposing of chips automatically.

AUTOMATED MATERIALS-HANDLING SUBSYSTEM

This subsystem deals with the storing, control, and retrieval of materials and work that (1) is not actually being processed, (2) finished parts, and (3) work in process. Materials handling is integrated into FMS. The design and development of the subsystem requires a consideration of the following characteristics of materials handling.

- The controlled path or travel of the workpiece.
- The transfer of workpieces from a conveyor line into a precisely piloted position for machining.
- The level(s) for placing machine tools in relation to materials conveyor shuttle levels.
- The need for fixtures and other nesting/holding devices for processing multiple workpieces.
- The advantages and/or problems associated with transporting parts by free or power-driven conveyors, shuttle cars, or other automatically-guided devices.
- Transporter traffic control to establish the number of transporters needed in order to minimize or eliminate machine time lost in waiting for tools or workpieces.

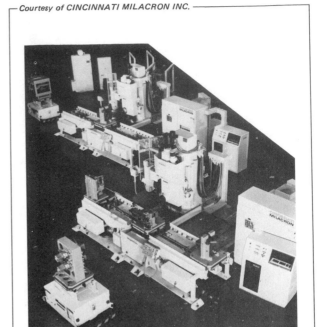

Figure 33-15 In-Line Palletized System for Work Loading and Work Retrieval

An in-line application of a palletized system for work fixturing, work loading, and work unloading is illustrated in Figure 33-15. The unmachined workpiece is fixtured (accurately positioned and secured) at the station in front of the machine tool. This in-line pallet arrangement is mounted on a wire-guided cart that is automatically deposited at the feed end of the machine. At the completion of the machining cycle, the machined part and pallet setup are then picked up at the other end of the line.

AUTOMATED TOOL-HANDLING SYSTEMS

Manual and computer-assisted loading and unloading of cutting tools and assemblies were described in earlier units on NC and CNC machine tools. In FMS applications, there is the added provision for the withdrawal and replacement of quick-change, cartridge-type, tool drums by a robotized subsystem.

AUTOMATED IN-PROCESS AND POST-PROCESS GAGING

In-process gaging refers to the taking of dimensional and form measurements and parts inspections for a limited number of part features *prior to the removal* of the workpiece. In-process gaging is important in correcting surface problems, maintaining form, and controlling dimensions within allowable tolerances.

New tools are brought up by command of the machine control unit (MCU) to remachine oversize surfaces. Undersize workpieces are rejected and the MCU signals a malfunction.

In-process gaging is carried on by linking a computer-controlled coordinate measuring machine (CMM) program into a DNC system (which remotely controls a group of NC machines). The in-process gaging of intricately detailed parts is exceedingly difficult and complex to achieve.

Post-process gaging, by contrast, does permit checking *all part details*. However, there are two major disadvantages. First, compensation for out-of-tolerance machining takes place *after* the part has been machined. Secondly, the part cannot be returned to the machine in FMS to be reworked.

AUTOMATED PRECISION SURFACE SENSING PROBES

A *sensing probe* is a multidirectional electronic switching device. When applied to a machining center, the sensing probe has the following components.

- A *body* fitted with a standard tapered shank to fit the machine spindle nose and a tool-storage matrix.
- An *interchangeable stylus* held in the body.
- A *probe-mounted inductive module* on the machine spindle connected to a *control circuit board* mounted in another location.

COMPONENTS OF A SURFACE SENSING PROBE

The major components of a precision surface sensing probe, generally used on machining centers, are shown in Figure 33–16. The probe is held in the storage matrix. After the workpiece is machined, the sensing probe is automatically transferred from the storage matrix onto the machine spindle (being interchangeable with the last used cutting tool). Any deflection of the probe stylus instantaneously records the surface location of a defect in machining or variation from an allowable tolerance.

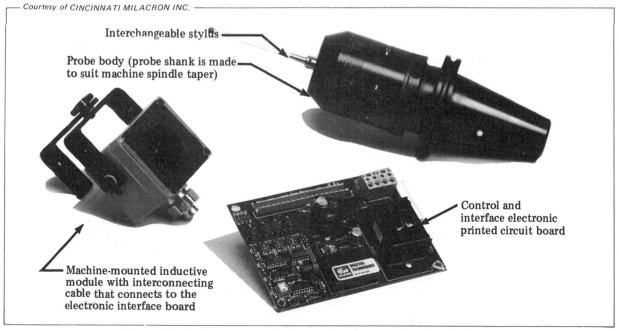

Courtesy of CINCINNATI MILACRON INC.

Interchangeable stylus

Probe body (probe shank is made to suit machine spindle taper)

Control and interface electronic printed circuit board

Machine-mounted inductive module with interconnecting cable that connects to the electronic interface board

Figure 33–16 Components of a Typical Precision Surface Sensing Probe

Staging Area

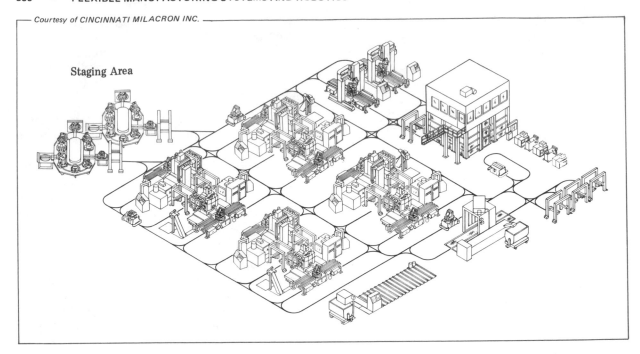

Figure 33–17 Graphic Display of a Computer-Directed Flexible Manufacturing System (FMS)

Surface sensing probes serve these functions:

- Measuring and checking a particular surface,
- Detecting and compensating for size or form variations,
- Making in-process alignments, and
- Sensing any drift (moving away from a normal condition in machining) which may produce out-of-tolerance parts.

AUTOMATED FEED, SPEED, AND/OR ADAPTATIVE CONTROL SUBSYSTEMS

Adaptative control provides feed-back signals to a machine control unit (MCU) for purposes of adjusting feeds and speeds. Sensory circuits and computations are used to compare *cutting torque* with specific cutting torque limits. Limits are based on prevailing tooling, material, and machine tool operations.

The adaptative control unit senses such factors as torsional tool overload, cutter wear, a broken or damaged cutting tool, surface irregu-larities, and dimensional variations. Feed and speed rates are adjusted upward or downward automatically as machining conditions require.

GRAPHIC DISPLAY OF A FLEXIBLE MANUFACTURING SYSTEM

Figure 33–17 visually displays a computer-directed flexible manufacturing system (FMS). This particular system consists of eight machining centers. Each machining center is equipped with a 90 tool-storage matrix to accommodate the range of parts which are to be machined and to automatically replace worn or damaged tools. Also, there are two coordinate measuring machines, one parts cleaning machine, and an automatic chip-removal system.

Workpieces and tools, mounted on pallet carrousels, are transported throughout the system under computer control. Additional carrousels are shown being loaded in the *staging area*. These carrousels are used for continuous parts processing, particularly during unmanned machining periods.

Safe Practices Applied to Robotics and Flexible Manufacturing Systems (FMS)

- Read the equipment manufacturer's specifications and Occupational Safety and Health Act safety guidelines and requirements before working with or around robotic and other automated machine tools and accessories.
- Establish the *danger zone of a robot* from the various zones of movement of the robot, the machine tools and control equipment, the materials and machining/working zones, emergency access areas, and work-exiting spaces.
- Check the automatic computer diagnostic data and sensor information and other preventive maintenance monitoring devices to determine and identify any reduced machine tool or robot functions.
- Make sure that the movements and functions of materials and tool handling equipment, automatic positioning devices, and specialized accessories are inspected and may be operated safely.
- Establish and follow a routine preventive maintenance schedule to reduce the possibility of malfunctioning and to ensure that each item is in safe, operable condition to fulfill specific designated functions.

- Analyze and study under direction of qualified technical personnel the potential hazards of all components that are totally integrated into a factory manufacturing system.
- Follow required safety practices in checking to ensure that automatic robot protective devices will initiate a safety command for immediate shut-down in the event of any possible danger.
- Determine in advance of any machine tool operation where manually-operated emergency stops and full-system shut-down switches are located and accessible.
- Report robot malfunctions and problems of transfer and tooling systems, grippers, protective monitoring devices, automated computer program and control equipment, and machine control units.
- Secure all guards and protective panels before operating any piece of equipment.
- Guard against and be continually alert for any unexpected or unintended motion.
- Establish a systematic approach to evaluate the effectiveness of safety devices to personally understand safe equipment operating procedures.
- Maintain records of hazardous control equipment and work procedures and take corrective action.

TERMS APPLIED TO ROBOTICS AND ADVANCED MANUFACTURING TECHNOLOGY AND SYSTEMS

Adaptive control	Determining fixed speeds and feeds for a machine control unit by feedback sensors rather than by programming.
Automated materials handling	A system in which completely automated equipment such as robots and wire-guided transport systems are employed to retrieve, load, unload, and store materials and workpieces.
Axes of motion (robots)	The number of linear, rotary, and combination movement axes in which a robot performs.
Cellular manufacturing	Manufacturing processes performed in a group technology machining cell for a part family.
Central processing unit	A component within a computer system that contains the circuitry necessary to control, schedule, perform arithmetical functions, and address data for memory storage or for instantaneous transmission to a machine control unit.

Computer-aided process planning (CAPP)	A planning system in which engineering data and process information are processed by computer to generate detailed plans for the manufacture of a particular part or a complete assembly.
Computer-integrated manufacturing (CIM)	The interfacing of CAD, CAM, NC, CNC, DNC, robotics, materials handling, and other engineering/process data gathering systems into a totally integrated manufacturing system.
Database management system	The collection, organization, management, and analysis of information and data in computer memory for storage and retrieval.
End effector	A work-performing tool that is attached to a robot arm or manipulator.
Flexible manufacturing system (FMS)	A computer-controlled, random-order, highly-automated manufacturing system wherein CADD, CAPP, CIM, robotics, materials handling, group technology, and other subsystems are interfaced into a flexible integrated factory manufacturing system.
Group technology	A subsystem in flexible integrated manufacturing through which parts are classified according to common characteristics and alphanumeric code designations into part families for computerized manufacturing purposes.
In-line, loop, ladder, and open-loop FMS configurations	Typical shop-floor layout plans for positioning machinery, accessories, and computer control units in relation to parts/materials delivery systems in a manufacturing cell.
Interactive graphics	The use of computer graphics input data to modify design characteristics or manufacturing technology and processes with graphic output display.
Off-line equipment	Equipment and other data processing system devices that are not directly controlled by a central processing unit.
Pallet system	A materials handling system consisting of a group of components that include work-holding pallets, positioning accessories, transfer mechanisms, pallet changers, rail-guided vehicles, and a control unit.
Teach pendant	A control panel used by an operator to guide, direct, and program a robot through a prescribed sequence of motions to perform specific tasks.
Tool center point (robot) TCP	A reference point on the robot hand from which all robot task measurements are programmed.

SUMMARY

- Flexible manufacturing systems consist of subsystems of computer-aided design and drafting (CADD), computer-assisted process planning (CAPP), computer integrated manufacturing (CIM), and other mainframe computer and machine control units.

  - FMS subsystems interface with group technology and cellular manufacturing, parts handling and transport systems, robotics systems, and other computer-aided subsystems.

- CIM systems interlock the computer capability required to manufacture a finished product, the part and tool design stages, and other successive processes, materials handling, inspection, marketing, cost accounting, and other database and records functions.

  - Computer-aided design and planning (CAPP) provide engineering graphics and other manufacturing database information for effective just-in-time manufacturing planning.

- Group technology manufacturing is based on engineering and machining data that is classified and coded through interactive computer graphics to establish a family-of-parts that have similar process characteristics.

  - Cellular manufacturing requires the clustering of a core of machines, quality control inspection accessories, computer-integrated materials handling units, and other equipment to perform a specific sequence of tasks within a group technology cell.

- Palletized systems permit the automated flow of parts to and away from conventional or cellular manufacturing processes. Palletized systems are designed for single machine tool or machining center operation or for in-line, loop, ladder, or open-field manufacturing layouts.

  - Robots are employed in conventional machine tool processes, cellular manufacturing, and flexible integrated manufacturing systems to perform work/parts handling tasks and processor functions.

- Electrical/hydraulic powered robots produce linear axes, rotary axes, and an infinite number of combination axes movements. Predictable path movements are electronically programmed and controlled using computer-aided and interfaced speeds, holding forces, load controls, and other function control units.

  - Computer control systems are used to program a robot through a sequence of tasks when the functions are verified as usable taught tasks that may be replicated for the automatic robot mode.

- Robots may be programmed manually in-line by the robot operator or off-line.

  - Factory automated subsystems in flexible manufacturing systems include CAD, CAPP, CIM, computer graphics, and other computer generated engineering and process program databases. Other subsystems deal with materials and tool handling, in-process and post-processor gaging, surface sensing probes, and automated feed, speed, and adaptive controls.

- All manufacturers' specifications and Federal Occupational Safety and Health Act requirements relating to the work envelope (zone of robot movement); zones of machine tool, accessories, and materials transfer systems; and access openings for workpiece feeding and removing of parts, must be strictly followed for safety.

UNIT 33 REVIEW AND SELF-TEST

1. Identify two functions that are provided by a computer graphics system.

2. Describe briefly the functions served by (a) CADD, (b) GEN, and (c) FMS systems.

3. State three advantages of an effective computer-aided process planning system (CAPP).

4. Make three statements about computer-integrated manufacturing functions (CIM) that relate to materials management for just-in-time manufacturing.

5. Identify any combination of four machining or fabricating processes which can be automated for computer-aided manufacturing.

6. a. Secure a Group Technology Classification Coding System Matrix Chart for concentric parts.
   b. Provide the specifications for a part that has the following group technology classification: **1 2 2 4 4 4 7 2** .

7. State three advantages of cellular manufacturing over conventional manufacturing.

8. a. Name four major components of a palletized materials work-handling system.
   b. Explain briefly how a palletized handling system works in a ladder configuration machining cell.

9. List the functions served by four subsystems of a fully-automated flexible manufacturing system (FMS).

10. a. Identify one function served by automated in-process gaging.
    b. State one advantage of in-process and one of post-process gaging.

11. List three uses of automated surface sensing probes.

12. Explain briefly the principle on which an adaptive subsystem automatically regulates machining speeds and feeds.

13. Give two examples of (a) handling applications and (b) processor applications for robots.

14. State four functions which are served by robots in multimachine manufacturing systems.

15. Tell briefly why the teach mode is important in programming robots.

16. Describe each of the following design features of robots.
    a. Degree of freedom   c. Compliance
    b. Repeatability   d. End effector

17. List four major danger zones for robotic applications in automated manufacturing.

APPENDIX

- **TROUBLESHOOTING CHARTS**

- **HANDBOOK TABLES**

- **INDEX**

TROUBLESHOOTING CHARTS

Two Charts are included as examples of the detailed information that is provided by machine tool and tooling producers. Chart 1 provides in capsule form problems commonly encountered by bar, chucker, and automatic screw machine operators relating to feeding, hole cutting and knurling, turret tooling, and cross slide tooling. Chart 2 deals with grinding wheel, machine tool, and work related problems encountered in cylindrical grinding.

Chart 1 Bar, Chucker, and Automatic Screw Machine Troubleshooting

Problem	Possible Cause	Corrective Action
Stocking and Rod Feeding Stock will not enter	—Chuck in closed position.	—Trip the chucking lever. Turn the handwheel until the chucking cam permits opening the chuck.
	—Bent bar stock.	—Cut off bent section or replace bar.
	—Misalignment of feed finger and chuck.	—Insert stock in the feed finger and push it through the spindle.
		—Place the chuck with the bar stock in its correct location in the chuck sleeve.
	—Master feed finger and out of position.	—Remove the feed tube. Inspect and repair damaged pads.
Chuck will not open	—Feed trip lever toe in raised position.	—Check the trip lever toe, spring, and the screw and pin in the lever toe.
	—Bent chuck fork, sheared spindle key, cracked chuck sleeve, broken clutch spring.	—Remove and replace defective parts.
Chuck will not close	—Drum cam not in open position.	—Turn the drum to the open position.
	—Feed trip lever toe in raised position.	—Check the trip lever toe, spring, and the screw and pin in the lever toe.
	—Broken chuck lever, chucking roll stud; worn or broken chuck lever fulcrum shoes, bent chuck levers; or worn chuck.	—Remove and replace the broken, worn, or bent parts.
Stock Stop Adjustment Workpiece length increases	—Stop is moving back in collet.	—Tighten the turret binding bolt.
	—Swing stop arm moves.	—Center the swing stop arm and the cutting-off tool. Tighten securely.
	—Feed slide movement is set longer than required length and there is excessive wear on the feeding finger.	—Reduce the amount of force of the feeding mechanism, which causes the stop to push back.
Variation in length	—Projection left because the cutting-off tool does not remove all of the teat.	—Reset the cutting-off tool on center.
	—Feed-out period on the lead cam is too short.	—Replace with the correct cam.
	—Worn roll and pin on lead cam lever.	—Remove and replace the worn parts.
	—Chuck or feed finger tension too loose (or worn).	—Adjust the chuck to the required tension or replace a worn feed finger.
Burred or scored end of work	—Stock feed cam lobe is not concentric with the center hole in the cam.	—Refinish the stock feed cam lobe.
	—Center hole in end of stock stop is too large.	—Use flat stock stops without center holes, or face the end of the stock stop.

Basic Hole Forming and Knurling

Drills and Drilling Hole larger diameter than the drill size	—Drill improperly centered in the holder.	—Adjust the outer body of the holder. Start the drill into the hole by hand feed. Tighten the holder screws.
	—Drill bushing out of line or drill shank is bent.	—Straighten the drill shank or replace both the bushing and the drill.
Holes drilled too rough	—Dull drill; feed too fast; insufficient or wrong coolant.	—Replace drill; reduce the rate of feed; use manufacturer's recommended cutting fluid.
Drill breakage	—Too high or too slow a starting speed; too fast a feed; too small lip clearance.	—Regrind with correct lip clearance; adjust speed and feed.
Drilled hole runs out	—Improper spotting, centering, or lengths of cutting edges.	— Use a spot drill that has a larger angle than the drill point angle; reset the drill on center; regrind the cutting edges evenly.
Reamers and Reaming Reamed hole larger than reamer size	—Reamer and forming tool cuts start together but the forming tool operation is completed ahead of reaming.	—Withdraw the reamer before the forming tool operation is completed.
Hole reamed too rough	—Excessive float in the holder.	—Reduce the amount the holder floats until the reamer enters without distorting the edge of the hole.
Hole reamed bell-mouthed	—Reamer bushing is misaligned with the workpiece axis.	—Check the bushing for straightness and the fit of the reamer shank. Replace any defective bushing.
Counterbores, Counterboring Hole counterbored too rough	—Dull counterbore; insufficient chip room; excessive clearance (producing chatter marks); excessive speed.	—Take general corrective actions for tool grinding and controlling the speed.
Threading Tools, Threading Rough threads	—Improperly ground or dull thread chasers; excessive speed; wrong and/or insufficient coolant.	—Take general corrective steps to correctly sharpen the chasers, reduce the cutting speed, and use the manufacturer's recommended cutting fluid.
	—Insufficient bearing.	—Set the cutting face at least 1/10th of the diameter of the workpiece ahead of center.
Die tears the thread or loads with chips	Hook or rake angle of the chasers too small.	—Regrind the chasers to increase the rake angle.
Die cutting edges wear rapidly or break off	—Too heavy bearing.	—Grind the chaser face to decrease the bearing. The face should be ahead of center.
	—Misalignment of die head.	—Adjust the die head to the correct alignment.
	—Die running into shoulder or chuck jaws.	—Use projection chasers; check the setup and chamfer angle.
Thread not long enough	—Slipping shaft or holding clutch fork on friction clutch; chuck tension.	—Adjust to apply a greater force against the movable friction clutch parts.
	—Loose feeding finger and chuck tension too loose.	—Check collet and bar sizes; adjust the chuck tension.
	—Insufficient cam rise.	—Select a cam with a lobe rise that permits thread cutting to length.
Thread too long	—Improper setting or tripping of self-opening die head.	—Check and reset the trip mechanism.
	—Cam rise too high.	—Use only a portion of the cam rise to produce the desired length.
Distorted and out-of-round threads	—Loose lead cam.	—Tighten the lead cam on the camshaft.
Tapered thread	—Tight (drag) or loose (misaligned) turret slide movement.	—Adjust the turret slide.
	—Incorrect threading alignment.	—Make adjustments to correct alignment.

Extensive tap breakage	—Tapping too close to the bottom of the hole; chips packing in the flutes; wrong coolant; tap drill too small.	—Take general corrective action for taps, tap drills, and cutting fluid.
	—Index dog trips before the tap is entirely withdrawn.	—Move the trip dog back to permit the completed withdrawal of the tap.

Knurls and Knurling

Poor knurl	—Knurl binds on the pin.	—Examine the pin. Stone any burrs.
	—Knurl not centered.	—Center each knurl and apply an equal force on multiple knurls.
Rough knurl	—Cam lobe rise too slow, causing chips to crowd and tear the knurl.	—Check and correct the cam design.
Improper tracking	—Knurls not of equal diameter or not mated.	—Check each knurl diameter; select knurl mates.

Turret Tooling

Knee Tools

Cutting with a taper	—Bar too small to use a knee tool; length of cut too long.	—Replace with a box tool, balance turning tool, or hollow mill.
	—Too much force required for the cut.	—Grind the correct clearance on the tool bit.
	—Tool not on center; too much feed; spindle speed too fast.	—Set tool on center or a few thousandths above; correct the feed and/or speed.
Cutting rough	—Incorrect rake (tool digs in); dull tool bit; improper feed and/or speed.	—Grind to correct rake, clearance, and front angles; set to correct speed and/or feed.
	—Incorrect coolant or rate of flow on cutting tool.	—Use manufacturer's recommended cutting fluid; correctly position nozzle and adjust flow rate.

Box Tools

Cutting rough	—Incorrect slide clearance (too little) and rake (too much).	—Regrind to correct clearance and rake angles. Check with an angle gage.
	—Back drag of the blade as it is withdrawn off the work.	—Loosen the back rest. Then, reset the blade to produce a 0.002″ (0.05mm) oversize diameter. Bring the back rest against the workpiece with sufficient force to turn the correct diameter.
Cutting taper on an irregular diameter	—Incorrect shank fit in turret hole.	—Check for burrs, nicks, and correct fit.
	—Back rest too tight on workpiece.	—Reset the back rest.
Eccentric diameters	—Support set ahead of the tool bit.	—Reset the support so that it follows the tool bit.

Balanced Turner

Cutting rough	—Front clearance too high; rake too deep.	—Regrind the tool bit to the correct rake and clearance angles.
Prominent shoulder steps	—First cutter advanced too far beyond the center.	—Draw the first cutter back.
	—Second cutter too far in back of the center.	—Push the second cutter out.

Hollow Mills

Cutting taper	—Improper alignment.	—Use floating holder.
Cutting rough	—Stock piled on inside of a lip.	—Reduce spindle RPM or flow on a larger quantity of cutting fluid.
		—Check back clearance of the hollow mill teeth.
	—Chipped, burned, or burred lip.	—Regrind to sharpen the mill.

Cutting too small	—Ring on mill tightened too much.	—Adjust the ring until the mill cuts to the correct size.
Ragged cut at start	—Stock not chamfered.	—Chamfer end of stock to permit the teeth to start cutting concentrically.

Turret and Cross Slide
Cross Slide

Does not go to stop	—Sheared pin in cross slide rack.	—Replace the shear pin. Relieve the tension on the stop screw before readjusting for the correct diameter.
	—Cam lobe cut down too far.	—Replace the cam with a properly machined one.
Cutting with chatter	—Tool too far below center; excessive tool rake; tool and work overhang; cracked cross slide; tool too wide for the cut.	—Reset tool to center height; mount tools as short as possible; replace damaged machine parts.
Irregular size formed diameter	—Worn roller or pin on cam rolls. —Cam lever roll out-of-round. —Insufficient dwell on the forming cam.	—Replace worn parts and reset the cam.
Form diameters fall outside specified tolerances	—Stock diameter too small to withstand cutting forces.	—Use back rest or support.
Cutting off with a burr	—Play in cutoff cross slide.	—Adjust the gib.
	—Cutting tool in too deep during the form turning operation, causing the workpiece to break off.	—Position the cross slide cams to either start forming sooner or cutting off later.

Turret

Clamp bolts do not hold the turret tools	—Turret tool clamp bushing too short; radius worn; or bolt too long.	—Replace the turret tool clamp bushing.
Play	—Turret disc nut loose.	—Tighten the turret disc nut enough to take up the play without causing the turret to bind.
Failure to lock securely after indexing	—Worn locking pin or broken spring.	—Remove and replace worn or broken parts.
Failure to index	—Turret trip lever in raised position.	—Drop the turret trip lever into the working position.
	—Short turret trip dog fails to adequately raise the clutch trip lever.	—Remove and replace the worn turret trip dog.
Locking	—Loose change roll stud.	—Tighten the change roll stud.
	—Incorrect location of the change roll stud.	—Change the two studs on the change roll disc that are used for double indexing to see whether the wrong one was removed. —Change the location of the stud.
Failure to advance the same distance each cycle	—Worn cam lever roll or pin; loose cam lead.	—Replace the worn roll or pin; tighten the lead cam.
	—Turret indexes before the lead cam lever reaches the end of the cam lobe.	—Reset the turret index trip dog to index just after the lead cam lever starts down the drop of the lead cam.

Chart 2 Common Cylindrical Grinding Problems, Probable Causes, and Corrective Action

Cylindrical Grinding Problem	Probable Cause	Corrective Action
Chatter Wheel	—Out-of-balance	—Run the wheel without coolant to remove the excess or unevenly distributed fluid in the wheel
		—Rebalance the wheel before and after truing and dressing
	—Out-of-round	—True the wheel face and sides before and after balancing
	—Too hard	—Change to coarser grit, finer grade, and a more open bond
	—Improperly dressed	—Check the mounting position and rigidity of the diamond dressing tool and its sharpness
Work and machine units	—Faulty work support or rotation	—Add work rests and position them for maximum support
		—Check the quality of the headstock and footstock work centers
	—Vibration	—Reduce the work speed and check the workpiece for balance
	—Faulty coolant	—Replace dirty and contaminated coolant solution
	—External vibration transmitted to cylindrical grinder	—Check machine leveling, secureness with which the machine is bolted to the floor, and use of vibration absorption material between the base and floor
	—Interference (operating clearance)	—Check the operating clearance between guards, work-holding devices, the grinding wheel, and the wheel head
	—Wheel head (unevenness of motion)	—Check the belt tension, pulleys, and motor drive for any conditions producing unevenness
		—Replace belts, where required, with belts that are of equal length and required shape
	—Headstock and center	—Adjust and correct the work speed, if needed
		—Check the fit of the work center
Work Defects Check marks	—Improper cutting action	—Dress the grinding wheel to act softer
		—Increase the coolant flow and position the nozzle for maximum cooling and cutting efficiency
	—Incorrect wheel	—Use a softer-grade wheel and check the correctness of grain size, abrasive, and bond
	—Incorrect dressing	—Use a sharp, uncracked, securely held, and correctly positioned diamond dresser
Burning and work discoloration	—Improper grinding	—Decrease the in-feed cutting rate
		—Retract the wheel before stopping the rotation of the workpiece
	—Improper wheel	—Use a softer wheel or dress to produce a softer cutting action
		—Increase the coolant flow
Surface imperfections	—Deep marks	—Change to a finer-grit wheel and check the abrasive wheel specifications for the workpiece material
	—Fine spiral marks	—Add work rests, if needed
		—Redress the wheel face parallel to the work axis
		—Reduce the traverse rate
	—Wide, varying depth, irregular marks	—Use a harder-grade wheel
	—Spots	—Remove glazed areas or oil spots on the wheel by turning and dressing
	—"Fish tails"	—Change to a cleaner coolant
		—Flush the wheel guards after dressing
	—Irregular marks	—Clean the machine to remove and flush away any loose chips or foreign particles
		—Check the condition of the blotters and the secureness of the grinding wheel

Cylindrical Grinding Problem	Probable Cause	Corrective Action
Improperly Operating Machine		
Dimensional inaccuracies	—Out-of-roundness	—Relap the work centers
		—Check the condition of the machine centers, accuracy of mounting, and the footstock force against the workpiece
	—Out-of-parallelism (taper)	—Recheck the alignment and secureness of the headstock, footstock, and table setting
	—Inconsistent work sizing	—Check the quality of the centers
		—Add work supports, if required
		—Check the grinding wheel for loading and balance
		—Take up backlash in the wheel head and carriage when readjusting stops for rapid and slow feeds
		—Check the traverse hydraulic system
		—Use a greater volume of clean coolant
Uneven traverse (in-feed) of the wheel head	—Scored ways on the carriage and wheel head	—Check the lubrication of the ways and the recommended oil for the lubricating and hydraulic systems
	—Leakage and gumminess of the hydraulic system fluids	—Check the valves, pistons, and system for oil leakage and gummy lubricant; flush the system and replace the lubricant
	—Unbalanced drive parts	—Replace an unbalanced drive motor, loose pulley, or uneven driving belts
Grinding Wheel Conditions		
Wheel defects	—Faulty wheel dressing	—Incline the dressing tool at least three degrees from the horizontal plane
		—Reduce the depth of the dressing cuts
		—Round off the wheel edges
	—Wheel acting too hard (problems of loading, glazing, chatter, burning, and so on)	—Increase the rate of in-feed, the RPM of the work, and the traverse rate
		—Decrease the wheel speed or width
		—Select a softer wheel grade and coarser grain size or dress the wheel to cut sharper
	—Wheel acting too soft (problems of wheel marks, reduced wheel life, and inconsistent work size)	—Decrease the in-feed and work and traverse speeds
		—Increase the wheel fpm (m/min) or cutting width of the abrasive wheel
		—Dress the wheel to act harder
		—Select a harder wheel, less fragile grain, or both
	—Wheel loading and glazing	—Use a coarser grain size with a more open bond or a softer wheel
		—Dress the wheel with a faster traversing feed and a deeper dressing cut
		—Soften the action of the wheel by controlling the cutting speed
		—Use less in-feed to prevent loading and more in-feed to correct glazing
		—Use a cleaner coolant and change the coolant composition to prevent gumming
Wheel breakage	—Radial breaks (two pieces)	—Bring the wheel up to operating speed and carefully move it to grinding position without jarring or striking the wheel
		—Discard a wheel that has been damaged in handling
	—Radial breaks (three or more pieces)	—Check the wheel for proper use of blotters, the presence of chips and foreign particles, good mounting fit, and the application of equal flange force
		—Reduce the in-feed; avoid jarring the wheel and work
		—Reduce the wheel speed to below the maximum rated fpm or m/min
		—Increase the coolant flow to prevent overheating

HANDBOOK TABLES

Table A–1 Machinability, Hardness, and Tensile Strength of Common Steels and Other Metals and Alloys

SAE Number	AISI Number	Tensile Strength (psi)	Hardness (Brinell)	Machinability Rating (percent)	SAE Number	AISI Number	Tensile Strength (psi)	Hardness (Brinell)	Machinability Rating (percent)
Carbon Steels					*Molybdenum Steels (continued)*				
1015	C1015	65,000	137	50	4140	A4140	90,000	187	56
1020	C1020	67,000	137	52	4150	A4150	105,000	220	54
×1020	C1022	69,000	143	62	×4340	A4340	115,000	235	58
1025	C1025	70,000	130	58	4615	A4615	82,000	167	58
1030	C1030	75,000	138	60	4640	A4640	100,000	201	69
1035	C1035	88,000	175	60	4815	A4815	105,000	212	55
1040	C1040	93,000	190	60					
1045	C1045	99,000	200	55					
1095	C1095	100,000	201	45	*Chromium Steels*				
					5120	A5120	73,000	143	50
Free-Cutting Steels					5140	A5140		174–229	60
×1113	B1113	83,000	193	120–140	52100	E52101	109,000	235	45
1112	B1112	67,000	140	100					
. . .	C1120	69,000	117	80	*Chromium-Vanadium Steels*				
					6120	A6120		179–217	50
Manganese Steels					6150	A6150	103,000	217	50
×1314	. . .	71,000	135	94					
×1335	A1335	95,000	185	70	*Other Alloys and Metals*				
					Aluminum (11S)		49,000	95	300–2,000
Nickel Steels					Brass, Leaded		55,000	RF 100	150–600
2315	A2317	85,000	163	50	Brass, Red or Yellow		25–35,000	40–55	200
2330	A2330	98,000	207	45	Bronze, Lead-Bearing		22–32,000	30–65	200–500
2340	A2340	110,000	225	40	Cast Iron, Hard		45,000	220–240	50
2345	A2345	108,000	235	50	Cast Iron, Medium		40,000	193–220	65
					Cast Iron, Soft		30,000	160–193	80
Nickel-Chromium Steels					Cast Steel (0.35 C)		86,000	170–212	70
3120	A3120	75,000	151	50	Copper (F.M.)		35,000	RF 85	65
3130	A3130	100,000	212	45	Low-Alloy, High-				
3140	A3140	96,000	195	57	Strength Steel		98,000	187	80
3150	A3150	104,000	229	50	Magnesium Alloys				500–2,000
3250	. . .	107,000	217	44	Malleable Iron				
					Standard		53–60,000	110–145	120
Molybdenum Steels					Pearlitic		80,000	180–200	90
4119	. . .	91,000	179	60	Pearlitic		97,000	227	80
×4130	A4130	89,000	179	58	Stainless Steel (12% Cr F.M.)		120,000	207	70

Table A-2 Allowances and Tolerances on Reamed Holes for General Classes of Fits

Class of Fit	*Allowances and Tolerances	*Nominal Diameters					
		Up to 1/2"	9/16" through 1"	1-1/16" through 2"	2-1/16" through 3"	3-1/16" through 4"	4-1/16" through 5"
A	High limit (+)	.0002	.0005	.0007	.0010	.0010	.0010
	Low limit (-)	.0002	.0002	.0002	.0005	.0005	.0005
	Tolerance	.0004	.0007	.0009	.0015	.0015	.0015
B	High limit (+)	.0005	.0007	.0010	.0012	.0015	.0017
	Low limit (-)	.0005	.0005	.0005	.0007	.0007	.0007
	Tolerance	.0010	.0012	.0015	.0019	.0022	.0024

Allowances and Tolerances for Forced Fits

Class of Fit	*Allowances and Tolerances	Up to 1/2"	9/16" through 1"	1-1/16" through 2"	2-1/16" through 3"	3-1/16" through 4"	4-1/16" through 5"
F	High limit (+)	.0010	.0020	.0040	.0060	.0080	.0100
	Low limit (+)	.0005	.0015	.0030	.0045	.0060	.0080
	Tolerance	.0005	.0005	.0010	.0015	.0020	.0020

Allowances and Tolerances for Driving Fits

Class of Fit	*Allowances and Tolerances	Up to 1/2"	9/16" through 1"	1-1/16" through 2"	2-1/16" through 3"	3-1/16" through 4"	4-1/16" through 5"
D	High limit (+)	.0005	.0010	.0015	.0025	.0030	.0035
	Low limit (+)	.0002	.0007	.0010	.0015	.0020	.0025
	Tolerance	.0003	.0003	.0005	.0010	.0010	.0010

Allowances and Tolerances for Push Fits

Class of Fit	*Allowances and Tolerances	Up to 1/2"	9/16" through 1"	1-1/16" through 2"	2-1/16" through 3"	3-1/16" through 4"	4-1/16" through 5"
P	High limit (-)	.0002	.0002	.0002	.0005	.0005	.0005
	Low limit (-)	.0007	.0007	.0007	.0010	.0010	.0010
	Tolerance	.0005	.0005	.0005	.0005	.0005	.0005

Allowances and Tolerances for Running Fits

Class of Fit	*Allowances and Tolerances	Up to 1/2"	9/16" through 1"	1-1/16" through 2"	2-1/16" through 3"	3-1/16" through 4"	4-1/16" through 5"
X	High limit (-)	.0010	.0012	.0017	.0020	.0025	.0030
	Low limit (-)	.0020	.0027	.0035	.0042	.0050	.0057
	Tolerance	.0010	.0015	.0018	.0022	.0025	.0027
Y (average machine work)	High limit (-)	.0007	.0010	.0012	.0015	.0020	.0022
	Low limit (-)	.0012	.0020	.0025	.0030	.0035	.0040
	Tolerance	.0005	.0010	.0013	.0015	.0015	.0018
Z (fine tool work)	High limit (-)	.0005	.0007	.0007	.0010	.0010	.0012
	Low limit (-)	.0007	.0012	.0015	.0020	.0022	.0025
	Tolerance	.0002	.0005	.0008	.0010	.0012	.0013

*These inch-standard measurements may be converted to equivalent metric-standard measurements.

Table A-3 American Standard Surface Texture Values

Dimensional Measurement	Measurement Values*							
	Microinches	μm	Microinches	μm	Microinches	μm	Microinches	μm
Roughness height (reference scales)	1 2 3 4 5 6 8†	0.03 0.05 0.08 0.10 0.13 0.15 0.20	13 16† 20 25 32† 40	0.33 0.41 0.51 0.63 0.81 1.02	50 63† 80 100 125† 160	1.27 1.60 2.03 2.54 3.18 4.06	200 250† 320 400 500† 600 800 1000†	5.08 6.35 8.13 10.16 12.70 15.24 20.32 25.40

Dimensional Measurement	Inches	mm	Inches	mm	Inches	mm	Inches	mm	Inches	mm
Standard roughness width cutoff	0.003	0.08	0.010	0.25	0.100	2.54	0.300	7.62	1.000	25.4

Dimensional Measurement	Inches	mm	Inches	mm	Inches	mm	Inches	mm
Waviness height (reference scales)	0.00002 0.00003 0.00005 0.00008	0.0005 0.0008 0.0013 0.0020	0.0001 0.0002 0.0003 0.0005 0.0008	0.0025 0.0050 0.0076 0.0127 0.0203	0.001 0.002 0.003 0.005 0.008	0.025 0.050 0.076 0.127 0.203	0.010 0.015 0.020 0.030	0.25 0.38 0.51 0.76

*SI values are rounded to the number of decimal places for each measurement.
†Values are recommended.

Table A-4 Conversion Table for Optical Flat Fringe Bands in Inch and SI Metric Standard Units of Measure

Number of Fringe Bands	Equivalent Measurement Value*			Number of Fringe Bands	Equivalent Measurement Value*		
	Microinches	Inches	Millimeters		Microinches	Inches	Millimeters
0.1	1.2	0.0000012	0.000029	6.0	69.4	0.0000694	0.001763
0.2	2.3	0.0000023	0.000059	7.0	81.0	0.0000810	0.002056
0.3	3.5	0.0000035	0.000088	8.0	92.5	0.0000925	0.002350
0.4	4.6	0.0000046	0.000118	9.0	104.1	0.0001041	0.002644
0.5	5.8	0.0000058	0.000147	10.0	115.7	0.0001157	0.002938
0.6	6.9	0.0000069	0.000176	11.0	127.2	0.0001272	0.003232
0.7	8.1	0.0000081	0.000206	12.0	138.8	0.0001388	0.003525
0.8	9.3	0.0000093	0.000235	13.0	150.4	0.0001504	0.003819
0.9	10.4	0.0000104	0.000264	14.0	161.9	0.0001619	0.004113
				15.0	173.5	0.0001735	0.004407
1.0	11.6	0.0000116	0.000294	16.0	185.1	0.0001851	0.004700
2.0	23.1	0.0000231	0.000588	17.0	196.6	0.0001966	0.004994
3.0	34.7	0.0000347	0.000881	18.0	208.2	0.0002082	0.005288
4.0	46.3	0.0000463	0.001175	19.0	219.8	0.0002198	0.005582
5.0	57.8	0.0000578	0.001469	20.0	231.3	0.0002313	0.005876

*For a general approximation, one band may be considered to be one microinch or 0.0003mm.

Table A–5 Microinch and Micrometer (μm) Ranges of Surface Roughness for Selected Manufacturing Processes

Roughness Height in Microinches and Micrometers (μm)*

Manufacturing Process	4000 (101.60)	3000 (76.20)	2000 (50.80)	1000 (25.40)	500 (12.70)	250 (6.35)	125 (3.18)	63 (1.60)	32 (0.81)	16 (0.41)	8 (0.20)	4 (0.10)	2 (0.05)	1 (0.03)	0.5 (0.01)
Flame cutting															
Snagging															
Sawing															
Planing, shaping															
Drilling															
Electrical discharge machining															
Milling (chemical)															
Milling (rough)															
Broaching															
Reaming															
Boring, turning (finish)															
Turning (rough)															
Barrel finishing															
Electrolytic grinding															
Burnishing (roller)															
Grinding (commercial)															
Grinding (finish)															
Honing															
Polishing															
Lapping															
Superfinishing															
Sand casting															
Hot rolling															
Forging															
Mold casting (permanent)															
Extruding															
Cold rolling (drawing)															
Die casting															

Code ■ General manufacturing (average) surface finish range

▒ Higher or lower range produced by using special processes

*Values rounded to nearest second place μm decimal

Table A–6 Screw Thread Forms and Formulas (Unified and American National)

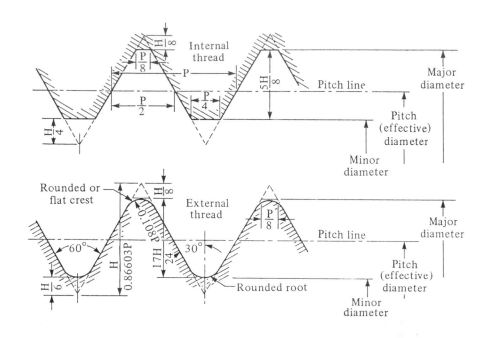

Formulas for Basic Dimensions of Unified and American National Systems
H = (height of sharp V-thread)
= 0.86603 × pitch

Pitch	$= \dfrac{1}{\text{number of threads per inch}}$	Crest truncation, external thread	$= 0.10825 \times \text{pitch} = \dfrac{H}{8}$
Depth, external thread	$= 0.61343 \times \text{pitch}$	Crest truncation, internal thread	$= 0.21651 \times \text{pitch} = \dfrac{H}{4}$
Depth, internal thread	$= 0.54127 \times \text{pitch}$	Root truncation, external thread	$= 0.14434 \times \text{pitch} = \dfrac{H}{6}$
Flat at crest, external thread	$= 0.125 \times \text{pitch}$	Root truncation, internal thread	$= 0.10825 \times \text{pitch} = \dfrac{H}{8}$
Flat at crest, internal thread	$= 0.250 \times \text{pitch}$	Addendum, external thread	$= 0.32476 \times \text{pitch}$
Flat at root, internal thread	$= 0.125 \times \text{pitch}$	Pitch diameter, external and internal	$= \text{major diameter} - 2 \text{ addendums}$ (addendum external thread)

Table A-7 ISO Metric Screw Thread Tap Drill Sizes in Millimeter and Inch Equivalents with Probable Hole Sizes and Percent of Thread

ISO Metric Tap Size	Recommended Metric Drill				Closest Recommended Inch Drill			
	Drill Size* (mm)	Inch Equivalent	Probable Hole Size (Inches)	Probable Percent of Thread	Drill Size	Inch Equivalent	Probable Hole Size (Inches)	Probable Percent of Thread
M1.6 × 0.35	1.25	0.0492	0.0507	69				
M1.8 × 0.35	1.45	0.0571	0.0586	69				
M2 × 0.4	1.60	0.0630	0.0647	69	#52	0.0635	0.0652	66
M2.2 × 0.45	1.75	0.0689	0.0706	70				
M2.5 × 0.45	2.05	0.0807	0.0826	69	#46	0.0810	0.0829	67
M3 × 0.5	2.50	0.0984	0.1007	68	#40	0.0980	0.1003	70
M3.5 × 0.6	2.90	0.1142	0.1168	68	#33	0.1130	0.1156	72
M4 × 0.7	3.30	0.1299	0.1328	69	#30	0.1285	0.1314	73
M4.5 × 0.75	3.70	0.1457	0.1489	74	#26	0.1470	0.1502	70
M5 × 0.8	4.20	0.1654	0.1686	69	#19	0.1660	0.1692	68
M6 × 1	5.00	0.1968	0.2006	70	#9	0.1960	0.1998	71
M7 × 1	6.00	0.2362	0.2400	70	15/64	0.2344	0.2382	73
M8 × 1.25	6.70	0.2638	0.2679	74	17/64	0.2656	0.2697	71
M8 × 1	7.00	0.2756	0.2797	69	J	0.2770	0.2811	66
M10 × 1.5	8.50	0.3346	0.3390	71	Q	0.3320	0.3364	75
M10 × 1.25	8.70	0.3425	0.3471	73	11/32	0.3438	0.3483	71
M12 × 1.75	10.20	0.4016	0.4063	74	Y	0.4040	0.4087	71
M12 × 1.25	10.80	0.4252	0.4299	67	27/64	0.4219	0.4266	72
M14 × 2	12.00	0.4724	0.4772	72	15/32	0.4688	0.4736	76
M14 × 1.5	12.50	0.4921	0.4969	71				
M16 × 2	14.00	0.5512	0.5561	72	35/64	0.5469	0.5518	76
M16 × 1.5	14.50	0.5709	0.5758	71				
M18 × 2.5	15.50	0.6102	0.6152	73	39/64	0.6094	0.6144	74
M18 × 1.5	16.50	0.6496	0.6546	70				
M20 × 2.5	17.50	0.6890	0.6942	73	11/16	0.6875	0.6925	74
M20 × 1.5	18.50	0.7283	0.7335	70				
M22 × 2.5	19.50	0.7677	0.7729	73	49/64	0.7656	0.7708	75
M22 × 1.5	20.50	0.8071	0.8123	70				
M24 × 3	21.00	0.8268	0.8327	73	53/64	0.8281	0.8340	72
M24 × 2	22.00	0.8661	0.8720	71				
M27 × 3	24.00	0.9449	0.9511	73	15/16	0.9375	0.9435	78
M27 × 2	25.00	0.9843	0.9913	70	63/64	0.9844	0.9914	70
M30 × 3.5	26.50	1.0433						
M30 × 2	28.00	1.1024						
M33 × 3.5	29.50	1.1614						
M33 × 2	31.00	1.2205						
M36 × 4	32.00	1.2598						
M36 × 3	33.00	1.2992						
M39 × 4	35.00	1.3780						
M39 × 3	36.00	1.4173						

Formula for Metric Tap Drill Size:

$$\frac{\text{basic major diameter}}{\text{(mm)}} - \frac{\% \text{ thread} \times \text{pitch (mm)}}{76.980} = \frac{\text{drilled hole size}}{\text{(mm)}}$$

Formula for Percent of Thread:

$$\frac{76.980}{\text{pitch (mm)}} \times \left(\frac{\text{basic major diameter}}{\text{(mm)}} - \frac{\text{drilled hole size}}{\text{(mm)}}\right) = \frac{\text{percent of}}{\text{thread}}$$

*Reaming is recommended to the drill size as given.
Table adapted with permission from TRW: Greenfield Tap & Die Division and Geometric Tool Division

Table A–8 Natural Trigonometric Functions (1° to 90°)

Angle	Sine	Cosine	Tangent	Angle	Sine	Cosine	Tangent
1°	.0175	.9998	.0175	46°	.7193	.6947	1.0355
2°	.0349	.9994	.0349	47°	.7314	.6820	1.0724
3°	.0523	.9986	.0524	48°	.7431	.6691	1.1106
4°	.0698	.9976	.0699	49°	.7547	.6561	1.1504
5°	.0872	.9962	.0875	50°	.7660	.6428	1.1918
6°	.1045	.9945	.1051	51°	.7771	.6293	1.2349
7°	.1219	.9925	.1228	52°	.7880	.6157	1.2799
8°	.1392	.9903	.1405	53°	.7986	.6018	1.3270
9°	.1564	.9877	.1584	54°	.8090	.5878	1.3764
10°	.1736	.9848	.1763	55°	.8192	.5736	1.4281
11°	.1908	.9816	.1944	56°	.8290	.5592	1.4826
12°	.2079	.9781	.2126	57°	.8387	.5446	1.5399
13°	.2250	.9744	.2309	58°	.8480	.5299	1.6003
14°	.2419	.9703	.2493	59°	.8572	.5150	1.6643
15°	.2588	.9659	.2679	60°	.8660	.5000	1.7321
16°	.2756	.9613	.2867	61°	.8746	.4848	1.8040
17°	.2924	.9563	.3057	62°	.8829	.4695	1.8807
18°	.3090	.9511	.3249	63°	.8910	.4540	1.9626
19°	.3256	.9455	.3443	64°	.8988	.4384	2.0503
20°	.3420	.9397	.3640	65°	.9063	.4226	2.1445
21°	.3584	.9336	.3839	66°	.9135	.4067	2.2460
22°	.3746	.9272	.4040	67°	.9205	.3907	2.3559
23°	.3907	.9205	.4245	68°	.9272	.3746	2.4751
24°	.4067	.9135	.4452	69°	.9336	.3584	2.6051
25°	.4226	.9063	.4663	70°	.9397	.3420	2.7475
26°	.4384	.8988	.4877	71°	.9455	.3256	2.9042
27°	.4540	.8910	.5095	72°	.9511	.3090	3.0777
28°	.4695	.8829	.5317	73°	.9563	.2924	3.2709
29°	.4848	.8746	.5543	74°	.9613	.2756	3.4874
30°	.5000	.8660	.5774	75°	.9659	.2588	3.7321
31°	.5150	.8572	.6009	76°	.9703	.2419	4.0108
32°	.5299	.8480	.6249	77°	.9744	.2250	4.3315
33°	.5446	.8387	.6494	78°	.9781	.2079	4.7046
34°	.5592	.8290	.6745	79°	.9816	.1908	5.1446
35°	.5736	.8192	.7002	80°	.9848	.1736	5.6713
36°	.5878	.8090	.7265	81°	.9877	.1564	6.3138
37°	.6018	.7986	.7536	82°	.9903	.1392	7.1154
38°	.6157	.7880	.7813	83°	.9925	.1219	8.1443
39°	.6293	.7771	.8098	84°	.9945	.1045	9.5144
40°	.6428	.7660	.8391	85°	.9962	.0872	11.4301
41°	.6561	.7547	.8693	86°	.9976	.0698	14.3007
42°	.6691	.7431	.9004	87°	.9986	.0523	19.0811
43°	.6820	.7314	.9325	88°	.9994	.0349	28.6363
44°	.6947	.7193	.9657	89°	.9998	.0175	57.2900
45°	.7071	.7071	1.0000	90°	1.0000	.0000	

Table A-9 Constants for Setting a 5″ Sine Bar (0°1′ to 10°60′)*

Minutes	0°	1°	2°	3°	4°	5°	6°	7°	8°	9°	10°
0	0.00000	0.08725	0.17450	0.26170	0.34880	0.43580	0.52265	0.60935	0.69585	0.78215	0.86825
1	.00145	.08870	.17595	.26315	.35025	.43725	.52410	.61080	.69730	.78360	.86965
2	.00290	.09015	.17740	.26460	.35170	.43870	.52555	.61225	.69875	.78505	.87110
3	.00435	.09160	.17885	.26605	.35315	.44015	.52700	.61370	.70020	.78650	.87255
4	.00580	.09310	.18030	.26750	.35460	.44155	.52845	.61510	.70165	.78790	.87395
5	0.00725	0.09455	0.18175	0.26895	0.35605	0.44300	0.52985	0.61655	0.70305	0.78935	0.87540
6	.00875	.09600	.18320	.27040	.35750	.44445	.53130	.61800	.70450	.79080	.87685
7	.01020	.09745	.18465	.27185	.35895	.44590	.53275	.61945	.70595	.79225	.87825
8	.01165	.09890	.18615	.27330	.36040	.44735	.53420	.62090	.70740	.79365	.87970
9	.01310	.10035	.18760	.27475	.36185	.44880	.53565	.62235	.70885	.79510	.88115
10	0.01455	0.10180	0.18905	0.27620	0.36330	0.45025	0.53710	0.62380	0.71025	0.79655	0.88255
11	.01600	.10325	.19050	.27765	.36475	.45170	.53855	.62520	.71170	.79795	.88400
12	.01745	.10470	.19195	.27910	.36620	.45315	.54000	.62665	.71315	.79940	.88540
13	.01890	.10615	.19340	.28055	.36765	.45460	.54145	.62810	.71460	.80085	.88685
14	.02035	.10760	.19485	.28200	.36910	.45605	.54290	.62955	.71600	.80230	.88830
15	0.02180	0.10905	0.19630	0.28345	0.37055	0.45750	0.54435	0.63100	0.71745	0.80370	0.88970
16	.02325	.11055	.19775	.28490	.37200	.45895	.54580	.63245	.71890	.80515	.89115
17	.02475	.11200	.19920	.28635	.37345	.46040	.54725	.63390	.72035	.80660	.89260
18	.02620	.11345	.20065	.28780	.37490	.46185	.54865	.63530	.72180	.80800	.89400
19	.02765	.11490	.20210	.28925	.37635	.46330	.55010	.63675	.72320	.80945	.89545
20	0.02910	0.11635	0.20355	0.29070	0.37780	0.46475	0.55155	0.63820	0.72465	0.81090	0.89685
21	.03055	.11780	.20500	.29220	.37925	.46620	.55300	.63965	.72610	.81230	.89830
22	.03200	.11925	.20645	.29365	.38070	.46765	.55445	.64110	.72755	.81375	.89975
23	.03345	.12070	.20795	.29510	.38215	.46910	.55590	.64255	.72900	.81520	.90115
24	.03490	.12215	.20940	.29655	.38360	.47055	.55735	.64400	.73040	.81665	.90260
25	0.03635	0.12360	0.21085	0.29800	0.38505	0.47200	0.55880	0.64540	0.73185	0.81805	0.90405
26	.03780	.12505	.21230	.29945	.38650	.47345	.56025	.64685	.73330	.81950	.90545
27	.03925	.12650	.21375	.30090	.38795	.47490	.56170	.64830	.73475	.82095	.90690
28	.04070	.12800	.21520	.30235	.38940	.47635	.56315	.64975	.73615	.82235	.90830
29	.04220	.12945	.21665	.30380	.39085	.47780	.56455	.65120	.73760	.82380	.90975
30	0.04365	0.13090	0.21810	0.30525	0.39230	0.47925	0.56600	0.65265	0.73905	0.82525	0.91120
31	.04510	.13235	.21955	.30670	.39375	.48070	.56745	.65405	.74050	.82665	.91260
32	.04655	.13380	.22100	.30815	.39520	.48210	.56890	.65550	.74190	.82810	.91405
33	.04800	.13525	.22245	.30960	.39665	.48355	.57035	.65695	.74335	.82955	.91545
34	.04945	.13670	.22390	.31105	.39810	.48500	.57180	.65840	.74480	.83100	.91690
35	0.05090	0.13815	0.22535	0.31250	0.39955	0.48645	0.57325	0.65985	0.74625	0.83240	0.91835
36	.05235	.13960	.22680	.31395	.40100	.48790	.57470	.66130	.74770	.83385	.91975
37	.05380	.14105	.22825	.31540	.40245	.48935	.57615	.66270	.74910	.83530	.92120
38	.05525	.14250	.22970	.31685	.40390	.49080	.57760	.66415	.75055	.83670	.92260
39	.05670	.14395	.23115	.31830	.40535	.49225	.57900	.66560	.75200	.83815	.92405
40	0.05820	0.14540	0.23265	0.31975	0.40680	0.49370	0.58045	0.66705	0.75345	0.83960	0.92545
41	.05965	.14690	.23410	.32120	.40825	.49515	.58190	.66850	.75485	.84100	.92690
42	.06110	.14835	.23555	.32265	.40970	.49660	.58335	.66995	.75630	.84245	.92835
43	.06255	.14980	.23700	.32410	.41115	.49805	.58480	.67135	.75775	.84390	.92975
44	.06400	.15125	.23845	.32555	.41260	.49950	.58625	.67280	.75920	.84530	.93120
45	0.06545	0.15270	0.23990	0.32700	0.41405	0.50095	0.58770	0.67425	0.76060	0.84675	0.93200
46	.06690	.15415	.24135	.32845	.41550	.50240	.58915	.67570	.76205	.84820	.93405
47	.06835	.15560	.24280	.32990	.41695	.50385	.59060	.67715	.76350	.84960	.93550
48	.06980	.15705	.24425	.33135	.41840	.50530	.59200	.67860	.76495	.85105	.93690
49	.07125	.15850	.24570	.33280	.41985	.50675	.59345	.68000	.76635	.85250	.93835
50	0.07270	0.15995	0.24715	0.33425	0.42130	0.50820	0.59490	0.68145	0.76780	0.85390	0.93975
51	.07415	.16140	.24860	.33570	.42275	.50960	.59635	.68290	.76925	.85535	.94120
52	.07565	.16285	.25005	.33715	.42420	.51105	.59780	.68435	.77070	.85680	.94260
53	.07710	.16430	.25150	.33865	.42565	.51250	.59925	.68580	.77210	.85820	.94405
54	.07855	.16580	.25295	.34010	.42710	.51395	.60070	.68720	.77355	.85965	.94550
55	0.08000	0.16725	0.25440	0.34155	0.42855	0.51540	0.60215	0.68865	0.77500	0.86110	0.94690
56	.08145	.16870	.25585	.34300	.43000	.51685	.60355	.69010	.77645	.86250	.94835
57	.08290	.17015	.25730	.34445	.43145	.51830	.60500	.69155	.77785	.86395	.94975
58	.08435	.17160	.25875	.34590	.43290	.51975	.60645	.69300	.77930	.86540	.95120
59	.08580	.17305	.26028	.34735	.43435	.52120	.60790	.69445	.78075	.86680	.95260
60	0.08725	0.17450	0.26170	0.34880	0.43580	0.52265	0.60935	0.69585	0.78215	0.86825	0.95405

5″ sine bar applications of constant values to different sizes of sine bars, plates, and chucks

Sine Bar Size (Inches)	Constant for 5″ Sine Bar
10	Multiply by 2
20	Multiply by 4
2-1/2	Multiply by 0.5
3	Multiply by 0.6
4	Multiply by 0.8

*Complete tables of sine bar constants from 0°1′ to 59°60′ are included in engineering and trade handbooks.

Table A-10 Suggested Starting Points for Cutting Speeds on Bar and Chucker Machines for Basic *External Cuts*

Machinability Class	Nature of Cut	Turning — Overhead and Cross Slide — Cutter Material — Carbide	Turning — Roller Turners — Cutter Material — High-Speed Steel	Turning — Roller Turners — Cutter Material — Carbide	Facing — Carbide	Forming — High-Speed Steel	Forming — Carbide	Cutting Off — High-Speed Steel	Cutting Off — Carbide
A	Roughing	490	150	490	490	150	490	120	500
A	Finishing	560			560				
B	Roughing	420	120	420	420	120	420	100	500
B	Finishing	490			490				
C	Roughing	320	90	320	320	90	320	80	400
C	Finishing	420			420				
D	Roughing	245	70	245	245	70	245	60	400
D	Finishing	320			320				
E	Roughing	220	50	220	220	50	220	50	300
E	Finishing	230			230				
F	Roughing	Max.			Max.	250	Max.	Max.	Max.
F	Finishing	Max.			Max.				

Table A-11 Suggested Starting Points for Cutting Speeds on Bar and Chucker Machines for Basic *Internal Cuts*

Machinability Class	Drilling (HSS Drills) — Regular Drill	Drilling (HSS Drills) — Core Drill — Cored Hole	Drilling (HSS Drills) — Core Drill — Drilled Hole	Boring (Carbide Cutters)		Reaming (HSS Reamers)*
A	120	120	150	Roughing	490	60
A				Finishing	560	
B	80	100	120	Roughing	420	50
B				Finishing	490	
C	60	80	90	Roughing	350	40
C				Finishing	420	
D	50	60	70	Roughing	260	30
D				Finishing	320	
E	50	50	50	Roughing	175	25
E				Finishing	230	
F	200	200	250	Roughing	Max.	120
F				Finishing	Max.	

*Finishing reamers that remove only 0.003″ (0.1mm) to 0.015″ (0.4mm) total stock

Table A–12 Recommended Cutting Speeds and Feeds for Various Depths of Cut on Common Metals (Single-Point Carbide Cutting Tools)

Metal	Depth of Cut (inches)	Feed per Revolution (inches)	Cutting Speed (sfpm)
Aluminum	.005–.015	.002–.005	700–1000
	.020–.090	.005–.015	450–700
	.100–.200	.015–.030	300–450
	.300–.700	.030–.090	100–200
Brass, Bronze	.005–.015	.002–.005	700–800
	.020–.090	.005–.015	600–700
	.100–.200	.015–.030	500–600
	.300–.700	.030–.090	200–400
Cast Iron (gray)	.005–.015	.002–.005	350–400
	.020–.090	.005–.015	250–350
	.100–.200	.015–.030	200–250
	.300–.700	.030–.090	75–150
Machine Steel	.005–.015	.002–.005	700–1000
	.020–.090	.005–.015	550–700
	.100–.200	.015–.030	400–550
	.300–.700	.030–.090	150–300
Tool Steel	.005–.015	.002–.005	500–700
	.020–.090	.005–.015	400–500
	.100–.200	.015–.030	300–400
	.300–.700	.030–.090	100–300
Stainless Steel	.005–.015	.002–.005	375–500
	.020–.090	.005–.015	300–375
	.100–.200	.015–.030	250–300
	.300–.700	.030–.090	75–175

Table A–13 ANSI Spur Gear Rules and Formulas for Required
Features of 20° and 25° Full-Depth Involute Tooth Forms

Required Feature	Rule	Formula*
Diametral pitch (P_d)	Divide the number of teeth by the pitch diameter	$P_d = \dfrac{N}{D}$
	Add 2 to the number of teeth and divide by the outside diameter	$P_d = \dfrac{N+2}{O}$
	Divide 3.1416 by the circular pitch	$P_d = \dfrac{3.1416}{P_c}$
Number of teeth (N)	Multiply the diametral pitch by the pitch diameter	$N = P_d \times D$
	Multiply the diametral pitch by outside diameter and then subtract 2	$N = (P_d \times O) - 2$
Pitch diameter (D)	Divide the number of teeth by the diametral pitch	$D = \dfrac{N}{P_d}$
	Multiply the addendum by 2 and subtract the product from the outside diameter	$D = O - 2A$
Outside diameter (O)	Add 2 to the number of teeth and divide by the diametral pitch	$O = \dfrac{N+2}{P_d}$
	Add 2 to the number of teeth and divide by the quotient of the number of teeth divided by the pitch diameter	$O = \dfrac{N+2}{N/D}$
Circular pitch (P_c)	Divide 3.1416 by the diametral pitch	$P_c = \dfrac{3.1416}{P_d}$
	Divide the pitch diameter by the product of 0.3183 times the number of teeth	$P_c = \dfrac{D}{0.3183 \times N}$
Addendum (A)	Divide 1.000 by the diametral pitch	$A = \dfrac{1.000}{P_d}$
Working depth (W^1)	Divide 2.000 by the diametral pitch	$W^1 = \dfrac{2.000}{P_d}$
Clearance (S)	Divide 0.250 by the diametral pitch	$S = \dfrac{0.250}{P_d}$
Whole depth of tooth (W^2)	Divide 2.250 by the diametral pitch	$W^2 = \dfrac{2.250}{P_d}$
Thickness of tooth (T)	Divide 1.5708 by the diametral pitch	$T = \dfrac{1.5708}{P_d}$
Center distance (C)	Add the pitch diameters and divide the sum by 2	$C = \dfrac{D^1+D^2}{2}$
	Divide one-half the sum of the number of teeth in both gears by the diametral pitch	$C = \dfrac{1/2(N^1+N^2)}{P_d}$

*Dimensions are in inches.

Table A–14 SI Metric Spur Gear Rules and Formulas for Required Features of 20° Full-Depth Involute Tooth Form

Required Feature	Rule	Formula*
Module (M)	Divide the pitch diameter by the number of teeth	$M = \dfrac{D}{N}$
	Divide the outside diameter by the number of teeth plus 2	$M = \dfrac{O}{N+2}$
Number of teeth (N)	Divide the pitch diameter by the module	$N = \dfrac{D}{M}$
Pitch diameter (D)	Multiply the module by the number of teeth	$D = M \times N$
Outside diameter (O)	Add the pitch diameter and twice the module	$O = D + 2M$
	Multiply the number of teeth plus 2 by the module	$O = M \times (N+2)$
Addendum (A)	Multiply the module by 1.000	$A = M \times 1.000$
Clearance (S)	Multiply the module by 0.250	$S = M \times 0.250$
Whole depth (W^2)	Multiply the module by 2.250	$W^2 = M \times 2.250$
Circular pitch (P_c)	Multiply the module by 3.1416	$P_c = M \times 3.1416$
	Divide the pitch diameter by the number of teeth and multiply by 3.1416	$P_c = \dfrac{D}{N} \times 3.1416$
Tooth thickness (T)	Multiply the module by 1.5708	$T = M \times 1.5708$
Center distance (C)	Multiply the module by the number of teeth in both gears and divide by 2	$C = \dfrac{M \times (N^1 + N^2)}{2}$
	Add the pitch diameters of the two gears and divide by 2	$C = \dfrac{D^1 + D^2}{2}$

*Dimensions are in millimeters.

Table A–15 Constants for Calculating Screw Thread Elements (Inches) for ISO Metric Screw Threads

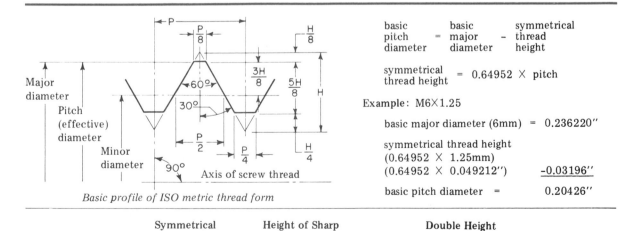

Basic profile of ISO metric thread form

$$\begin{matrix} \text{basic} \\ \text{pitch} \\ \text{diameter} \end{matrix} = \begin{matrix} \text{basic} \\ \text{major} \\ \text{diameter} \end{matrix} - \begin{matrix} \text{symmetrical} \\ \text{thread} \\ \text{height} \end{matrix}$$

$$\begin{matrix} \text{symmetrical} \\ \text{thread height} \end{matrix} = 0.64952 \times \text{pitch}$$

Example: M6×1.25

basic major diameter (6mm) = 0.236220″

symmetrical thread height
(0.64952 × 1.25mm)
(0.64952 × 0.049212″)   −0.03196″

basic pitch diameter =   0.20426″

Symmetrical Thread Height	Height of Sharp V Thread	Double Height Internal Thread
$2\left(\dfrac{3H}{8}\right) = \dfrac{0.64952P}{\text{(Inches)}}$	$H = 0.866025P$ (Inches)	$2\left(\dfrac{5H}{8}\right) = \dfrac{5H}{4}$; $\dfrac{1.08253P}{\text{(Inches)}}$

Table A–16 Grinding Relief or Clearance Angles on High-Speed Steel Cutting Tools*

	Wheel Diameter (inches)	Offset: Center of Cutter and Tip of Tooth Rest Below (or Above) The Center of The Grinding Wheel							
		Cutter Relief or Clearance Angles							
		4 Degrees		5 Degrees		6 Degrees		7 Degrees	
		inch	mm	inch	mm	inch	mm	inch	mm
Using Straight Grinding Wheels	3	0.104	2.64	0.131	3.33	0.157	3.99	0.183	4.65
	3¼	0.113	2.87	0.141	3.58	0.170	4.32	0.198	5.03
	3½	0.122	3.10	0.152	3.86	0.183	4.65	0.213	5.41
	3¾	0.131	3.33	0.163	4.14	0.196	4.98	0.227	5.76
	4	0.139	3.53	0.174	4.42	0.209	5.31	0.242	6.15
	4¼	0.150	3.81	0.185	4.70	0.222	5.64	0.259	6.58
	4½	0.157	3.99	0.195	4.95	0.235	5.97	0.274	6.96
	4¾	0.165	4.19	0.207	5.26	0.248	6.30	0.289	7.34
	5	0.174	4.42	0.218	5.54	0.261	6.63	0.305	7.75
	5¼	0.183	4.65	0.228	5.79	0.274	6.96	0.319	8.10
	5½	0.191	4.85	0.239	6.07	0.287	7.29	0.335	8.51
	5¾	0.200	5.08	0.250	6.35	0.300	7.62	0.350	8.89
	6	0.209	5.31	0.261	6.63	0.313	7.95	0.365	9.27
	6¼	0.218	5.54	0.272	6.91	0.326	8.28	0.381	9.68
	6½	0.226	5.74	0.283	7.19	0.339	8.61	0.396	10.06
	6¾	0.235	5.97	0.294	7.47	0.352	8.94	0.411	10.44
	7	0.244	6.20	0.305	7.75	0.365	9.27	0.426	10.82
	Cutter Diameter (inches)	Offset: Tip of Tooth Rest Below (or Above) Center Line of Cutter When Grinding Peripheral Cutter Teeth							
Using Cup Grinding Wheels	½	0.017	0.43	0.022	0.56	0.026	0.66	0.031	0.79
	¾	0.026	0.66	0.033	0.84	0.040	1.02	0.046	1.17
	1	0.035	0.89	0.044	1.12	0.053	1.35	0.061	1.55
	1¼	0.044	1.12	0.055	1.40	0.066	1.68	0.077	1.95
	1½	0.052	1.32	0.066	1.68	0.079	2.01	0.092	2.34
	1¾	0.061	1.55	0.076	1.93	0.092	2.34	0.108	2.74
	2	0.070	1.78	0.087	2.21	0.105	2.67	0.123	3.12
	2½	0.087	2.21	0.109	2.77	0.131	3.33	0.153	3.89
	2¾	0.096	2.44	0.120	3.05	0.144	3.66	0.168	4.27
	3	0.104	2.64	0.131	3.33	0.158	4.01	0.184	4.67
	3½	0.122	3.10	0.153	3.89	0.184	4.67	0.215	5.46
	4	0.139	3.53	0.174	4.42	0.210	5.33	0.245	6.22
	4½	0.157	3.99	0.197	5.00	0.237	6.02	0.276	7.01
	5	0.174	4.42	0.219	5.56	0.263	6.68	0.307	7.80
	5½	0.192	4.88	0.241	6.12	0.289	7.34	0.338	8.58
	6	0.207	5.26	0.262	6.65	0.315	8.00	0.368	9.35

*Adapted Courtesy of Norton Company

Table A–17 Standard Types of Diamond Grinding Wheels

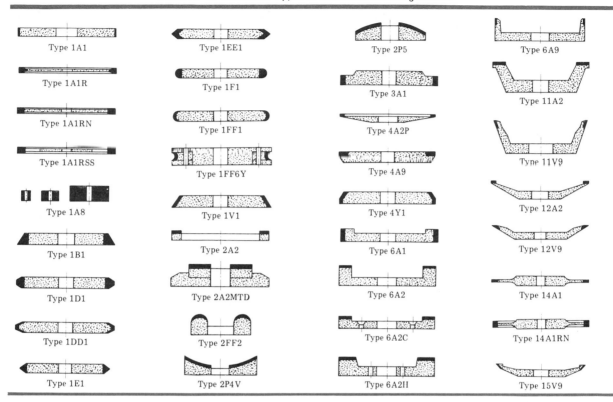

Table A–18 Conversion of Surface Speeds (sfpm) to Spindle Speeds (RPM) for Various Diameters of Grinding Wheels (1″ to 20″ or 25.4mm to 508.0mm)

| Wheel Diameter (mm) | Surface Speed in Feet per Minute (sfpm) | | | | | | | | | | Wheel Diameter (Inches) |
| | 4,000 | 4,500 | 5,500 | 6,500 | 7,500 | 9,500 | 12,500 | 14,200 | 16,000 | 17,000 | |
	Spindle Speeds in Revolutions per Minute (RPM)										
25.4	15,279	17,189	21,008	24,828	28,647	36,287	47,745	54,240	61,116	64,935	1
50.8	7,639	8,594	10,504	12,414	14,328	18,143	23,875	27,120	30,558	32,465	2
76.2	5,093	5,729	7,003	8,276	9,549	12,096	15,915	18,080	20,372	21,645	3
101.6	3,820	4,297	5,252	6,207	7,162	9,072	11,940	13,560	15,278	16,235	4
127.0	3,056	3,438	4,202	4,966	5,730	7,258	9,550	10,850	12,224	12,985	5
152.4	2,546	2,865	3,501	4,138	4,775	6,048	7,960	9,040	10,186	10,820	6
177.8	2,183	2,455	3,001	3,547	4,092	5,183	6,820	7.750	8,732	9,275	7
203.2	1,910	2,148	2,626	3,103	3,580	4,535	5,970	6,780	7,640	8,115	8
228.6	1,698	1,910	2,334	2,758	3,182	4,032	5,305	6,030	6,792	7,215	9
254.0	1,528	1,719	2,101	2,483	2,865	3,629	4,775	5,425	6,112	6,495	10
304.8	1,273	1,432	1,751	2,069	2,386	3,023	3,980	4,520	5,092	5,410	12
355.6	1,091	1,228	1,500	1,773	2,046	2,592	3,410	3,875	4,366	4,640	14
406.4	955	1,074	1,313	1,552	1,791	2,268	2,985	3,390	3,820	4,060	16
457.2	849	955	1,167	1,379	1,591	2,016	2,655	3,015	3,396	3,605	18
508.0	764	859	1,050	1,241	1,432	1,814	2,390	2,715	3,056	3,245	20

Table A–19 Selected Standard Shapes, Design Features, and General Toolroom Applications of Grinding Wheels (Based on ANSI B 74.2)

Standard Shapes, Dimensions*, and Grinding Surfaces	Design Features	Principal Applications	
		Nature of Grinding	Grinding Category
Type 1 straight wheel	Cylindrical with concentric bore	Peripheral	—Surface —Cylindrical —Tool and cutter —Offhand —Portable machine
Type 2 cylindrical wheel	Mounted on the diameter, in a chuck, or on a faceplate	Side	Surface (vertical spindle)
Type 5 straight wheel	Recessed on one side	Peripheral	—Cylindrical —Internal —Surface (horizontal spindle)
Type 6 straight-cup wheel	Equal wall thickness from back to grinding face	Side	—Tool and cutter —Portable machine
Type 7 straight wheel	Recessed two sides	Peripheral	—Cylindrical —Surface (horizontal spindle)
Type 11 flaring-cup wheel	Thickness back wall tapered outward to grinding face	Side	—Tool and cutter —Portable machine

Standard Shapes, Dimensions*, and Grinding Surfaces	Design Features	Principal Applications	
		Nature of Grinding	Grinding Category
 Type 12 dish wheel	Dressed on the side or "U" face	—Side —"U" face	Tool and cutter
 Type 13 saucer wheel	Saucer-shaped equal cross section	—Peripheral —Saw tooth shaping or sharpening	Tool and cutter
 Type 21 straight wheel (relieved)	Both sides relieved, leaving a flat area	Peripheral	Cylindrical
 Type 22 straight wheel (relieved and recessed)	One side relieved to a flat area and second side recessed	Peripheral	Cylindrical

Types 23, 24, 25, and 26 provide further modifications
of straight wheel relieved and recessed combinations

*Key for letter dimensions:

A	Radial width of flat at periphery	N	Depth of relief on one side
D	Outside diameter	O	Depth of relief on second side
E	Thickness at hole	P	Diameter of recess
F	Depth of recess on one side	R	Peripheral radius
G	Depth of recess on second side	T	Overall thickness
H	Hole (bore) diameter	U	Width of edge
J	Diameter of outside flat	W	Rim (wall) thickness at grinding face
K	Diameter of inside flat		

Table A–20 Applications of Wheel Types to Various Grinding Machines and Processes

Wheel	Centertype	Centerless	Internal	Horizontal Spindle	Vertical Spindle	Cutter Grinder	Floor Stand and Bench	High-Speed Hand Grinding	Saw Gumming	Cut Off	Weld Smoothing	Portable	Floor Stand	Swing Frame
	Precision Grinding — Cylindrical			Surface		Tool Cutter		Rough Grinding				Snagging		
Type 1	P	P	P	P	P	P	P	S	P	P	S	P	P	P
Type 2				P	P	S	S							
Type 3												P	P	
Type 4						S						P	P	
Type 5	P	P	P	P		S						P	S	
Type 6			S	S	P	P						P	P	
Type 7	P	P		P		S								
Type 11				S	S	P						P	P	
Type 12						P								
Type 13						S			P					
Type 16												S	P	
Type 17 and 17R												S	P	
Type 18 and 18R			S									S	P	
Type 19 and 19 R												S	P	
Type 20	P	S												
Type 21	P				S									
Type 22	P				S									
Type 24	P				S									
Type 25	P				S									
Type 27										S	P	P		
Plate mounted, nut inserted					S		P							
Mounted points			P					P				S	P	
Segments				P	P									

P = Principal use   S = Secondary use   *Courtesy of CINCINNATI MILACRON INC.*

Table A-21 Melting Points of Metals and Alloys in General Shop Use

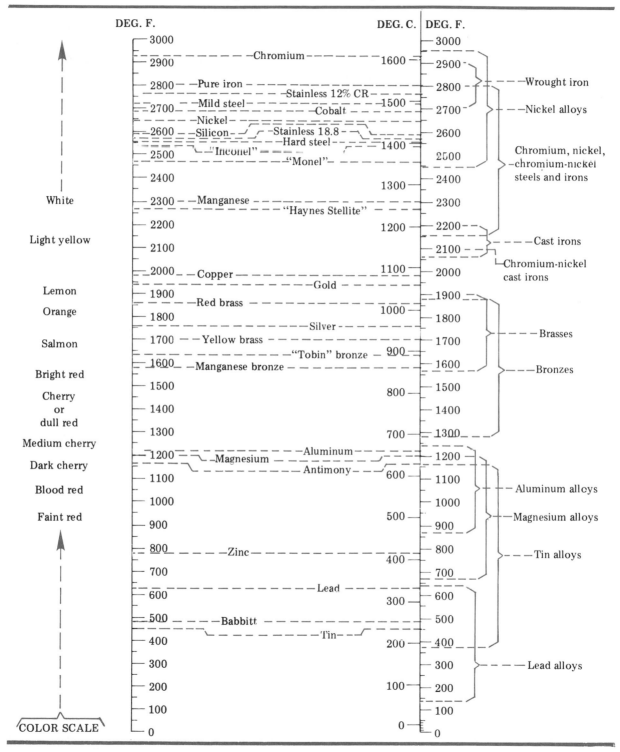

Table A–22 Equivalent Hardness Conversion Numbers for Steel

Rockwell C	Vickers	Shore Scleroscope	Brinell 10mm Ball, 3,000 kgf Load			Rockwell		Rockwell Superficial		
						A Scale, 60 kgf Load	D Scale, 100 kgf Load	15N Scale, 15 kgf Load	30N Scale, 30 kgf Load	45N Scale, 45 kgf Load
			Standard Ball	Hultgren Ball	Carbide Ball	Diamond Penetrator		Superficial Diamond Penetrator		
Equivalent Hardness Numbers for Conversion among the Systems										
68	940	97				85.6	76.9	93.2	84.4	75.4
67	900	95				85.0	76.1	92.9	83.6	74.2
66	865	92				84.5	75.4	92.5	82.8	73.3
65	832	91			739	83.9	74.5	92.2	81.9	72.0
64	800	88			722	83.4	73.8	91.8	81.1	71.0
63	772	87			705	82.8	73.0	91.4	80.1	69.9
62	746	85			688	82.3	72.2	91.1	79.3	68.8
61	720	83			670	81.8	71.5	90.7	78.4	67.7
60	697	81		613	654	81.2	70.7	90.2	77.5	66.6
59	674	80		599	634	80.7	69.9	89.8	76.6	65.5
58	653	78		587	615	80.1	69.2	89.3	75.7	64.3
57	633	76		575	595	79.6	68.5	88.9	74.8	63.2
56	613	75		561	577	79.0	67.7	88.3	73.9	62.0
55	595	74		546	560	78.5	66.9	87.9	73.0	60.9
54	577	72		534	543	78.0	66.1	87.4	72.0	59.8
53	560	71		519	525	77.4	65.4	86.9	71.2	58.6
52	544	69	500	508	512	76.8	64.6	86.4	70.2	57.4
51	528	68	487	494	496	76.3	63.8	85.9	69.4	56.1
50	513	67	475	481	481	75.9	63.1	85.5	68.5	55.0
49	498	66	464	469	469	75.2	62.1	85.0	67.6	53.8
48	484	64	451	455	455	74.7	61.4	84.5	66.7	52.5
47	471	63	442	443	443	74.1	60.8	83.9	65.8	51.4
46	458	62	432	432	432	73.6	60.0	83.5	64.8	50.3
45	446	60	421	421	421	73.1	59.2	83.0	64.0	49.0
44	434	58	409	409	409	72.5	58.5	82.5	63.1	47.8
43	423	57	400	400	400	72.0	57.7	82.0	62.2	46.7
42	412	56	390	390	390	71.5	56.9	81.5	61.3	45.5
41	402	55	381	381	381	70.9	56.2	80.9	60.4	44.3
40	392	54	371	371	371	70.4	55.4	80.4	59.5	43.1
39	382	52	362	362	362	69.9	54.6	69.9	58.6	41.9
38	372	51	353	353	353	69.4	53.8	79.4	57.7	40.8
37	363	50	344	344	344	68.9	53.1	78.8	56.8	39.6
36	354	49	336	336	336	68.4	52.3	78.3	55.9	38.4
35	345	48	327	327	327	67.9	51.5	77.7	55.0	37.2
34	336	47	319	319	319	67.4	50.8	77.2	54.2	36.1
33	327	46	311	311	311	66.8	50.0	76.6	53.3	34.9
32	318	44	301	301	301	66.3	49.2	76.1	52.1	33.7
31	310	43	294	294	294	65.8	48.4	75.6	51.3	32.5
30	302	42	286	286	286	65.3	47.7	75.0	50.4	31.3

Table A–22 Equivalent Hardness Conversion Numbers for Steel (Continued)

			Brinell			Rockwell		Rockwell Superficial		
Rockwell C	Vickers	Shore Scleroscope	10mm Ball, 3,000 kgf Load			A Scale, 60 kgf Load	D Scale, 100 kgf Load	15N Scale, 15 kgf Load	30N Scale, 30 kgf Load	45N Scale, 45 kgf Load
			Standard Ball	Hultgren Ball	Carbide Ball	Diamond Penetrator		Superficial Diamond Penetrator		
						Equivalent Hardness Numbers for Conversion among the Systems				
29	294	41	279	279	279	64.6	47.0	74.5	49.5	30.1
28	286	41	271	271	271	64.3	46.1	73.9	48.6	28.9
27	279	40	264	264	264	63.8	45.2	73.3	47.7	27.8
26	272	38	258	258	258	63.3	44.6	72.8	46.8	26.7
25	266	38	253	253	253	62.8	43.8	72.2	45.9	25.5
24	260	37	247	247	247	62.4	43.1	71.6	45.0	24.3
23	254	36	243	243	243	62.0	42.1	71.0	44.0	23.1
22	248	35	237	237	237	61.5	41.6	70.5	43.2	22.0
21	243	35	231	231	231	61.0	40.9	69.9	42.3	20.7
20	238	34	226	226	226	60.5	40.1	69.4	41.5	19.6

Note: Hardness numbers correspond to values for carbon and alloy steel (ASTM specifications E 140-71).

Table A–23 Recommended Heat Treatment Temperatures for Selected Grades and Kinds of Steel

AISI or SAE Number	Hardening Temperature Range		Annealing Temperature Range		Normalizing Temperature Range		Quenching Medium
	°F	°C	°F	°C	°F	°C	
1030	1550/1600	843/871	1525/1575	829/857	1625/1725	885/941	
1040	1500/1550	816/843	1475/1525	802/829	1600/1700	871/927	
1050	1475/1525	802/829	1450/1500	788/816	1550/1650	843/899	
1060	1450/1500	788/816	1425/1475	774/802	1500/1600	816/871	
1070	1450/1500	788/816	1425/1475	744/802	1500/1600	816/871	Water or brine
1080	1400/1450	760/788	1375/1425	746/774	1475/1575	802/857	
1090	1400/1450	760/788	1375/1425	746/774	1475/1575	802/857	
1132	1550/1600	843/871	1525/1575	829/857	1625/1725	885/941	
1140	1500/1550	816/843	1475/1525	802/829	1600/1700	871/927	
1151	1475/1525	802/829	1450/1500	788/816	1550/1650	843/899	
1330	1525/1575	829/857	1500/1550	816/843	1600/1700	871/927	Oil or water
1340	1500/1550	816/843	1475/1525	802/829	1550/1650	843/899	Oil or water
3140	1475/1525	802/829	1475/1525	802/829	1550/1650	843/899	Oil
4028	1550/1600	843/871	1525/1575	829/857	1600/1700	871/927	Oil or water
4042	1500/1550	816/843	1475/1525	802/829	1550/1650	843/899	Oil
4063	1475/1525	802/829	1450/1500	788/816	1550/1650	843/899	Oil
4130	1550/1600	843/871	1525/1575	829/857	1600/1700	871/927	Oil or water
4140	1525/1575	829/857	1500/1550	816/843	1600/1700	871/927	Oil
4150	1500/1550	816/843	1475/1525	802/829	1600/1700	871/927	Oil
4340	1500/1550	816/843	1500/1550	816/843	1600/1700	871/927	Oil

INDEX